**Series in Computational and Physical Processes
in Mechanics and Thermal Sciences**
*(Formerly the Series in Computational Methods in Mechanics and Thermal
Sciences)*

W. J. Minkowycz and E. M. Sparrow, *Editors*

HEAT CONDUCTION U

GREEN'S FUNCT

HEAT CONDUCTION USING GREEN'S FUNCTIONS

J. V. Beck

Heat Transfer Group
Department of Mechanical Engineering
Michigan State University

K. D. Cole

Mechanical Engineering Department
University of Nebraska, Lincoln

A. Haji-Sheikh

Mechanical Engineering Department
The University of Texas, Arlington

B. Litkouhi

Mechanical Engineering Department
Manhattan College, Riverdale, New York

⦿HEMISPHERE PUBLISHING CORPORATION
A member of the Taylor & Francis Group

London   Washington, DC   Philadelphia

HEAT CONDUCTION USING GREEN'S FUNCTIONS

1 2 3 4 5 6 7 8 9 0   E B E B   9 8 7 6 5 4 3 2 1

This book was set in Times Roman by Harper Graphics. The editors were Andrew N. Bartlett, Heather Jefferson, and Brenda Brienza; the production supervisor was Peggy M. Rote. Printing and binding by Edwards Brothers, Inc.

A CIP catalog record for this book is available from the British Library.

Library of Congress Cataloging-in-Publication Data

Heat conduction using Green's functions / J. V. Beck . . . [et al.].
     p.   cm.—(Series in computational methods in mechanics and thermal sciences)
   Includes bibliographical references and index.

   1. Heat—Conduction—Mathematics. 2. Green's functions.
I. Beck, J. V. (James Vere), date. II. Series.
TJ260.H3854   1992
621.402′2—dc20   91-28755
                                  CIP

ISBN 1-56032-096-6

CONTENTS

To my wife, Barbara; children, Sharon Beck (MSW, LCSW) and Douglas Beck (Ph.D.), and their spouses, Drew Foley and Sheila Beck; sister and her husband, Phyllis and Clarence Sawyer; and sister-in-law and her husband, Carole and Gilbert Becker

J.V.B.

To Mary, Allison, and Jonathan

K.D.C.

To Rosemary, Misty, and Michele

A.H.

To my wife, Mahshad, and children, Babak, Behrang, and Mehrie Litkouhi

B.L.

PREFACE

The purpose of this book is to simplify and organize the solution of heat conduction and diffusion problems and to make them more accessible. This is accomplished using the method of Green's functions, together with extensive tables of Green's functions and related integrals. The tables of Green's functions were first compiled as a supplement to a first-year graduate course in heat conduction taught at Michigan State University. The book was originally envisioned as a reference volume, but it has grown into a heat conduction treatise from a Green's function perspective.

There is enough material for a one-semester course in analytical heat conduction and diffusion. There are worked examples and student problems to aid in teaching. Because of the emphasis on Green's functions, some traditional topics such as Fourier series and Laplace transform methods are treated somewhat briefly; this material could be supplemented according to the interest of the instructor. The book can also be used as a supplementary text in courses on heat conduction, boundary value problems, or partial differential questions of the diffusion type.

We hope the book will be used as a reference for practicing engineers, applied mathematicians, physicists, geologists, and others. In many cases, a heat conduction or diffusion solution may be assembled from tabulated Green's functions rather than derived. The book contains the most extensive set of Green's functions and related integrals that are currently available for heat conduction and diffusion.

The book is organized on a geometric basis because each Green's function is associated with a unique geometry. For each of the three coordinate systems—Cartesian, cylindrical, and spherical—there is a separate appendix of Green's functions named Appendix X, Appendix R, and Appendix RS, respectively. Each of the Green's functions listed is identified by a unique alphanumeric character that begins with either X, R, or RS to denote the x, r, or the spherical r coordinate, respectively. It is important for the reader to know something about this numbering system to use the tables of Green's functions. A more detailed numbering system, which covers both Green's functions and temperature solutions, is discussed in Chapter 2. We find the numbering system very helpful in identifying exactly which solution is under discussion, and all of the solutions discussed in the text are listed in Appendix N indexed according to the numbering system.

The level of treatment is intended for senior and first-year graduate students in engineering and mathematics. We have emphasized solution of problems rather than theorems and proofs, which are generally omitted. A prerequisite is an undergraduate course in ordinary differential equations. A previous introduction to the method of separation of variables for partial differential equations is also important.

The first nine chapters of the book are written with senior engineering students in mind. The Introduction contains background information on heat conduction and brief derivations of the heat conduction equations. Chapters 1 through 5 introduce Green's functions for transient heat conduction in one-dimensional bodies. The Cartesian coordinate system is emphasized in this section as an aid to learning. Steady-state problems are treated as a special case of the transient solution in Sections 3.5 and 3.6. Chapters 6 through 9 are devoted to the solution of problems in the rectangular, cylindrical, and spherical coordinate systems. Transient problems are emphasized and steady problems are treated briefly in separate sections for each coordinate system (Sections 6.9, 8.7, and 9.8). Chapters 10 and 11 introduce the Galerkin-based Green's function method, which combines the efficient analysis of the Green's function method with the flexibility of geometry afforded by numerical methods. Chapter 12 introduces the unsteady surface element method, a numerical method that involves the matching of analytical solutions at the boundaries of bodies in contact.

No other book on Green's functions combines introductory material, worked examples, and extensive tables of Green's functions. Important books that contain some of this material include *Heat Conduction* by M. N. Ozisik (Wiley, New York, 1980), *Conduction of Heat in Solids* by H. S. Carslaw and J. C. Jaeger (Oxford, London, 1959), *Methods of Theoretical Physics* by P. M. Morse and H. Feshbach (McGraw-Hill, New York, 1953), *Elements of Green's Functions and Propagation* by G. Barton (Oxford, London, 1989), *Green's Functions and Transfer Functions Handbook* by A. G. Butkovskiy (Halsted Press, New York, 1982), *Application of Green's Functions in Science and Engineering* by M. D. Greenberg (Prentice-Hall, Englewood Cliffs, New Jersey, 1971), and *Green's Functions: Introductory Theory with Applications* by G. F. Roach (Van Nostrand Reinhold, New York, 1970).

James Beck would like to express his appreciation to the National Science Foundation for support over the years that has aided in the development of this work. Particularly important is the support related to the unsteady surface element method in which Dr. Ned Keltner of Sandia National Laboratories has also had a very influential part.

Kevin Cole would like to acknowledge support from the Engineering Foundation that has contributed to this project. Thanks also go to the many students in heat conduction classes that have read the manuscript and have made many suggestions over the years.

A. Haji-Sheikh would like to acknowledge support from the National Science Foundation, under the directorship of Win Aung and Richard O. Buckius, which was instrumental in the development of the Galerkin-based integral method. Special thanks also to Win Aung who recognized the potential of the Galerkin-based integral method even before the work began. Thanks also to my wife who spent many hours typing and proofreading the manuscript, and to David Lou, former chairman of the Mechanical Engineering Department at UTA, for his encouragement.

Special thanks to the staff at Hemisphere for their competent handling of an equation-filled book. The authors take full responsibility for any errors that may remain in the book, but because this book contains many new solutions we invite readers to send us any errors that they may find. Concerning errors please contact Kevin Cole, Department of Mechanical Engineering, P.O. Box 880656, University of Nebraska-Lincoln, Lincoln, NE 68588-0656 (402-472-5857). We will compile a list of errata and make it available to interested readers.

J. V. Beck
K. D. Cole
A. Haji-Sheikh
B. Litkouhi

NOMENCLATURE

a	geometrical dimension
A	area, m^2
b	geometrical dimension
B	Biot number, usually hL/k
c_p	specific heat, units J/(kg K)
ds_j	differential element of body surface S$_j$
dv'	differential element of volume
f	boundary term
f_j	basis function (Chapter 10)
$F(\mathbf{r})$	initial temperature distribution
g	energy generation per unit time and per unit volume
G	Green's function
h	heat transfer coefficient, units W/(m^{2}K)
i, j	indices
J	Joules, unit of energy
k	thermal conductivity, units W/(mK)
kg	kilogram, unit of mass
K	fundamental heat conduction solution
K	Kelvin, unit of absolute temperature
L	slab thickness, meters
m	fin effect parameter; also meter, unit of length
n_j	outward normal vector on surface S$_j$
N_m	norm (Chapter 4)

$\mathbf{q}$	heat flux, W/m^2
Q	heat energy, Joules
r	radial coordinate
$\mathbf{r}$	position vector
$\mathbf{r}'$	position vector; also, dummy variable
s	Laplace transform parameter (Chap 4)
S_i	surface
t	time, sec
T	temperature, Kelvin
T_o	initial temperature distribution
T^*	auxiliary solution
$\mathbf{u}, \mathbf{U}$	velocity, m/s
$\mathbf{V}, \mathbf{w}$	velocity, m/s
W	Watts, one Joule per second
x,y,z	Cartesian coordinates
$\mathbf{X}_\mathrm{m}$	eigenfunction (Chapter 4)

Greek Symbols

α	thermal diffusivity, W/m^2
β_m	eigenvalue
γ_n	eigenvalue (Chapter 10)
δ	Dirac delta function
ρ	density
σ	propagation speed for heat transfer, m/s
τ	time; also dummy variable
φ, Φ	angular coordinate (azimuth) for cylindrical and spherical geometries
θ	polar angle coordinate, spherical coordinate system
ψ_n	eigenvalue (Chapter 10)

INTRODUCTION

HEAT CONDUCTION BASICS

I.1 INTRODUCTION

This chapter gives the terminology and equations of heat conduction. The reader familiar with elementary heat conduction theory may be content to skim this material and begin with Chapter 1 where solution methods and Green's functions are introduced.

In this chapter, heat flux and thermal conductivity are defined in Section I.2 and the energy equation for homogeneous isotropic bodies is discussed in Sections I.3 and I.4. Boundary conditions and boundary value problems are introduced in Section I.5. The integral energy equation is derived in Section I.6. Heterogeneous bodies and anisotropic bodies are briefly discussed in Sections I.7 and I.8. Section I.9 discusses finite propagation of heat conduction.

I.2 HEAT FLUX AND TEMPERATURE

In a solid body that contains variations of temperature, heat flow proceeds from a region of high temperature to a region of lower temperature. The term heat flow is the rate of flow of energy (in Joules per second, or J/s) associated with the vibrational energy of atoms and molecules in the body. Heat flux is the rate of heat flow per unit area at any point in the body. Heat conduction theory is the relationship between heat flux and temperature in a solid body; it also applies to liquids and gases when there is no fluid motion.

Heat flux cannot be measured directly, but its effects can be indirectly observed. At the surface of a solid body the heat flux can sometimes be observed as an effect on the surroundings, such as the melting of ice, the warming of a well-stirred water bath, or the vaporization of water at a certain rate. Inside a solid body, the heat flux can be deduced only from the temperature distribution, and then only if the relationship between temperature and heat flux is thoroughly understood.

In a solid body with a steady temperature gradient, say along the direction of coordinate x, the following equation is used to relate the heat flux q_x to the temperature T:

$$q_x = -k \frac{\partial T}{\partial x} \tag{I.1}$$

where k is the thermal conductivity of the body with units W/(m K). The above equation was introduced by Fourier and has come to be called Fourier's law of heat conduction. Fourier's law applies to any body that is *homogeneous* (the same substance all the way through) and *isotropic* (heat flows equally well in any direction). The thermal conductivity may be a function of the temperature.

Heat flux (W/m^2) is a vector quantity. The direction of the heat flux is given by the temperature gradient—heat flows from high temperature to low temperature. The magnitude of the heat flux depends on the magnitude of the temperature gradient and on the thermal conductivity. In general, the heat flux vector has three components, one for each coordinate direction. For example, in the rectangular coordinate system (x, y, z) the three components of the heat flux are given by

$$q_x = -k \frac{\partial T}{\partial x} \qquad q_y = -k \frac{\partial T}{\partial y} \qquad q_z = -k \frac{\partial T}{\partial z} \tag{I.2}$$

I.3 DIFFERENTIAL ENERGY EQUATION IN RECTANGULAR COORDINATES

The energy equation is derived in this section for homogeneous isotropic bodies. The rectangular (x, y, z) coordinate system is used for simplicity.

The energy equation, also called the heat conduction equation, is based on the conservation of energy. Consider a small parallelpiped-shaped control volume in a stationary, homogeneous, and isotropic body. The control volume is located at point (x, y, z) in the body and has volume $dV = dx\, dy\, dz$. See Fig. I.1. The energy balance on the control volume is given by a form of the first law of thermodynamics:

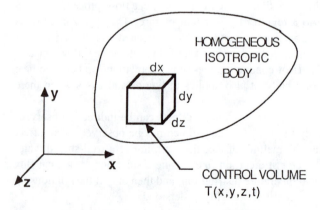

Figure I.1 Control volume.

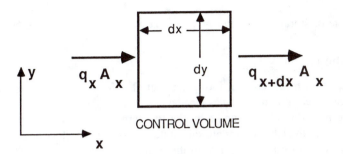

Figure I.2 Flow of heat in the x direction.

Net rate of heat flow in + rate of energy generation

$$= \text{rate of storage of energy} \quad (I.3)$$

Each term in this rate equation has units of energy/time (J/s or watts). The three terms in this equation will be examined one at a time.

Net rate of heat flow in. There are six faces on the control volume through which heat can enter or leave. Heat flux is positive in the positive coordinate directions, and each heat flux must be multiplied by the area of the face to give the correct units of watts. Figure I.2 shows the flow of heat in the x direction, where $q_x A_x$ has the units of watts. The *net* flow of heat is the difference between the inflow and the outflow $(q_x A_x - q_{x+dx} A_x)$. For all three directions and all six faces of the control volume,

Net rate of heat flow in

$$= (q_x - q_{x+dx})A_x + (q_y - q_{y+dy})A_y + (q_z - q_{z+dz})A_z \quad (I.4)$$

Rate of energy generation. Energy generation is energy that affects the temperature throughout the volume of the body. It is distinguished from energy that enters the body through the boundaries. Energy generation can come from electrical resistance heating inside the body, from chemical reaction (for example, epoxy and concrete generate heat when curing), or from absorption of radiation (nuclear, microwave, or other electromagnetic energy). The energy generation may vary from place to place in the body and it may vary with time. The energy generation may also be simply zero. It is given the symbol $g(x, y, z, t)$ with units W/m^3 (rate of energy generation per unit volume). For the control volume, then,

$$\text{Rate of energy generation} = g(x, y, z, t)dx \, dy \, dz \quad (I.5)$$

Rate of storage of energy. A change in the storage of energy is defined by a change in the specific internal energy (a thermodynamic quantity) which is given by $c\delta T$ for solid bodies. Here c is the specific heat [J/(kg K)] and δT is the change in temperature. The *rate* of specific energy storage (per unit mass) is given by the time derivative $c\partial T/\partial t$. The partial derivative on time is used because T also depends on position (x, y, z). Multiply the time rate-of-change of specific internal energy by the density and the volume to obtain watts:

$$\text{Rate of storage of energy} = \rho c \, \frac{\partial T}{\partial t} \, dx \, dy \, dz \tag{I.6}$$

I.3.1. Energy Equation

To place the energy equation in differential form, the control volume will be made arbitrarily small. Then, the heat flux at the faces located at $x + dx$, $y + dy$, and $z + dz$ can be related to the heat flux at x, y, and z by the first term of a Taylor series, according to Table I.1. When the table values are substituted into Eq. (I.4), the energy equation can be assembled from Eqs. (I.4)–(I.6) in the form

$$-\frac{dq_x}{dx} - \frac{dq_y}{dy} - \frac{dq_z}{dz} + g(x, y, z, t) = \rho c \, \frac{\partial T}{\partial t} \tag{I.7}$$

Now, applying Fourier's law yields

$$\frac{\partial}{\partial x} \left(k \, \frac{\partial T}{\partial x} \right) + \frac{\partial}{\partial y} \left(k \, \frac{\partial T}{\partial y} \right) + \frac{\partial}{\partial z} \left(k \, \frac{\partial T}{\partial z} \right) + g(x, y, z, t) = \rho c \, \frac{\partial T}{\partial t} \tag{I.8}$$

This is the energy equation for a homogeneous isotropic body. Properties c and k may depend upon the temperature and therefore may vary with position in the body.

I.3.2 Constant Properties

In the special case when the thermal conductivity does not depend on position (for example, when the temperature gradients are not too large), the energy equation can be written as

$$\frac{\partial^2 T}{\partial x^2} + \frac{\partial^2 T}{\partial y^2} + \frac{\partial^2 T}{\partial x^2} + \frac{1}{k} g(x, y, z, t) = \frac{1}{\alpha} \frac{\partial T}{\partial t} \tag{I.9}$$

where $\alpha = k/(\rho c)$ is the thermal diffusivity (m²/s). This form of the energy equation is extensively studied in this book.

I.4 OTHER COORDINATE SYSTEMS

The energy equation can be cast in other orthogonal coordinate systems; the cylindrical and spherical systems are listed in this section. A general vector form of Fourier's law is given by

Table I.1 One term of Taylor series for q

Direction	Flux	Area
x	$q_{x+dx} = q_x + \dfrac{\partial q_x}{\partial x} dx$	$A_x = dy \, dz$
y	$q_{y+dy} = q_y + \dfrac{\partial q_y}{\partial y} dy$	$A_y = dx \, dz$
z	$q_{z+dz} = q_z + \dfrac{\partial q_z}{\partial z} dz$	$A_z = dx \, dy$

$$\mathbf{q} = -k \nabla T \tag{I.10}$$

where ∇T is the gradient of the temperature and $\mathbf{q}$ is the heat flux vector. A vector form of the energy equation that is independent of coordinate system is given by (see Ozisik, 1980, pp. 4–6 for a derivation)

$$-\nabla \cdot \mathbf{q} + g(\mathbf{r}, t) = \rho c \frac{\partial T}{\partial t} \tag{I.11}$$

where $\nabla \cdot \mathbf{q}$ is the divergence of the heat flux. The energy equation in any coordinate system can be found by substituting the correct form of the divergence and gradient operators for that particular coordinate system.

I.4.1 Cylindrical Coordinate System

In the cylindrical coordinate system shown in Figure I.3a the energy equation is

$$\frac{1}{r} \frac{\partial}{\partial r} \left(kr \frac{\partial T}{\partial r} \right) + \frac{1}{r^2} \frac{\partial}{\partial \phi} \left(k \frac{\partial T}{\partial \phi} \right) + \frac{\partial}{\partial z} \left(k \frac{\partial T}{\partial z} \right) + g = \rho c \frac{\partial T}{\partial t} \tag{I.12a}$$

or for $k = $ constant

$$\frac{\partial^2 T}{\partial r^2} + \frac{1}{r} \frac{\partial T}{\partial r} + \frac{1}{r^2} \frac{\partial^2 T}{\partial \phi^2} + \frac{\partial^2 T}{\partial z^2} + \frac{g}{k} = \frac{1}{\alpha} \frac{\partial T}{\partial t} \tag{I.12b}$$

I.4.2 Spherical Coordinate System

In the spherical coordinate system shown in Fig. I.3b, the energy equation is

$$\frac{1}{r^2} \frac{\partial}{\partial r} \left(kr^2 \frac{\partial T}{\partial r} \right) + \frac{1}{r^2 \sin \theta} \frac{\partial}{\partial \theta} \left(k \sin \theta \frac{\partial T}{\partial \theta} \right)$$
$$+ \frac{1}{r^2 \sin^2 \theta} \frac{\partial}{\partial \phi} \left(k \frac{\partial T}{\partial \phi} \right) + g = \rho c \frac{\partial T}{\partial t} \tag{I.13a}$$

or for $k = $ constant,

$$\frac{1}{r} \frac{\partial^2 (rT)}{\partial r^2} + \frac{1}{r^2 \sin \theta} \frac{\partial}{\partial \theta} \left(\sin \theta \frac{\partial T}{\partial \theta} \right) + \frac{1}{r^2 \sin^2 \theta} \frac{\partial^2 T}{\partial \phi^2} + \frac{g}{k} = \frac{1}{\alpha} \frac{\partial T}{\partial t} \tag{I.13b}$$

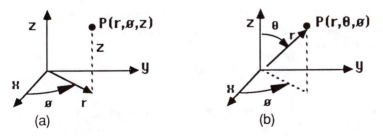

(a) (b)

Figure I.3 (a) Cylindrical coordinate system. (b) Spherical coordinate system.

I.5 BOUNDARY AND INITIAL CONDITIONS

This book is concerned with solutions to the energy equation as they apply to problems in engineering and physics. The mathematical form of the solutions (such as Green's functions) are determined by the boundary conditions, that is, the value of the temperature (or its derivative) at the boundaries of the heat conducting body. The combination of the energy equation, the specific boundary conditions, and the initial condition is called a *boundary value problem.*

Most of this book is concerned with orthogonal bodies, whose boundaries are located where one coordinate is a constant, such as $x = 0$ or $x = L$. Where possible, the coordinate system is chosen so that the body of interest may be treated as an orthogonal body. (Nonorthogonal bodies are discussed in Chapter 11.)

The number of boundary conditions required for a boundary value problem is determined by the form of the energy equation. For example, the two-dimensional energy equation in the rectangular coordinate system,

$$\frac{\partial^2 T}{\partial x^2} + \frac{\partial^2 T}{\partial y^2} + \frac{1}{k}\, g(x, y, t) = \frac{1}{\alpha}\frac{\partial T}{\partial t} \tag{I.14}$$

requires five conditions: two each for the boundaries on x and y and one initial condition. Boundary conditions typically have the form

$$k_i \frac{\partial T}{\partial n_i} + h_i T = f_i(r_i, t) \tag{I.15}$$

where all quantities are evaluated at the ith boundary and where r_i represents the location of the ith boundary in a specific coordinate system. Initial conditions have the form

$$T(r_i, t = 0) = F(r_i) \tag{I.16}$$

Boundary conditions and initial conditions are discussed in detail in Chapter 2.

I.6 INTEGRAL ENERGY EQUATION

In this section, the integral energy equation is derived for heat transfer in a solid. The solid may be moving but it may not change shape. There are no changes in the shape of the body during heating due to thermal expansion; the subject of thermal stresses is beyond our scope.

The derivation starts with a system, which is a body or portion of a body that is identified for study. The system may move and exchange energy with its surroundings. The first law of thermodynamics for a system can be written as

$$\frac{\delta Q}{dt} = \frac{\delta W}{dt} + \frac{dE}{dt} \tag{I.17}$$

where δW and δQ denote path-dependent quantities. Each term in Eq. (I.17) can be described in words by

$$\frac{\delta Q}{dt} = \text{Rate of heat addition to the system at the boundaries}$$

$$\frac{\delta W}{dt} = \text{Rate of work done by the system on its surroundings}$$

$$\frac{dE}{dt} = \text{Rate of energy accumulation inside the system}$$

If we neglect kinetic energy and potential energy compared to thermal energy, then the energy total of the system E is given by the internal energy of the system,

$$E = mu \tag{I.18}$$

where m is mass of the system in kilograms (kg), and u is the internal energy per unit mass, J/kg.

The next step is to relate the system to a control volume with the Reynolds transport theorem. (See Potter and Foss, 1975, p. 80, or White, 1974, p. 104.) The control volume is fixed in space and has fixed shape and fixed boundaries. At the moment of interest, time t, the system and the control volume occupy the same region. At a later time, $t + \Delta t$, the system has moved away from the fixed control volume. Refer to Figure I.4. A statement of the Reynolds transport theorem for the change of energy in the system is

$$\left.\frac{dE}{dt}\right|_{sys} = \frac{\partial}{\partial t} \int_{c.v.} u\rho \, dv + \int_{c.s.} \rho u (\mathbf{V} \cdot \hat{\mathbf{n}}) \, dA \tag{I.19}$$

where c.v. denotes the control volume, c.s. denotes the surface of the control volume (control surface), dv is an element of volume, ρ is density, $\mathbf{V}$ is velocity and $\hat{\mathbf{n}}$ is an

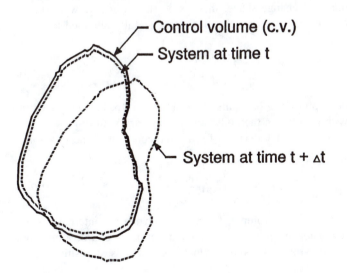

Control volume (c.v.)

System at time t

System at time t + Δt

Figure I.4 Relation between moving system and fixed control volume.

outward drawn unit normal vector. Equation (I.19) relates the energy in the system at time t to that in the control volume.

Next, replace dE/dt with the first law of thermodynamics, Eq. (I.17),

$$\left.\frac{\delta Q}{dt}\right|_{\text{sys}} - \left.\frac{\delta W}{dt}\right|_{\text{sys}} = \frac{\partial}{\partial t} \int_{\text{c.v.}} u\rho \, dv + \int_{\text{c.s.}} \rho u (\mathbf{V} \cdot \hat{\mathbf{n}}) \, dA \qquad (I.20)$$

The terms of Eq. (I.20) will next be examined separately. The first term of Eq. (I.20) relates to energy traveling across the control surface and can be given by

$$\frac{\delta Q}{dt} = \int_{\text{c.s.}} (-\mathbf{q} \cdot \hat{\mathbf{n}}) \, dA \qquad (I.21)$$

where $\mathbf{q}$ is the heat flux crossing the control surface in W/m^2. It can include conduction and radiation,

$$\mathbf{q} = \mathbf{q}_{\text{cond}} + \mathbf{q}_{\text{radiation}} \qquad (I.22)$$

but not any term caused by fluid flow for any element inside the body. When a body is "lumped" in some way so that a solid-fluid boundary is included in the control volume, then a convection-related term may enter. Otherwise, for any element in a solid body or porous body, the only two modes of heat transfer are conduction and radiation.

The $\delta W/dt$ term in Eq. (I.20) relates to the rate of work done *by* the system *on* the surroundings and could be composed of a number of parts,

$$\frac{\delta W}{dt} = \begin{cases} \text{Shaft work } + \text{ flow work } + \text{ viscous work} \\ + \text{ electrical work } + \text{ nuclear work } + \text{ chemical work,} \\ \text{all acting on the surroundings.} \end{cases}$$

For a solid body that does not change shape, there is no shaft work, flow work, or viscous work. The electrical, nuclear, and chemical work are all combined together as volume energy generation:

$$\frac{\delta W}{dt} = -\int_{\text{c.v.}} g \, dv \qquad (I.23)$$

The volume energy generation term g has units of W/m^3; $g > 0$ for heat produced in the body; g may vary with position in the body, and it may vary with time.

Next consider the third term of Eq. (I.20) for a fixed control volume in a solid $[\rho \neq \rho(t)]$,

$$\frac{\partial}{\partial t} \int_{\text{c.v.}} u\rho \, dv = \int_{\text{c.v.}} \rho \frac{\partial u}{\partial t} \, dv \qquad (I.24)$$

Let $v = \rho^{-1}$ where v is the specific volume. From thermodynamics, internal energy u can be a function of two independent thermodynamic quantities. Let u be a function of temperature T and specific volume v, both of which are functions of position $\mathbf{r}$ and time t, or

$$u = u(T(\mathbf{r}, t), v(\mathbf{r}, t))$$

Then using the chain rule for differentiation gives

$$\frac{\partial u}{\partial t} = \frac{\partial u}{\partial T}\bigg|_v \frac{\partial T}{\partial t} + \frac{\partial u}{\partial v}\bigg|_T \frac{\partial v}{\partial t} \tag{I.25}$$

In a solid, density is not a function of time, so that $\partial v/\partial t$ is equal to zero. Also, from the definition of the specific heat at constant volume,

$$c_v \equiv \frac{\partial u}{\partial T}\bigg|_v \tag{I.26}$$

In a solid, the specific heat at constant volume is the same as at constant pressure or

$$c_v = c_p = c \tag{I.27}$$

Substitute Eqs. (I.26) and (I.27) into Eq. (I.25)

$$\frac{\partial u}{\partial t} = c\frac{\partial T}{\partial t} \tag{I.28}$$

so the third term of Eq. (I.20) is given by

$$\int_{c.v.} \rho c \frac{\partial T}{\partial t}\, dv \tag{I.29}$$

Notice that the specific heat can be a function of position and temperature, $c = c(\mathbf{r}, T)$. In particular, note that c is *not* inside the derivative with respect to time in Eq. (I.29).

Then Eqs. (I.21), (I.23), and (I.29) can be substituted into Eq. (I.20) to give the **general form of the integral energy equation** for a solid,

$$\int_{c.s.} (-\mathbf{q}\cdot\hat{\mathbf{n}})\, dA + \int_{c.v.} g\, dv = \int_{c.v.} \rho c\frac{\partial T}{\partial t}\, dv + \int_{c.s.} \rho u\, \mathbf{V}\cdot\hat{\mathbf{n}}\, dA \tag{I.30}$$

This equation is valid for $\rho = \rho(\mathbf{r})$ and $c = c(\mathbf{r}, T)$.

Many forms of the heat conduction equation can be derived from this equation, including general partial differential equations and also lumped capacitance equations. If the control volume is taken to represent a thin element on either side of a boundary, Eq. (I.30) can be used to obtain boundary conditions.

I.7 HETEROGENEOUS BODIES

A body composed of two or more parts with different thermal conductivities is called a *heterogeneous* body (also called a nonhomogeneous body). Fourier's law may apply to each homogeneous part of such a body, but the interface where the conductivity changes must be treated with special techniques, two of which are discussed in this book. In Chapter 11 the Galerkin-based Green's function method is applied to a body with an inclusion. In Chapter 12 the surface element method is applied to two homogeneous bodies in thermal contact.

I.8 ANISOTROPIC BODIES

Many bodies of engineering interest do not conduct heat equally well in all directions and are called *anisotropic* bodies. Laminates, crystals, fiber/matrix composites, and wood are among the materials that have preferred directions in which heat flows more readily than others. For example, wood conducts heat along the grain more readily than across the grain.

I.8.1 Conductivity Matrix

For anisotropic bodies, a generalized form of Fourier's law is used that includes a thermal conductivity matrix. For example, in rectangular coordinates, the conductivity matrix is given by

$$
\begin{bmatrix}
k_{11} & k_{12} & k_{13} \\
k_{21} & k_{22} & k_{23} \\
k_{31} & k_{32} & k_{33}
\end{bmatrix}
\tag{I.31}
$$

and the heat flux vector is given by

$$
q_i = \sum_{j=1}^{3} k_{ij} \frac{\partial T}{\partial x_j}
\tag{I.32}
$$

The energy equation for anisotropic bodies contains cross derivatives and its solution is not covered here; refer to Carslaw and Jaeger (1959, p. 38) and Ozisik (1980, p. 611).

I.8.2 Orthotropic Bodies

The conductivity matrix depends on the orientation of the coordinate system in the body. If the coordinate system is parallel to three mutually perpendicular preferred directions of heat conduction, then the geometry is said to be *orthotropic* and the coordinate system lies along the principal axes of heat conduction. In an orthotropic body the conductivity matrix has a diagonal form,

$$
\begin{bmatrix}
k_{11} & 0 & 0 \\
0 & k_{22} & 0 \\
0 & 0 & k_{33}
\end{bmatrix}
\tag{I.33}
$$

The energy equation for orthotropic bodies does not contain any cross-derivatives and it can be transformed into the standard isotropic energy equation by a suitable choice of new spatial coordinates; see Ozisik (1980, pp. 631–632). Thus, all of the standard heat conduction solution methods apply to orthotropic bodies.

Wood is an example of an orthotropic body in the particular cylindrical coordinate system (r, ϕ, z) corresponding to the direction of the rays, rings, and axis of the tree (Carslaw and Jaeger, 1959, p. 41).

I.9 FINITE PROPAGATION OF HEAT CONDUCTION

Fourier's law of heat conduction describes heat transfer very accurately in most applications. However, it predicts that the response to a pulse of heat at one point in a body is instantaneous throughout the body. Of course, the size of the temperature response is vanishingly small far from the heat pulse, but the speed of propagation is infinite according to Fourier's law. Since infinite speed of propagation is not possible in real bodies, Fourier's law must be an approximation actually correct only after a very small time has elapsed. To correctly describe the finite propagation of heat for very short times, for very short distances, or for temperatures very near zero degrees Kelvin, another relation between temperature and heat flux is needed, and one relation is described here.

A relation between temperature and heat flux that allows for a finite speed of heat propagation is given by Ozisik and Vick (1984)

$$\frac{\alpha}{\sigma^2} \frac{\partial \mathbf{q}}{\partial t} + \mathbf{q} = \nabla \cdot (k \nabla T) \tag{I.34}$$

where σ is the propagation speed for heat transfer and α/σ^2 is the relaxation time for the heat flux to begin after a temperature gradient is imposed on the body. Conversely, the heat flow does not cease immediately after the temperature gradient is removed but dies away over a short period of time.

The energy equation that embodies the finite propagation of heat may be found by taking the divergence of Eq. (I.34),

$$\frac{\alpha}{\sigma^2} \frac{\partial}{\partial t} [\nabla \cdot \mathbf{q}] + \nabla \cdot \mathbf{q} = -k \nabla^2 T \tag{I.35}$$

Now solve the vector energy equation, Eq. (I.11), for $\nabla \cdot \mathbf{q}$,

$$\nabla \cdot \mathbf{q} = g(\mathbf{r}, t) - \rho c \frac{\partial T}{\partial t} \tag{I.36}$$

and substitute $\nabla \cdot \mathbf{q}$ into Eq. (I.35). After some rearranging, the result is

$$\nabla \cdot (k \nabla T) + \left[g(\mathbf{r}, t) + \frac{\alpha}{\sigma^2} \frac{\partial g}{\partial t} \right] = \rho c \frac{\partial T}{\partial t} + \frac{k}{\sigma^2} \frac{\partial^2 T}{\partial t^2} \tag{I.37}$$

This is the heat conduction equation that includes finite speed of heat propagation, and there are two additional terms that do not appear when Fourier's law is used in the energy equation. The second derivative of temperature with respect to time is a wave term and the wave speed is σ. This wave term is said to be hyperbolic in time, and Eq. (I.37) is sometimes called the hyperbolic heat conduction equation.

There is also a time derivative of the energy generation $g(\mathbf{r}, t)$ and this term is a consequence of the finite speed of heat propagation. In the limiting case of infinite propagation speed, Eq. (I.37) reduces to the classic diffusive energy equation.

Although Eq. (I.37) is now hyperbolic (rather than parabolic) in time, it is a linear equation if the wave speed is independent of the temperature, and Green's functions may be used to solve the hyperbolic heat conduction equation. However, a discussion of hyperbolic Green's functions is beyond the scope of the text.

REFERENCES

Carslaw, H. S., and Jaeger, J. C., 1959, *Conduction of Heat in Solids*, 2d ed., Oxford University Press, New York.

Ozisik, M. N., 1980, *Heat Conduction*, Wiley, New York.

Ozisik, M. N., and Vick, B., 1984, Propagation and Reflection of Thermal Waves in a Finite Medium, *Int. J. Heat Mass Transfer*, vol. 27, no. 10, pp. 1845–1854.

Potter, M. C., and Foss, J. F., 1975, *Fluid Mechanics*, p. 80, Wiley, New York.

White, F. M., 1974, *Viscous Fluid Flow*, p. 104, McGraw-Hill, New York.

CHAPTER ONE

INTRODUCTION TO GREEN'S FUNCTIONS

1.1 INTRODUCTION

Green's functions, named after the English mathematician and physicist, George Green (1773–1841), are very powerful tools for obtaining solutions of transient and steady-state linear heat conduction problems. They can also be applied for the solution of some convection problems and to many other phenomena which are described by the same type of equations. These problems usually involve solution of diffusion-type partial differential equations. A Green's function (GF) is a basic solution of a specific differential equation with homogeneous boundary conditions; it is a building block from which many useful solutions may be constructed. For transient heat conduction, a GF describes the temperature distribution caused by an instantaneous, local energy pulse.

This book contains an extensive set of exact GFs for the transient heat conduction equation, including Cartesian, cylindrical, and spherical coordinates. By utilizing these tabulated GFs, solutions of many transient heat conduction problems can be obtained in a straightforward and efficient manner. In many cases, the formal solutions can be *written directly* in terms of integrals which can be evaluated either exactly using integrals provided herein or approximately using an appropriate numerical method. Compared to the usual analytical methods, the GF method, using tabulated GFs, requires a lower level of mathematical ability than is frequently required for the solution of partial differential equations.

The GF method is related to other methods for solving heat conduction problems. The classic methods in heat conduction, including the method of separation of variables and the Laplace transform, are used to derive GFs (Chapter 4). Approximate methods of finding GFs developed by Haji-Sheikh (1988) and Haji-Sheikh and Lakshminarayanan (1987) are also discussed (see Chapter 10). In addition to solution procedures, the GF method also provides greater understanding of the nature of diffusion processes, including heat conduction and flow in porous media.

GFs have been used in the solution of transient heat conduction equations for many decades. This is demonstrated by the discussion of GFs in the classic books of Carslaw and Jaeger (1959) and Morse and Feshbach (1953). The purpose of this book is to provide a single text containing the following components: a lucid derivation of the GF solution equation; a systematic and practical approach to the solution of diffusion-type problems; and extensive compilation of GFs. Other books contain some of these components: Ozisik (1980) has a fine derivation of the GF solution equation; a book by Butkovskiy (1982) provides a catalog of many GFs; and Carslaw and Jaeger (1959) also list some GFs. Other important references on GFs are the books by Greenberg (1971), Roach (1970), and Stakgold (1979).

1.1.1 Advantages of the Green's Function Method

There is ample motivation for the use of GFs in linear transient heat conduction for basic (and other) geometries. One advantage of GFs is that they are flexible and powerful. The *same* GF for a given geometry and a given set of homogeneous boundary conditions is a building block for the temperature distribution resulting from (a) space-variable initial temperature distribution, (b) time- and space-variable boundary conditions, and (c) time- and space-variable volume energy generation.

A second advantage of the GF method is the systematic solution procedure. Many GFs have been derived and are tabulated in this book, so the derivation of the GF can be omitted in many cases. Eigenfunctions and eigenconditions need not be developed. In these cases the solutions can be written immediately in terms of the GFs. The saving of effort and reduced possibility of errors are particularly important for two- and three-dimensional geometries.

A third advantage is that two- and three-dimensional GFs can be found by simple multiplication of one-dimensional GFs for the rectangular coordinate system for most of the boundary conditions considered in this book, provided that the problem is linear, the body is homogeneous, and the geometry is "orthogonal." An orthogonal geometry is one for which any boundary is located where only one coordinate is a constant, such as $x = 0$ or $y = L$, but not along a boundary defined by, say $x + y = C$. The multiplicative property of one-dimensional GFs can result in great simplification in the derivation of the temperature for two- and three-dimensional orthogonal geometries as well as a very compact means of cataloging the GFs for these cases. For certain two-dimensional cases involving cylindrical coordinates, multiplication of the GFs can also be used.

A fourth advantage is that the GF solution equation has an alternative form which can improve the convergence of problems with nonhomogeneous boundary conditions. For heat conduction in finite bodies, the expression for temperature usually involves an infinite series. Slow convergence of these infinite series expressions can sometimes require evaluation of a very large number of terms of an infinite series to obtain accurate numerical values. For some problems having nonhomogeneous boundary conditions, the alternative formulation reduces the number of required terms.

A fifth advantage is that the GF function solution method can be time partitioned to reduce the number of terms of an infinite series that must be evaluated. Time partitioning is a general method that arises naturally from the GF method. The method of time partitioning can give accurate values for the temperature (or the GF) using only a few terms of the infinite series. Time partitioning is introduced in Chapter 5.

1.1.2 Overview of Chapter 1

The purposes of this chapter are (1) to introduce GFs, the Green's function solution equation (GFSE), and a fundamental heat conduction solution; (2) to discuss the concept of the auxiliary problem; (3) to provide some insight and motivation; and (4) to give the scope of the remainder of the book. More rigorous aspects, such as derivation of the GFSE, are deferred to later chapters. In Section 1.2 the use of GFs for an infinite body is considered. Examples of the uses of some GFs for calculation of temperature are provided. In Section 1.3, two interpretations of Green's functions are given. In Section 1.4, temperatures in semi-infinite bodies for boundary conditions of the first and second kinds are discussed. Section 1.5 provides a discussion of the auxiliary problem of GFs. Section 1.6 contains a list of properties common to all GFs. Finally, Section 1.7 outlines the scope of the remainder of the book.

1.2 TEMPERATURES IN AN INFINITE ONE-DIMENSIONAL BODY

In this section, the simplest GF is introduced for heat conduction in an infinite body. This GF is a fundamental solution of the transient heat conduction equation; rather than deriving this GF, it can be verified that it is a solution of the heat conduction equation. Also given without a derivation is the GF solution equation (GFSE) for a semi-infinite body; see Chapter 3 for derivations. Complete examples are given to show how GFs are used in the GFSE for the infinite body.

The temperature distribution in an infinite one-dimensional, constant-property body which has an initial temperature distribution $F(x)$, and volumetric energy generation $g(x, t)$ (with units of W/m^3), is described by

$$\frac{\partial^2 T}{\partial x^2} + \frac{1}{k} g(x, t) = \frac{1}{\alpha} \frac{\partial T}{\partial t} \qquad -\infty < x < \infty \qquad t > 0 \tag{1.1}$$

$$T(x, 0) = F(x) \tag{1.2}$$

where T is temperature (K), x is position (m), t is time (s), k is thermal conductivity [W/(m K)], and α is thermal diffusivity (m^2/s); α is equal to $k/\rho c$ where ρ is density in kg/m^3 and c is specific heat in J (kg K).

There are two terms in these equations that can cause transient variations in the temperature, namely, $g(x, t)/k$ and $F(x)$. These terms cause two independent effects that involve GFs in analogous manners. One of the purposes in the discussion below is to show similarity in the solutions involving $g(x, t)/k$ and $F(x)$.

1.2.1 Green's Function Solution Equation (GFSE)

The temperature $T(x, t)$, the solution of Eqs. (1.1) and (1.2) and derived in Chapter 3, is called a Green's function solution equation (GFSE)

$$T(x, t) = \int_{x' = -\infty}^{\infty} G(x, t|x', 0) \, F(x') \, dx'$$

$$+ \frac{\alpha}{k} \int_{\tau = 0}^{t} \int_{x' = -\infty}^{\infty} G(x, t|x', \tau) \, g(x', \tau) \, dx' \, d\tau \qquad (1.3)$$

where $G(x, t|x', \tau)$ is called a GF. Notice that the solution has two terms, one containing the initial condition F and the other containing the volumetric energy source g. Each of the two terms on the right side of Eq. (1.3) can be considered to be the solution of a problem, one with F and one with g. The two solutions are superimposed (i.e., added) for the complete solution.

1.2.2 Fundamental Heat Conduction Solution

Depending on the geometry and boundary conditions, there are many expressions for the GF $G(x, t|x', \tau)$. The particular form of $G(\cdot)$ for an infinite one-dimensional body is a fundamental heat conduction solution (Cannon, 1984),

$$G(x, t|x', \tau) = K(x - x', t - \tau)$$

$$\equiv [4\pi\alpha(t - \tau)]^{-1/2} \exp\left[-\frac{(x - x')^2}{4\alpha(t - \tau)}\right] \qquad t - \tau \geq 0$$

$$\equiv 0 \qquad t - \tau < 0 \qquad (1.4)$$

(In the numbering system introduced in Chapter 2, this is called the GF for the $X00$ case.) The fundamental heat conduction solution, $K(x - x', t - \tau)$, has several important properties:

First, $K(x - x', t - \tau)$ satisfies the heat conduction given by Eq. (1.1) for $g(x, t) = 0$ for $t - \tau$ greater than zero. See Problem 1.1.

Second, $K(x - x', t - \tau)$ is always equal to or greater than zero for $t - \tau$ greater than zero,

$$K(x - x', t - \tau) \geq 0 \qquad \text{for } t - \tau > 0 \qquad (1.5a)$$

Third, the integral of $K(x - x', t - \tau)$ from $x' = -\infty$ to ∞ is unity for all x values and for all times $t - \tau > 0$,

$$\int_{x' = -\infty}^{\infty} K(x - x', t - \tau) \, dx' = 1 \qquad t - \tau > 0 \qquad (1.5b)$$

and is equal to zero for times $t - \tau < 0$,

$$\int_{x' = -\infty}^{\infty} K(x - x', t - \tau) \, dx' = 0 \qquad t - \tau < 0 \qquad (1.5c)$$

Fourth, the value of $K(x - x', t - \tau)$ is unchanged if $x - x'$ is replaced by $x' - x$,

$$K(x - x', t - \tau) = K(x' - x, t - \tau) \qquad (1.5d)$$

Fifth, the limit of the integral of $\partial K / \partial x$ as x approaches x' from below is $\frac{1}{2}$

$$\lim_{x \uparrow x'} \int_0^t \frac{\partial K(x - x', t - \tau)}{\partial x} d\tau = \frac{1}{2} \tag{1.5e}$$

and approaching x' from above is $-\frac{1}{2}$

$$\lim_{x \downarrow x'} \int_0^t \frac{\partial K(x - x', t - \tau)}{\partial x} d\tau = -\frac{1}{2} \tag{1.5f}$$

Depending on the geometry and the boundary conditions, there are many expressions for the GF $G(x, t | x', \tau)$, but there is only one GF for the case of an infinite body, and a convenient form of it is given by Eq. (1.4).

It is instructive to examine a plot of $K(x - x', t - \tau)$; see Fig. 1.1 which shows $K(x - x, t - \tau)$ as a function of $x - x'$ for various values of $\alpha(t - \tau)$. As $\alpha(t - \tau)$ goes to zero, the $K(\cdot)$ function approaches the Dirac delta function. Each curve in Fig. 1.1 has the bell shape of the Gaussian distribution. At all times $t > \tau$, the area underneath a curve in Fig. 1.1 is unity as given by Eq. (1.5b). As times $t - \tau$ increase, the $K(\cdot)$ function spreads and the maximum decreases.

The temperature distribution in an infinite body ($-\infty < x < \infty$) for the initial temperature distribution $F(x)$ and the volumetric energy generation of $g(x, t)$ is found using Eq. (1.4) in Eq. (1.3). The result is

$$T(x, t) = \int_{-\infty}^{\infty} (4\pi\alpha t)^{-1/2} \exp\left[-\frac{(x - x')^2}{4\alpha t} \right] F(x') \, dx'$$

$$+ \frac{\alpha}{k} \int_{\tau=0}^t \int_{x'=-\infty}^{\infty} [4\pi\alpha(t - \tau)]^{-1/2}$$

$$\times \exp\left[-\frac{(x - x')^2}{4\alpha(t - \tau)} \right] g(x', \tau) \, dx' \, d\tau \tag{1.6}$$

Some examples of the use of Eq. (1.6) are given next.

Example 1.1 Find the temperature distribution for the case of

$$F(x) = T_1 \qquad \text{for } c < x < d$$

$$= 0 \qquad \text{otherwise}$$

$$g(x, t) = 0 \qquad \text{for all } x$$

SOLUTION The solution for T is obtained by using Eq. (1.6) with $F(x') = T_1$ for $c < x' < d$ and $F(x') = 0$ otherwise. Also $g(x', \tau) = 0$ for all x'. The result is

$$T(x, t) = \int_c^d K(x - x', t) T_1 dx'$$

$$= T_1 \int_c^d (4\pi\alpha t)^{-1/2} \exp\left[-\frac{(x - x')^2}{4\alpha t} \right] dx' \tag{1.7}$$

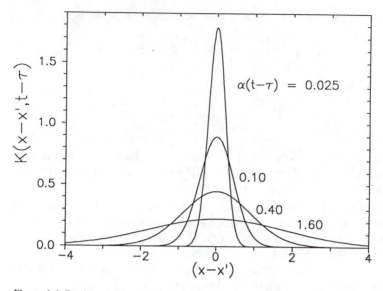

Figure 1.1 Fundamental heat conduction solution, $K(x - x', t - \tau)$.

Using the substitution $u = (x - x')/(4\alpha t)^{1/2}$, this integral can be written as

$$T(x, t) = \frac{T_1}{\pi^{1/2}} \int_{(x-d)/[(4\alpha t)^{1/2}]}^{(x-c)/[(4\alpha t)^{1/2}]} e^{-u^2} \, du \tag{1.8a}$$

$$T(x, t) = \frac{T_1}{2} \left\{ \mathrm{erf}\left[\frac{x - c}{(4\alpha t)^{1/2}}\right] - \mathrm{erf}\left[\frac{x - d}{(4\alpha t)^{1/2}}\right] \right\} \tag{1.8b}$$

$$T(x, t) = \frac{T_1}{2} \left\{ \mathrm{erfc}\left[\frac{x - d}{(4\alpha t)^{1/2}}\right] - \mathrm{erfc}\left[\frac{x - c}{(4\alpha t)^{1/2}}\right] \right\} \tag{1.8c}$$

where the error function, $\mathrm{erf}(\cdot)$, and the complementary error function, $\mathrm{erfc}(\cdot) = 1 - \mathrm{erf}(\cdot)$, are defined by

$$\mathrm{erf}(z) = \frac{2}{\pi^{1/2}} \int_0^z e^{-u^2} \, du \tag{1.9a}$$

$$\mathrm{erfc}(z) = \frac{2}{\pi^{1/2}} \int_z^\infty e^{-u^2} \, du \tag{1.9b}$$

These functions commonly occur in transient heat conduction. Some relations involving these functions are given in Appendix E. (See also Fig. 1.8.)

Equation (1.8) is plotted in Fig. 1.2 for $\alpha t/(d - c)^2 = 0.01, 0.05, 0.1, 0.5$, and 1 as a function of $(x - x_m)/(d - c)$ where x_m is the mean x value which is $(c + d)/2$. In this case, the temperature distribution can be written as

$$\frac{T}{T_1} = \frac{1}{2} \left[\mathrm{erfc}\left(\frac{x^+ - 0.5}{(4t^+)^{1/2}}\right) - \mathrm{erfc}\left(\frac{x^+ + 0.5}{(4t^+)^{1/2}}\right) \right] \tag{1.10}$$

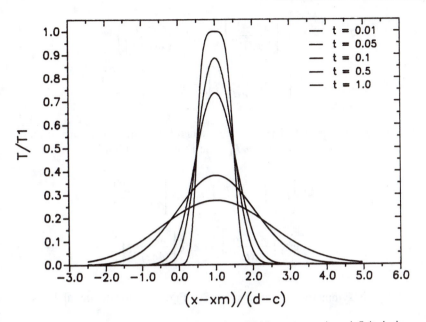

Figure 1.2 Temperature distribution for nonuniform initial temperature in an infinite body.

where $t^+ = \dfrac{\alpha t}{(d - c)^2}$

$x^+ = \dfrac{x - x_m}{d - c}$

The temperature distribution is affected first near the edges of the step change of the initial T distribution and, as the dimensionless time increases, the effect penetrates deeper.

Example 1.2 Find the temperature distribution for the case of $a < b < c < d$ and

$$F(x) = T_0 \qquad \text{for } a < x < b$$

$$= T_1 \qquad \text{for } c < x < d$$

$$= 0 \qquad \text{otherwise}$$

$$g(x, t) = 0 \qquad \text{for all } x$$

SOLUTION The solution can be found as in Example 1.1 by integrating over the two nonzero regions of $F(x)$ or by using Eq. (1.8c) as a building block (i.e., let $T_1 \to T_0$, $d \to b$, and $c \to a$). Using either procedure results in

$$T = \frac{T_0}{2}\left\{\text{erfc}\left[\frac{x-b}{(4\alpha t)^{1/2}}\right] - \text{erfc}\left[\frac{x-a}{(4\alpha t)^{1/2}}\right]\right\}$$

$$+ \frac{T_1}{2}\left\{\text{erfc}\left[\frac{x-d}{(4\alpha t)^{1/2}}\right] - \text{erfc}\left[\frac{x-c}{(4\alpha t)^{1/2}}\right]\right\} \quad (1.11)$$

Two interesting special cases can be obtained from Eq. (1.11). One of these is for $b \to -c$, $a \to -d$, and $T_0 \to T_1$. The resulting solution is

$$T = \frac{T_1}{2}\left\{\text{erfc}\left[\frac{x+c}{(4\alpha t)^{1/2}}\right] - \text{erfc}\left[\frac{x+d}{(4\alpha t)^{1/2}}\right]\right.$$

$$\left. + \text{erfc}\left[\frac{x-d}{(4\alpha t)^{1/2}}\right] - \text{erfc}\left[\frac{x-c}{(4\alpha t)^{1/2}}\right]\right\} \quad (1.12)$$

This solution is symmetric (about $x = 0$). See Fig. 1.3a for the initial T distribution.

Substitution of $-x$ for x in Eq. (1.12) and use of the Appendix E identity of $\text{erfc}(-z) = 2 - \text{erfc}(z)$ reveals the symmetry, and it can also be noted in Fig. 1.3a. This condition of symmetry can also be expressed mathematically by $\partial T/\partial x = 0$ at $x = 0$; $\partial T/\partial x = 0$ is sometimes called the insulation condition. In other words, the solution for a semi-infinite body ($x > 0$) which is insulated at $x = 0$ can be found from the infinite solution if the temperature distribution is made symmetric about $x = 0$.

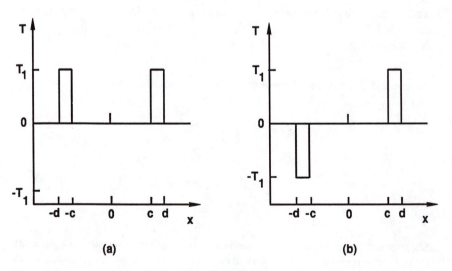

(a)   **(b)**

Figure 1.3 Initial temperature distribution for Example 1.2.

The other special case is for $b \rightarrow -c$, $a \rightarrow -d$, and $T_0 \rightarrow -T_1$ which has the solution

$$
T = \frac{T_1}{2} \left\{ \text{erfc} \left[\frac{x+c}{(4\alpha t)^{1/2}} \right] - \text{erfc} \left[\frac{x+d}{(4\alpha t)^{1/2}} \right] \right.
$$
$$
\left. - \text{erfc} \left[\frac{x-d}{(4\alpha t)^{1/2}} \right] + \text{erfc} \left[\frac{x-c}{(4\alpha t)^{1/2}} \right] \right\} \tag{1.13}
$$

This expression has the value of zero at $x = 0$ and is antisymmetric about the $x = 0$ axis. The zero temperature boundary condition is called the homogeneous isothermal condition. See Fig. 1.3b for the initial T distribution for this case.

Example 1.3 Find the temperature distribution for the case of

$$
F(x) = 0 \qquad \text{for all } x
$$

$$
g(x, t) = q_{x0} \delta(x - x_0)
$$

where q_{x0} has units of W/m^2, the same as those for heat flux. See Table 1.1 for properties of the Dirac delta function, $\delta(x - x_0)$. Also calculate the heat flux at x.

SOLUTION The solution for the temperature is obtained by using Eq. (1.6) with $F(x') = 0$,

$$
T(x, t) = \frac{\alpha}{k} \int_{\tau=0}^{t} \int_{x'=-\infty}^{\infty} [4\pi\alpha(t - \tau)]^{-1/2}
$$
$$
\times \exp\left[-\frac{(x - x')^2}{4\alpha(t - \tau)} \right] q_{x0} \delta(x' - x_0) \, dx' \, d\tau \tag{1.14}
$$

$$
T(x, t) = \frac{\alpha q_{x0}}{k} \int_{\tau=0}^{t} [4\pi\alpha(t - \tau)]^{-1/2} \exp\left[-\frac{(x - x_0)^2}{4\alpha(t - \tau)} \right] d\tau \tag{1.15}
$$

Table 1.1 Properties of Dirac delta function, $\delta(x - x')$

---

1. $\int_{-\infty}^{\infty} \delta(x - x') \, dx' = 1$

2. $\delta(x - x') = \begin{cases} \rightarrow \infty & \text{as } x \rightarrow x' \\ = 0 & \text{otherwise} \end{cases}$

3. $\int_{-\infty}^{\infty} F(x') \delta(x - x') \, dx' = F(x)$

4. $\dfrac{dH(t - \tau)}{dt} = \delta(t - \tau)$ where $H(\cdot)$ is the Heaviside unit step function

5. $\delta(t - \tau)$ has units of s^{-1}
    $\delta(x - x')$ has units of m^{-1}
    $\delta(\mathbf{r} - \mathbf{r}')$ has units such that $\int \delta(\mathbf{r} - \mathbf{r}') \, dv' = 1$

---

because the only contribution to the integral is at $x' = x_0$. Using integral 9 in Table F.6, Appendix F, gives, for Eq. (1.15),

$$T(x, t) = \frac{q_{x0}}{k} (\alpha t)^{1/2} \, \text{ierfc} \left[\frac{|x - x_0|}{(4\alpha t)^{1/2}} \right] \tag{1.16}$$

where $\text{ierfc}(z)$ is given by (see Appendix E)

$$\text{ierfc}(z) \equiv \int_z^\infty \text{erfc}(u) \, du = \pi^{-1/2} \exp(-z^2) - z \, \text{erfc}(z) \tag{1.17}$$

Notice that Eq. (1.16) is symmetric about $x = x_0$. The maximum temperature is finite and occurs at $x = x_0$ and is

$$T(x, t) = q_{x0} \left(\frac{t}{\pi k \rho c} \right)^{1/2} \tag{1.18}$$

1.3 TWO INTERPRETATIONS OF GREEN'S FUNCTIONS

Two different physical interpretations of $G(\cdot)$ can be found from Eq. (1.3) and are described below. The first physical interpretation of $G(\cdot)$ is the temperature distribution caused by a particular initial condition and the second interpretation is the temperature distribution for an instantaneous heat source.

The first physical interpretation is associated with the first term in Eq. (1.3) and is the solution $T(x, t)$ for the problem

$$\frac{\partial^2 T}{\partial x^2} = \frac{1}{\alpha} \frac{\partial T}{\partial t} \qquad -\infty < x < \infty \qquad t > 0 \tag{1.19a}$$

$$T(x, 0) = F(x) \tag{1.19b}$$

If the initial temperature distribution is zero everywhere except at x_0 where it is equal to F_0' times the Dirac delta function (see Table 1.1),

$$F(x) = F_0' \, \delta(x - x_0) \tag{1.20}$$

then the solution of Eq. (1.19a) and (1.20) is

$$T(x, t) = F_0' \, G(x, t | x_0, 0) \tag{1.21}$$

Hence, the GF $G(x, t | x_0, 0)$ can be interpreted as being the temperature distribution in the body that is the result of the initial temperature being zero everywhere except at point x_0 where there is a Dirac delta in the temperature distribution of magnitude $F_0' = 1$ K-m, where K-m denotes the units of Kelvin times meter. The units of $G(\cdot)$ and $K(\cdot)$ are both reciprocal length m^{-1}; the unit for $\delta(x - x_0)$ is also m^{-1}.

The second physical interpretation of a GF is the temperature caused by an instantaneous heat source at time t_0 and position x_0 and of strength $H\rho c$. For this case, the volumetric energy generation term in Eq. (1.1) becomes

$$g(x, t) = H\rho c \, \delta(x - x_0) \, \delta(t - t_0) \tag{1.22}$$

where H has the units of K-m; $\delta(x - x_0)$ has the unit m^{-1}; $\delta(t - t_0)$ has the units s^{-1}; and ρc has the units of J/m^3 K. These units are consistent with those of the volume energy generation g, which are W/m^3. The symbol g given by Eq. (1.22) represents the amount of energy that is released at $x = x_0$ and at $t = t_0$. It can be visualized as the energy associated with an instantaneous plane source in the direction normal to the x axis. It is also like an instantaneous laser sheet pulse being released at x_0 and at time t_0. For this case, the describing differential equation is

$$\frac{\partial^2 T}{\partial x^2} + \frac{1}{k} H\rho c \ \delta(x - x_0) \ \delta(t - t_0) = \frac{1}{\alpha} \frac{\partial T}{\partial t} \qquad -\infty < x < \infty \qquad t > 0 \qquad (1.23)$$

and the initial temperature distribution is zero,

$$T(x, 0) = 0 \qquad -\infty < x < \infty \qquad (1.24)$$

The solution for the temperature is zero until time $t = t_0$. After time t_0, the solution for $T(x, t)$ given by Eq. (1.3) is

$$T(x, t) = \frac{\alpha}{k} \int_{\tau=0}^{t} \int_{x'=-\infty}^{\infty} G(x, t|x', \tau) \ H\rho c$$

$$\times \ \delta(x' - x_0) \ \delta(\tau - t_0) \ dx' \ d\tau \qquad (1.25a)$$

$$T(x, t) = H \ G(x, t|x_0, t_0) \qquad (1.25b)$$

Notice that in using Eq. (1.3) for $g(x, t)$, it is necessary to replace x by x' and t by τ. The major point, however, is that GF is equal to the temperature rise for the instantaneous plane heat source given by Eq. (1.22) with $H = 1$.

These two alternate ways of thinking about GFs are important. In the first interpretation, the GF is equal to the temperature resulting from an initial temperature distribution that is zero everywhere except at the location of the Dirac delta function with strength of 1 K-m. In the second interpretation, the GF is equal to the temperature rise due to an instantaneous plane source with a strength of one K-m times ρc.

1.4 TEMPERATURES IN SEMI-INFINITE BODIES

A semi-infinite body is described by a body occupying the region $x \geq 0$. Although it represents an idealized body extending to positive infinity, it is a good model for many problems. A finite body of thickness L can be represented by a semi-infinite body, $0 < x < \infty$, when the boundary condition at $x = L$ does not influence the temperature distributions near $x = 0$. This happens for the small dimensionless times of $\alpha t/L^2 < 0.05$. Isothermal and insulation boundary conditions at $x = 0$ can be constructed from the infinite region solutions. The examples of Section 1.2 illustrate these points; also see Fig. 1.3.

Temperature solutions for a semi-infinite body with an isothermal surface and an insulated surface can be obtained using the fundamental heat conduction solution, given by Eq. (1.4). The homogeneous isothermal case is for the surface temperature (at $x = 0$) held at 0 degrees. A prescribed temperature at a boundary is called a boundary condition of the first kind. If the prescribed temperature is zero, the boundary condition is termed homogeneous. A prescribed heat flux at a surface is called a boundary condition of the second kind; if this heat flux is zero, the surface is said to be insulated and the boundary condition is also homogeneous. Both boundary conditions, the first and second kinds, are now considered by utilizing the concept of superposition which is valid because the problems are linear.

1.4.1 Boundary Condition of the First Kind

Consider a homogeneous boundary condition of the first kind (specified temperature) for a semi-infinite body,

$$\frac{\partial^2 T}{\partial x^2} = \frac{1}{\alpha}\frac{\partial T}{\partial t} \qquad 0 < x \leq \infty \tag{1.26}$$

$$T(x, 0) = F(x) \qquad 0 < x \leq \infty \tag{1.27}$$

$$T(0, t) = 0 \qquad t > 0 \tag{1.28}$$

See Fig. 1.4a for the geometry. The solution to this problem is the same as for an infinite body with the initial temperature $T(x, 0)$ equal to $F(x)$ for $x > 0$ and equal to $-F(-x)$ for $x < 0$; see Fig. 1.4b. Then the first term of Eq. (1.3) with $G(x, t|x', 0) = K(x - x', t)$ gives

$$T(x, t) = \int_{x'=0}^{\infty} K(x - x', t)\, F(x')\, dx'$$

$$- \int_{x'=-\infty}^{0} K(x - x', t) F(-x')\, dx' \tag{1.29a}$$

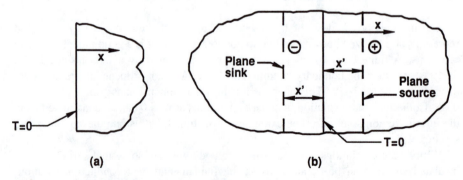

(a)   **(b)**

Figure 1.4 (a) Semi-infinite body with an isothermal boundary. (b) Semi-infinite body with $T = 0$ at $x = 0$ simulated by an infinite body with source at x' and sink at $-x'$.

In the second integral, replace $-x'$ by x'' to get

$$T(x, t) = \int_{x'=0}^{\infty} K(x - x', t) F(x') \, dx' - \int_{x''=0}^{\infty} K(x + x'', t) F(x'') \, dx''$$

$$= \int_{x'=0}^{\infty} [K(x - x', t) - K(x + x', t)] F(x') \, dx' \qquad (1.29b)$$

since x' and x'' are dummy variables. Notice that the domain of $0 \le x' \le \infty$ is included in the integral of Eq. (1.29b). This equation can be written in terms of a new GF,

$$T(x, t) = \int_{x'=0}^{\infty} G(x, t|x', 0) F(x') \, dx' \qquad (1.30a)$$

where the new GF is equal to

$$G(x, t|x', \tau) = K(x - x', t - \tau) - K(x + x', t - \tau) \qquad (1.30b)$$

$$G(x, t|x', \tau) = [4\pi\alpha(t - \tau)]^{-1/2} \left\{ \exp\left[-\frac{(x - x')^2}{4\alpha(t - \tau)} \right] \right.$$

$$\left. - \exp\left[-\frac{(x + x')^2}{4\alpha(t - \tau)} \right] \right\} \qquad t - \tau \ge 0 \qquad (1.30c)$$

This GF represents the physical problem of an instantaneous plane source of strength $H = 1$ m-K times ρc and at location x' and at time τ in a semi-infinite body with zero boundary conditions and zero initial conditions:

$$\frac{\partial^2 G}{\partial x^2} + \frac{1}{\alpha} \delta(x - x_0) \, \delta(t - t_0) = \frac{1}{\alpha} \frac{\partial G}{\partial t} \qquad 0 < x < \infty \qquad t > 0 \qquad (1.31)$$

$$G(0, t|x', \tau) = 0 \qquad G(\infty, t|x', \tau) = 0 \qquad (1.32)$$

$$G(x, 0|x', \tau) = 0 \qquad (1.33)$$

Equation (1.31) is obtained from Eq. (1.23) by replacing H by 1 and T by G. The presence of a sink at $x = -x'$ shown in Fig. 1.4b ensures that G is equal to zero at $x = 0$. The GF given by Eq. (1.30c) is plotted in Fig. 1.5. The curves are given for constant values of $\alpha(t - \tau)/x^2$ equal to 0.025, 0.05, 0.25, 1.0, and 4.0 versus x'/x; the same curves are obtained for fixed values of $\alpha(t - \tau)/x'^2$ versus x/x'. The GF is little affected by the isothermal boundary condition for $\alpha(t - \tau)/x^2 < 0.05$. For larger dimensionless times, the maximum G moves to larger x'/x values and its magnitude decreases.

1.4.2 Boundary Condition of the Second Kind

Next consider the case of the insulated surface (the boundary condition of the second kind). See Fig. 1.6a. This case can be treated in a similar manner as the homogeneous isothermal case. The differential equation, Eq. (1.26), and the initial condition, Eq. (1.27), are the same, but the boundary condition is

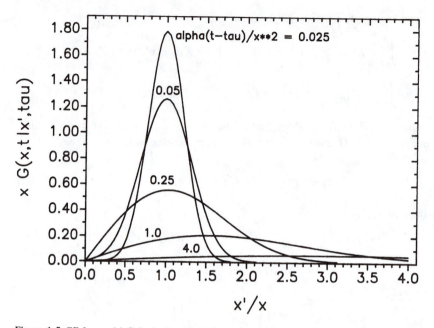

Figure 1.5 GF for semi-infinite body with isothermal condition of $G = 0$ at $x = 0$.

$$\frac{\partial T}{\partial x}\bigg|_{x=0} = 0 \tag{1.34}$$

which is a condition associated with symmetry about $x = 0$. (Other coordinate systems, such as radial, may not have symmetry for $\partial T/\partial r = 0$.)

The solution for the temperature can be obtained by using Eq. (1.3) (which is for $-\infty < x < \infty$) by making the initial temperature distribution symmetric, that is, equal to $F(x)$ for $x > 0$ and equal to $F(-x)$ for $x < 0$. Then using Eq. (1.3) with $G(x, t|x', 0) = K(x - x', t)$ gives

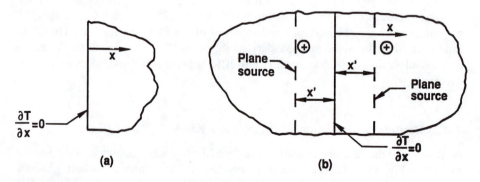

Figure 1.6 (a) Semi-infinite body with an insulated boundary. (b) Semi-infinite body with $\partial T/\partial x = 0$ at $x = 0$ simulated by an infinite body with source at x' and at $-x'$.

$$T(x, t) = \int_{x'=0}^{\infty} K(x - x', t) \, F(x') \, dx'$$

$$+ \int_{x'=-\infty}^{0} K(x - x', t) \, F(-x') \, dx' \qquad (1.35)$$

Replacing $-x'$ in the second integral by x'' and then combining into a single integral gives

$$T(x, t) = \int_{x'=0}^{\infty} K(x - x', t) \, F(x') \, dx'$$

$$+ \int_{x''=0}^{\infty} K(x + x'', t) \, F(x'') \, dx'' \qquad (1.36a)$$

$$T(x, t) = \int_{x'=0}^{\infty} [K(x - x', t) + K(x + x', t)] \, F(x') \, dx' \qquad (1.36b)$$

$$T(x, t) = \int_{x'=0}^{\infty} G(x, t|x', 0) \, F(x') \, dx' \qquad (1.36c)$$

where $G(\cdot)$ is given by

$$G(x, t|x', \tau) = K(x - x', t - \tau) + K(x + x', t - \tau) \qquad (1.37a)$$

$$G(x, t|x', \tau) = [4\pi\alpha(t - \tau)]^{-1/2} \left\{ \exp\left[-\frac{(x - x')^2}{4\alpha(t - \tau)} \right] \right.$$

$$\left. + \exp\left[-\frac{(x + x')^2}{4\alpha(t - \tau)} \right] \right\} \qquad t - \tau \geq 0 \qquad (1.37b)$$

This expression is the GF for a semi-infinite body insulated at $x = 0$. This solution can be also visualized as the result of superimposing two sources, one at $x = x'$ and the other at $x = -x'$. See Fig. 1.6b. The GF given by Eq. (1.37) is shown in Fig. 1.7 which shows $xG(\cdot)$ versus x'/x for $x \neq 0$; if $x = 0$, the $G(\cdot)$ function given by Eq. (1.37b) is twice as large as the GF shown in Fig. 1.1. As for the boundary condition of the first kind, the GF in Fig. 1.7 is unaffected by the $\partial T/\partial x = 0$ boundary condition at $x' = 0$ for $\alpha(t - \tau)/x^2 < 0.05$. Unlike that case, however, the maximum G moves to $x'/x = 0$ as the dimensionless time increases. Moreover, this case has G values (for the same x and t's) that are always as large or larger than the $G = 0$ at $x' = 0$ case, Fig. 1.5; the effect is most noticeable for $\alpha(t - \tau)/x^2 = 0.25$ to 4.0.

The method of deriving the GF given by Eq. (1.37) is related to the method of images for deriving the GFs, which is discussed in greater depth in Chapter 4.

Example 1.4 Find the temperature distribution for the problem

$$\frac{1}{\alpha} \frac{\partial^2 T}{\partial x^2} = \frac{\partial T}{\partial t} \qquad x > 0 \qquad t > 0$$

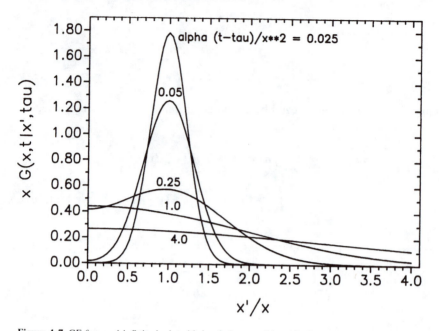

Figure 1.7 GF for semi-infinite body with insulation condition of $\partial G/\partial x = 0$ at $x = 0$.

$$T(0, t) = 0 \qquad T(\infty, t) \to T_0 = \text{constant}$$

$$T(x, 0) = T_0 \qquad x > 0$$

SOLUTION This problem has the boundary condition of the first kind and the solution is given by Eq. (1.30a) with $G(\cdot)$ given by Eq. (1.30b, c).

$$T(x, t) = \int_{x'=0}^{\infty} [K(x - x', t) - K(x + x', t)] T_0 \, dx'$$

$$= T_0 \left[\frac{1}{2} \operatorname{erfc}\left(\frac{x - x'}{(4\alpha t)^{1/2}}\right) \bigg|_{x'=0}^{\infty} + \frac{1}{2} \operatorname{erfc}\left(\frac{x + x'}{(4\alpha t)^{1/2}}\right) \bigg|_{x'=0}^{\infty} \right]$$

$$= T_0 \left[\left(1 - \frac{1}{2} \operatorname{erfc} \frac{x}{(4\alpha t)^{1/2}}\right) + \frac{1}{2}\left(0 - \operatorname{erfc} \frac{x}{(4\alpha t)^{1/2}}\right) \right]$$

$$= T_0 \left[1 - \operatorname{erfc} \frac{x}{(4\alpha t)^{1/2}} \right] = T_0 \operatorname{erf} \frac{x}{(4\alpha t)^{1/2}} \tag{1.38}$$

This solution is plotted in Fig. 1.8 versus $z = x/(4\alpha t)^{1/2}$; also shown are $\operatorname{erf}(z)$ and $\operatorname{erfc}(z)$. The variation of temperature is most pronounced for $x/(4\alpha t)^{1/2}$ less than 1.0.

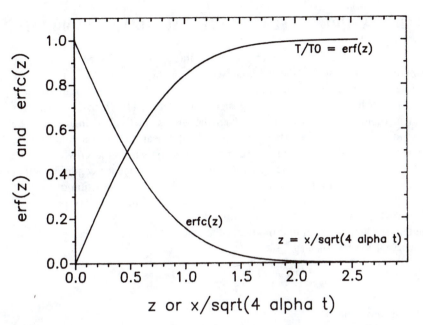

Figure 1.8 Error function (erf) and complementary error function (erfc).

Example 1.5 Find the temperature distribution for the problem

$$\frac{1}{\alpha}\frac{\partial^2 T}{\partial x^2} = \frac{\partial T}{\partial t} \qquad x > 0 \qquad t > 0$$

$$T(0, t) = T_0 \qquad T(\infty, t) \to 0$$

$$T(x, 0) = 0 \qquad x > 0$$

SOLUTION This problem does not have a homogeneous isothermal boundary condition but a related problem does. Define the new variable,

$$T' = T_0 - T$$

so that

$$\frac{1}{\alpha}\frac{\partial^2 T'}{\partial x^2} = \frac{\partial T'}{\partial t} \qquad x > 0 \qquad t > 0$$

$$T'(0, t) = 0 \qquad T'(\infty, t) \to T_0$$

$$T'(x, 0) = T_0 \qquad x > 0$$

the T' solution is given by Eq. (1.38) so that the solution is

$$\frac{T}{T_0} = \text{erfc}\left[\frac{x}{(4\alpha t)^{1/2}}\right] \qquad (1.39)$$

which also is shown in Fig. 1.8 as the erfc(z) curve.

1.5 TEMPERATURES IN FLAT PLATES: AUXILIARY PROBLEMS

The construction of the GF by superposition of the plane sources and sinks in an infinite body, as discussed in the previous section for the geometry of semi-infinite body, can also be extended to the finite geometry of the flat plate. This approach is an application of the *method of images* (Carslaw and Jaeger, 1959, p. 273) which is discussed in more detail in Chapter 4. Even though the method of images can be employed to construct the GFs (from the fundamental heat conduction solution) for the geometry of the flat plate, there are many cases for which the GFs cannot be obtained by this method, in particular, cases that involve boundary conditions other than the first and second kinds. A more general approach for construction of the GFs is through the use of an auxiliary problem. The concept of the auxiliary problem is now illustrated through the following discussion.

The temperature problem that motivates the study of the one-dimensional GF for the geometry of the flat plate is

$$\frac{\partial^2 T}{\partial x^2} + \frac{1}{k} g(x, t) = \frac{1}{\alpha} \frac{\partial T}{\partial t} \tag{1.40}$$

with boundary conditions

$$k_i \frac{\partial T}{\partial n_i}\bigg|_{x_i} + h_i T\big|_{x_i} = f_i(t) \tag{1.41}$$

where n_i is an outward normal from the body at the boundary. The subscript i is either 1 or 2 to represent the two boundaries. Thus, x_1 and x_2 are the locations of the left and right boundaries, respectively. The initial condition is

$$T(x, 0) = F(x) \tag{1.42}$$

The boundary condition, Eq. (1.41), represents three different kinds of boundary conditions by the choice of k_i, h_i, and f_i. These three boundary conditions are commonly studied and are called the first, second, and third kinds.

The first kind of boundary condition (also called the Dirichlet condition) is obtained from Eq. (1.41) by setting $k_i = 0$ and $h_i = 1$ to get the prescribed surface temperature

$$T(x_i, t) = f_i(t) \tag{1.43}$$

where $f_i(t)$ can also be simply zero. The second kind of boundary condition (also called the Neumann condition) is prescribed surface heat flux

$$k_i \frac{\partial T}{\partial n_i}\bigg|_{x_i} = f_i(t) \tag{1.44}$$

which becomes an insulated boundary if $f_i(t) = 0$.

The third kind of boundary condition is a convective boundary condition (also called the Robin condition) given by Eq. (1.41), where $f_i(t)$ is usually $h_i T_\infty$. The most familiar form of this boundary condition is then

$$-k_i \left. \frac{\partial T}{\partial n_i} \right|_{x_i} = h_i(T|_{x_i} - T_\infty) \tag{1.45}$$

where T_∞ is the constant or time-varying ambient temperature.

1.5.1 The Auxiliary Problem

The GF associated with the temperature given by Eqs. (1.40) through (1.42) is the solution to the *auxiliary equation*,

$$\frac{\partial^2 G}{\partial x^2} + \frac{1}{\alpha} \delta(x - x') \, \delta(t - \tau) = \frac{1}{\alpha} \frac{\partial G}{\partial t} \tag{1.46a}$$

subject to the *homogeneous* boundary conditions

$$k_i \left. \frac{\partial G}{\partial n_i} \right|_{x_i} + h_i G|_{x_i} = 0 \qquad i = 1, 2 \tag{1.46b}$$

and zero initial condition

$$G(x, t|x', \tau) = 0 \qquad \text{when } t < \tau \tag{1.46c}$$

[Equation (1.46a) is similar to Eq. (1.23) with $T \to G$ and $H \to 1$.] The auxiliary equation for any GF is identical to the original heat conduction equation except for the energy generation term, which is a Dirac delta function at location x' and at time τ. The one-dimensional GF G, defined by Eq. (1.46) has units of m^{-1}. This is apparent from the units of the energy generation term in Eq. (1.46a) [$\delta(x - x') \, \delta(t - \tau)/\alpha$, which has units of m^{-3}]. The homogeneous boundary conditions for the auxiliary equation are the same kinds as for the original problem.

1.6 PROPERTIES COMMON TO GREEN'S FUNCTIONS

The properties common to GF for heat conduction are summarized below.

1. G obeys the auxiliary equation.
2. G is a solution of the heat conduction problem having the same geometry but having homogeneous boundary conditions of the same kind as the original heat conduction problem.
3. $G \geq 0$ in the domain R for $t - \tau \geq 0$. $G = 0$ in the domain R for $t - \tau < 0$, the causality relation.
4. $G(x, t|x', \tau) = G(x', -t|x, -\tau)$, the reciprocity relation.
5. The time dependence of G is always $t - \tau$, so a one-dimensional GF could be written $G(x, x', t - \tau)$.
6. In rectangular coordinates, G has units of: m^{-1} for one-dimensional problems; m^{-2} for two-dimensional problems; and m^{-3} for three-dimensional problems.

Every GF is a solution to an auxiliary equation with homogeneous boundary conditions. The GF is always positive or zero, because it is the temperature caused by a positive heat pulse. The causality relation relates to the idea that the GF is the response at time t and location x to a pulse of heat occurring at time τ and at location x'. In a real (or causal) system, there can be no response before the pulse of heat occurs.

The reciprocity relation can be understood from the auxiliary equation, Eq. (1.46). Exchanging x and x' in the auxiliary equation leaves the sign of the solution unchanged because of the second derivative with respect to x. However, exchanging t and τ changes the sign of the solution, because of the first derivative with respect to t. Spatial orientation has no preferred direction in heat conduction, but time does have a preferred direction.

1.7 SCOPE OF BOOK

This section provides a preview of the remainder of this book. There are many unique parts of this book, including a new numbering system, extensive tables of GFs and an approximate GF solution procedure using a Galerkin-based analysis.

Chapter 2 provides a numbering system that is used in most of the rest of this book. Both GFs and temperature solutions are classified and tabulated using the system. An important advantage of the heat conduction numbering system is that existing solutions can be readily cataloged and recalled. This book contains an extensive set of GFs that are tabulated using this numbering system; see for example Appendices X, R, and RS. Also Appendix N gives the location of the temperature solutions in this book, indexed by the numbering system.

Chapter 3 gives two derivations of the GFSE, one for one-dimensional cases and the other for general multidimensional coordinates. These GFSEs are expressions for the temperature distribution in terms of various integrals of involving GFs. Chapter 3 also contains a derivation for an alternative form of the GFSE which can aid in obtaining solutions having better convergence properties than the GFSE for some problems having nonhomogeneous boundary conditions. Chapter 4 describes three methods of deriving Green's functions. The two interpretations of the GF given in Chapter 1 are utilized in obtaining the GFs.

Chapter 5 introduces the new concept of time partitioning. The GFs for finite bodies can be given in at least two different forms, one of which is for small values of $t - \tau$ and the other for all values of $t - \tau$. Although the one for large $t - \tau$ is valid for all times, its use for nonhomogeneous problems with the GFSE can result in infinite-series expressions that converge slowly. The small time solution, which comes from the semi-infinite solutions, requires only a few terms for its contribution. By utilizing the two types of GFs in the same solution, expressions can be derived that are computationally more efficient.

Chapter 6 discusses the use of the GFSE for the solution of heat conduction problems in Cartesian coordinates. One-, two-, and three-dimensional cases are covered. Chapter 7 covers radial heat flow and Chapter 8 covers two- and three-dimensional cases in cylindrical coordinates. Chapter 9 discusses temperature solutions in radial spherical coordinates.

Chapter 10 introduces the use of an important approximate method using Galerkin-based GFs. Although it is approximate, the answers provided for finite bodies can be extremely accurate, if a sufficient number of terms are used. Chapter 11 is a companion chapter that discusses various applications of these GFs. There is a computer program, COND, based on these Galerkin-base GFs for one-dimensional problems in Cartesian, cylindrical radial, and spherical radial coordinates. A large number of boundary conditions are considered and also the cases of homogeneous bodies and two-layer bodies (including a finite contact conductance) are treated. Temperatures and GFs can be plotted and evaluated. The use of these Galerkin-based GFs is particularly attractive for radial cylindrical coordinates, rather than the classical Bessel functions.

Chapter 12 discusses the unsteady surface element method. This method is useful when two or more bodies of distinctly different geometries are connected together. An example is the intrinsic thermocouple problem in which a cylindrical rod is attached to a semi-infinite body. In such a problem, the time domains of interest, based on the radius of the rod, can vary from 0.01 to 1,000,000. Numerical methods such as finite differences and finite elements have difficulty treating problems of this type.

At the end of the book are extensive tables of GFs, relevant integrals and data bases of transient heat conduction solutions. Also included is a description of the versatile one-dimensional computer program, COND, mentioned in connection with Chapters 10 and 11.

REFERENCES

Beck, J. V., and Keltner, N. R., 1987, Green's Function Partitioning Method Applied to Foil Heat Flux Gages, *J. Heat Transfer*, vol. 109, pp. 274–280.

Butkovskiy, A. G., 1982, *Green's Functions and Transfer Functions Handbook*, Halsted Press (div. of Wiley), New York.

Cannon, J. R., 1984, *The One-Dimensional Heat Equation*, Addison-Wesley, Menlo Park, Calif.

Carslaw, H. S., and Jaeger, J. C., 1959, *Conduction of Heat in Solids*, 2d ed., Oxford, New York.

Greenberg, M. D., 1971, *Application of Green's Functions in Science and Engineering*, Prentice-Hall, Englewood Cliffs, N.J.

Haji-Sheikh, A., and Lakshminarayanan, R., 1987, Integral Solution of Diffusion Equation: Part 2-Boundary Conditions of Second and Third Kinds, *ASME J. Heat Transfer*, vol. 109, pp. 557–562.

Haji-Sheikh, A., 1988, Heat Diffusion in Heterogeneous Media Using Heat-Flux-Conserving Basis Functions, *ASME J. Heat Transfer*, vol. 110, pp. 276–282.

Luikov, A. V., 1968, *Analytical Heat Diffusion Theory*, Academic Press, New York.

Mikhailov, M. D., and Ozisik, M. N., 1984, *Unified Analysis and Solutions of Heat and Mass Transfer*, Wiley, New York.

Morse, P. M., and Feshbach, H., 1953, *Methods of Theoretical Physics*, McGraw-Hill, New York.

Ozisik, M. N., 1980, *Heat Conduction*, Wiley, New York.

Potter, M. C., and Goldberg, J., 1987, *Mathematical Methods*, 2d ed., Prentice-Hall, Englewood Cliffs, N.J.

Roach, G. F., 1970, *Green's Functions, Introductory Theory with Applications*, Van Nostrand Reinhold, New York.

Stakgold, I., 1979, *Green's Functions and Boundary Value Problems*, Wiley-Interscience, New York.

PROBLEMS

1.1 By substituting $K(x - x', t - \tau)$ into Eq. (1.19a) for $T(x, t)$, verify that $K(x - x', t)$ is a solution. What is the initial condition?

1.2 Using the approximations of

$$f(x) = \begin{cases} A & -A^{-1} < x < A^{-1} \\ 0 & x^2 > A^{-2} \end{cases}$$

for

$$\frac{2}{\pi^{1/2}} \exp(-x^2)$$

and A equal to $\pi^{1/2}/2$, find approximate expressions for erf(z), erfc(z) and ierfc(z).

1.3 Find the value of A in Problem 1.2 that minimizes the least-squares function

$$S = \int_0^{1/A} [2\pi^{-1/2} \exp(-x^2) - A]^2 \, dx + \int_{1/A}^{\infty} [2\pi^{-1/2} \exp(-x^2)]^2 \, dx$$

(Answer: 0.9194.)

1.4 Verify Eq. (1.17) using integration by parts.

1.5 Investigate the behavior of the approximation of erfc(x) given by

$$\pi^{-1/2} \exp(-x^2) \left[\frac{1}{x} - \frac{1}{2x^3} + \frac{1 \cdot 3}{2^2 x^5} - \frac{1 \cdot 3 \cdot 5}{2^3 x^7} + \cdots \right]$$

for a given $x > 1$ as the number of terms is increased. Verify that the error is less in absolute value than the last term retained.

1.6 Find the temperature distribution in a semi-infinite body with the initial temperature given by

$$T = x \quad \text{for } 0 < x \le 1 \quad \text{and} \quad T = 0 \quad \text{for } x > 1$$

The surface temperature at $x = 0$ is maintained at zero temperature. (Tables 5.1–5.3 may be helpful.)

1.7 Find the temperature in a semi-infinite body with the initial temperature given by

$$T = x^2 + T_0 \quad \text{for } 0 < x \le L \quad \text{and} \quad T = T_0 \quad \text{for } x > L$$

The surface at $x = 0$ is insulated. (Tables 5.1–5.3 may be helpful.)

1.8 The temperature due to a specified heat flux boundary condition (nonhomogeneous boundary condition of the second kind) in a semi-infinite body may be found by using a planar heat source located at the surface. Find the temperature resulting from a volumetric heat source given by

$$g(x, t) = q_0 \, \delta(x - 0)$$

Also, find the heat flux through the point $x = a$ inside the body.

1.9 Derive the below expression for the heat flux at x starting with $T(x, t)$ given by Eq. (1.16),

$$q(x, t) = \frac{q_{x0}}{2} \text{sign}(x - x_0) \, \text{erfc} \left[\frac{|x - x_0|}{(4\alpha t)^{1/2}} \right]$$

NUMBERING SYSTEM IN HEAT CONDUCTION

2.1 INTRODUCTION

The number of exact solutions in transient heat conduction and diffusion is extremely large and is growing. These solutions are needed for thermal modeling of various devices, as test cases for finite difference/element programs, and as influence functions for the unsteady surface element method (see Chapter 12). Solutions are given in many different papers, government reports, and industry reports. Because of the lack of organization of the solutions, it was frequently easier to rederive a solution than to search for it. With the advent of large and inexpensive computer memories, the development of specialized data bases has become practical, and they exist in medicine, law, and many other fields. Data bases are used by expert systems, a form of artificial intelligence. One paper on using artificial intelligence for heat transfer problems is by Sharma and Minkowycz (1982). In developing such systems, it is very helpful to have a numbering system to organize the information.

The purpose of this chapter is to propose a numbering system for heat conduction and diffusion. Such a system not only simplifies construction of a computer data base but it makes deriving new solutions more simple and locating solutions less tedious. A related presentation of number systems is given by Beck and Litkouhi (1988) and other discussions are given in Beck (1984, 1986).

The numbering system covers basic geometries such as plates, cylinders, and spheres. Irregular geometries such as plates with several randomly spaced holes are not covered in the numbering system. This book deals mainly with solutions for temperature-independent thermal properties, but the numbering system can be employed for nonlinearities caused by temperature-variable properties.

The numbering system is specifically developed for transient diffusion and heat conduction. The same concepts, however, are applicable to other fields, such as convective heat transfer, fluid mechanics, and wave phenomena. Steady state is covered because it is included by the more general transient notation.

The plan of this chapter is first to give the numbering system for geometry and boundary conditions in Section 2.2. Section 2.3 provides boundary condition modifiers to describe the time and/or space variations of the nonhomogeneous term at a boundary. Section 2.4 gives an initial temperature distribution numbering system, and Section 2.5 provides a numbering system to treat interfaces between bodies. Section 2.6 gives a numbering system for the volumetric energy generation term $g(x, t)$, and then Section 2.7 gives some examples of the numbering system. The chapter concludes with Section 2.8, further discussion of advantages of the numbering system.

We recognize that not all readers will share our enthusiasm for the heat conduction numbering system. However, it is important that readers have some knowledge of the numbering system in order to use the extensive appendices of Green's functions (GFs) in this book. Most of the book will be accessible to the reader with a working knowledge of Section 2.2 on the numbering system for geometry and boundary conditions. Some readers may prefer to read Section 2.2 and then jump ahead to Chapter 3 on the Green's function solution equation (GFSE). Later these readers can return to Chapter 2 to learn more about the numbering system as the need arises.

2.2 GEOMETRY AND BOUNDARY CONDITION NUMBERING SYSTEM

For the rectangular coordinate system, the symbol X is used to denote the x coordinate; Y is used to denote the y direction; and Z is used to denote the z direction. For a two-dimensional problem involving x and y coordinates, X and Y are used; for a three-dimensional problem, X, Y, and Z are used. The three-dimensional equation for transient conduction with constant, isotropic thermal conductivity k is

$$k \left(\frac{\partial^2 T}{\partial x^2} + \frac{\partial^2 T}{\partial y^2} + \frac{\partial^2 T}{\partial z^2} \right) = \rho c \frac{\partial T}{\partial t} \tag{2.1}$$

For the cylindrical coordinates, r, ϕ, x, the symbol R is for r, Φ is for the angle ϕ, and X is for the axial coordinate. For constant k, the three-dimensional equation is

$$k \left[\frac{1}{r} \frac{\partial}{\partial r} \left(r \frac{\partial T}{\partial r} \right) + \frac{1}{r^2} \frac{\partial^2 T}{\partial \phi^2} + \frac{\partial^2 T}{\partial x^2} \right] = \rho c \frac{\partial T}{\partial t} \tag{2.2}$$

For spherical coordinates, r, ϕ, θ, the symbols are RS, Φ, Θ, respectively. The symbol RS is used to denote radial in the spherical direction. The angle ϕ for both the cylindrical and spherical coordinates goes from 0 to 2π.

Six different boundary conditions are given and are numbered 0, 1, 2, 3, 4, and 5. See Table 2.1.

The *first* kind of boundary condition is the prescribed temperature at boundary i,

$$T(\mathbf{r}_i, t) = f_i(\mathbf{r}_i, t) \tag{2.3}$$

where $f_i(\mathbf{r}_i, t)$ is the space- and time-dependent surface temperature. For a one-dimensional case at $x = 0$, $f_i(\cdot)$ can be a function of time only, such as $T(0, t) =$

Table 2.1 Types of boundary conditions

Notation	Name of boundary condition	Description of boundary condition
0	Zeroth kind (natural)	No physical boundary
1	Dirichlet	Prescribed temperature, Eq. (2.3)
2	Neumann	Prescribed heat flux, Eq. (2.4)
3	Robin	Convective condition, Eq. (2.6)
4	Fourth kind (Carslaw)	Thin film, no convection, Eq. (2.7)
5	Fifth kind (Jaeger)	Thin film, convection, Eq. (2.8)

$f_1(t)$. For a two-dimensional case with coordinates x, y, at $x = x_1$, $T(x_1, y, t) = f_1(y, t)$.

The *second* kind of boundary condition is prescribed heat flux,

$$k \left. \frac{\partial T}{\partial n_i} \right|_{\mathbf{r}_i} = f_i(\mathbf{r}_i, t) \tag{2.4}$$

where n_i is an outward pointing normal. For a one-dimensional case of boundaries at $x_1 = 0$ and $x_2 = L$, $n_1 = -x$ and $n_2 = x$; the boundary conditions are

$$-k \left. \frac{\partial T}{\partial x} \right|_{x=0} = f_1(t) \qquad k \left. \frac{\partial T}{\partial x} \right|_{x=L} = f_2(t) \tag{2.5a,b}$$

and $f_1(t)$ and $f_2(t)$ are heat fluxes directed toward the surfaces.

The *third* kind is a convective boundary condition,

$$k \left. \frac{\partial T}{\partial n_i} \right|_{\mathbf{r}_i} + h_i T|_{\mathbf{r}_i} = f_i(\mathbf{r}_i, t) \tag{2.6}$$

where h_i is the heat transfer coefficient and $f_i(\mathbf{r}_i, t)$ is usually equal to $h_i T_\infty$ with T_∞ being the ambient temperature, but $f_i(\mathbf{r}_i, t)$ can also include a prescribed heat flux.

The *fourth* kind is for a thin film at a surface with a prescribed heat flux $f_i(\cdot)$,

$$k \left. \frac{\partial T}{\partial n_i} \right|_{\mathbf{r}_i} = f_i(\mathbf{r}_i, t) - (\rho cb)_i \left. \frac{\partial T}{\partial t} \right|_{\mathbf{r}_i} \tag{2.7}$$

The product $(\rho cb)_i$ is for the film at the ith surface, and b_i is its thickness. A physical example of this type of boundary condition is heat transfer into a large ceramic object with a thin metal coating on the surface. The temperature distribution in the metal coating may be neglected across the small thickness b_i because the thermal conductivity of the metal is large compared to the ceramic, but storage of thermal energy in the metal coating may not be neglected. This boundary condition can also describe a surface film composed of a well-stirred fluid with heat capacity of $(\rho c_p b)_i$.

The *fifth* kind of boundary condition is for a thin film permitting heat losses from the film by convection,

$$k \left. \frac{\partial T}{\partial n_i} \right|_{\mathbf{r}_i} + h_i T = f_i(\mathbf{r}_i, t) - (\rho cb)_i \left. \frac{\partial T}{\partial t} \right|_{\mathbf{r}_i} \tag{2.8}$$

The boundary condition of the fifth kind is physically identical to the fourth kind except that instead of a specified heat flux on the thin film at the surface there is a specified heat transfer coefficient h.

Another important case is the *zeroth* kind. It is for conditions for which there is no physical boundary; it is sometimes called a natural boundary condition. It includes several cases, one of which is in the rectangular coordinates when a boundary extends to infinity. For example, a semi-infinite body that is convectively heated at $x = 0$ is denoted $X30$. Another case is for the center of radial cylindrical and spherical bodies that are solid. A solid cylinder with a prescribed surface heat flux is denoted $R02$. The case associated with a convective boundary condition at $r = a$ and a spherical domain outside $r = a$ is denoted $RS30$. Another case is for a thin annular ring which is denoted $\Phi00$.

Cases included by this numbering system are organized in Figs. 2.1–2.3; notice that the structural arrangement of each of these cases is different, with the radial coordinate having the largest number of distinct cases and the angular, the least. Figure 2.1 is for the Cartesian coordinate x and includes 21 distinct cases; others such as $X12$ can be listed but these can be found by a simple change of coordinates (i.e., $x \rightarrow L - x$, where L is the plate thickness). Notice that the cylindrical radial case shown in Fig. 2.2 includes 26 cases because the $RI0$ ($I = 1, \ldots, 5$) geometries are quite different from the $R0I$ geometries, the former being the infinite region bounded internally by the radius $r = a$ and the later for solid cylinders of radius a. For annular geometries with boundary radii of a and b, neither I nor J in RIJ are equal to zero. The spherical radial cases $RSIJ$ is similar to Fig. 2.2 with R replaced by RS. For the cylindrical coordinate ϕ and small changes in r, a ring is obtained; cases are displayed in Fig. 2.3. The special case in Fig. 2.3 is for a complete ring. There are neither $\Phi0I$ nor $\phi I0$ cases with $I \neq 0$. Except for the $\Phi00$ case, the ΦIJ cases in Fig. 2.3 have similar mathematical solutions as the corresponding XIJ cases of Fig. 2.1.

There are three special finite-body cases in Fig. 2.1 which (usually) have no steady state, namely $X22$, $X42$ and $X44$. There are five such special cases in Fig. 2.2 and

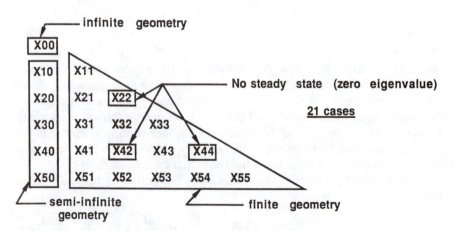

Figure 2.1 Distinct cases for one-dimensional Cartesian geometries.

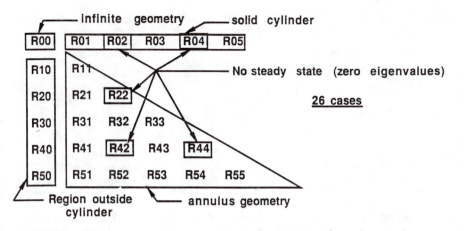

Figure 2.2 Distinct cases for one-dimensional cylindrical radial geometries.

four in Fig. 2.3. Mathematically, these cases are associated with zero eigenvalues. From a physical perspective, these cases do not have a steady state for time-independent values of $f_i(\cdot)$ in Equations (2.4) or (2.7) (unless there is the special case of zero net heat added). The $\Phi00$ case is unique since there are no physical boundaries; however, in this case (and the special finite bodies cases) there is no steady state for a constant volume source in the respective bodies.

For the infinite and semi-infinite geometries of Figs. 2.1 and 2.2, i.e., the first column in both figures, steady state is not usually attained in finite times.

2.3 BOUNDARY CONDITION MODIFIERS

The boundary conditions of the first through fifth kinds are denoted as indicated in Section 2.2 but the time and/or space variation must also be specified. This means

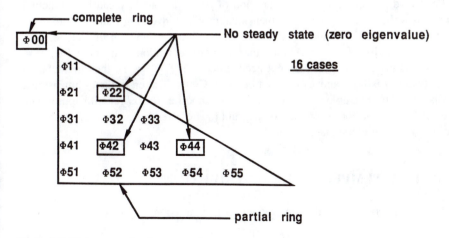

Figure 2.3 Distinct cases for ring geometries.

Table 2.2 Types of time- and space-variable function at boundary conditions

Notation	Time-variable boundary function	Notation	Space-variable boundary function (two-dimensional)
B-	Arbitrary $f(t)$	Bx-	Arbitrary $f(x)$
B0	$f(t) = 0$		
B1	$f(t) = C$		
B2	$f(t) = Ct$	Bx2	$f(x) = Cx$
B3	$f(t) = Ct^p$	Bx3	$f(x) = Cx^p, p > 1$
B4	$f(t) = \exp(-at)$	Bx4	$f(x) = \exp(-ax)$
B5	Step changes in $f(t)$	Bx5	Step changes in $f(x)$
B6	$\sin(\omega t + E), \cos(\omega t + E)$	Bx6	$\sin(\omega x + E), \cos(\omega x + E)$

that the function $f_i(\mathbf{r}_i, t)$ in Eqs. (2.3), (2.4), (2.6)–(2.8) must be described. For one-dimensional cases, f_i can be only a function of time. The one-dimensional case is first considered and then the two- and three-dimensional cases are discussed.

For one-dimensional cases, the function $f_i(t)$ includes zero (denoted B0), constant with time (B1) (actually a step increase at $t = 0$), linear with time (B2), some power other than 1 of t (B3), exponentials (B4), two or more step changes (B5), and sinusoids (B6). See Table 2.2. Only the basic cases are given specific notation. Solutions permitting an arbitrary time variation are indicated by a dash (–).

For one-, two-, or three-dimensional bodies, the geometry and boundary condition descriptors are followed by the boundary condition modifier BIJ. An example is X12B14 where the B14 indicates that the boundary condition of the first kind (prescribed T) at $x = 0$ is nonzero constant and the boundary condition of the second kind (prescribed q) at $x = L$ has an exponential dependence on time. In general, two indices follow B but there are exceptions. Only one index is needed when there is a boundary condition of the zeroth kind such as X20B1 or R03B1, where the B1's describe the nonzero boundary conditions. If both boundaries are of the zeroth kind (e.g., X00, R00 and Φ00), then the B modifier is not used.

For two-dimensional cases the variation of $f(\cdot)$ at a boundary can be a function of space as well as time. For a two-dimensional problem involving x and y coordinates and at a y surface, $f(\cdot)$ could be a function of x alone, a function of t alone, or a function of x and t. If $f = f(x)$, then the boundary condition is denoted $BxI, I = 2, \ldots, 6$ (since $I = 0$ and 1 are not needed here). If $f = f(x, t)$, then the notation $B(xItJ)$ (where I is for x and J for t) can be used. Generalization to three-dimensional cases is direct; for example, $f = f(x, z, t)$ has the modifier $B(xIzJtK)$ with appropriate values of I, J, and K corresponding to x, z, and t. The parentheses are used to enclose notation for a single boundary.

2.4 INITIAL TEMPERATURE DISTRIBUTION

The initial temperature distribution is given in general coordinates by

$$T(\mathbf{r}, 0) = F(\mathbf{r}) \tag{2.9}$$

and for a one-dimensional case with x being the coordinate,

$$T(x, 0) = F(x) \tag{2.10}$$

A numbering system for $F(\cdot)$ is given that is analogous to that for the boundary conditions. The letter T is followed by digits 0, 1, . . . , 7, as shown in Table 2.3. The coordinate r in Table 2.3 represents any single space coordinate such as r, x, or ϕ. Figure 2.4 displays some one-dimensional cases and gives the numbers including the notation for the initial temperature distribution. For two- and three-dimensional cases, see Figs. 2.5 and 2.6 which are discussed in Section 2.6. For steady state problems, the initial condition index T and the associated digit are not used.

2.5 INTERFACE DESCRIPTORS

The numbering system also includes composite bodies. The interface conditions are denoted in a manner similar to the boundary conditions. For perfect contact, a capital C is used for the interface. For example, a plate perfectly bonded to another one, with prescribed temperatures on either side is denoted, $X1B$-$CX1B$-T-, for arbitrary time-variation of the surface temperatures and arbitrary initial temperature distribution.

For other conditions the letter C is followed by a single digit; see Table 2.4. The notation $C2$ is used to denote a perfect contact with a heat source at the interface; since heat flux is involved, it is analogous to the boundary condition of the second kind, hence the use of 2. The notation $C3$ is used to denote an imperfect contact at location r_i with a contact conductance of h_c at the interface (analogous to the boundary condition of the third kind)

$$-k \left. \frac{\partial T}{\partial n_i} \right|_{r_i^-} = h_c(T_{r_i^-} - T_{r_i^+}) = -k \left. \frac{\partial T}{\partial n_i} \right|_{r_i^+} \tag{2.11}$$

The $C4$ case is for a thin film (or well-stirred fluid) in perfect contact at the interface,

$$-k \left. \frac{\partial T}{\partial n_i} \right|_{r_i^-} = (\rho cb)_i \left. \frac{\partial T}{\partial t} \right|_{r_i^+} -k \left. \frac{\partial T}{\partial n_i} \right|_{r_i^+} \tag{2.12}$$

where $(\rho cb)_i$ is for the thin film or well-stirred fluid.

Table 2.3 Types of space-variable initial conditions

Notation	Single space-variable initial condition
T-	Arbitrary $F(r)$
$T0$	$F(r) = 0$
$T1$	$F(r) = C$
$T2$	$F(r) = Cr$
$T3$	$F(r) = Cr^p$, p not 0 or 1
$T4$	$F(r) = \exp(-ar)$
$T5$	Step changes in $F(r)$
$T6$	$\sin(\omega r + E)$, $\cos(\omega r + E)$
$T7$	Dirac delta function, $\delta(r - r_0)$

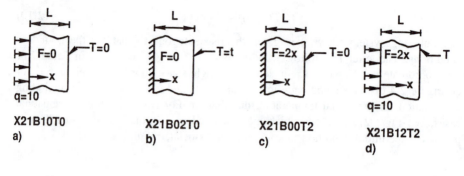

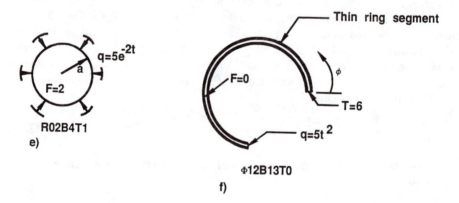

Figure 2.4 Some one-dimensional examples of numbering system.

2.6 NUMBERING SYSTEM FOR $g(x, t)$

A notation for the geometry and for the boundary conditions is given in previous sections. In this section, extensions to the numbering system are given to classify the volumetric source term $g(x, t)$.

The notation for the volumetric source term $g(x, t)$ is indicated by a capital G followed by up to four modifiers to denote the x and t dependence. The notation is $GxItJ$, where xI represents the x dependence, and tJ represents the time dependence of the volume source term. The values I and J can assume the values $0, 1, 2, \ldots,$ 7, or the dash (−) to represent different functions. See Table 2.5 for a listing of notation for the source term.

Several examples of the notation for the source term are presented below. For a source term of the form

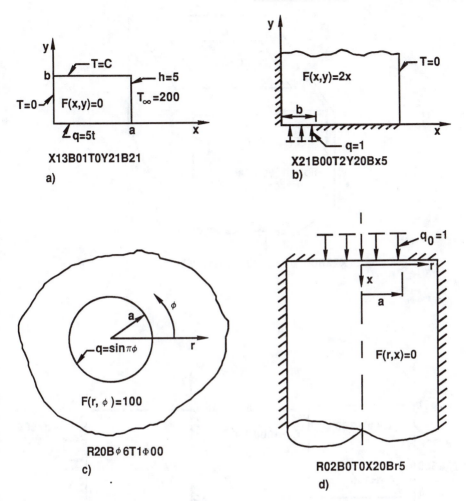

Figure 2.5 Two-dimensional examples of numbering system.

$$g(x, t) = 10xt,$$

the notation is $Gx2t2$. A source term of the form

$$g(x, t) = 10x$$

is denoted $Gx2t1$, or simply $Gx2$, since the modifier $t1$ is not needed to state that $g(x, t)$ does not depend on time. An even simpler case is

$$g(x, t) = 2$$

which is denoted $Gx1t1$ or more simply, $G1$. For the case where $g(x, t)$ is composed of a sum of several terms, such as

$$g(t) = a_0 + a_1 t + a_2 t^2$$

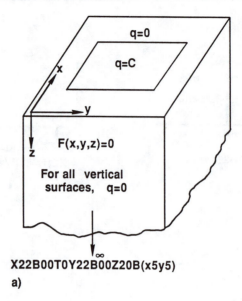

X22B00T0Y22B00Z20B(x5y5)

a)

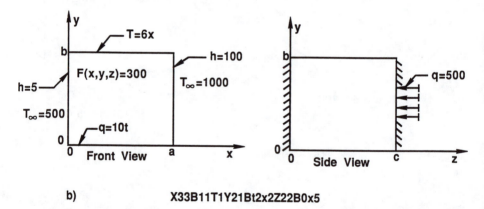

b)   X33B11T1Y21Bt2x2Z22B0x5

Figure 2.6 Three-dimensional examples of numbering system.

Table 2.4 Types of interface conditions

Notation	Description of interface condition
C	Perfect contact
$C2$	Perfect contact with source at interface
$C3$	Finite contact conductance
$C4$	Thin film at interface, perfect contact

Table 2.5 Notation for time-variable source terms

Notation	Time variation
$Gt-$	Arbitrary, $g(t)$
$Gt0$	$g(t) = 0$
$Gt1$	$g(t) = C$
$Gt2$	$g(t) = Ct$
$Gt3$	$g(t) = Ct^p, p \neq 0$ or 1
$Gt4$	$g(t) = \exp(-at)$
$Gt5$	Step changes in $g(t)$
$Gt6$	$\sin(\omega t + E)$, $\cos(\omega t + E)$
$Gt7$	Dirac delta function, $\delta(t - t_0)$

the notation is $Gx1t(1,2,3)$ or $Gt(1,2,3)$. Due to linearity of the heat conduction problem, the solution to the problem with source term $Gt(1,2,3)$ can be found as the sum of three problems,

$$Gt(1,2,3) = Gt1 + Gt2 + Gt3$$

2.7 EXAMPLES OF NUMBERING SYSTEM

The proposed numbering system can be used to describe a very large number of cases. Some one-dimensional cases are shown in Fig. 2.4. The first four cases of Fig. 2.4 are for the same basic case of $X21$. Figure 2.4a depicts a plate with a constant heat flux at $x = 0$ (boundary condition of the second kind) and $T = 0$ at $x = L$ (condition of the first kind). The initial temperature is zero. The number for this case is $X21B10T0$ where the 1 following B is for $q = C$ at $x = 0$ and the 0 following $B1$ is for the $T = 0$ condition at $x = L$. See Table 2.2. The problem of Fig. 2.4b has an insulated surface at $x = 0$, a linear time variation of temperature at $x = L$ and a zero initial temperature; its number is $X21B02T0$. The two in $B02$ is for the linear time variation at $x = L$. Figure 2.4c has $f = 0$ at both boundaries but the initial temperature is a linear function of x and thus is denoted $X21B00T2$. The case shown by Fig. 2.4d includes all the nonzero f_i and F values of Figs. 2.4a, b, and c.

A cylindrical radial case is shown in Fig. 2.4e. Depicted is a solid cylinder with a heat flux of exponential form at $r = a$ and the initial temperature is a constant. Figure 2.4f is for a segment of a thin ring.

Some two-dimensional cases are illustrated in Fig. 2.5. A rectangular plate is shown in Fig. 2.5a. The number description in the x direction is similar to that for a one-dimensional case and it is then followed by the one in the y direction. Since the initial temperature is known to be zero, it is redundant to repeat this information with the y direction notation. Another two-dimensional case is shown in Fig. 2.5b; it is for a plate that is finite in the x direction and semi-infinite in the y direction. For the x direction, the boundary conditions are of the second and first kinds and are homogeneous, but the initial temperature distribution is linear with x; thus this part of the notation is $X21B00T2$. For the y direction, there is a step increase in q at $x = 0$ and a step decrease at $x = b$, and there is no physical boundary for large y. Hence, the notation in the y direction is $Y20Bx5$ where the $Bx5$ notation is for the steps in q in the x direction at the $y = 0$ boundary. There is no y direction dependence of the initial temperature so it is omitted in the notation.

A case of a body outside the cylindrical radius of $r = a$ is shown by Fig. 2.5c. There is a sinusoidal variation with ϕ of the surface heat flux and the initial temperature distribution is constant. The notation is $R20B\phi6T1\Phi00$. The $B\phi6$ describes the boundary condition at $r = a$ and no index is needed for $r \rightarrow \infty$ where there is no physical boundary.

Figure 2.5d displays a semi-infinite cylinder that is insulated at all surfaces except at the center at the top where a circular heat flux is applied. The initial temperature is zero. The number for this case is $R02B0T0X20Br5$ where the $Br5$ notation is used because the heat flux is not constant with r but can be considered to have a step increase at $r = 0$ and a step decrease $r = a$. If the heat flux were over the circular region shown and also varied as ct in time, $Br5$ would be replaced by $B(r5t2)$ where the parentheses are used to denote that both conditions apply at the same boundary.

The numbering system readily extends to three-dimensional cases such as given in Fig. 2.6. The first case is for a semi-infinite rod that is insulated on all surfaces except there is a constant heat flux over a rectangular region at $z = 0$. The case of a rectangular block is shown in Fig. 2.6b, where front and side views are shown.

2.8 ADVANTAGES OF NUMBERING SYSTEM

There are several types of advantages of the numbering system. The first relates to a data base of conduction solutions. The second relates to an algebra that can be given for linear problems. The last major advantage relates to use of the method in conjunction with GFs to obtain solutions for linear problems; full explanation is deferred until after Chapter 3.

2.8.1 Data Base in Transient Heat Conduction

One of the obvious advantages of a numbering system is that it facilitates the organizing of a data base. A structure is provided that makes the storage of solutions easier. Also important is that it greatly reduces the effort in locating solutions. Instead of relying on imprecise verbal titles of papers (or abstracts) to describe a particular problem, a search based on the notation given herein can be much more direct and less prone to miss related solutions.

The numbering system has been utilized to catalog most of the solutions of Carslaw and Jaeger. An example of a portion of a data base for some solutions is given in Table 2.6. A more complete tabulation is available from the first author. Table 2.6 gives numbers of some one-dimensional cases from Carslaw and Jaeger (1959). The first column contains the number; the second and third columns give the page and equation numbers of the reference; and the last column contains some comments.

Table 2.6 Some one-dimensional cases in Carslaw and Jaeger (1959)

Number	Page	Equation	Comments
X00T5	54	3	$T(x, 0) = T_0$, $-a < x < a$; $T(x, 0) = 0$, $|x| > a$
X10B1T0	60	10	
X10B3T0	305	6	$T(0, t) = T_0 t^{n/2}$, $n = 1, 2, \ldots$
X11B00T1	96	6	
R01B0T1	199	5	
R01B1T0	331	3	Small time solution

2.8.2 Algebra for Linear Cases

For linear cases, several kinds of algebraic manipulations are possible. This brief discussion can include only a few possibilities.

One case involves boundary conditions of the zeroth, first, and third kinds and the uniform initial temperature distribution. An example is

$$[X10B1T0|_{T(0,t)=T_0}] = T_0[1 - (X10B0T1|_{T(x,0)=1})] \tag{2.13}$$

where T_0 is a constant.

In addition to relating boundary conditions and the initial temperature, the notation suggests a method of superimposing solutions. The number of nonzero values of the indices following B and T give the number of superposition problems that can be formed; this is the number of "forcing" terms. An example is provided by the first four cases of Fig. 2.4. The Fig. 2.4d case is the sum of the first three cases,

$$X21B12T2 = X21B10T0 + X21B02T0 + X21B00T2 \tag{2.14}$$

Notice that $B12$ contains two nonzero digits and $T2$ contains one; hence, the case of Fig. 2.4d can be given as the sum of three problems. The same superposition principles can be used for the two-dimensional problem of Fig. 2.5a.

Another type of superposition is possible for more than one forcing term at a boundary. An example is for the Fig. 2.4a case with

$$q = 10 + 5t \tag{2.15}$$

The temperature solution can be written as

$$T|_{q=10+5t} = 10 [X21B10T0|_{q=1}] + 5 [X21B20T0|_{q=t}] \tag{2.16}$$

Another aspect of the algebra for the numbering system is that it can aid in identifying the number of explicit dimensions of a problem. A plate is a three-dimensional object but the temperature distribution can be an explicit function of only one or two coordinates. Boundary conditions of the zeroth, second, and fourth kinds have the potential of reduction in the number of dimensions while the first, third, and fifth kinds do not. However, for reduction in the number of the dimensions, both boundaries in a given direction must be homogeneous and there cannot be any explicit dependence of the initial temperature or g in that direction.

As an example, consider the case of a cube which is at zero initial temperature and there is no volumetric energy source. At time zero, each surface is heated with a constant heat flux (which may or may not be the same for each face). The number for this case is $X22B11Y22B11Z22B11T0$ and the solution is equal to the sum of six one-dimensional problems,

$$X22B11Y22B11Z22B11T0 = X22B10T0 + X22B01T0 + Y22B10T0$$

$$+ Y22B01T0 + Z22B10T0 + Z22B01T0 \quad (2.17)$$

This reduction of dimensions on the right side of Eq. (2.17) is because the typical three-dimensional problem of $X22B10Y22B00Z22B00T0$ reduces to

$$X22B10Y22B00Z22B00T0 = X22B10Y22B00T0 = X22B10T0 \quad (2.18)$$

Note that the $Y22B00$ and the $Z22B00$ conditions have boundary conditions of the second kind and are homogeneous.

An example that does not reduce in the same manner is for a cube initially at $T = 0$ and subjected to a step increase in temperature on each surface (i.e., a constant temperature with time and over the surface). The number and algebra for this case are

$$X11B11Y11B11Z11B11T0$$

$$= X11B10Y11B00Z11B00T0 + X11B01Y11B00Z11B00T0$$

$$+ X11B00Y11B10Z11B00T0 + X11B00Y11B01Z11B00T0$$

$$+ X11B00Y11B00Z11B10T0 + X11B00Y11B00Z11B10T0 \quad (2.19)$$

Each of these problems is three-dimensional although simplifications in the solutions result because the problems are similar. If each surface of the cube is subjected to the same temperature condition (or even convective boundary condition), the GF solution leads to further simplifications. For example, if the cube is initially at temperature T_0, and suddenly immersed in a fluid at $T_\infty = 0$ with the same h on each surface, the temperature distribution is given by

$$X33B00Y33B00Z33B00T1$$

$$= T_0 [X33B00T1|_{F=1}][Y33B00T1|_{F=1}][Z33B00T1|_{F=1}] \quad (2.20)$$

This is related to the multiplication of solutions associated with Heisler charts in undergraduate heat transfer textbooks.

The possibilities of uses of this numbering system for transient heat conduction and diffusion are large and can be considerably expanded beyond what is outlined in this book.

REFERENCES

Beck, J. V., 1984, Green's Function Solution for Transient Heat Conduction Problems, *Int. J. Heat Mass Transfer*, vol. 27, pp. 1235–1244.

Beck, J. V., 1986, Green's Functions and Numbering System for Transient Heat Conduction, *AIAA J.*, vol. 24, pp. 327–333.

Beck, J. V., and Keltner, N. R., 1982, Transient Thermal Contact of Two Semi-Infinite Bodies Over a Circular Area, *Spacecraft Radiative Transfer and Temperature Control, Progr. Astronaut. Aeronaut.*, vol. 33, pp. 66–82.

Beck, J. V., and Keltner, N. R., 1985, Green's Function Partitioning Method Applied to Foil Heat Flux Gages, ASME Paper No. 85-HT-56.

Carslaw, H. S., and Jaeger, J. C., 1959, *Conduction of Heat in Solids*, 2 ed., Oxford University Press, New York.

Keltner, N. R., and Beck, J. V., 1981, Unsteady Surface Element Method, *J. Heat Transfer*, vol. 103, pp. 759–764.

Sharma, A., and Minkowycz, W. J., 1982, KNOWTRAN: An Artificial Intelligence System for Solving Heat Transfer Problems, *Int. J. Heat Mass Transfer*, vol. 25, pp. 1279–1289.

PROBLEMS

2.1 Give the numbering system designation for Example 1.1 of Chapter 1.

2.2 Give two numbers for Eq. (1.12) that are valid for $x > 0$.
(Answer: $X00T5$ and $X20B0T5$)

2.3 Give two numbers for Eq. (1.13) that are valid for $x > 0$.

2.4 Give the numbering system designation for Example 1.3 of Chapter 1.

2.5 Give the number for Eq. (1.21). Also give the number for the initial condition given by Eq. (1.25).

2.6 Give the numbering system designation for Example 1.4 of Chapter 1.

2.7 Give the numbering system designation for Problem 1.8 of Chapter 1.

2.8 Give the number for the problem with the same geometry and boundary condition shown in Fig. 1.6a with the initial temperature being a constant and with a constant volumetric energy source.

2.9 Give the number for Fig. 2.4d with $F = 6$, $q = 2$. At $x = L$, $T = 5 + 2 \sin 4t$.

2.10 Using the numbering system for conduction, give the numbers for the following one-dimensional cases, each of which is the partial differential equation,

$$\frac{\partial C}{\partial t} = D \frac{\partial^2 C}{\partial x^2} \qquad 0 < x < L \qquad t > 0$$

(*a*) $C(0, t) = C_0$, $C(L, t) = 0$, $C(x, 0) = 6 \sin 2\pi x/L$.
(*b*) $- \partial C/\partial x = 0$ at $x = 0$, $C(L, t) = C_0$, $C(x, 0) = C_1$.
(*c*) $C(0, t) = 3 + 4t^2$, $C(L, t) = \cos 2t$, $C(x, 0) = \cos 2x$.

2.11 Write the describing differential equation, boundary conditions, and initial condition for the problem denoted $X24B21G1T0$.

2.12 For the partial differential equation for cylindrical heat flow with volume energy generation, give the numbers for the following cases.

(*a*) A solid cylinder is initially at a uniform temperature and is suddenly plunged into a fluid at a temperature of T_∞; where $g = 0$.

(*b*) A hollow cylinder is initially at a uniform temperature is insulated at the inner surface and is heated by a constant heat flux at the outer surface; $g = 5$.

(*c*) The region is that outside the radius of $r = a$ and a constant heat flux exists at $r = a$. The initial temperature is T_0 and $g = 0$.

(*d*) The geometry is the same as shown in Fig. 2.5d but q at $x = 0$ is $\sin \pi r/a$ for $r < a$ and zero for larger values of r. The initial temperature is a function of r and ϕ.

2.13 Use the numbering system algebra to construct six problems, each of which has only one nonzero "forcing" term and the sum of which adds to the original three-dimensional problem. How many of these problems reduce to less than three-dimensional problems?

THREE

DERIVATION OF THE GREEN'S FUNCTION SOLUTION EQUATION

3.1 INTRODUCTION

The Green's function solution equation (GFSE) for transient heat conduction is derived in this chapter in several forms. First, the one-dimensional form for rectangular co-ordinates is derived for boundary conditions of the first, second, and third kinds. This form is easy to understand and examples are included to demonstrate how the equation is applied. Second, the GFSE is derived in a general three-dimensional form that applies to rectangular, cylindrical, and spherical coordinates. An even more general form of the GFSE is derived in Chapter 10; it covers the case of nonhomogeneous materials. Third, an alternative form particularly appropriate for nonhomogeneous boundary conditions is given. Fourth, a steady-state form is given and, finally, the GFSE is given for moving solids.

This chapter contains background material that, although important, is not essential to the application of the Green's functions (GF) method. One can begin with the GFSE, choose the correct GF, evaluate the integrals, and find the solution for temperature. However, an understanding of the GFSE will lead to a greater understanding of the GFs themselves.

This chapter covers the derivation of the one-dimensional GFSE in Section 3.2 and a general vector-based form in Section 3.3. Section 3.4 contains an alternative form of the GFSE (AGFSE) which may be helpful for nonhomogeneous boundary conditions when slow convergence is obtained. Section 3.5 covers the m^2T term which is associated with fins. Section 3.6 covers the steady-state GSFE as a limit of the transient case. Finally, Section 3.7 contains a derivation of the GFSE for moving solids.

3.2 DERIVATION OF THE ONE-DIMENSIONAL GREEN'S FUNCTION SOLUTION EQUATION

The one-dimensional GFSE for rectangular coordinates is derived in this section. The one-dimensional form of the GFSE is free of vector calculus, so one can gain intuition about the GF method with a minimum of notation. The derivation makes use of the properties of GFs, and the result is an expression for the temperature that fully exploits the linear property of the heat conduction equation.

The boundary value problem for the temperature in a one-dimensional rectangular geometry is given in Section 1.5, by Eqs. (1.40) through (1.52) as

$$\frac{\partial^2 T}{\partial x^2} + \frac{1}{k} g(x, t) = \frac{1}{\alpha} \frac{\partial T}{\partial t} \qquad t > 0 \tag{3.1}$$

$$k_i \frac{\partial T}{\partial n_i}\bigg|_{x_i} + h_i T|_{x_i} = f_i(t) \qquad t > 0 \qquad \text{and } i = 1, 2 \tag{3.2}$$

$$T(x, 0) = F(x) \tag{3.3}$$

This is the problem that we are trying to solve with the GF method. In general, Eq. (3.2) describes convection boundary conditions (boundary conditions of the third kind), but temperature or heat flux boundary conditions may be obtained by taking $k_i = 0$ or $h_i = 0$, respectively, on surfaces $i = 1$ or $i = 2$.

The derivation of the GFSE begins with the auxiliary boundary value problem for the GF that corresponds to the above temperature problem. The auxiliary boundary value problem is very similar to the boundary value problem for the temperature with two important differences: first, the energy generation term in the differential equation for the GF is a Dirac delta function; and second, the boundary conditions and the initial conditions for the GF are homogeneous. The auxiliary boundary value problem was previously discussed in Section 1.5 and is given by

$$\frac{\partial^2 G}{\partial x^2} + \frac{1}{\alpha} \delta(x - x') \delta(t - \tau) = \frac{1}{\alpha} \frac{\partial G}{\partial t} \qquad t > \tau \tag{3.4a}$$

$$k_i \frac{\partial G}{\partial n_i}\bigg|_{x_i} + h_i G|_{x_i} = 0 \qquad i = 1, 2 \tag{3.4b}$$

$$G(x, t = 0|x', \tau) = 0 \qquad t < \tau \tag{3.4c}$$

Next, the reciprocity relation (Section 1.6)

$$G(x, t|x', \tau) = G(x', -\tau|x, -t)$$

is applied to the auxiliary equation (3.4a) to give

$$\frac{\partial^2 G}{\partial x'^2} + \frac{1}{\alpha} \delta(x' - x) \delta(t - \tau) = -\frac{1}{\alpha} \frac{\partial G}{\partial \tau} \tag{3.5}$$

Notice the minus sign on the time derivative. The next step is to write the original heat conduction equation for T in terms of x' and τ. That is, write Eq. (3.1) with a simple change of variables: replace x by x' and replace t by τ to give

$$\frac{\partial^2 T}{\partial x'^2} + \frac{1}{k} g(x', \tau) = \frac{1}{\alpha} \frac{\partial T}{\partial \tau} \tag{3.6}$$

Multiply Eq. (3.6) by $G(x, t|x', \tau)$, multiply Eq. (3.5) by $T(x', \tau)$, and then subtract Eq. (3.5) from Eq. (3.6) to get

$$G \frac{\partial^2 T}{\partial x'^2} - T \frac{\partial^2 G}{\partial x'^2} + \frac{G}{k} g(x', \tau) - \frac{T}{\alpha} \delta(x' - x) \delta(t - \tau) = \frac{1}{\alpha} \frac{\partial (TG)}{\partial \tau} \tag{3.7}$$

Integrate Eq. (3.7) with respect to x' over the domain $0 \le x' \le L$, and integrate with respect to τ from 0 to $t + \epsilon$, where ϵ is a small positive number. The result is

$$\int_{\tau=0}^{t+\epsilon} d\tau \int_{x'=0}^{L} \left(G \frac{\partial^2 T}{\partial x'^2} - T \frac{\partial^2 G}{\partial x'^2} \right) dx' + \frac{1}{k} \int_{\tau=0}^{t+\epsilon} d\tau \int_{x'=0}^{L} g(x', \tau) G(x, t|x', \tau) dx'$$

$$- \frac{1}{\alpha} T(x, t) = \frac{1}{\alpha} \int_{x'=0}^{L} \left[T G \right]_{\tau=0}^{\tau=t+\epsilon} dx' \tag{3.8}$$

Note that the properties of the Dirac delta function give the term $T(x' = x, \tau = t)$ on the left-hand side of this equation. This equation can be solved for $T(x, t)$ to give

$$T(x, t) = - \int_{x'=0}^{L} \left[T G \right]_{\tau=0}^{\tau=t+\epsilon} dx'$$

$$+ \frac{\alpha}{k} \int_{\tau=0}^{t+\epsilon} d\tau \int_{x'=0}^{L} g(x', \tau) G(x, t|x', \tau) dx'$$

$$+ \alpha \int_{\tau=0}^{t+\epsilon} d\tau \int_{x'=0}^{L} \left(G \frac{\partial^2 T}{\partial x'^2} - T \frac{\partial^2 G}{\partial x'^2} \right) dx' \tag{3.9}$$

This is the GFSE for one-dimensional rectangular coordinates. The three terms on the right-hand side of Eq. (3.9) will next be examined and simplified one at a time.

The first term of Eq. (3.9) can be simplified by observing that $G(x, t|x', t + \epsilon) = 0$ from the causality relation. That is, G is zero because $t - \tau = t - (t + \epsilon) = -\epsilon < 0$; there is zero response before the impulse occurs. Also, $T(x', 0)$ can be replaced by the initial condition, given by Eq. (3.3). Thus, the first term of the GF equation represents the effect of the initial condition, and it is written

$$\int_{x'=0}^{L} F(x') G(x, t|x', 0) dx' \tag{3.10}$$

The second term in Eq. (3.9) is the effect of the volume energy generation. This term will not be simplified any further at this point.

The third term of Eq. (3.9) can be simplified with integration by parts. (The analogous step in the three-dimensional derivation involves Green's theorem.) Consider just the integral on x' from this third term, and integrate by parts to get

$$\int_{x'=0}^{L} \left(G \frac{\partial^2 T}{\partial x'^2} - T \frac{\partial^2 G}{\partial x'^2} \right) dx' = G \frac{\partial T}{\partial x'} \Big|_{x'=0}^{x'=L} - \int_{x'=0}^{L} \frac{\partial G}{\partial x'} \frac{\partial T}{\partial x'} dx'$$

$$- T \frac{\partial G}{\partial x'} \Big|_{x'=0}^{x'=L} + \int_{x'=0}^{L} \frac{\partial T}{\partial x'} \frac{\partial G}{\partial x'} dx'$$

$$= G \frac{\partial T}{\partial x'} \Big|_{x'=0}^{x'=L} - T \frac{\partial G}{\partial x'} \Big|_{x'=0}^{x'=L} \tag{3.11}$$

Note that the two integrals in Eq. (3.11) cancel.

If the boundary conditions are of the second or third kinds, then the boundary conditions for T and G can be used to evaluate $\partial T/\partial x'$ and $\partial G/\partial x'$ at the boundaries. Equations (3.2) and (3.4b) can be written as

$$\frac{\partial G}{\partial n_i} \Big|_{x'=x_i} = - \frac{h_i}{k_i} G \Big|_{x'=x_i} \tag{3.12}$$

$$\frac{\partial T}{\partial n_i} \Big|_{x'=x_i} = \frac{f_i(\tau)}{k_i} - \frac{h_i}{k_i} T \Big|_{x'=x_i} \tag{3.13}$$

The notation n_i is for the *outward normal* from the body. Substitute these boundary conditions into Eq. (3.11) to get

$$G \frac{\partial T}{\partial x'} \Big|_{x'=0}^{x'=L} - T \frac{\partial G}{\partial x'} \Big|_{x'=0}^{x'=L} = \left[\frac{f_i(\tau)}{k_i} G - \frac{h_i}{k_i} T G \right]_{x'=L} - \left[-\frac{f_i(\tau)}{k_i} G + \frac{h_i}{k_i} T G \right]_{x'=0}$$

$$- \left(-\frac{h_i}{k_i} T G \right)_{x'=L} + \left(\frac{h_i}{k_i} T G \right)_{x'=0}$$

$$= \sum_{i=1}^{2} \frac{f_i(\tau)}{k_i} G \Big|_{x'=x_i} \tag{3.14}$$

Note that the terms that involve T cancel. The summation over $i = 1, 2$ is meant to cover all the possibilities for the boundary conditions of one-dimensional bodies. The total number of boundary terms is two for a finite body ($0 \le x \le L$). (The derivation also applies for semi-infinite and finite bodies. The semi-infinite one-dimensional body requires only one boundary term, and the infinite body does not require any boundary terms.)

If the boundary conditions are of the first kind, Eq. (3.11) takes a different form. At the boundaries, $G = 0$ and $T = f_i(t)$, so that

$$G \frac{\partial T}{\partial x'} \Big|_{x'=0}^{x'=L} - T \frac{\partial G}{\partial x'} \Big|_{x'=0}^{x'=L} = - \sum_{j=1}^{2} f_j(\tau) \frac{\partial G}{\partial n_j} \Big|_{x'=x_j} \tag{3.15}$$

Again, the summation over $j = 1, 2$ is used to represent the contribution from both boundaries.

The last step in the derivation of the GF equation is to take the limit of Eq. (3.9) as $\epsilon \to 0$. Then, $t + \epsilon$ can be replaced by t in the equation, without altering the conclusions that are drawn from $\epsilon > 0$. Finally, Eq. (3.9) is combined with the simplified terms given by Eqs. (3.10), (3.11), (3.14), and (3.15) to give the desired result

$$
T(x, t) = \int_{x'=0}^{L} G(x, t|x', 0) \, F(x') \, dx' \quad \text{(for the initial condition)}
$$

$$
+ \frac{\alpha}{k} \int_{\tau=0}^{t} d\tau \int_{x'=0}^{L} g(x', \tau) \, G(x, t|x', \tau) \, dx' \quad \text{(for energy generation)}
$$

$$
+ \alpha \int_{\tau=0}^{t} d\tau \sum_{i=1}^{2} \left[\frac{f_i(\tau)}{k_i} G(x, t|x_i, \tau) \right] \quad \begin{array}{l}\text{(for boundary conditions of}\\ \text{the second and third kinds)}\end{array}
$$

$$
- \alpha \int_{\tau=0}^{t} d\tau \sum_{i=1}^{2} \left[f_i(\tau) \frac{\partial G}{\partial n_i} \bigg|_{x'=x_i} \right] \quad \begin{array}{l}\text{(for boundary conditions}\\ \text{of the first kind only)} \quad (3.16)\end{array}
$$

This is the desired GFSE which applies to one-dimensional transient heat conduction in the rectangular coordinate system. The one-dimensional body is assumed to be homogeneous and to have constant properties (independent of temperature and position).

Each term in the GFSE must have the units of temperature. In the first term, $F(x')$ has units of temperature, so the product $G\,dx'$ must be dimensionless for the units to be correct, therefore the one-dimensional GF has units of m^{-1}. In the second term, $g(x', \tau)$ has units of W/m^3, so the product $(\alpha/k)g(x', \tau) \, d\tau$ has units of temperature, as it should. In the third term, $f_i(\tau)$ has units of W/m^2 (heat flux), so the product $(\alpha/k_i) \, f_i(\tau) \, d\tau$ has the units of temperature. Finally, in the fourth term, $f_i(\tau)$ has units of temperature, so the product $\alpha(\partial G/\partial n_i)d\tau$ is dimensionless.

In the usual cases discussed in this book, the boundary terms $f_i(t)$ are known. There are special cases when $T(x, t)$ is known from measurements, and $f_i(t)$ is the unknown. This is called the inverse heat conduction problem (Beck et al., 1985). In this case, Eq. (3.16) is considered to be an integral equation because the unknown, $f_i(t)$, is inside the integral.

Each $G(\cdot)$ term in Eq. (3.16) represents the same GF, which is mathematically unique for each set of boundary conditions. For example, in a geometry with $X12$ boundary conditions, the correct GF to use in Eq. (3.16) is the $X12$ GF, as in the following example.

Example 3.1 For the geometry shown in Fig. 3.1, the boundary conditions for $T(x, t)$ are

$$
T(0, t) = T_0 \tag{3.17}
$$

$$
-k \frac{\partial T}{\partial x} \bigg|_{x=L} = q(t) \tag{3.18}
$$

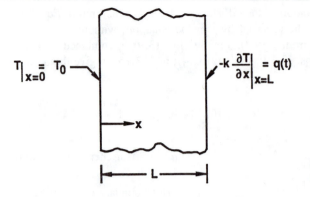

Figure 3.1 Slab body geometry for Example 3.1: $X12$ case.

This is an example of the $X12$ geometry. The initial condition is

$$T(x, 0) = F(x) \tag{3.19}$$

and there is no energy generation in this case. If the $X12$ GF is assumed to be a known function named $G_{X12}(x, t|x', \tau)$, what is the appropriate form of the GFSE?

SOLUTION The GFSE is a sum of the various effects that contribute to the temperature $T(x, t)$. The contribution of the initial condition is given by the first term from Eq. (3.16),

$$\int_{x'=0}^{L} G_{X12}(x, t|x', 0) F(x') \, dx' \tag{3.20}$$

The boundary condition at $x = 0$ is of the first kind. This boundary condition contributes to the temperature according to the last term of Eq. (3.16) where $f_i(\tau) = T_0$. This term is

$$- \alpha \int_{\tau=0}^{t} \left(- T_0 \frac{\partial G_{X12}}{\partial x'} \bigg|_{x'=0} \right) d\tau \tag{3.21}$$

Notice the minus sign that appears because $\partial/\partial n_i' = - \partial/\partial x'$ at $x = 0$; the outward pointing normal n_i is in the minus x direction for the $x = 0$ surface.

The boundary condition at $x = L$ is of the second kind. This boundary condition contributes to the temperature according to the third term of Eq. (3.16), where $f_i(\tau) = q(\tau)$,

$$\alpha \int_{\tau=0}^{t} \frac{q(\tau)}{k} G_{X12}(x, t|L, \tau) \, d\tau \tag{3.22}$$

The temperature $T(x, t)$ is the sum of these three effects, or

$$T(x, t) = \int_{x'=0}^{L} G_{X12}(x, t|x', 0) F(x') \, dx'$$

$$+ \alpha \int_{\tau=0}^{t} T_0 \left. \frac{\partial G_{X12}}{\partial x'} \right|_{x'=0} d\tau$$

$$+ \alpha \int_{\tau=0}^{t} \frac{q(\tau)}{k} G_{X12}(x, t|L, \tau) \, d\tau \qquad (3.23)$$

which is the GFSE for this example.

Notice that $G_{X12}(x, t|x', \tau)$ in each term is evaluated at the time or location appropriate to that term in the GF equation. For example, in the initial condition term, G_{X12} is evaluated at $\tau = 0$. In the term for the left-side boundary condition, $\partial G_{X12}/\partial x'$ is evaluated at $x' = 0$.

Example 3.2 The one-dimensional semi-infinite body shown in Fig. 3.2 has a convection boundary condition given by

$$-k \left. \frac{\partial T}{\partial x} \right|_{x=0} = h \left(T_\infty - T|_{x=0} \right) \qquad (3.24)$$

where T_∞ is the ambient temperature. This is the $X30$ geometry. The volume heat generation is given by $g(x, t) = g_c$, where g_c is a constant. The heat conduction equation is thus given by

$$\frac{\partial^2 T}{\partial x^2} + \frac{1}{k} g_c = \frac{1}{\alpha} \frac{\partial T}{\partial t} \qquad (3.25)$$

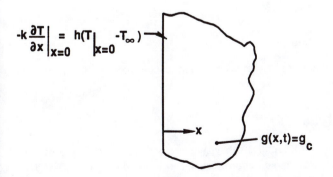

Figure 3.2 Semi-infinite body with convection at the boundary and internal energy generation. Geometry for Example 3.2

The initial condition is

$$T(x, 0) = F(x) \tag{3.26}$$

If the $X30$ GF is assumed to be a known function denoted $G_{X30}(x, t|x', \tau)$, what is the appropriate form of the GFSE?

SOLUTION There are three terms that contribute to the temperature $T(x, t)$: the initial condition, the volume heat generation, and the convection boundary condition. The effect of the boundary at infinity does not require an explicit term, because it is already included in the correct GF, denoted $G_{X30}(\cdot)$. The temperature for this case is given by the GFSE

$$T(x, t) = \int_{x'=0}^{\infty} G_{X30}(x, t|x', 0) F(x') \, dx'$$

$$+ \frac{\alpha}{k} \int_{\tau=0}^{t} \int_{x'=0}^{\infty} g_c \, G_{X30}(x, t|x', \tau) \, dx' \, d\tau$$

$$+ \alpha \int_{\tau=0}^{t} \frac{hT_{\infty}}{k} G_{X30}(x, t|0, \tau) \, d\tau \tag{3.27}$$

Note that the integrals on x' in the first two terms are evaluated over the entire body, $0 \le x' \le \infty$. This is an extension of Eq. (3.16) to the semi-infinite case. (A similar extension to the $X00$, or infinite body case, is to evaluate the x' integral over $-\infty \le x' \le \infty$).

3.3 GENERAL FORM OF THE GREEN'S FUNCTION SOLUTION EQUATION

In this section the GFSE will be derived in a general form for an additional term in the heat conduction equation (the m^2T term) and for two additional boundary conditions. This general form of the GFSE can be applied to three-dimensional geometries in any orthogonal coordinate system. The rectangular, cylindrical, or spherical coordinate systems are treated in this book.

3.3.1 Temperature Problem

The partial differential equation that describes transient, multidimensional, linear heat conduction in a homogeneous isotropic body is,

$$\nabla^2 T + \frac{1}{k} g(\mathbf{r}, t) - m^2 T = \frac{1}{\alpha} \frac{\partial T}{\partial t} \quad \text{in region } R \text{ and } t > 0 \tag{3.28}$$

The thermal conductivity k and thermal diffusivity α are both constant with position, time, and temperature. Any orthogonal coordinate $\mathbf{r}$ can be used in Eq. (3.28). The $g(\mathbf{r}, t)$ term represents space- and time-variable volume energy generation.

The m^2T term could represent side heat losses for a fin; m^2 can be a function of $\mathbf{r}$ but not t. (The m^2T term is not needed for the three-dimensional treatment of a fin.)

If there is a component of volume energy generation g that is linearly proportional to temperature, it should be included in the m^2T term which could then encompass the effects of electric heating and dilute chemical reactions; in such cases m^2 could be either positive or negative. An example of transient conduction involving the m^2T term is given in Section 3.5.

The initial temperature distribution is expressed by

$$T(\mathbf{r}, 0) = F(\mathbf{r}) \tag{3.29}$$

The boundary conditions for Eq. (3.28) have the general form

$$k_i \frac{\partial T}{\partial n_i} + h_i T = f_i(\mathbf{r}_i, t) - (\rho cb)_i \frac{\partial T}{\partial t} \qquad t > 0 \tag{3.30}$$

where the temperature T and its derivatives are evaluated at the boundary surface S_i, and $\mathbf{r}_i$ denotes the boundary. The spatial derivative $\partial/\partial n_i$ denotes differentiation along an *outward* drawn normal to the boundary surface S_i, $i = 1, 2, \ldots, s$. The heat transfer coefficient, h_i, and $(\rho cb)_i$ can vary with position on S_i but are independent of temperature and time. The boundary condition given by Eq. (3.30) includes the possibility of a high conductivity surface film of thickness b_i. There is a negligible temperature gradient through the film and there is no heat flux parallel to the surface inside the film. Five different boundary conditions can be obtained from Eq. (3.30) by setting $k_i = 0$ or k, $h_i = 0$ or h, and also $b = 0$ or nonzero.

Figure 3.3 shows some examples of boundary conditions of the first and second kinds. Figures 3.4 and 3.5 show some examples of boundary conditions of the third kind and fifth kind, respectively. The five different boundary conditions are discussed in Chapter 2.

3.3.2 Derivation of the Green's Function Solution Equation

The GFSE is derived using Eqs. (3.28)–(3.30) and also an auxiliary problem for an instantaneous heat source inside the body. The solution to the auxiliary problem is GF $G(\mathbf{r}, t|\mathbf{r}', \tau)$, where the instantaneous source is located at position $\mathbf{r}'$ and at time τ; $\mathbf{r}$ is the location at which the temperature is observed at time t. There can be a nonzero response at $\mathbf{r}$ only if $t - \tau > 0$. The auxiliary problem has homogeneous boundary conditions and a zero initial temperature.

The derivation of the general GFSE begins with the reciprocity relation of GF,

$$G(\mathbf{r}, t|\mathbf{r}', \tau) = G(\mathbf{r}', -\tau|\mathbf{r}, -t) \tag{3.31}$$

substituted into the auxiliary equation, resulting in

$$\nabla_0^2 G + \frac{1}{\alpha} \delta(\mathbf{r} - \mathbf{r}') \delta(t - \tau) - m^2 G = -\frac{1}{\alpha} \frac{\partial G}{\partial \tau} \qquad t > \tau \tag{3.32}$$

$$G(\mathbf{r}', -\tau|\mathbf{r}, -t) = 0 \qquad t < \tau \tag{3.33}$$

$$k_i \frac{\partial G}{\partial n_i'} + h_i G = (\rho cb)_i \frac{\partial G}{\partial \tau} \qquad t > \tau \tag{3.34}$$

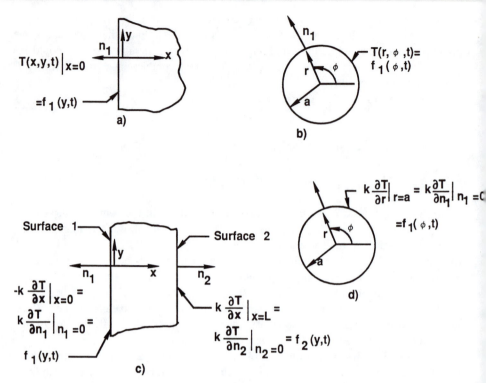

Figure 3.3 Examples of boundary conditions of the first and second kinds: (*a*) first kind of boundary condition at $x = 0$; (*b*) first kind of boundary condition at $r = a$; (*c*) second kind of boundary condition at $x = 0$ and L, rectangular coordinates; (*d*) second kind of boundary condition at $r = a$, cylindrical coordinates.

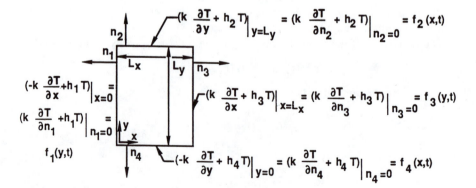

Figure 3.4 Examples of convection boundary conditions (third kind) on rectangular body.

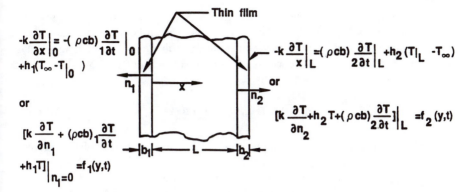

$$-k\frac{\partial T}{\partial x}\Big|_0 = -(\rho cb)\frac{\partial T}{1\partial t}\Big|_0$$
$$+h_1(T_\infty - T|_0)$$

or

$$[k\frac{\partial T}{\partial n_1} + (\rho cb)_1\frac{\partial T}{\partial t}$$
$$+h_1 T]\Big|_{n_1=0} = f_1(y,t)$$

$$-k\frac{\partial T}{\partial x}\Big|_L = (\rho cb)\frac{\partial T}{2\partial t}\Big|_L + h_2(T|_L - T_\infty)$$

or

$$[k\frac{\partial T}{\partial n_2} + h_2 T + (\rho cb)\frac{\partial T}{2\partial t}]\Big|_L = f_2(y,t)$$

Figure 3.5 Examples of the film boundary condition (fifth kind).

where ∇_0^2 is the Laplacian operator for the $\mathbf{r}'$ coordinates and the minus sign on the left side in Eq. (3.32) is a result of Eq. (3.31), with t being replaced by $-\tau$. Next, the temperature equation (3.28), can be written in terms of $\mathbf{r}'$ and τ as

$$\nabla_0^2 T + \frac{1}{k} g(\mathbf{r}', \tau) - m^2 T = \frac{1}{\alpha} \frac{\partial T}{\partial \tau} \tag{3.35}$$

Multiply Eq. (3.35) by G, multiply Eq. (3.32) by T, and subtract Eq. (3.32) from Eq. (3.35) to get

$$(G\nabla_0^2 T - T\nabla_0^2 G) + \frac{g(\mathbf{r}', \tau)}{k} G - \frac{1}{\alpha} \delta(\mathbf{r} - \mathbf{r}')\delta(t - \tau) T = \frac{1}{\alpha} \frac{\partial(GT)}{\partial \tau} \tag{3.36}$$

Integrate this equation with respect to $\mathbf{r}'$ over the total region R, and integrate with respect to τ from 0 to $t + \epsilon$, where ϵ is an arbitrarily small positive number. This yields

$$\int_{\tau=0}^{t+\epsilon} \int_R \alpha (G\nabla_0^2 T - T\nabla_0^2 G) \, dv' \, d\tau + \int_{\tau=0}^{t+\epsilon} \int_R \frac{\alpha}{k} G g(\mathbf{r}', \tau) \, dv' \, d\tau - T(\mathbf{r}, t)$$
$$= \int_R \Big[G T \Big]_{\tau=0}^{t+\epsilon} dv' \tag{3.37}$$

where dv' is a volume element in the region R. By rearranging the above equation, the temperature distribution in the body is

$$T(\mathbf{r}, t) = -\int_R \Big[G T \Big]_{\tau=0}^{t+\epsilon} dv' + \int_{\tau=0}^{t+\epsilon} \int_R \frac{\alpha}{k} G g(\mathbf{r}', \tau) \, dv' \, d\tau$$
$$+ \int_{\tau=0}^{t+\epsilon} \int_R \alpha (G\nabla_0^2 T - T\nabla_0^2 G) \, dv' \, d\tau \tag{3.38}$$

The left-hand side of this equation is the temperature distribution in the body at location $\mathbf{r}$ and at time t. The right-hand side of this equation is now examined term by term.

The first term on the right hand side of Eq. (3.38) can be simplified because $G(\mathbf{r}, t | \mathbf{r}', t + \epsilon) = 0$ by the causality relation; the effect cannot begin before the instantaneous source. Also, at $\tau = 0$, the temperature distribution $T(\mathbf{r}', 0)$ is the initial temperature distribution $F(\mathbf{r})$. Hence, the first right-side term of Eq. (3.38) becomes

$$\int_R G(\mathbf{r}, t | \mathbf{r}', 0) \, F(\mathbf{r}') \, dv' \tag{3.39}$$

For transient heat conduction in a body, this is the effect of the initial temperature distribution on the transient temperature distribution.

The second term on the right side of Eq. (3.38) arises from the volume energy generation $g(\mathbf{r}, t)$. This term will not be simplified further.

The third term on the right-hand side of Eq. (3.38) represents the contribution of all the boundary conditions. This term can be simplified with Green's theorem to change the volume integral to a surface integral. The result is

$$\int_{\tau=0}^{t+\epsilon} \int_R \alpha \, (G \nabla_0^2 T - T \nabla_0^2 G) \, dv' \, d\tau$$

$$= \int_{\tau=0}^{t+\epsilon} \sum_{i=1}^{s} \int_{S_i} \alpha \left(G \left. \frac{\partial T}{\partial n_i'} \right|_{\mathbf{r}'=\mathbf{r}_i'} - T \left. \frac{\partial G}{\partial n_i'} \right|_{\mathbf{r}'=\mathbf{r}_i'} \right) ds_i' \, d\tau \tag{3.40}$$

where $\partial/\partial n_i'$ denotes differentiation along an outward drawn normal to the boundary surface S_i and ds_i is an area element of S_i.

The integrand from Eq. (3.40) can be expressed in terms of the boundary conditions of the heat conduction equation and the auxiliary GF equation. If the boundary conditions are of the second, third, fourth, or fifth kind, then the boundary conditions for T and G can be used to evaluate $\partial T/\partial n_i'$ and $\partial G/\partial n_i'$ at the boundaries. Equations (3.30) and (3.34) can be written as

$$\left. \frac{\partial G}{\partial n_i'} \right|_{\mathbf{r}'=\mathbf{r}_i'} = - \frac{h_i}{k_i} G |_{\mathbf{r}'=\mathbf{r}_i'} + \frac{(\rho c b)_i}{k_i} \frac{\partial G}{\partial \tau} \tag{3.41}$$

$$\left. \frac{\partial T}{\partial n_i'} \right|_{\mathbf{r}'=\mathbf{r}_i'} = \frac{f_i(\mathbf{r}_i', \tau)}{k_i} - \frac{h_i}{k_i} T |_{\mathbf{r}'=\mathbf{r}_i'} - \frac{(\rho c b)_i}{k_i} \frac{\partial T}{\partial \tau} \tag{3.42}$$

Multiplying the boundary condition Eq. (3.42) by the GF, multiplying Eq. (3.41) by the temperature and subtracting yields,

$$\left[G \left. \frac{\partial T}{\partial n_i'} \right|_{\mathbf{r}'=\mathbf{r}_i'} - T \left. \frac{\partial G}{\partial n_i'} \right|_{\mathbf{r}'=\mathbf{r}_i'} \right] = \frac{f_i(\mathbf{r}_i', \tau)}{k_i} G - \frac{(\rho c b)_i}{k_i} \left(T \frac{\partial G}{\partial \tau} + G \frac{\partial T}{\partial \tau} \right)$$

$$= \frac{f_i(\mathbf{r}_i', \tau)}{k_i} G - \frac{(\rho c b)_i}{k_i} \frac{\partial (GT)}{\partial \tau} \tag{3.43}$$

Replace Eq. (3.43) into Eq. (3.40) to obtain for boundary conditions of the second through fifth kinds:

$$\int_{\tau=0}^{t+\epsilon} \sum_{i=1}^{s} \int_{S_i} \alpha \left(G \frac{\partial T}{\partial n_i'} \bigg|_{r'=r_i} - T \frac{\partial G}{\partial n_i'} \bigg|_{r'=r_i} \right) ds_i' \, d\tau$$

$$= \alpha \int_{\tau=0}^{t+\epsilon} \sum_{i=1}^{s} \int_{S_i} \frac{f_i(r_i', \tau)}{k_i} G(r, t|r', \tau) \, ds_i' \, d\tau$$

$$+ \alpha \sum_{i=1}^{s} \int_{S_i} \frac{(\rho c b)_i}{k_i} G(r, t|r', 0) F(r') \, ds_i' \qquad (3.44)$$

Note that the integral over τ has been evaluated for the term $\partial(GT)/\partial\tau$.

For a boundary condition of the first kind the right side of Eq. (3.40) takes a different form. At the boundary, G is zero and T is $f_i(r_i, t)$ for boundary conditions of the first kind. Then, the right-hand side of Eq. (3.40) becomes

$$- \alpha \int_{\tau=0}^{t+\epsilon} \sum_{j=1}^{s} \int_{S_j} f_j(r_j', \tau) \frac{\partial G}{\partial n_j'} \bigg|_{r'=r_j} ds_j' \, d\tau \qquad (3.45)$$

for boundary conditions of the first kind.

The final step in the derivation of the GFSE is to take the limit of Eq. (3.38) as $\epsilon \to 0$. Then, $t + \epsilon$ can be replaced by t in the equation without altering the conclusions drawn from $\epsilon > 0$. The derivation is completed by combining Eq. (3.38) with the simplified terms given by Eqs. (3.39), (3.44), and (3.45) to give the important *general GFSE for heat conduction* for homogeneous bodies:

$$T(r, t) = T_{in}(r, t) + T_g(r, t) + T_{b.c.}(r, t) \qquad (3.46a)$$

which contains three terms, one for the initial conditions, one for the volumetric energy source, and one for the nonhomogeneous boundary conditions. The initial temperature contribution term is

$$T_{in}(r, t) = \int_R G(r, t|r', 0) F(r') \, dv' \qquad \text{(for all boundary conditions)}$$

$$+ \alpha \sum_{i=1}^{s} \int_{S_i} \frac{(\rho c b)_i}{k_i} G(r, t|r', 0) F(r') \, ds_i' \qquad \begin{array}{l}\text{(for boundary conditions}\\\text{of the fourth and}\\\text{fifth kinds)}\end{array}$$

$$(3.46b)$$

The term for the volumetric energy generation inside the body is

$$T_g(r, t) = \int_{\tau=0}^{t} \int_R \frac{\alpha}{k} G(r, t|r', \tau) g(r', \tau) \, dv' \, d\tau \qquad (3.46c)$$

The term for the boundary conditions contains two types of expressions, one for boundary conditions of the second through fifth kinds and the other is for boundary conditions of the first kind. The term for the boundary conditions is

$$T_{\text{b.c.}}(\mathbf{r}, t) = \alpha \int_{\tau=0}^{t} \sum_{i=1}^{s} \int_{S_i} \frac{f_i(\mathbf{r}'_i, \tau)}{k_i} G(\mathbf{r}, t | \mathbf{r}'_i, \tau) \, ds'_i \, d\tau$$

(for boundary conditions of the second through fifth kinds)

$$- \alpha \int_{\tau=0}^{t} \sum_{j=1}^{s} \int_{S_j} f_j(\mathbf{r}_j, \tau) \frac{\partial G}{\partial n'_j} \bigg|_{\mathbf{r}'=\mathbf{r}'_j} ds'_j \, d\tau$$

(for boundary conditions of the first kind only)   (3.46d)

This equation has two parts because the boundary condition of the first kind must be treated in a different manner than the others.

Equation (3.46) applies to any orthogonal coordinate system if the correct form for ds and dv are used. See Table 3.1 for the differential elements ds_i, and dv for rectangular, cylindrical, and spherical coordinates systems.

Table 3.1 Quantities ds'_i and dv' for the transient GFSE for three coordinate systems

Example of geometry	Coordi- nates	ds'_i	dv'	Units of $G*$	
					Rectangular coordinates
Slab	x	$1^\dagger$	dx'	m^{-1}	
Rectangle	x, y	dx' or dy'	$dx'dy'$	m^{-2}	
Parallelpiped	x, y, z	$dx'dy'$, $dx'dz'$ or $dy'dz'$	$dx'dy'dz'$	m^{-3}	
					Cylindrical coordinates
Infinite cylinder	r	$^\dagger 2\pi r_i$	$2\pi r' dr'$	m^{-2}	
Thin shell	ϕ	$^\dagger \delta$ (thin-shell thickness)	$\delta a d\phi'$ (a = shell radius)	m^{-2}	
Finite cylinder	r, z	$2\pi r_i dz'$ or $2\pi r' dr'$	$2\pi r' dr' dz'$	m^{-3}	
Wedge	r, ϕ	dr' or $r_i d\phi'$	$r' dr' d\phi'$	m^{-2}	
					Spherical coordinates
Sphere	r	$^\dagger 4\pi r_i^2$	$4\pi (r')^2 dr'$	m^{-3}	
Conical section of sphere	r, θ	$2\pi (r')2dr' \sin \theta_i$, or $2\pi r_i^2 \sin' d\theta'$	$2\pi (r')^2 dr' \sin \theta' d\theta'$	m^{-3}	

*Units of G are such that $G \, dv'$ is dimensionless for heat conduction. Refer to the initial condition term of the GFSE.

†No integral on S_i.

X218295

The total number of terms considered between the i and j summations is exactly s, that is, the heat flux boundary conditions (second, third, fourth, and fifth kinds) and temperature boundary conditions (first kind) are mutually exclusive on a given boundary. For a one-dimensional boundary, $0 \leq s \leq 2$; for a two-dimensional geometry, $0 \leq s \leq 4$; and, for a three-dimensional geometry, $0 \leq s \leq 6$. The number of boundary conditions s includes only conditions at "real" boundaries; it does not include a boundary condition at $x \to \infty$ for a semi-infinite body, for example.

Equation (3.46) is the main result of this chapter and is a general form of the GFSE. See Chapter 10 for a general form that applies to nonhomogeneous bodies.

Example 3.3 Consider a two-dimensional rectangular region starting at $x = a_1$ and extending to $x = a_2$ in the x direction and starting at $y = b_1$ and going to $y = b_2$ in the y direction. See Fig. 3.6.

(*a*) Formulate and discuss the problem for boundary conditions of zeroth, first, second, and third kinds.

(*b*) Give the appropriate form of the GFSE for this problem.

SOLUTION

a. Formulation of the problem. The describing partial differential equation is

$$k \left(\frac{\partial^2 T}{\partial x^2} + \frac{\partial^2 T}{\partial y^2} \right) + g(x, y, t) = \rho c \frac{\partial T}{\partial t} \tag{3.47}$$

and the boundary conditions are either of the zeroth, first, second, or third kinds. For the boundary condition of the zeroth kind, there is actually no boundary. For the X20Y10 case, for example, it is convenient to set $a_1 = 0$ and $b_1 = 0$ and to note that $a_2 \to \infty$ and $b_2 \to \infty$, and hence no boundary source term, $f(\cdot)$, enters for $x = a_2 \to \infty$ and $y = b_2 \to \infty$.

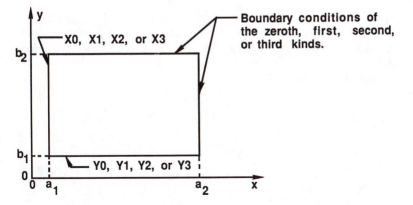

Figure 3.6 Two-dimensional rectangular body, geometry for Example 3.3.

The boundary condition at $x = a_1$ can be written as

$$- k_{a1} \frac{\partial T(a_1, y, t)}{\partial x} + h_{a1} T(a_1, y, t) = f_{a1}(y, t) \tag{3.48}$$

for boundary conditions of the first, second, and third kinds. For boundary conditions of the first kind, the k_{a1}, h_{a1}, and $f_{a1}(\cdot)$ terms are

$$k_{a1} = 0 \qquad h_{a1} = 1 \qquad f_{a1}(y, t) = T_{a1}(y, t) \tag{3.49}$$

where $T_{a1}(y, t)$ is the prescribed temperature history at $x = a_1$. For the boundary condition of the second kind, the values are

$$k_{a1} = k \qquad h_{a1} = 0 \qquad f_{a1}(y, t) = q_{a1}(y, t) \tag{3.50}$$

where k is the thermal conductivity of the solid, and $q_{a1}(\cdot)$ is the prescribed heat flux at $x = a_1$. For the boundary condition of the third kind, these terms in Eq. (3.48) are

$$k_{a1} = k \qquad h_{a1} = h_{a1}(y) \qquad f_{a1}(y, t) = h_{a1}(y) T_{\infty a1}(y, t) \tag{3.51}$$

where $h_{a1}(y)$ is the heat transfer coefficient at $x = a_1$ and $T_{\infty a1}(y, t)$ is the ambient temperature at $x = a_1$. In general, the usual GF approach permits h to be a function of position but not time.

In Eq. (3.49), the $f(\cdot)$ function represents prescribed temperatures at a boundary. Functions $f(\cdot)$ can depend on time and position. For boundary conditions of the second kind, $f(\cdot)$ in Eqs. (3.50) is a prescribed heat flux q. For a boundary condition of the third kind, $f(\cdot)$ in Eqs. (3.51) is a prescribed variation of hT_∞ where h can be a function of position (not time) and T_∞ can be a function of time and position. The boundary conditions at the $x = a_2$ surface are similar to the one at $x = a_1$

$$k_{a2} \frac{\partial T(a_2, y, t)}{\partial x} + h_{a2} T(a_2, y, t) = f_{a2}(y, t) \tag{3.52}$$

This equation also applies for the first, second, and third kind of boundary conditions by suitable choice of k_{a2}, h_{a2}, and f_{a2} in a manner similar to that in Eqs. (3.49)–(3.51).

The boundary condition at $y = b_1$ is

$$- k_{b1} \frac{\partial T(x, b_1, t)}{\partial y} + h_{b1} T(x, b_1, t) = f_{b1}(x, t) \tag{3.53}$$

and the boundary condition at $y = b_2$ is

$$k_{b2} \frac{\partial T(x, b_2, t)}{\partial y} + h_{b2} T(x, b_2, t) = f_{b2}(x, t) \tag{3.54}$$

In order to complete the statement of the problem, the initial temperature distribution is needed,

$$T(x, y, 0) = F(x, y) \tag{3.55}$$

where $F(x, y)$ is the temperature distribution at $t = 0$.

b. Two-Dimensional GFSE. The two-dimensional GFSE can be written as

$$
\begin{aligned}
T(x, y, t) = & \int_{x'=a_1}^{a_2} \int_{y'=b_1}^{b_2} G_{XIJ}(x, t|x', 0) \, G_{YMN}(y, t|y', 0) F(x', y') \, dx' \, dy' \\
& + \frac{\alpha}{k} \int_{\tau=0}^{t} \int_{x'=a_1}^{a_2} \int_{y'=b_1}^{b_2} G_{XIJ}(x, t|x', \tau) \, G_{YMN}(y, t|y', \tau) \\
& \times g(x', y', \tau) \, dy' \, dx' \, d\tau' + I_{x'=a_1} \\
& + I_{x'=a_2} + I_{y'=b_1} + I_{y'=b_2}
\end{aligned} \tag{3.56}
$$

The notation G_{XIJ} refers to the GF specific to the rectangular coordinate type of boundary condition on the boundaries $x = a_1$ and $x = a_2$. Similarly, YMN refers to the GF for the type of boundary conditions at $y = b_1$ and b_2. The last four terms denoted I in Eq. (3.56) depend on the type of boundary condition, of the zeroth, first, second or third kinds. There are four I terms, one for each boundary. For a boundary condition of the zeroth kind, the associated I term is equal to zero. For boundary condition of type 1, specified temperature, at $x' = a_1$, the $I_{x'=a_1}$ term for the boundary at $x' = a_1$ is

$$
\begin{aligned}
I_{x'=a_1} = & \ \alpha \int_{\tau=0}^{t} \int_{y'=b_1}^{b_2} \left(- \frac{\partial G_{XIJ}(x, t|a_1, \tau)}{\partial n'} \right) \\
& \times G_{YMN}(y, t|y', \tau) f_{x1}(y', \tau) \, dy' \, d\tau
\end{aligned} \tag{3.57a}
$$

and for boundary conditions of second or third kinds term $I_{x'=a_1}$ is

$$
\begin{aligned}
I_{x'=a_1} = & \ \frac{\alpha}{k} \int_{\tau=0}^{t} \int_{y'=b_1}^{b_2} G_{XIJ}(x, t|a_1, \tau) \\
& \times G_{YMN}(y, t|y', \tau) f_{x1}(y', \tau) \, dy' \, d\tau
\end{aligned} \tag{3.57b}
$$

For the $I_{x'=a_2}$ term, the same expressions as given in Eq. (3.57a) are used with a_1 in $G_{XIJ}(\cdot)$ replaced by a_2, $f_{a1}(\cdot)$ by $f_{a2}(\cdot)$. For $x' = a_1$ in Eq. (3.57b), $\partial n'$ is $-\partial x'$, while for $x' = a_2$, $\partial n'$ is $\partial x'$.

The terms $I_{y'=b_1}$ and $I_{y'=b_2}$ may be found in a manner similar to Eq. (3.57).

Example 3.4 Give the appropriate form of the GFSE for a three-dimensional rectangular parallelepiped region of $a_1 \leq x \leq a_2$, $b_1 \leq y \leq b_2$, $c_1 \leq z \leq c_2$, with the initial temperature of $F(x, y, z)$.

SOLUTION The three-dimensional GFSE can be written as

$$
T(x, y, z, t) = \int_{x'=a_1}^{a_2} \int_{y'=b_1}^{b_2} \int_{z'=c_1}^{c_2} G_{XIJ}(x, t|x', 0) G_{YKL}(y, t|y', 0)
$$

$$
\times\ G_{ZMN}(z, t|z', 0) F(x', y', z')\ dx'\ dy'\ dz'
$$

$$
+ \frac{\alpha}{k} \int_{\tau=0}^{t} \int_{x'=a_1}^{a_2} \int_{y'=b_1}^{b_2} \int_{z'=c_1}^{c_2}
$$

$$
\times\ G_{XIJ}(x, t|x', \tau) G_{YKL}(y, t|y', \tau)
$$

$$
\times\ G_{ZMN}(z, t|z', \tau) g(x', y', z', \tau)\ dx'\ dy'\ dz'
$$

$$
+ I_{x'=a_1} + I_{x'=a_2} + I_{y'=b_1} + I_{y'=b_2} + I_{z'=c_1} + I_{z'=c_2} \quad (3.58)
$$

where the I's can be found as in Example 3.3.

3.4 ALTERNATIVE GREEN'S FUNCTION SOLUTION EQUATION

In some cases, the use of GFs for nonhomogeneous boundary conditions can yield slowly converging solutions. Some of these cases can be modified to produce better-behaved solutions by using an alternative GFSE (AGFSE). A brief derivation is given in this section and a more complete derivation is given in Section 10.3.

The derivation begins with a known solution, $T^*(\mathbf{r}, t)$, to the problem

$$
\nabla^2 T^* - m^2 T^* = -\frac{g^*(\mathbf{r}, t)}{k} \quad \text{in region } R \quad (3.59)
$$

with the general boundary condition of

$$
k_i \frac{\partial T^*}{\partial n_i}\bigg|_{\mathbf{r}_i} + h_i T^* = f_i(\mathbf{r}_i, t) - (\rho c b)_i \frac{\partial T^*}{\partial t}\bigg|_{\mathbf{r}_i} \quad (3.60)
$$

Notice that the boundary conditions are nonhomogeneous and contain the same prescribed source term $f_i(\mathbf{r}, t)$ that is in the $T(\mathbf{r}, t)$ problem. In addition, Eq. (3.59) contains the arbitrary source term of $g^*(\mathbf{r}, t)$, which in some cases is set equal to zero and in others a particular choice, such as $g^* = g$, simplifies the problem; it does not have to correspond to $g(\mathbf{r}, t)$.

Let the solution to the usual transient heat conduction problem be made equal to

$$
T(\mathbf{r}, t) = T^*(\mathbf{r}, t) + T'(\mathbf{r}, t) \quad (3.61)
$$

Then a solution is desired for $T'(\mathbf{r}, t)$,

$$T'(\mathbf{r}, t) = T(\mathbf{r}, t) - T^*(\mathbf{r}, t) \tag{3.62}$$

which must satisfy

$$\nabla^2 T' + \frac{1}{k}[g(\mathbf{r}, t) - g^*(\mathbf{r}, t)] - m^2 T' - \frac{1}{\alpha}\frac{\partial T^*}{\partial t} = \frac{1}{\alpha}\frac{\partial T'}{\partial t} \qquad \text{in } R \tag{3.63}$$

with the initial condition

$$T'(\mathbf{r}, 0) = F(\mathbf{r}) - T^*(\mathbf{r}, 0) \tag{3.64}$$

and the general boundary condition

$$k_i \frac{\partial T'}{\partial n_i} + h_i T' = -(\rho c b)_i \frac{\partial T'}{\partial t} \tag{3.65}$$

which is now homogeneous. From the above, it can be seen that the solution to the $T'(\mathbf{r}, t)$ problem can be obtained by using the GFSE given by Eq. (3.46) but using a modified initial condition, a modified volume energy generation term, and homogeneous boundary conditions. Using Eq. (3.46) for $T'(\mathbf{r}, t)$ and then using Eq. (3.61) yields the AGFSE for $T(\mathbf{r}, t)$:

$$T(\mathbf{r}, t) = T^*(\mathbf{r}, t) + \int_R G(\mathbf{r}, t|\mathbf{r}', 0)[F(\mathbf{r}') - T^*(\mathbf{r}', 0)] \, dv'$$

$$+ \alpha \sum_{i=1}^{s} \int_{S_i} \frac{(\rho c b)_i}{k_i} G(\mathbf{r}, t|\mathbf{r}', 0)[F(\mathbf{r}_i') - T^*(\mathbf{r}_i', 0)] \, ds_i'$$

(for boundary conditions of the fourth and fifth kinds only)

$$+ \frac{\alpha}{k} \int_{\tau=0}^{t} \int_R G(\mathbf{r}, t|\mathbf{r}', \tau) \left[g(\mathbf{r}', \tau) - g^*(\mathbf{r}', \tau) \right.$$

$$\left. - \rho c \frac{\partial T^*(\mathbf{r}', \tau)}{\partial \tau} \right] dv' \, d\tau \tag{3.66}$$

Example 3.5 Consider the problem of a plate with the boundary and initial conditions

$$T(0, t) = T_0$$

$$T(L, t) = T_0 + (T_L - T_0) \sin \omega t$$

$$T(x, 0) = T_0$$

where T_0 and T_L are constants and ω is the frequency of oscillation of the temperature at $x = L$. Solve this problem using the standard GFSE and AGFSE.

SOLUTION The standard form of the GFSE is used first. In this solution (and the alternative form) it is convenient to solve the problem

$$T'(0, t) = T(0, t) - T_0 = 0$$

$$T'(L, t) = (T_1 - T_0) \sin \omega t$$

$$T'(x, 0) = 0$$

By solving this problem rather than the $T(x, t)$ problem, the nonhomogeneous boundary condition and nonzero initial conditions are replaced by the easier zero conditions. For this problem, the solution using Eq. (3.46) for $T'(x, t)$ has a nonzero term only for the boundary condition at $x = L$,

$$
\begin{aligned}
T'(x, t) &= -\alpha \int_0^t \frac{\partial G_{X11}(x, t|L, \tau)}{\partial n'} f(\tau)\, d\tau \\
&= \alpha \int_0^t \frac{2\pi}{L^2} \sum_{m=1}^{\infty} e^{-m^2\pi^2\alpha(t-\tau)/L^2} m(-1)^m \\
&\quad \times \sin\left(m\pi\frac{x}{L}\right)(T_L - T_0)\sin \omega\tau\, d\tau \\
&= (T_L - T_0)\frac{2\pi a}{\omega L^2} \sum_{m=1}^{\infty} \frac{m(-1)^m \sin(m\pi x/L)}{D_m^2 + 1} \\
&\quad \times (e^{-m^2\pi^2\alpha t/L^2} + D_m \sin \omega t - \cos \omega t) \qquad (3.67a)
\end{aligned}
$$

where

$$D_m = \frac{m^2\pi^2\alpha}{\omega L^2} \qquad (3.67b)$$

Here the derivative of the GF, $\partial G_{X11}/\partial n'$, has been taken from Appendix X, Eq. (X11.12). The integral on τ is given by

$$
\int_0^t e^{-m^2\pi^2\alpha(t-\tau)/L^2}\sin \omega\tau\, d\tau = \frac{1}{\omega(D_m^2 + 1)}(e^{-m^2\pi^2\alpha t/L^2} + D_m \sin \omega t - \cos \omega t) \qquad (3.68)
$$

The expression given by Eq. (3.67) contains two parts, a "steady-state" part and a transient part. The steady-state part persists in time and is periodic. The expression is not a rapidly convergent one, however. Notice that there is a term in the numerator proportional to m^3 and in the denominator to m^4; this results in terms that are proportional to m^{-1}. Series with terms that are proportional to m^{-1} typically converge very slowly, if at all. An indication of difficulty is observed for the location of $x = L$, because $\sin m\pi = 0$ but this value gives $T'(L, t) = 0$ which is not equal to the given boundary condition. This seeming contradiction is related to the convergence problem.

Consider now the use of the AGFSE. The $T^*(x, t)$ solution is obtained by solving Eq. (3.59) in the form

$$\frac{\partial^2 T^*}{\partial x^2} = 0 \tag{3.69}$$

and the boundary conditions

$$T^*(0, t) = T_0 \qquad T^*(L, t) = T_0 + (T_L - T_0) \sin \omega t \tag{3.70}$$

The solution for $T^*(x, t)$ is

$$T^*(x, t) = T_0 + (T_L - T_0) \frac{x}{L} \sin \omega t \tag{3.71}$$

Now Eq. (3.66) is used. The first integral has no contribution because

$$F(x') - T^*(x', 0) = T_0 - (T_0 + 0) = 0$$

The second integral is not present because the boundary conditions are not the fourth or fifth kinds. Then, Eq. (3.66) gives

$$T(x, t) = \left[T_0 + \left(T_L - T_0 \right) \frac{x}{L} \sin \omega t \right]$$

$$- \frac{\alpha}{k} \int_0^t \int_{x'=0}^L G_{X11}(x, t|x', \tau) \rho c \frac{\partial T^*(x', \tau)}{\partial \tau} \, d\tau \, dx'$$

$$= T_0 + (T_L - T_0) \frac{x}{L} \sin \omega t$$

$$- \frac{2}{L} \int_{\tau=0}^t \int_{x'=0}^L \sum_{m=1}^\infty e^{-m^2\pi^2\alpha(t-\tau)/L^2} \sin \left(m\pi \frac{x}{L} \right)$$

$$\times \sin \left(m\pi \frac{x'}{L} \right) (T_L - T_0) \frac{x'}{L} \omega \cos \omega\tau \, dx' \, d\tau$$

$$= T_0 + (T_L - T_0) \frac{x}{L} \sin \omega t$$

$$+ \frac{2}{\pi} (T_L - T_0) \sum_{m=1}^\infty \frac{\sin (m\pi x/L)(-1)^m}{m(D_m^2 + 1)}$$

$$\times \left(D_m \cos \omega t + \sin \omega t - D_m e^{-\frac{m^2\pi^2\alpha t}{L^2}} \right) \tag{3.72}$$

where $D_m = m^2\pi^2 \alpha(\omega L^2)$. In contrast with Eq. (3.67), which has terms proportional to m^{-1}, Eq. (3.72) has terms proportional to m^{-3} for large m. Eq. (3.72) converges rapidly and has no convergence problems; it also gives the correct result at $x = L$.

In general, the alternative GFSE is preferred over the standard form for nonhomogeneous boundary conditions when the large time form of the GF is used. This is particularly true for boundary conditions of the first kind and when results near the boundaries are needed. Notice, for this example, however, that two integrations were required for the alternative form but only one for the standard form. The large-time GFs have the time- and space-dependent components in separate terms, such as

$$\exp\left(-\frac{\beta_m^2\alpha(t-\tau)}{L^2}\right) \qquad \text{and} \qquad \sin\frac{\beta_m x}{L}$$

while the short-time GFs have the t and x together, such as

$$\exp\left[-\frac{(2mL+x-x')^2}{4\alpha(t-\tau)}\right]$$

When the short-time GFs for nonhomogeneous boundary conditions are used, the standard form may be better than the alternative form of the GFSE because fewer integrations are needed.

Another way to solve these problems is to use time partitioning, Section 5.3. In general, the standard or alternative forms of the GFSE should be tried first. Time-partitioning is needed more often for multidimensional problems than for one-dimensional ones. The methods of acceleration of series, Section 5.5, can also be used.

3.5 FIN TERM m^2T

The fin approximation may be applied in geometries with one dimension that is thin and if the temperature distribution in the thin-axis direction is approximately uniform (lumped). In this case, the energy equation may be simplified by replacing the diffusion term corresponding to the thin-axis direction by the term m^2T, called the fin term. In general, the fin parameter m can be a function of position $\mathbf{r}$, but not a function of time. The fin term can also be used to represent volume heat generation that is proportional to temperature, such as electric heating or dilute chemical reactions.

The GF method applies to fin problems even though the GFSE for the transient temperature, Eq. (3.46), does not explicitly involve the m^2 term. In the GFSE there are terms for the boundary conditions, the energy generation, and the initial condition but there is no term for fins. The dependence of the solution on m^2 is hidden in the GF $G(\mathbf{r}, t|\mathbf{r}, \tau)$ so that a different GF must be found when the fin term is used in the differential equation. The dependence of the GF on the fin term m^2 may be seen explicitly in the auxiliary equation, Eq. (3.32).

In this section, transient and steady GFs are discussed for the special case of a *spatially constant* fin term. In the case when m^2 is not spatially constant, the GF may be quite complicated if it can be found at all; in this event, the Galerkin-based GF method discussed in Chapter 10 is recommended.

3.5.1 Transient Fin Problems

All the transient GFs listed in this book are for the $m^2 = 0$ case. However, these same transient GFs can also be used for the case of a *spatially constant* m^2 when the following transformation is applied to the temperature.

Let $W(\mathbf{r}, t)$ be a new dependent variable, related to $T(\mathbf{r}, t)$ by

$$T(\mathbf{r}, t) = W(\mathbf{r}, t) \exp(-m^2 \alpha t) \qquad (3.73)$$

where m^2 is constant. Substitute this relation into the heat conduction equation, Eq. (3.28), and multiply the equation by $e^{+m^2 \alpha t}$. The result is

$$\nabla^2 W + \frac{1}{k} g(\mathbf{r}, t) e^{m^2 \alpha t} = \frac{1}{\alpha} \frac{\partial W}{\partial t} \qquad (3.74)$$

The m^2 term has canceled out so the transformed variable $W(\mathbf{r}, t)$ may be found using GFs that do not involve the $m^2 T$ term. Then the transformation can be inverted to find the original temperature $T(\mathbf{r}, t)$. The transformation does not work on steady-state problems at all because the time derivative is involved in canceling the $m^2 T$ term. For steady-state problems with the fin term, a separate set of GFs must be used; see Section 3.5.2 on steady problems.

Transient problems that involve the fin term are quite complex, and although the transformation allows a familiar set of transient GFs to be applied to these problems, the complexity of the solution has not been removed but has been shifted to the energy generation term and the boundary conditions. The boundary conditions for the transformed variable W involve the term $e^{+m^2 \alpha t}$.

The initial condition and boundary conditions for $W(\mathbf{r}, t)$ can be found by carefully applying the transformation. The initial condition for W is given by

$$W(\mathbf{r}, 0) = F(\mathbf{r}) e^0 = F(\mathbf{r}) \qquad (3.75)$$

which is unchanged. The boundary conditions will be examined according to kind. The boundary condition of the first kind is

$$T(\mathbf{r}, t) = f_i(\mathbf{r}, t) \qquad (3.76)$$

Using the relationship that defines the new variable W, the boundary condition of the first kind becomes

$$W(\mathbf{r}_i, t) = f_i(\mathbf{r}, t) \exp(m^2 \alpha t) \qquad (3.77)$$

The boundary condition of the second kind becomes

$$k \frac{\partial W}{\partial n_i}\bigg|_{\mathbf{r}_i} = f_i(\mathbf{r}_i, t) \exp(m^2 \alpha t) \qquad (3.78)$$

The boundary condition of the third kind becomes

$$k \frac{\partial W}{\partial n_i}\bigg|_{\mathbf{r}_i} = h_i[f_i(\mathbf{r}_i, t) \exp(m^2 \alpha t) - W(\mathbf{r}, t)] \qquad (3.79)$$

The boundary conditions of the fourth and fifth kinds are more affected. Using Eq. (3.73) in Eq. (3.30) gives

$$k_i \frac{\partial W}{\partial n_i}\bigg|_{\mathbf{r}_i} + [h_i - (\rho c b)_i m^2 \alpha] W(\mathbf{r}, t) = f_i(\mathbf{r}, t) \exp(m^2 \alpha t) - (\rho c b)_i \frac{\partial W}{\partial t}\bigg|_{\mathbf{r}_i} \quad (3.80)$$

Notice the extra coefficient $(\rho c b)_i m^2 \alpha$ that appears with h_i in this equation.

In summary, for the case of the transient heat conduction equation with the $m^2 T$ term for m^2 *constant*, the GFs for the transformed variable $W(\mathbf{r}, t)$ are exactly the same as for the $m^2 = 0$ case, but the energy generation term is now multiplied by $e^{+m^2 \alpha t}$, and the boundary conditions are different. For boundary conditions of the fourth and fifth kinds, h_i is replaced by $h_i - (\rho c b)_i m^2 \alpha$ at the ith boundary. In addition, for each of the five types of boundary conditions, $f_i(\mathbf{r}, t)$ in Eq. (3.46) is replaced by $f_i(\mathbf{r}, t) \exp(m^2 \alpha t)$. After the GF solution for $W(\mathbf{r}, t)$ is obtained, $T(\mathbf{r}, t)$ is simply obtained by multiplying $W(\mathbf{r}, t)$ by $\exp(-m^2 \alpha t)$ as given in Eq. (3.73).

Example 3.6: X11 case with fin term. A thin fin of uniform cross section is initially at temperature T_∞ and the $x = 0$ end of the fin is suddenly set to temperature T_0. Derive the one-dimensional fin equation and find the GF solution for the temperature in the fin if the heat transfer coefficient for side heat losses is constant and the $x = L$ end of the fin is maintained at T_∞.

SOLUTION

a. Differential equation. The fin geometry is shown in Fig. 3.7. The fin has thickness $\delta \ll L$ so that the temperature varies only in the x direction. The differential equation for the fin may be found by considering the control volume of length dx at location x. The energy balance for the control volume given by the integral energy equation (I.30) could be used.

$$w\delta[q(x) - q(x + dx)] - (2w\,dx)(T - T_\infty) = \rho c(w\delta\,dx)\frac{\partial T}{\partial t} \quad (3.81)$$

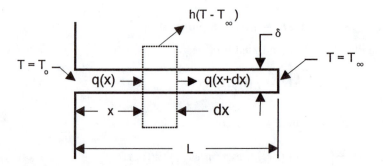

Figure 3.7 One-dimensional fin with constant cross section for Examples 3.6 and 3.7.

where w is the width of the fin, $q(\cdot)$ is heat flux (W/m^2), and h is the constant heat transfer coefficient. Divide the energy balance by the volume of the control volume ($w\delta\, dx$) to get

$$-\frac{q(x + dx) - q(x)}{dx} - \frac{2h}{\delta}(T - T_\infty) = \rho c \frac{\partial T}{\partial t} \tag{3.82}$$

The heat flux terms may be replaced by a derivative in the limit as $dx \to 0$. Replace the heat flux terms by Fourier's law $q(x) = -k\, \partial T/dx$ to give

$$k\frac{\partial^2 T}{\partial x^2} - \frac{2h}{\delta}(T - T_\infty) = \rho c \frac{\partial T}{\partial t} \tag{3.83}$$

Finally, divide by k and introduce a new variable $\Theta(x, t) = (T - T_\infty)$ to make the equation homogeneous:

$$\frac{\partial^2 \Theta}{\partial x^2} - m^2\Theta = \frac{1}{\alpha}\frac{\partial \Theta}{\partial t} \tag{3.84}$$

where now $m^2 = 2h/(\delta k)$ with units m^{-2}. This is the differential equation for a fin of uniform cross section. For the problem at hand the boundary conditions are

$$\Theta(x, 0) = T_0 - T_\infty$$

$$\Theta(0, t) = 0$$

$$\Theta(L, t) = 0$$

b. Green's function solution. The boundary value problem for $\Theta(x, t)$ may be transformed according to Eq. (3.73) for the GF solution. The transformed boundary value problem for $W(x, t)$ is given by

$$\frac{\partial^2 W}{dx^2} = \frac{1}{\alpha}\frac{\partial W}{\partial t}$$

$$W(x, 0) = (T_0 - T_\infty)e^{m^2\alpha t}$$

$$W(0, t) = 0 \tag{3.85}$$

$$W(L, t) = 0$$

The transient temperature is driven by the boundary condition at $x = 0$ and the solution is given by the GF method as

$$W(x, t) = \alpha \int_{\tau=0}^{t} (T_0 - T_\infty)e^{m^2\alpha\tau}\frac{\partial G_{X11}}{\partial x'}\bigg|_{x'=0} d\tau \tag{3.86}$$

Note that the boundary condition is introduced into the integral as a function of dummy variable τ. The function G_{X11} and its derivative is given in Appendix X so the solution is

$$W(x, t) = \alpha \int_{\tau=0}^{t} (T_0 - T_\infty) e^{m^2\alpha\tau} \frac{2\pi}{L}$$

$$\times \sum_{n=1}^{\infty} e^{-n^2\pi^2\alpha(t-\tau)/L^2} n \sin \frac{n\pi x}{L} d\tau \tag{3.87}$$

Be careful to distinguish fin parameter m from the summation index n, and to distinguish integration variable τ from time t. The integral on τ may be carried out to give the transformed solution:

$$W(x, t) = (T_0 - T_\infty) \frac{2\pi}{L} \sum_{n=1}^{\infty} \left(e^{m^2\alpha t} - e^{-n^2\pi^2\alpha t/L^2} \right) n$$

$$\times \sin \left(\frac{n\pi x}{L} \right) (m^2 + n^2\pi^2)^{-1}$$

Finally the temperature in the fin may be found by the inverse transform $\Theta = W \exp(-m^2\alpha t)$, or,

$$\Theta(x, t) = T(x, t) - T_\infty = (T_0 - T_\infty) \frac{2\pi}{L}$$

$$\times \sum_{n=1}^{\infty} \left(1 - e^{-m^2\alpha t} e^{-n^2\pi^2\alpha t/L^2} \right) n$$

$$\times \sin \left(\frac{n\pi x}{L} \right) (m^2 + n^2\pi^2)^{-1} \tag{3.88}$$

In the limit as $t \to \infty$, the series converges to the steady-state solution, but the series converges slowly (like $1/n^2$). A better form of the steady solution can be found by using a steady GF directly as shown in Section 3.6.

3.5.2 Steady Fin Problems in One Dimension

The W transformation discussed in Section 3.5.1 does not apply to steady fin problems because the W transformation relies on the time derivative $\partial T/\partial t$ to cancel the fin term from the differential equation. Many steady fin solutions exist in the literature and methods other than GFs may be appropriate.

Steady fin problems may be solved with the steady GF method if the steady-fin GF can be found. An example of a steady fin problem is given in the next section. A list of steady-fin GFs in rectangular coordinates is given in Appendix X, Table X.2, for the special case $m^2 = $ constant.

3.6 STEADY HEAT CONDUCTION

In this section, steady GFs are presented through their relationship with the transient GFs. The steady-state GFSE is stated in a general form.

3.6.1 Relationship between Steady and Transient Green's Functions

The steady GF is the limit as $t \to \infty$ of the time integral of the transient GF:

$$G(\mathbf{r}|\mathbf{r}') = \lim_{t \to \infty} \int_{\tau=0}^{t} \alpha\, G(\mathbf{r}, t|\mathbf{r}', \tau)\, d\tau \qquad (3.89)$$

This relationship may be regarded as the definition of the steady GF and it is one way to find the steady GF if the transient GF is known. For two- and three-dimensional geometries, this relationship is useful; refer to Section 4.6.4 on the limit method. For one-dimensional geometries, it is better to find the steady GF directly from the auxiliary equation for G; refer to Section 4.6.1 for this procedure.

The limit in Eq. (3.89) does not exist for all geometries and, therefore, the steady GF does not exist for all geometries. The existence of a steady-state solution depends on several factors, but geometries with insulated boundaries (boundary conditions of type 2) may not have steady solutions.

In Eq. (3.89), the transient GF is multiplied by the term $\alpha\, d\tau$ with units (m^2), so the steady GF has different units than the transient GFs which depend on the dimensionality of the geometry under discussion. The relationship between units of steady and transient GF in Cartesian coordinates are given in Table 3.2.

3.6.2 Steady Green's Function Solution Equation

In this section, the steady GFSE is stated in a general form. The steady GFSE may be derived as the limit of the transient GFSE as $t \to \infty$ because the steady temperature is simply the transient temperature in the limit as $t \to \infty$. The steady GFSE may also be derived directly from the boundary value problem for the temperature and from the auxiliary equation for the GF in a manner parallel to that for the transient GFSE presented in Section 3.3; this derivation is given as a problem at the end of the chapter.

The partial differential equation that describes steady, multidimensional, linear heat conduction is

$$\nabla^2 T + \frac{1}{k} g(\mathbf{r}) - m^2 T = 0 \qquad \text{in region } R \qquad (3.90)$$

where ∇^2 is the Laplacian operator in the appropriate coordinate system. The thermal conductivity k is constant with position and temperature. The $m^2 T$ term could represent side heat losses for a fin; in general m^2 can be a function of $\mathbf{r}$. (The $m^2 T$ term is not needed for the three-dimensional treatment of fins.)

Table 3.2 Units of steady and transient GFs in Cartesian coordinates

Geometry	Units of transient GF	Units of steady GF
One dimension	m^{-1}	m
Two dimensions	m^{-2}	1 (dimensionless)
Three dimensions	m^{-3}	m^{-1}

The steady boundary conditions for Eq. (3.90) have the general form

$$k_i \frac{\partial T}{\partial n_i} + h_i T = f_i(\mathbf{r}_i) \tag{3.91}$$

where the temperature T and its derivatives are evaluated at the boundary surface S_i, and $\mathbf{r}_i$ denotes the location of the boundary. The spatial derivative $\partial/\partial n_i$ denotes differentiation along an *outward* drawn normal to the boundary surface S_i, $i = 1, 2, \ldots, s$. The heat transfer coefficient h_i can vary with position on S_i but is independent of temperature. Three different boundary conditions can be obtained from Eq. (3.91) by setting $k_i = 0$ or k, and by setting $h_i = 0$ or h. Boundary conditions of type 4 or 5 involve energy storage $\partial T/\partial t$ and therefore do not appear in steady problems.

The steady GF satisfies the auxiliary equation

$$\nabla^2 G + \delta(\mathbf{r} - \mathbf{r}') - m^2 G = 0 \tag{3.92}$$

$$k_i \frac{\partial G}{\partial n_i'} + h_i G = 0 \tag{3.93}$$

If the GF is known for a geometry, the steady temperature may be found from the steady-state GFSE:

$$T(\mathbf{r}) = \int_R \frac{1}{k} G(\mathbf{r}|\mathbf{r}') \, g(\mathbf{r}') \, dv' \qquad \text{(for internal energy generation)}$$

$$+ \sum_{i=1}^{s} \int_{S_i} \frac{f_i(\mathbf{r}_i')}{k_i} G(\mathbf{r}|\mathbf{r}_i') \, ds_i' \qquad \begin{array}{l}\text{(for boundary conditions of}\\ \text{the second and third kind)}\end{array}$$

$$- \sum_{j=1}^{s} \int_{S_j} f_j(\mathbf{r}_j) \frac{\partial G}{\partial n_j'}\bigg|_{\mathbf{r}'=\mathbf{r}_j'} \, ds_j' \qquad \begin{array}{l}\text{(for boundary condition}\\ \text{of the first kind only)}\end{array} \tag{3.94}$$

Next, a steady example is given that includes the fin term. Other examples of steady heat transfer are given in Sections 6.9, 8.7, and 9.8.

Example 3.7: Steady fin of constant cross section with specified temperatures on the ends. Find the steady temperature in a fin with equation and boundary conditions given by

$$\frac{d^2 T}{dx^2} - m^2(T - T_\infty) = 0 \qquad 0 < x < L$$

$$T(x = 0) = T_0 \tag{3.95}$$

$$T(x = L) = T_\infty$$

where T_0 and T_∞ are constant temperatures, and $m^2 = h/(k\delta)$, as shown in Fig. 3.7. This is a fin of constant cross section and the number of this case is X11.

SOLUTION The temperature equation may be made homogeneous with the new variable $\Theta(x) = T(x) - T_\infty$, to give

$$\frac{d^2\Theta}{dx^2} - m^2\Theta = 0 \qquad \Theta(x = 0) = T_0 - T_\infty \qquad \Theta(x = L) = 0 \quad (3.96)$$

The GF for this case is given in Appendix Table X.2 as

$$G_{X11}(x|x') = \begin{cases} \sinh\left[m(L - x')\right] \dfrac{\sinh mx}{m \sinh mL} & x < x' \\[4mm] \sinh\left[m(L - x)\right] \dfrac{\sinh mx'}{m \sinh mL} & x > x' \end{cases} \quad (3.97)$$

The temperature solution is given by the boundary condition term of the steady GFSE, Eq. (3.94), or

$$T(x) - T_\infty = - (T_0 - T_\infty) \left.\frac{dG}{dn'}\right|_{x'=0} = (T_0 - T_\infty) \left.\frac{dG}{dx'}\right|_{x'=0} \quad (3.98)$$

Because the boundary term is evaluated at $x' = 0$, the $x > x'$ form of the above GF must be used to give

$$T(x) - T_\infty = (T_0 - T_\infty) \frac{\sinh m(L - x)}{\sinh mL} \quad (3.99)$$

where the derivative was carried out using the following:

$$d(\sinh z)\, dz = \cosh z \text{ and } \cosh(0) = 1$$

The shape of the temperature distribution is a decreasing exponential. This particular example may be solved directly from Eq. (3.96) with independent solutions cosh and sinh, but the GF method also applies to fin problems with internal energy generation.

3.7 MOVING SOLIDS

3.7.1 Introduction

Moving solid problems occur in many cases in heat conduction. These problems can be the result of a solid moving past a heating condition, such an extruded wire moving out of a die and being cooled by convection and radiation. Another case is a physically fixed solid with a moving heat source, such as a moving laser source on the surface of a plate. A third case can result from a moving surface, such as the ablating surface of a reentry heat shield. In each of these cases, it frequently is convenient to formulate the problem so that the coordinate system is attached to the heat source which causes a velocity term to appear in the partial differential equation of heat conduction. The equations usually must be derived using a control volume approach as discussed in the Introduction.

These problems can be one-, two-, or three-dimensional. An example of a one-dimensional problem is a moving circular die that is convectively cooled and lumped in the radial direction. That is, the temperature is only a function of the axial coordinate (not radial also) and possibly time. The describing equation can be given as

$$k \frac{\partial^2 T}{\partial x^2} - \frac{2h}{a} (T - T_\infty) = \rho c \left(\frac{\partial T}{\partial t} + U \frac{\partial T}{\partial x} \right) \tag{3.100}$$

where the thermal conductivity is assumed to be independent of temperature, the coordinate system is fixed at the die with the wire moving at a velocity of U in the positive x direction, and a is the wire radius. It is possible to have a steady state (actually, called a quasi-steady state) in this problem with respect to the die. In that case, the time derivative disappears in Eq. (3.100). The expression "quasi-steady state" is used because the temperature at any location fixed in the body varies with time, even though the temperature at a location fixed with respect to the die does not depend on time.

Another problem is for a small laser beam heating the surface of a plate. One way to visualize the problem is for the beam to be stationary and the plate to be moving in the x, y, and z directions with velocities of U_1, U_2, and U_3, respectively. Another way is to visualize that the beam is moving in the $-U_1$, $-U_2$, and $-U_3$ directions, in other words, just opposite to the previous way. In both cases, the coordinate system is fixed on the beam. The describing equation can be given as

$$k \left(\frac{\partial^2 T}{\partial x^2} + \frac{\partial^2 T}{\partial y^2} + \frac{\partial^2 T}{\partial z^2} \right) = \rho c \left(\frac{\partial T}{\partial t} + U_1 \frac{\partial T}{\partial x} + U_2 \frac{\partial T}{\partial y} + U_3 \frac{\partial T}{\partial z} \right) \tag{3.101}$$

The velocities U_1, U_2, and U_3 are assumed to be known. Again, a quasi-steady state exists for the coordinates fixed on the beam and the velocities being steady, although the temperature varies with time for a fixed point in the plate. To simplify the problem, assume that the beam is moving in the negative x direction while the plate is fixed (or equivalently, the beam is fixed and the plate is moving in the positive direction), then the equation becomes

$$k \left(\frac{\partial^2 T}{\partial x^2} + \frac{\partial^2 T}{\partial y^2} + \frac{\partial^2 T}{\partial z^2} \right) = \rho c \left(\frac{\partial T}{\partial t} + U_1 \frac{\partial T}{\partial x} \right) \tag{3.102}$$

A further simplification occurs when the velocity U_1 is sufficiently large that the U_1 term in Eq. (3.102) is much larger than the second derivative with respect to the x term, resulting in the second derivative in the x term being negligible. If, further, there is a quasi-steady state, then Eq. (3.102) simplifies to

$$k \left(\frac{\partial^2 T}{\partial y^2} + \frac{\partial^2 T}{\partial z^2} \right) = \rho c U_1 \frac{\partial T}{\partial x} \tag{3.103}$$

This equation is interesting because it is the same parabolic type as the heat conduction equation, but now the time is replaced by x/U_1. This is an important point, but it is not the main thrust of this section.

3.7.2 Three-Dimensional Formulation

The emphasis in this section is to develop a method to treat moving solid problems in a manner that the same GF and GFSE can be used, with appropriate modifications. We consider the describing equation

$$k\left(\frac{\partial^2 T}{\partial x^2} + \frac{\partial^2 T}{\partial y^2} + \frac{\partial^2 T}{\partial z^2}\right) = \rho c\left(\frac{\partial T}{\partial t} + V\frac{\partial T}{\partial x}\right) \tag{3.104}$$

where V is the velocity in the positive direction of the solid through a fixed control volume. (A more general equation is considered in the problems at the end of this chapter.) The boundary conditions can be of the first kind such as

$$T(0, y, z, t) = T_{x1}(y, z, t) \tag{3.105}$$

$$T(L, y, z, t) = T_{x2}(y, z, t) \tag{3.106}$$

and the second and third kinds,

$$-k\frac{\partial T}{\partial x}\bigg|_{x=0} = h_{x1}[T_{x\infty 1}(y, z, t) - T(0, y, z, t)] + q_{x1}(y, z, t) \tag{3.107a}$$

$$-k\frac{\partial T}{\partial x}\bigg|_{x=L} = h_{x2}[T(L, y, z, t) - T_{x\infty 2}(y, z, t)] - q_{x2}(y, z, t) \tag{3.107b}$$

or equivalently,

$$-k\frac{\partial T}{\partial x}\bigg|_{x=0} + h_{x1}T(0, y, z, t) = T_{x\infty 1}(y, z, t) + q_{x1}(y, z, t)$$

$$= f_{x1}(y, z, t) \tag{3.108a}$$

$$k\frac{\partial T}{\partial x}\bigg|_{x=L} + h_{x2}T(L, y, z, t) = h_{x2}T_{x\infty 2}(y, z, t) + q_{x2}(y, z, t)$$

$$= f_{x2}(y, z, t) \tag{3.108b}$$

Notice the definition of f_{x1} and f_{x2} implied by these equations.

The initial condition is

$$T(x, y, z, 0) = F(x, y, z) \tag{3.109}$$

These equations and boundary conditions are transformed using

$$T(x, y, z, t) = W(x, y, z, t)\exp\left(\frac{Vx}{2\alpha} - \frac{V^2}{4\alpha}t\right) \tag{3.110}$$

where $W(x, y, z, t)$ is the velocity transformation and is described by

$$k\left(\frac{\partial^2 W}{\partial x^2} + \frac{\partial^2 W}{\partial y^2} + \frac{\partial^2 W}{\partial z^2}\right) = \rho c\frac{\partial W}{\partial t} \tag{3.111}$$

with boundary conditions of the first kind

$$W(0, y, z, t) = T_{x1}(y, z, t)e^{V^2t/(4\alpha)} \tag{3.112}$$

$$W(L_x, y, z, t) = T_{x2}(y, z, t)e^{-VL_x/(2\alpha) + (V^2t)/(2\alpha)} \tag{3.113}$$

or boundary conditions of the second or third kinds,

$$-k\frac{\partial W}{\partial x}\bigg|_{x=0} + h_{xe1}W|_{x=0} = f_{x1}(y, z, t)e^{V^2t/(2\alpha)} \tag{3.114}$$

$$h_{xe1} = h_{x1} - \frac{kV}{2\alpha} \tag{3.115}$$

$$k\frac{\partial W}{\partial x}\bigg|_{x=L} + h_{xe2}W|_{x=L} = f_{x2}(y, z, t)e^{-VL/(2\alpha) + V^2t/(4\alpha)} \tag{3.116}$$

$$h_{xe2} = h_{x2} + \frac{kV}{2\alpha} \tag{3.117}$$

Notice the effective heat transfer coefficient definitions in Eqs. (3.115) and (3.117). Also notice that the boundary condition of the second kind turns into one of the third kind; this means that the G_{X2-} and G_{X-2} GFs are transformed to the G_{X3-} and G_{X-3} GFs with the effective h values being $-kV/2\alpha$ and $kV/2\alpha$, respectively.

The initial condition for W is obtained from Eqs. (3.109) and (3.110); the result is

$$W(x, y, z, 0) = F(x, y, z) \exp\left(-\frac{Vx}{2\alpha}\right) \tag{3.118}$$

This concludes the formulation of the W problem. It now remains to obtain the solution to the W problem and then to use Eq. (3.110) to get the T solution.

The GFSE can be written as

$$T(x, y, z, t) = \exp\left(\frac{Vx}{2\alpha} - \frac{V^2t}{4\alpha}\right)[W_{in}(x, y, z, t)$$
$$+ W_{bc1}(x, y, z, t) + W_{bc2,3}(x, y, z, t)] \tag{3.119}$$

where $W_{in}(\cdot)$ is for the initial condition, $W_{bc1}(\cdot)$ is for boundary conditions only of the first kind, and $W_{bc2,3}(\cdot)$ is for boundary conditions of the second and third kinds. It is important to note that there can only be one boundary condition at a given boundary, but it can be of the first or second or third kinds. The second and third kinds are treated in a similar manner. The boundary condition of the zeroth kind (no physical boundary) does not have an explicit term in Eq. (3.119).

Each of the W terms in Eq. (3.119) is now considered separately. The expression for $W_{in}(\cdot)$ is

$$W_{in}(x, y, z, t) = \int_{x'=0}^{L} \int_{y'} \int_{z'} G_{X--}(x, t|x', 0) G_{Y--}(y, t|y', 0)$$

$$\times G_{Z--}(z, t|z', 0) e^{-Vx'/(2\alpha)} F(x', y', z') \, dx' \, dy' \, dz' \quad (3.120)$$

The dashes in $X--$ can be 1, 2, or 3; the second dash could also be 0, but then the upper limit L must be changed to infinity. The first dash in $Y--$ and $Z--$ can be 1, 2, or 3, while the second dash can be 0, 1, 2, or 3. If the problem is two-dimensional with x and y coordinates, then the dependence on z and z' disappears. If the problem is one-dimensional with x being the only coordinate, Eq. (3.120) becomes

$$W_{in}(x, t) = \int_{x'=0'}^{L} G_{X--}(x, t|x', 0) e^{-Vx'/(2\alpha)} F(x') \, dx' \quad (3.121)$$

Consider next the boundary conditions of the first kind. There could be all boundary conditions of this kind in a given problem or none might be present. Also the problem might be only one- or two-dimensional. For reasons of brevity and clarity, only the $x = 0$ and $x = L$ boundaries are explicitly considered to be of the first kind for the three-dimensional case, resulting in the $W_{bc1}(x, y, z, t)$ expression of

$W_{bc1}(x, y, z, t)$

$$= \alpha \int_{\tau=0}^{t} \int_{y'} \int_{z'} \frac{\partial G_{X1-}(x, t|0, \tau)}{\partial x'} G_Y(y, t|y', \tau) G_Z(z, t|z', \tau)$$

$$\times T_{x1}(y', z', \tau) e^{V^2\tau/(4\alpha)} \, d\tau \, dy' \, dz'$$

$$- \alpha \int_{\tau=0}^{t} \int_{y'} \int_{z'} \frac{\partial G_{X-1}(x, t|L, \tau)}{\partial x'} G_Y(y, t|y', \tau) G_Z(z, t|z', \tau)$$

$$\times T_{x2}(y', z', \tau) e^{V^2\tau/(4\alpha)} \, d\tau \, dy' \, dz' \quad (3.122)$$

where the Y and Z notation subscripts have omitted the $--$ symbols. Recall that boundary conditions of the second kind have been transformed to those of the third kind. If there are boundary conditions of the first kind at the y boundaries as well as at the x boundaries, then in addition to the two terms in Eq. (3.122), two more terms are added with the integration now on x', z' and τ, the x' derivative replaced with one with respect to y', and the appropriate boundary temperature used. For a one-dimensional problem in the x direction, Eq. (3.122) reduces to

$$W_{bc1}(x, t) = \alpha \int_{\tau=0}^{t} \frac{\partial G_{X1-}(x, t|0, \tau)}{\partial x'} T_{x1}(\tau) e^{V^2\tau/(4\alpha)} \, d\tau$$

$$- \alpha \int_{\tau=0}^{t} \frac{\partial G_{X-1}(x, t|L, \tau)}{\partial x'} T_{x2}(\tau) e^{VL/(4\alpha) + V^2\tau/(4\alpha)} \, d\tau \quad (3.123)$$

Consider next boundary conditions of the second and/or third kinds. Again for brevity, only the x-direction boundary conditions are treated. The result for $W_{bc2,3}$ (x, y, z, t) is

$$W_{bc2,3}(x, y, z, t) = \frac{\alpha}{k} \int_{\tau=0}^{t} \int_{y'} \int_{z'} G_{X3\text{-}}(x, t|0, \tau) G_Y(y, t|y', \tau)$$

$$\times\, G_Z(z, t|z', \tau) f_{x1}(y', z', \tau) e^{V^2\tau/(4\alpha)}\, d\tau\, dy'\, dz'$$

$$+ \frac{\alpha}{k} \int_{\tau=0}^{t} \int_{y'} \int_{z'} G_{X\text{-}3}(x, t|L, \tau) G_Y(y, t|y', \tau)$$

$$\times\, G_Z(z, t|z', \tau) f_{x2}(y', z', \tau) e^{-VL/(2\alpha)\,+\,V^2\tau/(4\alpha)}\, d\tau\, dy'\, dz' \qquad (3.124)$$

The GFs used above, $[G_{X\text{-}\text{-}}(x, t|x', \tau), G_{Y\text{-}\text{-}}(y, t|y', \tau), G_{Z\text{-}\text{-}}(z, t|z', \tau)]$, are tabulated in the appendices and can be used, along with the eigenconditions. There are some changes, however. The boundary condition of the second kind is transformed to the third kind, while the first and third kinds remain the same. For both the second and third kinds, however, the h_1 and h_2 values are replaced by other values. At $x = 0$, for boundary conditions of the second kind (having $G_{X2\text{-}}$), G becomes $G_{X3\text{-}}$, and the h_1 values become

$$h_1 \rightarrow h_1 - \frac{kV}{2\alpha} = -\frac{kV}{2\alpha} \qquad (3.125)$$

where h_1 on the right is zero for boundary conditions of the second kind. For $x = L$ with $G_{X\text{-}2}$, G goes to $G_{X\text{-}3}$ and h_2 goes to

$$h_2 \rightarrow h_2 + \frac{kV}{2\alpha} = \frac{kV}{2\alpha} \qquad (3.126)$$

Hence, at $x = 0$ for positive values of V, the effective h is decreased while it is increased at $x = L$. If the velocity is in the negative direction, these relations are changed.

Example 3.8 A large body is initially at the temperature T_i, and then its surface at $x = 0$ is suddenly decreased to zero. The body is porous and a fluid is flowing through so that the describing partial differential equation is

$$k\frac{\partial^2 T}{\partial x^2} = \rho c \left(\frac{\partial T}{\partial t} + V\frac{\partial T}{\partial x} \right) \qquad (3.127)$$

The body can be considered to be semi-infinite $(0 < x < \infty)$ since it is said to be large. The boundary and initial conditions are

$$T(0, t) = 0 \qquad (3.128)$$

as $x \rightarrow \infty$:
$$T(x, t) \rightarrow T_i \qquad (3.129)$$

$$T(x, 0) = T_i \qquad (3.130)$$

SOLUTION Only the initial condition gives a contribution so that Eqs. (3.119) and (3.120) are needed. The number of this case is $X10B0T1V$. The equations become

$$T(x, t) = \exp\left(\frac{Vx}{2\alpha} - \frac{V^2t}{4\alpha}\right)$$

$$\times \int_{x'=0}^{\infty} G_{X10}(x, t|x', 0)e^{-Vx'/(2\alpha)} T_i \, dx' \tag{3.131}$$

The $G_{X10}(x, t|x', 0)$ GF can be found in Appendix X and is equal to

$$G_{X10}(x, t|x', 0) = (4\pi\alpha t)^{-1/2}\left\{\exp\left[-\frac{(x - x')^2}{4\alpha t}\right]\right.$$

$$\left. - \exp\left[-\frac{(x + x')^2}{4\alpha t}\right]\right\} \tag{3.132}$$

Integrals of the type

$$I_1 = \int_{x'=0}^{\infty} \exp\left[-\frac{(x - x')^2}{4\alpha t}\right] \exp\left(-\frac{Vx'}{2\alpha}\right) dx'$$

$$= (\pi\alpha t)^{1/2} \exp\left(\frac{V^2t}{4\alpha} - \frac{Vx}{2\alpha}\right)$$

$$\times \text{erfc}\left[\frac{(\alpha t)^{1/2}V}{2\alpha} - \frac{x}{(4\alpha t)^{1/2}}\right] \tag{3.133}$$

and another integral of the same type is needed with x replaced by $-x$. This integral can be evaluated by completing the square or by using integral 1 in Table F.6, Appendix F. Then using Eq. (3.133) in Eq. (3.131) gives

$$T(x, t) = \frac{T_i}{2}\left\{\text{erfc}\left[\frac{(\alpha t)^{1/2}V}{2\alpha} - \frac{x}{(4\alpha t)^{1/2}}\right]\right.$$

$$\left. - e^{Vx/\alpha} \text{erfc}\left[\frac{(\alpha t)^{1/2}V}{2\alpha} + \frac{x}{(4\alpha t)^{1/2}}\right]\right\} \tag{3.134}$$

For the case of positive V and $t \to \infty$, the steady-state temperature $T(x, \infty)$ goes to zero, while for a negative $V \, (= -U)$ and $t \to \infty$, $T(x, \infty)$ goes to

$$T(x, \infty) = T_i\left[1 - \exp\left(-\frac{Ux}{\alpha}\right)\right] \tag{3.135}$$

where $\text{erfc}(-\infty) = 2$ is used. Equation (3.135) is also valid for steady-state ablation in which a solid is being decomposed at its heated surface by intense heating and is moving at a constant velocity; x would be measured from the ablating surface and T and T_i would be interpreted as the temperature differences from the ablation temperature.

REFERENCES

Beck, J.V., Blackwell, E., and St. Clair, C.R., Jr., 1985, *Inverse Heat Conduction*, Wiley, New York.

PROBLEMS

Note: Unless otherwise requested, the explicit forms of the GFs are not needed; simply using the notation $G_{X12}(\cdot)$, for example, is sufficient.

3.1 For a vector $\mathbf{A}$, Green's theorem is usually stated

$$\iiint \nabla \cdot \mathbf{A} \, dV = \iint \mathbf{A} \cdot \mathbf{n} \, dS$$

where $\mathbf{n}$ is the outward normal. Use this form of Green's theorem to establish the following identities:

(a) $\iiint \{\Phi \nabla^2 \Phi + |\nabla \Phi|^2\} \, dV = \iint \Phi(\nabla \Phi) \cdot \mathbf{n} \, dS$

(b) $\iiint \{\Psi \nabla^2 \Phi - \Phi \nabla^2 \Psi\} \, dV = \iint [\Psi(\nabla \Phi) \cdot \mathbf{n} - \Phi(\nabla \Psi) \cdot \mathbf{n}] \, dS$

3.2 Demonstrate for XIJ ($I, J = 1, 2, 3,$ and 4) that

$$T(x, t) = \int_{x'=0}^{L} G_{XIJ}(x, t | x', 0) F(x') \, dx'$$

is the solution to the equation

$$\alpha \frac{\partial^2 T}{\partial x^2} = \frac{\partial T}{\partial t}$$

with the initial condition of $T(x, t) = F(x)$ and appropriate homogeneous boundary conditions. Use Eq. (3.4a) in your solution.

3.3 A plate has the boundary conditions given by

$$T(0, t) = T_0(t) \qquad \text{and} \qquad T(L, t) = T_i$$

and the initial condition $T(x, 0) = T_i$. Give the solution for the temperature in terms of the appropriate GF. Only one integral should be in the solution.

3.4 A semi-infinite region, $0 \leq x \leq \infty$, is initially at temperature $F(x)$. For times $t > 0$, boundary surface at $x = 0$ is kept at zero temperature and heat is generated within the solid at the rate of $g(x, t)$. Give the expression for the temperature distribution in terms of GFs.

3.5 A semi-infinite region, $0 \leq x \leq \infty$, is initially at zero temperature. For times $t > 0$, boundary surface at $x = 0$ is heated by a constant heat flux q_0. Heat is generated within the solid at the rate of $g_0 = $ constant from $x = a$ to b. Give the GFSE expression for the temperature distribution.

3.6 Give the GF solution to the problems in Problem 2.10.

3.7 Give the Green's function solution for determining the temperature in a concrete driveway (modeled as a one-dimensional semi-infinite solid) that is exposed to a convective surface heating condition with heat transfer coefficient h_s, plus a net radiative heat input of $q(t)$. The ambient temperature is assumed to be varying with time and is given by $T_\infty(t)$. At time zero, there is a nonuniform initial temperature-distribution given by $F(x)$.

3.8 Give the GF solution to the problem denoted $X23B10Y13B00T$-G- and also give the describing differential equation, boundary, and initial conditions.

3.9 A plane wall is suddenly subjected to a step change in temperature at $x = 0$ to temperature of 100°C and the initial temperature is 50°C. The $x = L$ boundary is exposed to a convection condition with an h of 10 W/m² °C and a fluid temperature of $50 + 50 \sin(5t)$°C. Obtain three different expressions for the temperature distribution in terms of the appropriate G_X (which should not be given explicitly). The three different expressions are found by different treatments of the initial condition.

3.10 A cube is initially at the temperature $F(x, y, z)$ and the surfaces are exposed to a fluid at temperature T_∞, which is a constant, and a heat transfer coefficient h. Give an expression using GF for $T(x, y, z, t)$.

3.11 A solid cylinder of radius a in a nuclear reactor is initially at the temperature $F(r)$. It is cooled by a fluid at $T_\infty(t)$ and has a heat transfer coefficient of h. Give a mathematical statement of the problem and also the number using the number system of Chapter 2. Find the solution in terms of GFs.

3.12 Solve Problem 3.11 also with a volumetric heat source due the nuclear reactions of $g(r) = g_0 \exp[-(a - r)/R]$ where R is a constant.

3.13 The alternative GFSE involves the quantity T^*, defined by

$$\nabla^2 T^* - m^2 T^* = \frac{1}{k}g^*(\mathbf{r}, t)$$

Give a physical interpretation of T^*, then in one-dimensional rectangular coordinates find a general solution for T^* for the following cases for $m^2 = 0$:

    (a) $g^*(\mathbf{r}, t) = g_1$, a constant
    (b) $g^* = x$
    (c) $g^* = e^{-ax}$

3.14 Using the notation $G(r, \theta, \phi, t|r', \theta', \phi', \tau)$ for the GF, write the GFSE for the temperature in an infinite body in spherical polar coordinates. The initial condition is $F(r, \theta, \phi)$ and the volume energy generation is $g(r, \theta, \phi, t)$.

3.15 Using this name $G(r, \phi, z, t|r', \phi', z', \tau)$ for the GF, write the GFSE for the temperature in a half cylinder, $0 \le r \le a$, $0 \le \phi \le \pi$, $0 \le z \le L$. The boundary conditions are homogeneous, the initial condition is $F(r, \phi, z)$ and the volume energy generation is $g(r, \phi, z, t)$.

3.16 Repeat the derivation of Section 3.3 for the same problem but the right-hand side replaced by

$$\frac{1}{\alpha} u(\mathbf{r}) \frac{\partial T}{\partial t}$$

The function $u(\mathbf{r})$ could represent a velocity term for a flow problem if the second derivative in the flow direction were dropped and t were replaced by the coordinate in the flow direction. Show that the GFSE is the same as Eq. (3.46) except $u(\mathbf{r}')$ is also inside the first integral of Eq. (3.46b).

3.17 An orthotropic plate is a model for aligned-fiber composite materials. For a two-dimensional orthotropic body, the thermal conductivity has two components (and only two), such as k_x and k_y for the x and y directions, respectively. Consider the problem of

$$k_x \frac{\partial^2 T}{\partial x^2} + k_y \frac{\partial^2 T}{\partial y^2} = \rho c \frac{\partial T}{\partial t}$$

$$-k_x \frac{\partial T}{\partial x}\bigg|_{x=0} = q_{x0}(y, t) \qquad T(a, y, t) = T_a(y, t) \qquad T(x, 0, t) = 0$$

$$-k_y \frac{\partial T}{\partial y}\bigg|_{y=b} = h_{yb}[T(x, b, t) - T_\infty(x, t)]$$

The objective is to obtain a GFSE for this case by using the transformation given below.

    (a) By using the transformation $y' = y(k_y/k_x)^{1/2}$, show that the problem can be transformed to

$$k_x \frac{\partial^2 T}{\partial x^2} + k_x \frac{\partial^2 T}{\partial y'^2} = \rho c \frac{\partial T}{\partial t}$$

$$-k_x \frac{\partial T}{\partial x}\bigg|_{x=0} = q_{x0}(y', t) \qquad T(a, y', t) = T_a(y', t) \qquad T(x, 0, t) = 0$$

$$-k_x \frac{\partial T}{\partial y'}\bigg|_{y'=b'} = h'_{yb}[T(x, b', t) - T_\infty(x, t)]$$

where $b' = b(k_y/k_x)^{1/2}$ and $h'_{yb} = h_{yb}(k_x/k_y)^{1/2}$.

    (b) By comparing the above problem with those previously given, obtain a GFSE. (It is not necessary to completely rederive the GFSE.) Leave in a form that does not contain the GFs in explicit form.

    (c) Give the GF(s) for this problem.

3.18 Derive the steady-state GFSE, Eq. (3.94), from first principles.

3.19 Derive Eq. (3.100) using the control volume equation in the Introduction.

3.20 Using the relationship between steady and unsteady GF, (Eq. 3.89), show how the unsteady GFSE reduces to the steady GFSE in the limit as $t \to \infty$.

3.21 Repeat Example 3.7 with added constant energy generation in the body: $g(x, t) = g_0$.

3.22 Repeat Example 3.7 with the boundary condition at $x = L$ given by

$$k \frac{\partial T(x = L)}{\partial x} + h[T(x = L) - T_\infty] = 0$$

3.23 Show that if $m = ax$ in the equation

$$\frac{\partial^2 T}{\partial x^2} - m^2 T = \frac{1}{\alpha} \frac{\partial T}{\partial t}$$

that the W transformation (Eq. 3.73) does not eliminate the $m^2 T$ term.

3.24 Give the solution in terms of GFs for the moving long circular die described by Eq. (3.100) for T_∞ equal to a constant and the boundary condition at $x = 0$ of $T = T_0$. The initial temperature is $F(x)$.

3.25 Give the solution using GFs for the problem denoted $X23B11T$-V.

3.26 Use the alternative GF solution equation to obtain $T(x, t)$ for

$$\alpha \frac{\partial^2 T}{\partial x^2} = \frac{\partial T}{\partial t} + V \frac{\partial T}{\partial x}$$

$$-k \frac{\partial T}{\partial x} = h(T(0, t) - T_\infty) \qquad \text{at } x = 0$$

$$T = 0 \quad \text{at } x = L \qquad T(x, 0) = 0$$

3.27 The GF for the hyperbolic energy equation is defined by

$$\nabla^2 G - \frac{1}{\alpha} \frac{\partial G}{\partial t} - \frac{1}{\sigma^2} \frac{\partial^2 G}{\partial t^2} = -\frac{\delta(\mathbf{r} - \mathbf{r}') \, \delta(t - \tau)}{\alpha}$$

Derive the GFSE for the hyperbolic energy equation in the infinite body.

3.28 Show that the equation

$$\frac{\partial^2 T}{\partial x^2} + \frac{\partial^2 T}{\partial y^2} + \frac{\partial^2 T}{\partial z^2} + \frac{g(x, y, z, t)}{k} - m^2 T = \frac{1}{\alpha} \left[\frac{\partial T}{\partial t} + u \frac{\partial T}{\partial x} + v \frac{\partial T}{\partial y} + w \frac{\partial T}{\partial z} \right]$$

by using the transformation

$$T(x, y, z, t) = W(x, y, z, t) \exp \left[\frac{ux}{2\alpha} - \left(\frac{u^2}{4\alpha} + m^2 \alpha \right) t \right] \exp \left[\frac{vy}{2\alpha} - \frac{v^2 t}{4\alpha} \right] \exp \left[\frac{wz}{2\alpha} - \frac{w^2 t}{4\alpha} \right]$$

can be written as

$$\frac{\partial^2 W}{\partial x^2} + \frac{\partial^2 W}{\partial y^2} + \frac{\partial^2 W}{\partial z^2} + \frac{G(x, y, z, t)}{k} = \frac{1}{\alpha} \frac{\partial T}{\partial t}$$

where G is defined to be

$$G = g(x, y, z, t) \exp \left[-\frac{ux}{2\alpha} + \left(\frac{u^2}{4\alpha} + m^2 \alpha \right) t \right] \exp \left[-\frac{vy}{2\alpha} + \frac{v^2 t}{4\alpha} \right] \exp \left[-\frac{wz}{2\alpha} + \frac{w^2 t}{4\alpha} \right]$$

METHODS FOR OBTAINING GREEN'S FUNCTIONS

4.1 INTRODUCTION

Although the Green's function (GF) approach represents a powerful and flexible method for solving heat conduction and diffusion problems, it is necessary to have mathematical expressions for the GFs. However, many GFs are known; Appendixes X, R, and RS provide listings of GFs in a systematic form for rectangular, cylindrical, and spherical coordinates, respectively. The purpose of this chapter is to demonstrate several methods of obtaining *exact* expressions for the GFs. Galerkin-based GFs for composite bodies and other difficult cases are discussed in Chapters 10 and 11. Once the GF is known for a given problem, the general solution of the problem can be written down immediately using the GF solution equations given in Chapter 3; integrations may still be needed, but the equations can be performed numerically (as shown in Chapter 5), if not analytically.

For many problems involving finite bodies, the GF expressions have two different forms: the small-time GF and the large-time GF. Various solution techniques are used to determine the different forms. The small-time and large-time forms of the GF are mathematically equivalent and both apply for $t \geq 0$; however, depending on the practical applications, one may be preferred to the other. Applications of the small-time and large-time Green's functions are discussed in more detail in Chapter 5 where the concept of the time partitioning is introduced.

In Chapter 1, we saw that the appropriate GF for a given problem is the solution to the corresponding homogeneous auxiliary problem. Consequently, the GFs themselves can be found by classic mathematical methods. In this chapter, three different approaches for obtaining the GFs are discussed and illustrated through various examples. The first method uses sources and sinks in an infinite body for construction of the GF in a finite planar body. This method, which is known as the method of images, is illustrated in Section 4.2. The next method utilizes the Laplace transform. Many small-time GFs are derived from the Laplace transform solutions of the heat conduction equation. This approach is discussed in Section 4.3. The third method uses the separation of variables technique. Many large-time GFs are obtained through this procedure. The method of separation of variables (and its relation to the GF) is discussed in Section 4.4. Section 4.5 shows that certain two- and three-dimensional GFs can be found by simple multiplication of the corresponding one-dimensional GFs. Finally, Section 4.6 covers steady-state GFs and their relationship with transient GFs.

4.2 METHOD OF IMAGES

The method of images for rectangular coordinates is based on the construction of a GF for a finite body from the GF for an infinite body (the fundamental heat conduction solution). A disadvantage of this method is that it can be readily applied only to problems with boundary conditions of the zeroth, first, and second kinds. Section 1.3 shows how the method of images can be used to find the GFs for the geometry of a semi-infinite body with isothermal and insulated surface conditions. To demonstrate more fully the application of this method, this method is employed in this section to obtain the GFs for four different flat plate cases. They are denoted $X11$, $X12$, $X21$, and $X22$ in the heat conduction numbering system. Planar sources and sinks are considered here, but line and point sources can also be used in two- or three-dimensional geometries.

The temperature solutions for each of the cases mentioned above with an initial temperature of $F(x)$ and homogeneous boundary conditions is given by

$$T(x, t) = \int_{x'=0}^{L} G(x, t \mid x', 0) F(x') \, dx' \qquad (4.1)$$

The integration is over the domain 0 to L. Four $G(\cdot)$ functions can be constructed by superimposing the plane source solution for an *infinite* body (the fundamental heat conduction solution). See Fig. 4.1 for the location of these plane sources (which are denoted by the plus signs) or sinks (which are denoted by the minus signs). The physical locations of the sources or sinks are at positions included by the equations

$$z^- = 2nL + x - x' \qquad n = \ldots, -2, -1, 0, 1, 2, \ldots \qquad (4.2a)$$

$$z^+ = 2nL + x + x' \qquad n = \ldots, -2, -1, 0, 1, 2, \ldots \qquad (4.2b)$$

One of the simplest cases to visualize is the $X22$ case which has two insulated boundary conditions; these boundary conditions can be modeled by symmetric images or reflections. The result is a series of sources (not sinks) at the z^- and z^+ locations given by Eqs. (4.2a,b). As a consequence, the $X22$ GF has only positive components as given in Table 4.1.

Another case is denoted $X11$ and is shown at the top of Fig. 4.1. Notice that adjacent images across any boundary, (at $x = 0$, $\pm L$, $\pm 2L$, . . .) must have the opposite sign to the adjacent one in order to have a zero contribution at the common boundary. This leads to the distribution of signs shown in the $X11$ case in Fig. 4.1 and the $X11$ GF given in Table 4.1. The same procedure is followed in the $X12$ and $X21$ cases shown in Fig. 4.1. The boundaries at $x = 0$, $\pm 2L$, $\pm 4L$, . . . are repeated as are those at $x = \pm L$, $\pm 3L$, . . . ; as a consequence, the symmetric condition

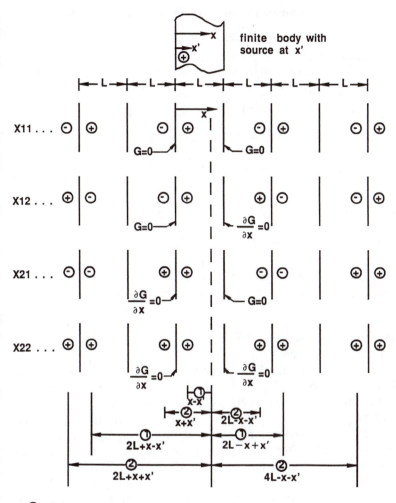

Figure 4.1 Location of sources ($+$) and sinks ($-$) for finite-body GFs created from infinite-body GFs.

(boundary condition of the second kind) has the same sign on both sides of a boundary and the antisymmetric condition (boundary condition of the first kind) is modeled by a source on one side and a sink on the other.

The cases shown in Fig. 4.1 have the GFs that are tabulated in Table 4.1 as the last five cases, with the last case being a general form containing all of the previous four cases. There are summations that extend from $n = -\infty$ to $n = +\infty$, but only a few terms are needed for small dimensionless times; this is discussed further in the next paragraph. A more extensive table of GFs for Cartesian coordinates is given in Appendix X.

Table 4.1 Green's functions formed from fundamental heat conduction solution

Case	Green's function
$X00$	$K(x - x', t - \tau) = [4\pi\alpha(t - \tau)]^{-1/2}\exp[-(x - x')^2/4\alpha(t - \tau)]$
$X10$	$K(x - x', t - \tau) - K(x + x', t - \tau)$
$X20$	$K(x - x', t - \tau) + K(x + x', t - \tau)$
$X11$	$\displaystyle\sum_{n=-\infty}^{\infty} [K(2nL + x - x', t - \tau) - K(2nL + x + x', t - \tau)]$
$X12$	$\displaystyle\sum_{n=-\infty}^{\infty} (-1)^n [K(2nL + x - x', t - \tau) - K(2nL + x + x', t - \tau)]$
$X21$	$\displaystyle\sum_{n=-\infty}^{\infty} (-1)^n [K(2nL + x - x', t - \tau) + K(2nL + x + x', t - \tau)]$
$X22$	$\displaystyle\sum_{n=-\infty}^{\infty} [K(2nL + x - x', t - \tau) + K(2nL + x + x', t - \tau)]$
XIJ	$\displaystyle\sum_{n=-\infty}^{\infty} (-1)^{(I+J)n} [K(2nL + x - x', t - \tau) + (-1)^I K(2nL + x + x', t - \tau)], I, J = 1, 2$

It is instructive to see how many terms in the $X11$, $X12$, $X21$, and $X22$ cases are needed for small dimensionless times, $\alpha(t - \tau)/L^2$. Consider the typical term, $K(2nL + x \pm x', t - \tau)$, which is plotted in Figs. 4.2 and 4.3. Results for the dimensionless time of 0.025 are plotted in the first figure and for the dimensionless time of 0.1 in the second figure. The function $K(\cdot)$ is plotted versus $(x - x')/L$ or $(x + x')/L$, where $(x - x')/L$ can vary from -1 to $+1$, and $(x + x')/L$ can vary from 0 to 2. For $\alpha(t - \tau)/L^2 = 0.025$, the maximum K value is almost 2. See Fig. 4.2. For terms with values at least 0.0001 (0.005% of the maximum), the $n = 0$ term is needed for $(x - x')/L$ between -1 and 1, and for the $(x + x')/L$ term for 0 to 1. The $n = -1$ term is needed only for $(x + x')/L$ between 1 and 2. For the larger time of $\alpha(t - \tau)/L^2 = 0.1$, Fig. 4.3 shows that for terms being less than 0.005% of the maximum, the $K(\cdot)$ terms for $(x - x')/L$ are needed for $n = 0$ (region of -1 to 1), $n = 1$ (region of -1 to 0), and $n = -1$ (region of 0 to 1). The $K(\cdot)$ terms for $(x + x')/L$ are needed for $n = 0$ (region of 0 to 2) and $n = -1$ (region of 0 to 2). For other criteria regarding the magnitude of terms that are neglected, the number of required terms could be greater or smaller. The major point is that for small dimensionless times such as $\alpha(t - \tau)/L^2 < 0.025$, only two terms are needed for $K(\cdot)$, one for $n = 0$ and the other for $n = -1$.

4.3 LAPLACE TRANSFORM DERIVATION OF GREEN'S FUNCTIONS

The Laplace transformation is a powerful tool in the solution of linear ordinary and partial differential equations, and has accordingly been applied to many heat conduction

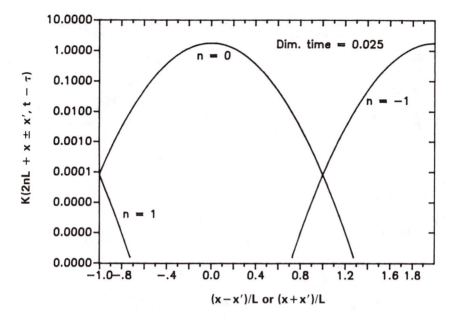

Figure 4.2 Function $K(2nL + x \pm x', t - \tau)$, a component of the early-time GF, at dimensionless time $\alpha(t - \tau)/L^2 = 0.025$.

problems (Carslaw and Jaeger, 1959; Arpaci, 1966; Luikov, 1968; Ozisik, 1980). The method is particularly well suited for the solution of one-dimensional time-dependent problems. The process of solution consists of three main steps. First, the time variable is removed from the problem by means of Laplace transformation, resulting in a simpler equation than the original equation. Next, the new equation is solved in the transformed space; and finally, the solution of the new equation is transformed back to obtain the solution to the original problem. Since this chapter is intended mainly to illustrate various approaches for obtaining the Green's functions (GFs), a detailed treatment of the Laplace transform method will not be given here. For a more comprehensive presentation of the application of the Laplace transform method to heat conduction problems, the reader may refer to Chapters 12, 13, and 15 of Carslaw and Jaeger (1959).

In this section, we first present a brief description of the Laplace transformation and a short list of its properties. An example problem is given next, to demonstrate the application of the method to a typical heat conduction problem by employing a table of transform pairs. Finally, the method is utilized for the determination of the GFs through the use of two examples.

4.3.1 Definition

Consider a function $f(t)$ for $t \geq 0$. This function can be multiplied by e^{-st} and integrated with respect to t from zero to infinity. Then, if the resulting integral exists, it is a function of the parameter s; that is,

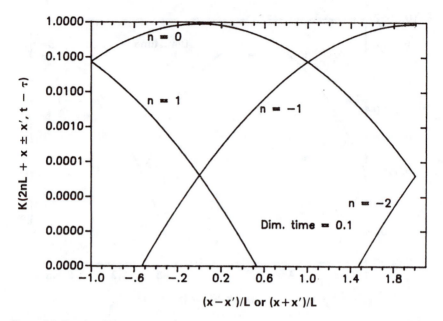

Figure 4.3 Function $K(2nL + x \pm x', t - \tau)$, a component of the early-time GF, at dimensionless time $\alpha(t - \tau)/L^2 = 0.10$.

$$\bar{f}(s) = \int_0^\infty e^{-st} f(t) \, dt \tag{4.3}$$

The function $\bar{f}(s)$ is called the Laplace transform of the function $f(t)$, and is denoted by $L[f(t)]$. The original function $f(t)$ is called the inverse transform of $\bar{f}(s)$ and is denoted by

$$f(t) = \mathscr{L}^{-1}[\bar{f}(s)] \tag{4.4}$$

Both functions $f(t)$ and $\bar{f}(s)$ are called a Laplace transform pair, and knowledge of either one enables the other to be recovered.

An important step in the process of solving a problem by Laplace transforms is that of inverting the transform to obtain the solution to the original problem. Fortunately, extensive tables of transform pairs are available in the literature, which can directly be utilized for the solution of many problems. However, in practice it might not be always the case. In those cases that the transform functions are not available in the table of transforms, the appropriate inversion can be achieved by integration in the complex domain. (For more details, see Ozisik, 1980.)

To better illustrate the Laplace transformation defined in Eq. (4.3), a few simple examples are provided;

Example 1. Let $f(t) = 1$ when $t \geq 0$, then

$$\bar{f}(s) = \mathscr{L}[1] = \int_0^\infty (1) \, e^{-st} \, dt = -\frac{1}{s} e^{-st} \Big]_0^\infty = \frac{1}{s} \qquad \text{for } s > 0 \tag{4.5}$$

Example 2. Let $f(t) = t$ when $t > 0$, then

$$\bar{f}(s) = \mathcal{L}[t] = \int_0^\infty te^{-st}\, dt = \frac{1}{s^2} \qquad \text{for } s > 0 \qquad (4.6)$$

Example 3. Let $f(t) = e^{at}$ when $t \geq 0$, then

$$\bar{f}(s) = \mathcal{L}[e^{at}] = \int_0^\infty e^{at} e^{-st}\, dt$$

$$= \int_0^\infty e^{(a-s)t}\, dt = \frac{1}{s-a} \qquad \text{for } s > a \qquad (4.7)$$

4.3.2 Properties of the Laplace Transform

Some important properties of the Laplace transform that are useful in the solution of heat conduction problems are linear property, shifting property, multiplication by t^n, division by t, transform of integrals, change of scale, and transform of convolution.

a. *Linear property.* For c_1 and c_2 arbitrary constants,

$$\mathcal{L}[c_1 f(t) + c_2 g(t)] = c_1 \mathcal{L}[f(t)] + c_2 \mathcal{L}[g(t)] = c_1 \bar{f}(s) + c_2 \bar{g}(s) \qquad (4.8a)$$

b. *Shifting property.* For a being any constant,

$$\mathcal{L}[e^{at} f(t)] = \bar{f}(s - a) \qquad (4.8b)$$

c. *Multiplication by* t^n. *For n any positive integer,*

$$\mathcal{L}[t^n f(t)] = (-1)^n \frac{d^n \bar{f}(s)}{ds^n} = (-1)^n \bar{f}^{(n)}(s) \qquad (4.8c)$$

d. *Division by* t.

$$\mathcal{L}\left[\frac{f(t)}{t}\right] = \int_s^\infty \bar{f}(s')\, ds' \qquad (4.8d)$$

e. *Transform of derivatives.* If $n > 0$ is an integer and $\lim f(t)e^{-st} = 0$ as $t \to \infty$, then for $t > 0$,

$$\mathcal{L}[f^{(n)}(t)] = s^n \bar{f}(s) - s^{n-1} f(0) - s^{n-2} f'(0) - \cdots - f^{(n-1)}(0) \qquad (4.8e)$$

f. *Transform of integrals.* If $\lim e^{-st} \int_0^t f(u)\, du = 0$ as $t \to \infty$, then

$$\mathcal{L}\left[\int_0^t f(u)\, du\right] = \frac{1}{s} \bar{f}(s) \qquad (4.8f)$$

g. *Change of scale.* If a is any positive constant, then

$$\mathcal{L}[f(at)] = \frac{1}{a} \bar{f}\left(\frac{s}{a}\right) \qquad (4.8g)$$

h. *Transform of convolution.* If $\bar{f}(s)$ is $\mathcal{L}[f(t)]$ and $\bar{g}(s)$ is $\mathcal{L}[g(t)]$, then

$$\mathscr{L}\left[\int_0^t f(u)g(t-u)\,du\right] = \mathscr{L}\left[\int_0^t f(t-u)g(u)\,du\right] = \bar{f}(s) \cdot \bar{g}(s) \quad (4.8h)$$

Example 4.1: Heat conduction in a semi-infinite body with specified surface temperature—X10B1T0 case. Consider a semi-infinite body initially at zero temperature subjected to a constant surface temperature T_0, for times $t > 0$. There is no volume energy generation in the body. Using the Laplace transform method, find the transient temperature distribution in the body.

SOLUTION The differential equation and the boundary and initial conditions for this problem are given as

$$\frac{\partial^2 T(x, t)}{\partial x^2} = \frac{1}{\alpha} \frac{\partial T(x, t)}{\partial t} \quad (4.9)$$

$$T(0, t) = T_0 \quad (4.10a)$$

$$\lim T(x, t) \to 0 \quad \text{as } x \to \infty \quad (4.10b)$$

$$T(x, 0) = 0 \quad (4.10c)$$

The first step in the solution is to find the Laplace transform of the differential equation (4.9) with respect to t; that is,

$$\mathscr{L}\left[\frac{\partial^2 T(x, t)}{\partial x^2}\right] = \frac{1}{\alpha} \mathscr{L}\left[\frac{\partial T(x, t)}{\partial t}\right] \quad (4.11)$$

The use of the properties of Laplace transform yields

$$\mathscr{L}\left[\frac{\partial^2 T(x, t)}{\partial x^2}\right] = \frac{d^2 \bar{T}(x, s)}{dx^2} \quad (4.12a)$$

$$\mathscr{L}\left[\frac{\partial T(x, t)}{\partial t}\right] = s\bar{T}(x, s) - sT(x, 0)$$

$$= s\bar{T}(x, s) \quad \text{since } T(x, 0) = 0 \quad (4.12b)$$

where

$$\bar{T}(x, s) = \mathscr{L}[T(x, t)] = \int_0^\infty e^{-st} T(x, t)\,dt \quad (4.12c)$$

Thus, Eq. (4.11) can be written as

$$\frac{d^2 \bar{T}(x, s)}{dx^2} - \frac{s}{\alpha} \bar{T}(x, s) = 0 \quad (4.13)$$

Similarly, the Laplace transform of the boundary conditions, Eqs. (4.10a, b), yields

$$\bar{T}(0, s) = \mathscr{L}[T_0] = \frac{T_0}{s} \quad (4.14a)$$

$$\overline{T}(x, s) \rightarrow 0 \qquad \text{as } x \rightarrow \infty \qquad (4.14b)$$

Equation (4.13) is an ordinary differential equation for $\overline{T}(x, s)$ with the only independent variable being x. The solution of this equation with the boundary conditions given by Eqs. (4.14a, b) may be written as

$$\overline{T}(x, s) = \frac{T_0}{s} e^{-x\sqrt{s/\alpha}} \qquad (4.15)$$

The final step is now to transform $\overline{T}(x, s)$ back to obtain the solution for $T(x, t)$; that is,

$$T(x, t) = \mathscr{L}^{-1}\left[\frac{T_0}{s} e^{-x\sqrt{s/\alpha}}\right] \qquad (4.16)$$

Equation (4.16) can be inverted simply by utilizing a table of transform pairs (e.g., see Beyer, 1976, CRC Standard Mathematical Tables) to obtain

$$T(x, t) = T_0 \operatorname{erfc}\left[\frac{x}{(4\alpha t)^{1/2}}\right] \qquad (4.17)$$

This is the same solution as given by Eq. (6.16) which was obtained by the GF method.

4.3.3 Derivation of Green's Functions

The short-time GF for many heat conduction problems are derived from the Laplace transform solution of the corresponding auxiliary equation with homogeneous boundary conditions. As discussed in Chapter 1, the auxiliary equation for a given problem is identical to the original heat conduction equation for that problem except for the heat generation term, which is a unit instantaneous heat source modeled by a Dirac delta function. The homogeneous boundary conditions for the auxiliary problem must be of the same kind as the original problem. Determination of the GFs by the method of Laplace transform is best illustrated through the use of examples.

Example 4.2: Semi-infinite body with convection—$X30$ case. Find the GF for the problem of a semi-infinite body with the convective boundary condition at the surface.

SOLUTION This is the $X30$ case. The GF associated with this problem is the solution to the following auxiliary equation:

$$\frac{\partial^2 G}{\partial x^2} + \frac{1}{\alpha} \delta(x - x') \delta(t - 0) = \frac{1}{\alpha}\frac{\partial G}{\partial t} \qquad t \geq 0 \qquad x > 0 \qquad (4.18)$$

subject to the homogeneous boundary conditions of

$$-k\frac{\partial G(0, t|x', 0)}{\partial x} + hG(0, t|x', 0) = 0 \qquad t \geq 0 \qquad (4.19a)$$

$$G(\infty, t|x', 0) = 0 \qquad t \geq 0 \tag{4.19b}$$

and initial condition

$$G(x, t|x', 0) = 0 \qquad t < 0 \tag{4.19c}$$

Notice that the second term in Eq. (4.18) represents a unit instantaneous plane source at location x' released at time $\tau = 0$. Consequently, $G(x, t|x', 0)$ is the X30 GF for $\tau = 0$. Once the appropriate expression for $G(x, t|x', 0)$ is determined, then GF for $\tau \neq 0$ can be found by replacing t by $t - \tau$ in that expression.

In the Laplace transform approach, the auxiliary problem given by Eq. (4.18) is subdivided into two problems. One gives the solution due to the instantaneous plane source at location x' and at time τ for an infinite one-dimensional body (the fundamental heat conduction solution), and the other satisfies the given initial and boundary conditions. Hence, $G(x, t|x', 0)$ is written as

$$G(x, t|x', 0) = K(x - x', t - 0) + v(x, t) \tag{4.20}$$

where K is the fundamental heat conduction solution for $\tau = 0$, given by

$$K(x - x', t - 0) = (4\pi\alpha t)^{-1/2} \exp\left[-\frac{(x - x')^2}{4\alpha t}\right] \tag{4.21}$$

and $v(x, t)$ satisfies the one-dimensional heat conduction equation in the semi-infinite region; that is,

$$\frac{\partial^2 v(x, t)}{\partial x^2} = \frac{1}{\alpha}\frac{\partial v(x, t)}{\partial t} \qquad \text{for } t > 0 \qquad \text{and} \qquad 0 < x < \infty \tag{4.22a}$$

with the initial condition of

$$v(x, 0) = 0 \tag{4.22b}$$

and v should be such that the boundary conditions, Eqs. (4.19a, b), are satisfied.

Taking the Laplace transform of K and v in Eqs. (4.21) and (4.22a) results in the following equations,

$$\overline{K} = \frac{1}{2(\alpha s)^{1/2}} \exp\left[-\left(\frac{s}{\alpha}\right)^{1/2}|x - x'|\right] \tag{4.23}$$

$$\frac{d^2\overline{v}}{dx^2} - \frac{s}{\alpha}\overline{v} = 0 \tag{4.24}$$

where the overbar denotes the Laplace transform

$$\mathcal{L}\{F(t)\} = \int_0^\infty e^{-st}F(t)\,dt = \overline{F} \tag{4.25}$$

and s is the Laplace transform parameter. The general solution of Eq. (4.24) may be written as

$$\overline{v}(x, s) = A \exp\left[\left(\frac{s}{\alpha}\right)^{1/2}x\right] + B \exp\left[-\left(\frac{s}{\alpha}\right)^{1/2}x\right] \tag{4.26}$$

Now, taking the Laplace transform of Eq. (4.20) and substituting the values for $\overline{K}$ and $\overline{v}$ from Eqs. (4.23) and (4.26) into the transformed equation yields

$$
\overline{G}(x, s|x'\ 0) = \frac{1}{2(\alpha s)^{1/2}} \exp\left[-\left(\frac{s}{\alpha}\right)^{1/2}|x - x'|\right]
$$
$$
+ A \exp\left[\left(\frac{s}{\alpha}\right)^{1/2} x\right] + B \exp\left[-\left(\frac{s}{\alpha}\right)^{1/2} x\right] \quad (4.27)
$$

The constants A and B in Eq. (4.27) are determined from the boundary conditions Eqs. (4.19a, b). The Laplace transform of these equations are

$$
-k\frac{\partial\overline{G}(0, s|x', 0)}{\partial x} + h\overline{G}(0, s|x', 0) = 0 \quad (4.28a)
$$

$$
\overline{G}(\infty, s|x', 0) = 0 \quad (4.28b)
$$

Then, by introducing the transformed conditions Eqs. (4.28a, b) into Eq. (4.27), the constants A and B are

$$
A = 0 \quad (4.29a)
$$

$$
B = \frac{1}{2(\alpha s)^{1/2}} \frac{(s/\alpha)^{1/2} - H}{(s/\alpha)^{1/2} + H} \exp\left[-\left(\frac{s}{\alpha}\right)^{1/2} x'\right] \quad (4.29b)
$$

where $H = h/k$. Substituting Eqs. (4.29a, b) back into Eq. (4.27) yields

$$
\overline{G}(x, s|x', 0) = \frac{1}{2(\alpha s)^{1/2}} \left\{\exp\left[-\left(\frac{s}{\alpha}\right)^{1/2}|x - x'|\right]\right.
$$
$$
+ \exp\left[-\left(\frac{s}{\alpha}\right)^{1/2}(x + x')\right] - \frac{2H}{[(s/\alpha)^{1/2} + H]}
$$
$$
\left.\times \exp\left[-\left(\frac{s}{\alpha}\right)^{1/2}(x + x')\right]\right\} \quad (4.30)
$$

From a Laplace transform table (such as in the appendix of Carslaw and Jaeger, 1959), the inverse transform of Eq. (4.30) gives the solution of $G(x, t|x', 0)$; that is,

$$
G(x, t|x', 0) = \frac{1}{2(\pi\alpha t)^{1/2}} \left\{\exp\left[-\frac{(x - x')^2}{4\alpha t}\right] + \exp\left[-\frac{(x + x')^2}{4\alpha t}\right]\right\}
$$
$$
- \frac{h}{k} \exp\left[\frac{h}{k}(x + x') + \alpha\frac{h^2 t}{k^2}\right] \mathrm{erfc}\left[\frac{(x + x')}{(4\alpha t)^{1/2}} + \frac{h}{k}(\alpha t)^{1/2}\right]
$$
$$
(4.31)
$$

which is the $X30$ GF for $\tau = 0$. The $X30$ GF for $\tau \neq 0$ can now be determined by replacing t by $t - \tau$ in Eq. (4.31); that is,

$$G_{X30}(x, t|x', \tau) = \frac{1}{[4\pi\alpha(t - \tau)]^{1/2}} \left\{ \exp\left[-\frac{(x - x')^2}{4\alpha(t - \tau)} \right] \right.$$

$$+ \exp\left[-\frac{(x + x')^2}{4\alpha(t - \tau)} \right] \right\}$$

$$- \frac{h}{k} \exp\left[\frac{h}{k}(x + x') + \alpha\frac{h^2(t - \tau)}{k^2} \right]$$

$$\times \operatorname{erfc}\left\{ \frac{x + x'}{[4\alpha(t - \tau)]^{1/2}} + \frac{h}{k}[\alpha(t - \tau)]^{1/2} \right\} \quad (4.32)$$

This equation is tabulated in Appendix X.

Example 4.3: Region outside a spherical cavity with convection—RS30 case. Find the GF for the infinite region outside a spherical cavity of radius a with a convective boundary condition. This is the $RS30$ case.

SOLUTION The GF is the solution to the auxiliary equation,

$$\frac{1}{r}\frac{\partial^2(rG)}{\partial r^2} + \frac{1}{\alpha}\delta(\mathbf{r} - \mathbf{r}')\,\delta(t - 0) = \frac{1}{\alpha}\frac{\partial rG}{\partial t} \qquad a < r < \infty \qquad t \geq 0 \quad (4.33)$$

Here $\delta(\mathbf{r} - \mathbf{r}')$ has units m^{-3}. The homogeneous boundary conditions are

$$-k\frac{\partial G(a, t|r', 0)}{\partial r} + hG(a, t|r', 0) = 0 \qquad t \geq 0 \quad (4.34a)$$

$$G(\infty, t|r', 0) = 0 \qquad t \geq 0 \quad (4.34b)$$

and the initial condition is

$$G(r, t|r', 0) = 0 \qquad t < 0 \quad (4.34c)$$

Equations (4.33) and (4.34) represent the problem of an infinite region outside the spherical cavity of $r = a$ (initially at zero temperature) subject to a unit instantaneous spherical surface source at $r = r'$ released at time $\tau = 0$ with a homogeneous convective boundary condition at $r = a$.

Again, in a manner similar to that used in the previous example, the solution for $G(r, t|r', 0)$ is subdivided into two parts in the following form:

$$G(r, t|r', 0) = K_s(r - r', t - 0) + v(r, t) \quad (4.35)$$

where K_s is the fundamental heat conduction solution for radial flow in the spherical region; it is the GF for the $RS00$ case (see Appendix RS) and is given by

$$K_s(r - r', t - 0) = \frac{1}{8\pi rr'(\pi\alpha t)^{1/2}} \left\{ \exp\left[-\frac{(r - r')^2}{4\alpha t} \right] \right.$$

$$- \exp\left[-\frac{(r + r')^2}{4\alpha t} \right] \right\} \quad (4.36)$$

and its Laplace transform is given by

$$\overline{K}_s = \frac{1}{8\pi r r'(\alpha s)^{1/2}} \left\{ \exp\left[-\left(\frac{s}{\alpha}\right)^{1/2} |r - r'| \right] \right.$$

$$\left. - \exp\left[-\left(\frac{s}{\alpha}\right)^{1/2} (r + r') \right] \right\} \tag{4.37}$$

The temperature v in this case satisfies the heat conduction equation for one-dimensional heat flow in the region outside the spherical cavity $r = a$; that is,

$$\frac{\partial^2 [rv(r, t)]}{\partial r^2} = \frac{1}{\alpha} \frac{\partial [rv(r, t)]}{\partial t} \qquad \text{for } t > 0 \qquad \text{and} \qquad a < r < \infty \tag{4.38a}$$

with the initial condition of

$$v(r, 0) = 0 \tag{4.38b}$$

The Laplace transform of Eq. (4.38a) yields

$$\frac{d^2(r\overline{v})}{dr^2} - \frac{s}{\alpha} r\overline{v} = 0 \qquad \text{for } a < r < \infty \tag{4.39}$$

which has the general solution of the form

$$\overline{v}(r, s) = \frac{A}{r} \exp\left[\left(\frac{s}{\alpha}\right)^{1/2} r \right] + \frac{B}{r} \exp\left[-\left(\frac{s}{\alpha}\right)^{1/2} r \right] \tag{4.40}$$

Taking the Laplace transform of Eq. (4.35) and substituting the values for $\overline{K}$ and $\overline{v}$ from Eq. (4.37) and (4.38) into the result gives

$$\overline{G}(r, s|r', 0) = \frac{1}{8\pi r r'(\alpha t)^{1/2}} \left\{ \exp\left[-\left(\frac{s}{\alpha}\right)^{1/2} |r - r'| \right] \right.$$

$$\left. - \exp\left[-\left(\frac{s}{\alpha}\right)^{1/2} (r + r') \right] \right\}$$

$$+ \frac{A}{r} \exp\left[\left(\frac{s}{\alpha}\right)^{1/2} r \right] + \frac{B}{r} \exp\left[-\left(\frac{s}{\alpha}\right)^{1/2} r \right] \tag{4.41}$$

Equation (4.41) must satisfy the boundary conditions Eqs. (4.34a, b). The Laplace transforms of these equations are

$$-\frac{\partial \overline{G}(a, s|r', 0)}{\partial r} + H\overline{G}(a, s|r', 0) = 0 \tag{4.42a}$$

$$\overline{G}(\infty, s|r', 0) = 0 \tag{4.42b}$$

where h/k is denoted H. It follows from Eq. (4.42b) that

$$A = 0 \tag{4.43a}$$

Then, from Eq. (4.42a), one can show that

$$B = \frac{1}{8\pi r'(\alpha t)^{1/2}} \left\{ \exp\left[-\left(\frac{s}{\alpha}\right)^{1/2} r'\right] - \exp\left[-\left(\frac{s}{\alpha}\right)^{1/2}(r' - 2a)\right] \right.$$

$$\left. + \frac{2(s/\alpha)^{1/2}}{(s/\alpha)^{1/2} + \dfrac{1}{a} + H} \exp\left[-\left(\frac{s}{\alpha}\right)^{1/2}(r' - 2a)\right] \right\} \tag{4.43b}$$

Substituting the values for A and B from Eqs. (4.43a, b) into Eq. (4.41) yields,

$$\overline{G}(r, s|r', 0) = \frac{1}{8\pi r r'(\alpha t)^{1/2}} \left\{ \exp\left[-\left(\frac{s}{\alpha}\right)^{1/2} |r - r'|\right] \right.$$

$$- \exp\left[-\left(\frac{s}{\alpha}\right)^{1/2}(r + r' - 2a)\right]$$

$$+ \frac{2(s/\alpha)^{1/2}}{(s/\alpha)^{1/2} + \dfrac{1}{a} + H}$$

$$\left. \times \exp\left[-\left(\frac{s}{\alpha}\right)^{1/2}(r + r' - 2a)\right] \right\} \tag{4.44}$$

which is the Laplace transform of $G(r, t|r', 0)$. Taking the inverse transform of Eq. (4.44) (see Laplace transform table in the appendix of Carslaw and Jaeger, 1959) and by replacing t by $t - \tau$ gives

$$G_{RS30}(r, t|r', \tau) = \frac{1}{8\pi r r'[\alpha\pi(t - \tau)]^{1/2}} \left(\exp\left[-\frac{(r - r')^2}{4\alpha(t - \tau)}\right] \right.$$

$$+ \exp\left[-\frac{(r + r' - 2a)^2}{4\alpha(t - \tau)}\right]$$

$$- \frac{k + ah}{ak} [4\pi\alpha(t - \tau)]^{1/2}$$

$$\times \exp\left[\alpha(t - \tau)\left(\frac{k + ah}{ak}\right)^2 \right.$$

$$+ \frac{k + ah}{ak}(r + r' - 2a)\right]$$

$$\times \operatorname{erfc}\left\{ \frac{(r + r' - 2a)}{2[\alpha(t - \tau)]^{1/2}} \right.$$

$$\left. \left. + \frac{k + ah}{ak}[\alpha(t - \tau)]^{1/2}\right\} \right) \tag{4.45}$$

which is the $RS30$ Green's function; it is included in Appendix RS.

In the example problems considered above, the inversion of the transformed solutions were obtained directly from a table of Laplace transforms. However, note that there are cases (e.g., in finite geometries such as plates, cylinders, and spheres) for which the transformed solution $\overline{G}$ does not appear in the Laplace transform tables. In such cases the original solution G is derived by the use of the series expansion or the usual inversion theorem for the Laplace transformation. The series expansion approach is often less complicated and more useful than the use of the inversion theorem, particularly for small times. In this approach, usually the transformed solution $\overline{G}$ is expanded into a series of ascending powers of $1/s$ which converges rapidly for large values of s (corresponding to small values of t). Then, the resulting series is inverted term by term using the Laplace transform table. For another discussion of this approach, see Chapter 12 of Carslaw and Jaeger (1959).

4.4 METHOD OF SEPARATION OF VARIABLES FOR DERIVATION OF GREEN'S FUNCTIONS

The method of separation of variables can be used to find the GFs through the relationship between the GF and the Dirac delta function. Chapter 1 showed that GFs are proportional to the temperature rise in a body driven by a Dirac delta function initial temperature distribution. The method of separation of variables provides a straightforward method for solving finite-body problems with arbitrary initial temperature distributions. Once the temperature $T(x, t)$ is known for an arbitrary space-variable initial temperature $F(x)$, then the GF can be found from $T(x, t)$ because an arbitrary initial temperature includes the Dirac delta function as a special case.

In this section, several one-dimensional flat plate GFs are found using the method of separation of variables. The flat plate with the temperature fixed at both sides ($X11$) is used in a full discussion of the method and the flat plate with one insulated boundary ($X21$) is discussed in an example. A more general derivation of GFs using the separation of variables method is given by Beck (1984) for the flat plate with boundary conditions of the first, second, third, fourth, or fifth kinds.

4.4.1 Plate with Temperature Fixed at Both Sides ($X11$)

One of the simplest cases to consider using the method of separation of variables is for prescribed temperatures of zero at both boundaries of a plate. The describing partial differential equation, boundary conditions, and initial conditions are given by

$$\frac{\partial^2 T}{\partial x^2} = \frac{1}{\alpha} \frac{\partial T}{\partial t} \qquad 0 < x < L \tag{4.46}$$

$$T(0, t) = 0 \qquad T(L, t) = 0 \tag{4.47a, b}$$

$$T(x, 0) = F(x) \tag{4.48}$$

Note that the boundary conditions and the partial differential equation are both homogeneous. This case has the notation $X11B00T-$.

Since the thermal diffusivity α is a constant, the differential equation can be solved by adding many solutions, each of which satisfies the differential equation. This is also called superimposing solutions. Let

$$T(x, t) = \sum_{n=1}^{\infty} T_n(x, t) \tag{4.49}$$

where the solutions $T_n(x, t)$ satisfy Eq. (4.46). That is, when $T_n(x, t)$ is substituted into Eq. (4.46), an identity results. In addition, each $T_n(x, t)$ solution satisfies the homogeneous boundary conditions given by Eqs. (4.47a, b). A $T_n(x, t)$ solution for a given n does not usually satisfy the initial condition given by Eq. (4.48).

The procedure continues by assuming that

$$T_n(x, t) = \mathbf{X}(x)\,\Theta(t) \tag{4.50}$$

where $\mathbf{X}(x)$ is a function of only x, and where $\Theta(t)$ is a function of only t. In other words, $T_n(x, t)$ is chosen to be a product of two functions, one that depends only on x and the other that depends only on t. The variables have been separated in Eq. (4.50), hence the name separation of variables technique. Replacing T in Eq. (4.46) by T_n gives

$$\frac{\partial^2 T_n}{\partial x^2} = \frac{1}{\alpha}\frac{\partial T_n}{\partial t} \tag{4.51}$$

and substituting Eq. (4.50) into Eq. (4.51) gives

$$\alpha\,\frac{d^2\mathbf{X}}{dx^2}\,\Theta = \mathbf{X}\,\frac{d\Theta}{dt} \tag{4.52}$$

Dividing Eq. (4.52) by $\alpha T_n(x, t)$ yields

$$\frac{1}{\mathbf{X}}\frac{d^2\mathbf{X}}{dx^2} = \frac{1}{\alpha\Theta}\frac{d\Theta}{dt} \tag{4.53}$$

This equation states that a function of x is equal to a function of t. This equality can only be true if the functions are both simply the same constant. For that reason, let both sides be equal to the negative (real) quantity of $-\lambda^2$,

$$\frac{1}{\mathbf{X}}\frac{d^2\mathbf{X}}{dx^2} = \frac{1}{\alpha\Theta}\frac{d\Theta}{dt} = -\lambda^2 \tag{4.54}$$

Another choice is a positive constant λ^2, but as is shown below, a positive constant gives meaningless results. (This assumes that λ is restricted to real and not imaginary values.) In some cases, the constant may be equal to zero.

Two *ordinary* differential equations now must be solved,

$$\frac{d^2\mathbf{X}}{dx^2} + \lambda^2\mathbf{X} = 0 \tag{4.55a}$$

$$\frac{d\Theta}{dt} + \alpha\lambda^2\Theta = 0 \tag{4.55b}$$

The general solutions of these equations are

$$\mathbf{X} = C_1 \sin \lambda x + C_2 \cos \lambda x \tag{4.56a}$$

$$\Theta = C_3\, e^{-\lambda^2 \alpha t} \qquad (4.56b)$$

Notice that $\mathbf{X}$ is a sum of two periodic functions. Also Θ is a decaying exponential function. Note that if $-\lambda^2$ were replaced by λ^2, the solution for Θ would result in explosive growth over time—clearly not physically reasonable. (Again, if λ is allowed to be imaginary, different conclusions are possible.) For large times, the solution of the problem given by Eqs. (4.46)–(4.48) must tend toward zero. Consequently, the constant in Eq. (4.54) must be $-\lambda^2$, where the negative sign is both necessary and important.

At this point, it has been assured that $T_n(x, t)$ satisfies the partial differential equation. Next, $T_n(x, t)$ must satisfy the two (homogeneous) boundary conditions. From the boundary condition at $x = 0$, we have

$$T_n(0, t) = \mathbf{X}(0)\, \Theta(t) = 0 \qquad (4.57)$$

Since $\Theta(t)$ is an arbitrary function of time, it cannot be set equal to zero without causing $T_n(x, t)$ to be zero for all values of t; such a trivial solution clearly cannot satisfy the nonzero initial conditions which will be examined shortly. Hence, $\mathbf{X}(0) = 0$, and from Eq. (4.56a) it is necessary that

$$\mathbf{X}(0) = 0 = C_1 \cdot 0 + C_2 \cdot 1 \qquad (4.58)$$

which yields

$$C_2 = 0 \qquad (4.59)$$

Next consider the boundary condition at $x = L$ which gives

$$T_n(L, t) = \mathbf{X}(L)\, \Theta(t) = 0 \qquad (4.60)$$

and again since $\Theta(t)$ cannot be always zero, the result is

$$\mathbf{X}(L) = 0 = C_1 \sin \lambda L \qquad (4.61)$$

Consequently, the *eigencondition* is

$$\sin \lambda_n L = 0 \qquad (4.62)$$

which can occur at only certain values, namely,

$$\lambda_n L = n\pi \qquad n = \ldots, -2, -1, 0, 1, 2, \ldots$$

All of these n values are not needed, however. The negative values do not give independent eigenfunctions (sin $\lambda_n x$ is called an eigenfunction), since

$$\sin(-\lambda_n L) = -\sin(-\lambda_n L) \qquad (4.63)$$

Also the $n = 0$ value makes no contribution in this case since sin $(0) = 0$. Hence, the eigenvalues λ_n are given by

$$\lambda_n = \frac{n\pi}{L} \qquad n = 1, 2, 3, \ldots \qquad (4.64)$$

Usually the eigenvalues in this book are made dimensionless. Let the dimensionless eigenvalues be denoted β_n where for this case

$$\beta_n = n\pi \qquad n = 1, 2, 3, \ldots \tag{4.65}$$

and the eigenfunction is

$$\sin \frac{\beta_n x}{L} \tag{4.66}$$

At this point the differential equation and the two homogeneous boundary conditions for $T_n(x, t)$ have been satisfied. The next step is to bring the two parts of $T_n(x, t)$ together to find

$$T_n(x, t) = C_1 \sin \frac{\beta_n x}{L} C_3 e^{-\beta_n^2 \alpha t/L^2} = A_n \sin \frac{\beta_n x}{L} e^{-\beta_n^2 \alpha t/L^2} \tag{4.67}$$

where A_n is a constant that depends on n. Introduce this form of T_n into Eq. (4.49) to get

$$T(x, t) = \sum_{n=1}^{\infty} A_n \sin \frac{\beta_n x}{L} e^{-\beta_n^2 \alpha t/L^2} \tag{4.68}$$

The remaining condition to satisfy is the initial condition, Eq. (4.48). This condition is nonzero, unlike the boundary conditions. Using the value of $t = 0$ in Eq. (4.68) and the value of $T(x, 0) = F(x)$ gives

$$F(x) = \sum_{n=1}^{\infty} A_n \sin \frac{\beta_n x}{L} \tag{4.69}$$

The objective is now to determine values of the constants A_n, $n = 1, 2$, etc. A result from the theory of Fourier series is that the sine functions are *orthogonal*, which can be stated as

$$\int_{x=0}^{L} \sin \frac{\beta_n x}{L} \sin \frac{\beta_m x}{L} \, dx = \begin{cases} \dfrac{L}{2} & m = n \neq 0 \\ 0 & m = n \end{cases} \tag{4.70}$$

for the β_n values of $n\pi$, $n = 1, 2, \ldots$. This orthogonality condition provides a very powerful tool for determining one value of A_n at a time. Multiplying both sides of Eq. (4.69) by $\sin(\beta_m x/L) \, dx$ and integrating from $x = 0$ to L yields

$$\int_{x=0}^{L} F(x) \sin \frac{\beta_m x}{L} \, dx = \int_{x=0}^{L} \sum_{n=1}^{\infty} A_n \sin \frac{\beta_n x}{L} \sin \frac{\beta_m x}{L} \, dx \tag{4.71}$$

Now, according to the orthogonality condition, Eq. (4.70), there is a nonzero term on the right hand side of Eq. (4.71) only when $m = n$. In other words, the orthogonality condition just picks out one term in the summation to give

$$\int_{x=0}^{L} F(x) \sin \frac{\beta_m x}{L} \, dx = \frac{A_m L}{2} \tag{4.72}$$

Another way to think of this procedure is to imagine that m is a particular value such as 2. If $m = 2$, then the right side of Eq. (4.71) is

$$\int_{x=0}^{L} A_1 \sin \frac{\beta_1 x}{L} \sin \frac{\beta_2 x}{L} \, dx + \int_{x=0}^{L} A_2 \sin^2 \frac{\beta_2 x}{L} \, dx$$

$$+ \int_{x=0}^{L} A_3 \sin \frac{\beta_3 x}{L} \sin \frac{\beta_2 x}{L} \, dx + \cdots$$

Only the second term (when $\beta_m = \beta_n = \beta_2$) yields a nonzero value, namely, $A_2 L/2$. See Eq. (4.70). Solving Eq. (4.72) for A_m yields

$$A_m = \frac{2}{L} \int_{x=0}^{L} F(x) \sin \frac{\beta_m x}{L} \, dx \qquad (4.73)$$

where $m = 1, 2, \ldots$, and the m subscript in Eq. (4.73) could be replaced by another index symbol, such as n.

Normally, the separation of variables procedure terminates at this point with the observation that A_m (with $m \to n$) in Eq. (4.73) can be used to obtain the A_n values for Eq. (4.68). This gives the complete solution, since the partial differential equation, the two homogeneous boundary conditions, and the initial condition are all satisfied. Since our objective is to obtain a GF, further steps are added. Introducing A_m (with $m \to n$) from Eq. (4.73) and changing x to x' in Eq. (4.68) results in

$$T(x, t) = \sum_{n=1}^{\infty} \frac{2}{L} \int_{x'=0}^{L} F(x') \sin \frac{\beta_n x'}{L} \sin \frac{\beta_n x}{L} \, dx' \, e^{-\beta_n^2 \alpha t/L^2} \qquad (4.74)$$

Taking the integral outside and rearranging gives

$$T(x, t) = \int_{x'=0}^{L} \left[\frac{2}{L} \sum_{n=1}^{\infty} e^{-\beta_n^2 \alpha t/L^2} \sin \frac{\beta_n x}{L} \sin \frac{\beta_n x'}{L} \right] F(x') \, dx' \qquad (4.75a)$$

$$T(x, t) = \int_{x'=0}^{L} G_{X11}(x, t|x', 0) F(x') \, dx' \qquad (4.75b)$$

Notice that the expression inside the brackets in Eq. (4.75a) is the $X11$ GF, evaluated at $\tau = 0$. The $X11$ GF for $\tau \neq 0$ can be found by replacing $(t - 0)$ by $(t - \tau)$ inside the brackets to obtain

$$G_{X11}(x, t|x', \tau) = \frac{2}{L} \sum_{n=1}^{\infty} e^{-\beta_n^2 \alpha (t-\tau)/L^2} \sin \frac{\beta_n x}{L} \sin \frac{\beta_n x'}{L} \qquad (4.76)$$

where the β_n values are

$$\beta_n = n\pi \qquad n = 1, 2, \ldots \qquad (4.77)$$

A few more comments are appropriate regarding this result. It is stated in Chapter 1 that the GF can be interpreted as the temperature rise in the body caused by a Dirac delta function of unit value at position x_0 and time $t_0 = 0$. Since $F(x)$ is arbitrary, let $F(x)$ be the impulse of $T_0 L \, \delta(x' - x_0)$. Then, integrating Eq. (4.75b) gives

$$T(x, t) = T_0 L \, G_{X11}(x, t|x_0, 0) \qquad (4.78)$$

That is, the temperature rise is equal to the GF for the source located at x_0 and $t_0 =$

0 with strength T_0L (the units of T_0L are K m). The symbol x_0 in Eq. (4.78) could be replaced by x' to denote that the source is at x'.

The $G_{X11}(x, t|x', \tau)$ function is found by replacing t in Eq. (4.76) by $t - \tau$ and limiting the time domain to $0 \leq \tau \leq t$. The $G_{X11}(\cdot)$ function satisfies the boundary conditions of $G_{X11}(0, t|x', \tau) = 0$ and $G_{X11}(L, t|x', \tau) = 0$. Also note that the $G_{X11}(\cdot)$ function given by Eq. (4.76) is unchanged by interchanging x and x'. In other words, if the value of a GF at x is known for a source at x', then the same value applies to the GF at x' for a source at x; $G(\cdot)$ is symmetric in x and x'.

It is instructive to examine a plot of $G_{X11}(\cdot)$ for several dimensionless times $\alpha(t - \tau)/L^2$ and several values of x'/L. See Figs. X.1–X.5 in Appendix X. For small time values such as $\alpha(t - \tau)/L^2 < 0.025$ and x' not near the boundary, $G_{X11}(\cdot)$ is approximated by $G_{X10}(\cdot)$. See Section 4.2 and the short time expression given in Table 4.1. As the time $\alpha(t - \tau)/L^2$ becomes larger, the effects of the boundaries increase. The $G_{X11}(\cdot)$ function approaches zero for $\alpha(t - \tau)/L^2 > 0.5$.

The $G_{X11}(\cdot)$ expression given by Eq. (4.76) and that in Table 4.1 are both exact and give the same numerical values, but the former only needs a few terms for large $\alpha(t - \tau)/L^2$ values, while the latter needs only a few terms for small $\alpha(t - \tau)/L^2$ values. In general, the larger time expression, Eq. (4.76) is easier to manipulate mathematically.

Example 4.4: Plate insulated on both sides—$X22$ case. Find the GF for a plate insulated at both $x = 0$ and at $x = L$ with the separation of variables method.

SOLUTION The boundary value problem for an arbitrary initial condition is given by

$$\frac{\partial^2 T}{\partial x^2} = \frac{1}{\alpha} \frac{\partial T}{\partial t} \qquad 0 < x < L \qquad t > 0 \tag{4.79}$$

$$\left.\frac{\partial T}{\partial x}\right|_{x=0} = 0 \qquad \left.\frac{\partial T}{\partial x}\right|_{x=L} = 0 \tag{4.80}$$

$$T(x, 0) = F(x) \tag{4.81}$$

The solution procedure is similar to that for the $X11$ case, and Eqs. (4.49) through (4.56) also apply to this case. The boundary condition at $x = 0$ is different, however, and yields

$$\left.\frac{\partial T_n(x, t)}{\partial x}\right|_{x=0} = \left.\frac{d\mathbf{X}}{dx}\right|_{x=0} \Theta(t) = 0 \tag{4.82}$$

and thus $d\mathbf{X}/dx = 0$ at $x = 0$ to give, from Eq. (4.56a),

$$\left.\frac{d\mathbf{X}}{dx}\right|_{x=0} = 0 = B\lambda \cos(0) - C\lambda \sin(0)$$

$$= B \cdot 1 - C \cdot 0 \tag{4.83}$$

and thus $B = 0$. Repeating this procedure at $x = L$ gives

$$\left. \frac{dX}{dx} \right|_{x=L} = 0 = -C\lambda \sin \lambda L \tag{4.84}$$

and thus the eigencondition is

$$\sin \lambda_n L = 0 \tag{4.85}$$

Notice that $n = 0$ is included because the eigenfunction for this case, $\cos(\beta_n x/L)$ ($n = 1, 2, 3, \ldots$), reduces to unity for $n = 0$. The $X(x)$ function (the eigenfunction) now becomes

$$X(x) = \begin{cases} C_n \cos \dfrac{\beta_n x}{L} & n = 1, 2, \ldots \\ C_0 \cdot 1 & n = 0 \end{cases} \tag{4.86}$$

where Eq. (4.56) is used with Eq. (4.85) and with $C = 0$. At this point the partial differential equation for $T_n(x, t)$ and the two homogeneous boundary conditions are satisfied.

Using the relation that $T(x, t)$ is the sum of the $T_n(x, t)$ values gives

$$T(x, t) = \sum_{n=0}^{\infty} A_n e^{-\beta_m \alpha t/L^2} n \cos \frac{\beta_n x}{L} \tag{4.87}$$

Using the initial condition, Eq. (4.81), yields

$$F(x) = \sum_{n=0}^{\infty} A_n \cos \frac{\beta_n x}{L} \tag{4.88}$$

which is a Fourier cosine series. The A_n's can be found by multiplying Eq. (4.88) by $\cos(\beta_m x/L)\, dx$ and integrating over the domain, which is $0 < x < L$,

$$\int_{x=0}^{L} F(x) \cos \frac{\beta_m x}{L} \, dx = \int_{x=0}^{L} \sum_{n=0}^{\infty} A_n \cos \frac{\beta_n x}{L} \cos \frac{\beta_m x}{L} \, dx \tag{4.89}$$

For the β_n values of $n\pi$, the orthogonality relation involving the cosine function is

$$\int_{x=0}^{L} \cos \frac{\beta_n x}{L} \cos \frac{\beta_m x}{L} \, dx = \begin{cases} 0 & m \neq n \\ L & m = n = 0 \\ \dfrac{L}{2} & m = n \neq 0 \end{cases} \tag{4.90}$$

Utilizing this relation in Eq. (4.89) gives

$$A_0 = \frac{1}{L} \int_{x=0}^{L} F(x) \, dx \tag{4.91a}$$

$$A_m = \frac{2}{L} \int_{x=0}^{L} F(x) \cos \frac{\beta_m x}{L} \, dx \qquad m = 1, 2, \ldots \tag{4.91b}$$

As noted in connection with Eq. (4.73), the subscript m in Eq. (4.91b) can be replaced by another index such as n.

The complete solution to this $X22$ problem posed by Eqs. (4.79)–(4.81) is given by Eq. (4.87) with $A_m (m \rightarrow n)$ given by Eq. (4.91b). However, the purpose here is to demonstrate that a GF can be derived with separation of variables theory. Hence, introduce Eq. (4.91b) with $m \rightarrow n$ and $x \rightarrow x'$ into Eq. (4.87) to get

$$T(x, t) = \frac{1}{L} \int_{x=0}^{L} F(x') \, dx' + \sum_{n=1}^{\infty} \frac{2}{L} \int_{x=0}^{L} F(x') \cos \frac{\beta_n x'}{L} \, dx'$$

$$\times \ e^{-\beta_m \alpha t/L^2} n \cos \frac{\beta_n x}{L} \tag{4.92}$$

$$T(x, t) = \int_{x=0}^{L} \left[\frac{1}{L} + \frac{2}{L} \sum_{n=1}^{\infty} e^{-\beta_m \alpha t/L^2} n \cos \frac{\beta_n x}{L} \cos \frac{\beta_n x'}{L} \right] F(x') \, dx' \tag{4.93}$$

in which the term in brackets is the $G_{X22}(x, t | x', \tau)$ GF evaluated at $\tau = 0$. That is, Eq. (4.93) can be written as

$$T(x, t) = \int_{x=0}^{L} G_{X22}(x, t | x', 0) \, F(x') \, dx' \tag{4.94}$$

where $G_{X22}(x, t | x', \tau)$ is found from the bracketed term in Eq. (4.93) by replacing $(t - 0)$ by $(t - \tau)$ for $\tau \leq t$,

$$G_{X22}(x, t | x', \tau) = \frac{1}{L} + \frac{2}{L} \sum_{n=1}^{\infty} e^{-\beta_n \alpha(t-\tau)/L^2} \cos \frac{\beta_n x}{L} \cos \frac{\beta_n x'}{L} \tag{4.95a}$$

$$\beta_n = n\pi \qquad n = 1, 2, \ldots \tag{4.95b}$$

Notice that the $X22$ GF in Eq. (4.95a) has one more explicit term than the $X11$ GF in Eq. (4.76). The $n = 0$ term is not zero in the $X22$ case because $\cos(\beta_n x/L)$ is not zero for $n = 0$. The summation terms of the $X11$ and $X22$ GFs are quite similar. Both summations contain two trigonometric functions with arguments $\beta_n x/L$ and $\beta_n x'/L$. The eigenvalues are identical for the two summations. Both summations contain the factor $\exp [-\beta_n^2 \alpha t/L^2]$. For "large" values of dimensionless time, such as $\alpha(t - \tau)/L^2 \geq 1$, this exponential factor causes the summations in Eqs. (4.76) and (4.95) to approach zero in value. That is, $G_{X11}(\cdot)$ goes to zero and $G_{X22}(\cdot)$ goes to $1/L$ for large values of $\alpha(t - \tau)/L^2$.

A compact list of one-dimensional GFs based on the separation of variables approach is contained in Tables 4.2 and 4.3. These are best for "large" times; a complete compilation for both large and small times are given in Appendix X. A brief list of eigenvalues for some flat plate geometries involving convection boundary conditions (3rd kind) are given in Table 4.4.

4.5 PRODUCT SOLUTION FOR GREEN'S FUNCTIONS

The solution of certain two- and three-dimensional transient heat conduction problems can be obtained very simply as the product of one-dimensional transient solutions. In

Table 4.2 Eigenfunctions for Green's functions given by $G(x, t | x', \tau) = \dfrac{X_0(x)}{N_0}$ $+ \displaystyle\sum_{m=1}^{\infty} e - \beta_m^2 \alpha(t - \tau)/L^2 \dfrac{X_m(x) X_m(x')}{N_m}$

Number	Eigenfunctions, $X_m(x)$	A_1	A_2
$X1J, J = 1, 2, 3, 4, 5$	$\sin \beta_m x/L$	1	0
$X2J, J = 1, 2, 3, 4, 5$	$\cos \beta_m x/L$	0	1
$X31$	$\sin \beta_m(L - x)/L$	1	0
$X32$	$\cos \beta_m(L - x)/L$	0	1
$X33, X34, X35$	$B_1 \sin (\beta_m x/L) + \beta_m \cos (\beta_m x/L)$	B_1	β_m
$X4J, J = 1, 2, 3, 4, 5$	$-C_1 \beta_m \sin (\beta_m x/L) + \cos (\beta_m x/L)$	$-C_1\beta_m$	1
$X5J, J = 1, 2, 3, 4, 5$	$(B_1 - C_1\beta_m^2) \sin (\beta_m x/L) + \beta_m \cos (\beta_m x/L)$	$B_1 - C_1\beta_m^2$	β_m

Special cases:
  For $X22$, $X24$, $X42$, and $X44$: $X_0(x) = 1$
  For all other cases $X_0(x) = 0$

$$B_i = h_i L/k, \quad C_i = (\rho cb)_i/\rho cL, \quad i = 1, 2$$

Table 4.3 Eigenvalues and norms for Green's functions obtained using the method of separation of variables

Eigenvalues are positive roots of:

$$\tan \beta_m = \frac{\beta_m[K_1(B_2 - C_2\beta_m^2) + K_2(B_1 - C_1\beta_m^2)]}{K_1 K_2 \beta_m^2 - (B_1 - C_1\beta_m^2)(B_2 - C_2\beta_m^2)}$$

(K, B and C are defined below.)

   Simple cases:
     for $X11$ and $X22$, $\beta_m = m\pi$, $m = 1, 2, \ldots$
     for $X12$ and $X21$, $\beta_m = (2m - 1)\pi/2$, $m = 1, 2, \ldots$

   See Appendix X for some approximate relations.

Norms for $m = 1, 2, \ldots$

$$N_m = L\left(\frac{1}{2}(A_1^2 + A_2^2) + A_2^2(C_1 + C_2)\right.$$

$$\left. + \frac{\tan \beta_m}{1 + \tan^2 \beta_m}\left\{\frac{1}{2\beta_m}(A_2^2 - A_1^2) + 2C_2A_1A_2 + \tan \beta_m\left[C_2(A_1^2 - A_2^2) + \frac{1}{\beta_m}A_1A_2\right]\right\}\right)$$

(A_1 and A_2 are given in Table 4.2.)

   Simple cases: $N_m = L/2$ for $X11$, $X12$, $X21$, and $X22$.
   Special cases: $N_0 = (1 + C_1 + C_2)L$ for $X22$, $X24$, $X42$, and $X44$ for $\beta_0 = 0$.

Use XIJ; $I, J = 1, 2, 3, 4, 5$:

I	K_1	B_1	C_1	J	K_2	B_2	C_2
1	0	1	0	1	0	1	0
2	1	0	0	2	1	0	0
3	1	B_1	0	3	1	B_2	0
4	1	0	C_1	4	1	0	C_2
5	1	B_1	C_1	5	1	B_2	C_2

and where $K_i = k_i/k$, $B_i = h_i L/k$, $C_i = (\rho cb)_i/\rho cL$, $i = 1, 2$.

Table 4.4 Some eigenvalues for $X13$, $X31$, $X23$, $X32$ and $X33$

Eigenvalues	B	β_1	β_2	β_3	
Of $\tan \beta_m = -\beta_m/B$ for $X13$ and $X31$	0	1.5708	4.7124	7.8540	(also $X12$ and $X21$)
	0.1	1.6320	4.7335	7.8667	
	1	2.0288	4.9132	7.9787	
	10	2.8628	5.7606	8.7083	
	100	3.1105	6.2211	9.3317	
	∞	3.1416	6.2832	9.4248	(also $X11$)
Of $\tan \beta_m = B/\beta_m$ for $X23$ and $X32$	0	0	3.1416	6.2832	(also $X22$)
	0.1	0.3111	3.1731	6.2991	
	1	0.8603	3.4256	6.4373	
	10	1.4289	4.3058	7.2281	
	100	1.5552	4.6658	7.7764	
	∞	1.5708	4.7124	7.8540	(also $X12$ and $X21$)
Of $\tan \beta_m = 2\beta_m B/(\beta_m^2 - B^2)$ for $X33$	0	0	3.1416	6.2832	(also $X22$)
with $B_1 = B_2$	0.1	0.4435	3.2040	6.3149	
	1	1.3065	3.6918	6.5854	
	10	2.6277	5.3073	8.0671	
	100	3.0800	6.1601	9.2405	
	∞	3.1416	6.2832	9.4248	(also $X11$)

this section, certain two- and three-dimensional GFs are shown to be products of one-dimensional GFs in the rectangular and cylindrical coordinate systems. Product solutions are not permitted in the spherical coordinate system. Product solutions are not generally possible for steady heat conduction.

4.5.1 Rectangular Coordinates

In rectangular coordinates, one-dimensional transient GFs can be multiplied together to form two- and three-dimensional GFs under the following restrictions: (1) the boundary conditions are of the type 0, 1, 2, or 3 (types 4 and 5 are not permitted); (2) if boundary conditions of the third type are present, the heat transfer coefficient h_i must be a constant for a given surface s_i.

The following discussion of product solutions begins with product solutions for *temperature* due to arbitrary initial conditions. Then, a particular initial condition, the Dirac delta function, is used to show that GFs also form product solutions. A two-dimensional case is demonstrated, but the procedure can be repeated to treat three-dimensional cases.

Arbitrary initial conditions. Consider first the temperature due to an arbitrary initial condition in a two-dimensional body described by rectangular coordinates. The boundary conditions are homogeneous and volume energy generation is zero. That is, consider the following heat conduction problem:

$$\frac{\partial^2 T}{\partial x^2} + \frac{\partial^2 T}{\partial y^2} = \frac{1}{\alpha} \frac{\partial T}{\partial t} \tag{4.96a}$$

$$\frac{T(x, y, t = 0)}{T_0} = F^+(x, y) \tag{4.96b}$$

$$k_j \frac{\partial T}{\partial n_j} + h_j T = 0 \qquad j = 1, 2, \ldots, s \tag{4.96c}$$

where T_0 is a characteristic temperature, and s represents the number of boundary conditions ($0 \leq s \leq 4$ for the two-dimensional case). The convection heat transfer coefficient h_j must be a constant. Only boundary conditions of types 0, 1, 2, or 3 are treated.

Suppose that the dimensionless initial condition, $F^+(x, y)$, can be written as a product of two functions, one a function of x and the other a function of y:

$$F^+(x, y) = F_1^+(x) F_2^+(y) \tag{4.97}$$

Then, the following statement is true: the solution of the two-dimensional heat conduction problem defined by Eqs. (4.96a, b, c), can be written as the product of two functions

$$\frac{T(x, y, t)}{T_0} = T_1(x, t) T_2(y, t) \tag{4.98}$$

where T_1 and T_2 are dimensionless, and are defined by the following one-dimensional heat conduction problems:

x direction:
$$\frac{\partial^2 T_1}{\partial x^2} - \frac{1}{\alpha} \frac{\partial T_1}{\partial t} = 0 \tag{4.99a}$$

$$T_1(x, t = 0) = F_1^+(x) \tag{4.99b}$$

$$k_i \frac{\partial T_1}{\partial n_i}\bigg|_{x = x_i} + h_i T_1|_{x = x_i} = 0 \qquad i = 1, 2 \tag{4.99c}$$

y direction:
$$\frac{\partial^2 T_2}{\partial y^2} - \frac{1}{\alpha} \frac{\partial T_2}{\partial t} = 0 \tag{4.100a}$$

$$T_2(y, t = 0) = F_2^+(y) \tag{4.100b}$$

$$k_i \frac{\partial T_2}{\partial n_i}\bigg|_{y = y_i} + h_i T_2|_{y = y_i} = 0 \qquad i = 1, 2 \tag{4.100c}$$

Note that $i, j = 1, 2$ defines the two boundaries for each finite geometry. However, semi-infinite and infinite geometries are also allowed.

The above statement is proved by direct substitution of the product solution, Eq. (4.98), into Eq. (4.96a, b, c). First, consider Eq. (4.96a), the differential equation,

$$T_2 \frac{\partial^2 T_1}{\partial x^2} + T_1 \frac{\partial^2 T_2}{\partial y^2} - \frac{1}{\alpha} \left(T_2 \frac{\partial T_1}{\partial t} + T_1 \frac{\partial T_2}{\partial t} \right) = 0 \tag{4.101}$$

which can be written as

$$T_2 \left(\frac{\partial^2 T_1}{\partial x^2} - \frac{1}{\alpha} \frac{\partial T_1}{\partial t} \right) + T_1 \left(\frac{\partial^2 T_2}{\partial y^2} - \frac{1}{\alpha} \frac{\partial T_2}{\partial t} \right) = 0 \qquad (4.102)$$

This equation is satisfied because it is the sum of the one-dimensional heat conduction equations, Eq. (4.99a) and Eq. (4.100a).

Next, consider the initial condition, Eq. (4.96b). Direct substitution of the product solution gives

$$T_1(x, 0) \, T_2(y, 0) = F^+(x, y) \qquad (4.103)$$

and the initial condition has a product form given by Eq. (4.97) to give

$$T_1(x, 0) \, T_2(y, 0) = F_1^+(x) \, F_2^+(y) \qquad (4.104)$$

This equation is satisfied because it is the product of Eq. (4.99b) and Eq. (4.100b). There are no unusual restrictions on the functions F_1^+ and F_2^+ (they may be zero, piecewise continuous functions, etc.).

Finally, consider the boundary condition Eq. (4.96c). Direct substitution of the product solution gives

$$k_j \frac{\partial (T_1 T_2)}{\partial n_j} + h_j (T_1 T_2) = 0 \qquad (4.105)$$

There are two possibilities for the normal vector n_j in a two-dimensional rectangular coordinate system. The first possibility is for n_j parallel to the x direction, in which case Eq. (4.105) becomes

$$T_2 \left(k_j \frac{\partial T_1}{\partial n_j} + h_j T_1 \right) = 0 \qquad (4.106)$$

This equation is satisfied because it is Eq. (4.99a) multiplied by T_2. The second possibility is for n_j parallel to the y direction, in which case Eq. (4.105) is identical to Eq. (4.100c) multiplied by T_1. This concludes the proof of product solutions for temperature due to arbitrary initial conditions given by Eq. (4.97).

Dirac delta function initial condition. Next, consider a specific initial condition, the Dirac delta function, given by

$$F^+(x, y) = L^2 \, \delta(x - x') \, \delta(y - y') \qquad (4.107)$$

where the length L may have any desired significance; it is used to make $F^+(x, y)$ dimensionless. The dimensionless initial condition, Eq. (4.107), can be written as a product,

$$F^+(x, y) = L\delta(x - x') \cdot L\delta(y - y') \qquad (4.108)$$

Then, the temperature $T(x, y)$ in a two-dimensional body that obeys Eq. (4.96a) and boundary conditions given by Eq. (4.96c) can also be written in product form (Eq. 4.98):

$$\frac{T(x, y, t)}{T_0} = T_1(x, t) \, T_2(y, t) \qquad (4.109)$$

Chapter 1 showed that the temperature, $T(r, t)$, caused by a Dirac delta function initial condition is equivalent to a GF multiplied by a constant:

$$T(r, t) = T_0 L^m G(r, t|r', 0) \qquad (4.110)$$

where $m = 1, 2$, or 3 for one-, two-, or three-dimensional bodies; $G(\cdot)$ is the GF; T_0 is a characteristic temperature; and L is a characteristic length (for dimensional consistency).

Now, each of the functions $T_1(x, t)$ and $T_2(y, t)$ in Eq. (4.109) can also be written in the form of GFs given in Eq. (4.110),

$$T_1(x, t) = L \, G_1(x, t|x', 0) \qquad (4.111a)$$

$$T_2(y, t) = L \, G_2(y, t|y', 0) \qquad (4.111b)$$

Replace Eqs. (4.110) and (4.111) into Eq. (4.109) to obtain

$$G(x, y, t|x', y', 0) = G(x, t|x', 0) \cdot G(y, t|y', 0) \qquad (4.112)$$

Finally, the time dependence of all GFs is $(t - \tau)$, so that in general, $(t - 0)$ can be replaced by $(t - \tau)$ to give

$$G(x, y, t|x', y', \tau) = G(x, t|x', \tau) \cdot G(y, t|y', \tau) \qquad (4.113)$$

That is, the GF for the two-dimensional boundary value problem given in Eq. (4.96) is the product of the one-dimensional GFs associated with the boundary value problems given in Eq. (4.99) and (4.100).

In general, one-dimensional GFs multiply in rectangular coordinates to give two-dimensional GFs. Recall that product solutions are limited to boundary conditions of types 0, 1, 2, and 3. A repeated application of this analysis can be carried out to show the three-dimensional GF in rectangular coordinates can be found from a product of three one-dimensional GFs; that is, $G_{XYZ} = G_X \cdot G_Y \cdot G_Z$.

4.5.2 Cylindrical Coordinates

In cylindrical coordinates (r, ϕ, z), product solutions of GFs are allowed under the following restrictions: (1) the boundary conditions are of the type 0, 1, 2, or 3 (types 4 and 5 are not permitted); (2) if boundary conditions of the third type are present, the heat transfer coefficient h_i must be a constant for a given surface s_i; (3) a GF that depends only on the z coordinate is multiplied by another GF that does *not* depend on the z coordinate.

For example, let G_R, G_Φ, and G_Z represent one-dimensional GFs, let G_{RZ}, $G_{R\Phi}$, and $G_{\Phi Z}$ represent all possible two-dimensional GFs, and let $G_{R\Phi Z}$ represent the three-dimensional GF in cylindrical coordinates. Then, if the boundary conditions meet restrictions (1) and (2), the following product solutions are allowed in cylindrical coordinates:

$$G_{RZ} = G_R \cdot G_Z \qquad (4.114a)$$

$$G_{\Phi Z} = G_\Phi \cdot G_Z \qquad (4.114b)$$

$$G_{R\Phi Z} = G_{R\Phi} \cdot G_Z \tag{4.114c}$$

Note that the GF $G_{R\Phi}$ cannot be found by a product solution.

4.6 STEADY GREEN'S FUNCTIONS

Under steady-state conditions the heat conduction equation reduces to the Poisson equation. Much has been written about the Poisson equation in the fields of electrostatics, elasticity, diffusion, and heat transfer. Many books on theoretical physics contain an overview of solution methods to the Poisson equation and its special case, the Laplace equation; one is Morse and Feshbach (1953). The method of GFs is only one of many solution methods, and we have chosen a unified treatment of GFs at the expense of completeness. Although we do not present other methods, we do not mean to imply that other methods are not important. For example, the use of complex variables and conformal transformations is an enormously powerful method for two-dimensional problems.

In some ways the steady GFs are more difficult to apply than the transient GFs. The steady GFs behave very differently in one, two, and three dimensions. Unlike the transient GFs, the one-dimensional steady GF may not be multiplied to find two- or three-dimensional solutions; the steady GF for each geometry must be found separately.

There are sometimes two forms of the steady GF, depending on the method used to derive it. For example, in a two-dimensional rectangle, the method of images produces a GF with terms containing log functions, and the limit method produces a GF containing sine and cosine functions. If two forms of a solution exist, they are different expansions of the same unique solution and the two expansions may have different convergence properties that can be used to advantage.

In this section, four methods to obtain the steady GFs are discussed. Our attention is focused on those methods related to the transient solutions, and there is no attempt at a complete treatment.

4.6.1 Integration of the Auxiliary Equation: The Source Solutions

For one-dimensional cases the auxiliary equation for the steady GF can be solved directly by integration. The solution for the point source, the line source, and the plane source in the infinite body will be examined to demonstrate the method. The source solutions are important in certain numerical methods, such as the boundary element method. For the present discussion, the source solutions are useful in understanding the functional form of steady GFs before considering the added complexity of boundary conditions. The distinction between the source solutions and the GFs is important: the source solution satisfies the auxiliary equation alone and may or may not satisfy homogeneous boundary conditions, but the GF satisfies a *boundary value problem* which includes the auxiliary equation and homogeneous boundary conditions.

Point source (three dimensions). The point source solution is the steady temperature induced at location **r** by a point heat source at location **r**′. The point source solution depends only on the distance $(\mathbf{r} - \mathbf{r}')$, so the appropriate coordinate system is spherical polar coordinates. The point-source solution satisfies the equation

$$\frac{1}{r^2} \frac{d}{dr} \left(r^2 \frac{dG}{dr} \right) = -\delta(\mathbf{r} - \mathbf{r}') \qquad (4.115)$$

where $\delta(\mathbf{r} - \mathbf{r}')$ has units of (meters)$^{-3}$. The solution to Eq. (4.115) is:

$$G(\mathbf{r}|\mathbf{r}') = \frac{1}{4\pi |r - r'|} \qquad (4.116)$$

The point-source solution is given the symbol $G(\mathbf{r}|\mathbf{r}')$ because it is also a GF: it satisfies the homogeneous boundary condition $G(r \rightarrow \infty) = 0$. The point source solution is singular at $|\mathbf{r} - \mathbf{r}'| = 0$. In rectangular coordinates the point source solution may be written

$$G(x, y, z|x', y', z') = \frac{1}{4\pi} [(x - x')^2 + (y - y')^2 + (z - z')^2]^{-1/2} \quad (4.117)$$

Derivation of the point source solution. The point source may be found by integrating the differential equation (4.115). For the moment, let the source be located at $\mathbf{r}' = 0$ to simplify the analysis. We can translate the source back to $\mathbf{r}' \neq 0$ later. The Dirac delta function $\delta(\mathbf{r})$ is zero everywhere except at $\mathbf{r} = 0$, so except at this point, G should satisfy the Laplace equation in spherical polar coordinates

$$\frac{1}{r^2} \frac{d}{dr} \left(r^2 \frac{dG}{dr} \right) = 0$$

Integrating once:

$$r^2 \frac{dG}{dr} = C_1 \qquad \frac{dG}{dr} = \frac{C_1}{r^2}$$

integrating again gives,

$$G = -\frac{C_1}{r} + C_2$$

The constant C_2 may have any value to satisfy the Laplace equation and, if we take $C_2 = 0$, it will also satisfy the GF boundary condition $G \rightarrow 0$ at $r \rightarrow \infty$. The constant C_1 may be found to have the value $-1/(4\pi)$ by replacing G back into the differential equation (4.115) and integrating both sides of the equation over all space. The nature of the Dirac delta function allows us to equivalently integrate over a small sphere P centered at $r = 0$ with arbitrary small radius σ, because the integrand is zero for any integral that does not include the location of the Dirac delta function:

$$\int_P \nabla^2 G \, dv = -\int_P \delta(\mathbf{r}) \, dv$$

Here dv is the differential volume. The right-hand side yields, with the sifting property of the Dirac delta function (note the units of $\delta(\mathbf{r})$ here),

$$\int_P \nabla^2 G \, dv = -1$$

The left-hand side may be simplified with the divergence theorem to give the integral over the surface of sphere P:

$$\int_{r=\sigma} d\mathbf{S} \cdot \nabla G = -1$$

The value of ∇G in spherical coordinates evaluated at $r = \sigma$ may be substituted to give

$$\int_{r=\sigma} ds \, \frac{C_1}{\sigma^2} = -1$$

Finally, the integral may be evaluated to give the surface area of the sphere,

$$\frac{4\pi \, \sigma^2 C_1}{\sigma^2} = -1$$

or, $C_1 = -1/(4\pi)$, which completes the derivation for $r' = 0$: $G(r|0) = 1/(4\pi r)$. Finally, the point source may be translated to arbitrary location $r' \neq 0$ by noting that $G(r|0) > 0$, and since a change of coordinate system should not change the sign of G, the vector magnitude is required: $G(\mathbf{r}|\mathbf{r}') = 1/(4\pi|\mathbf{r} - \mathbf{r}'|)$.

Line source (two dimensions). The cylindrical coordinate system is appropriate for the line source. The two-dimension differential equation for the line source in cylindrical coordinates is

$$\frac{1}{r} \frac{d}{dr} \left(r \frac{dG_0}{dr} \right) = -\delta(\mathbf{r} - \mathbf{r}') \tag{4.118}$$

The notation G_0 is used for the line source to distinguish it from a GF. Here $\delta(\mathbf{r} - \mathbf{r}')$ has units of (meters)$^{-2}$. The solution to Eq. (4.118) is

$$G_0(r|r') = \frac{-1}{2\pi} \ln |\mathbf{r} - \mathbf{r}'| \tag{4.119}$$

where $|\mathbf{r} - \mathbf{r}'|$ is a vector magnitude in cylindrical coordinates. Strictly speaking, Eq. (4.119) has an error in the units because the argument of the log function should be dimensionless; however, in physical use the line source always has the form $\ln (a/|\mathbf{r} - \mathbf{r}'|)$ where a has the units of meters. In rectangular coordinates, the line source may be written

$$G_0(x, y|x', y') = -\frac{1}{2\pi} \ln \{[(x - x')^2 + (y - y')^2]^{1/2}\}$$

$$= -\frac{1}{4\pi} \ln [(x - x')^2 + (y - y')^2] \tag{4.120}$$

Unlike the point source solution, the line source is not a GF because it does not satisfy the homogeneous boundary condition $G \to 0$ at $r \to \infty$; the log function increases without bound as $r \to \infty$.

Even though the line source is not a GF, it can be used to construct temperature solutions in the infinite body. For example, the line source is important in the method of images (Section 4.6.3).

The line source is important in numerical methods such as the boundary element method. The boundary element method in two dimensions involves a distribution of line sources on a closed curve in the infinite body. The closed curve is broken into line segments called boundary elements, and the distribution of the line sources on the boundary elements is chosen to satisfy boundary conditions on the closed curve. The temperature in the body is evaluated by numerical summation over all the boundary elements, in effect superimposing the temperature induced by each source distribution. This is equivalent to the GF procedure of integrating over the volume to account for volume energy generation. For an introduction to the method see Brebbia and Walker (1978).

Plane source (one dimension). The steady plane source solution is described by the one-dimension steady-state heat equation

$$\frac{d^2 G_0}{dx^2} = -\delta(x - x') \tag{4.121}$$

where $\delta(x - x')$ has units of $(\text{meters})^{-1}$. The solution for G_0 is

$$G_0(x|x') = -\tfrac{1}{2} |x - x'| \tag{4.122}$$

The notation G_0 is used for the plane source solution because it is not a GF, and the reason is that the function $G_0(x|x')$ does not satisfy homogeneous boundary conditions for a GF. The plane source solution blows up at $x \to \infty$, and in fact it blows up proportional to $|x - x'|$, which is faster than the line source solution which blows up like $\ln |\mathbf{r} - \mathbf{r}'|$.

The plane source solution may be derived by integrating the differential equation (4.121) directly, but a little care is required. Since the heated plane divides the infinite body into two regions, the differential equation is integrated in two different regions and then the two solutions are linked by a jump condition at $(x - x') = 0$. The plane source solutions in the one-dimensional slab geometries $X11$, $X12$, and $X13$ may also be derived this way if in addition the correct homogeneous boundary conditions are applied.

Derivation of the plane source solution. The plane source solution is the response to a plane heat source of unit strength. Suppose the plane heat source is located at $x' = 0$. Later the plane source can be translated to an arbitrary location $x' \neq 0$. Let $G_1(x)$ be the solution to Eq. (4.121) for $(x > 0)$ and let $G_2(x)$ be the solution to Eq. (4.121) for $(x < 0)$. Since, for $(x \neq 0)$, the Dirac delta function $\delta(x)$ on the right-hand side of Eq. (4.121) is zero, the functions G_1 and G_2 may be found by integrating the now-homogeneous equation

$$\frac{d^2 G_i}{dx^2} = 0 \qquad i = 1 \text{ or } 2$$

to give solutions

$$G_1(x) = Ax + B \qquad G_2(x) = Cx + D$$

where A, B, C, and D are constants that must be determined. Two conditions for evaluating these four constants come from putting function G_1 and G_2 back together at $x = 0$. The two solutions G_1 and G_2 must agree at the location of the heat source $x = 0$:

$$G_1(x = 0) = G_2(x = 0)$$

from which $B = D$. Another condition comes from substituting G_1 and G_2 into Eq. (4.121) and integrating over the entire body. This is equivalent to performing an energy balance for a control volume that contains only the plane heat source of unit strength, because the only contribution to the integral occurs at the location of the Dirac delta function. The integral of Eq. (4.121) is

$$\left. \frac{dG_2}{dx} \right|_{x=0} - \left. \frac{dG_1}{dx} \right|_{x=0} = -1 \qquad (4.123)$$

which shows that the plane heat source causes a unit jump in the slope of the solution. Substitute the general form of G_1 and G_2 into Eq. (4.123) to obtain $C - A = -1$.

Two more conditions are needed to evaluate the constants, but only one more condition may be found from symmetry about the location of the plane heat source: $G_1(x) = G_2(-x)$. In terms of the constants this gives $A = -C$. Equation (4.123) may now be solved to give

$$A = \tfrac{1}{2} \qquad C = -\tfrac{1}{2}$$

and the temperature distribution is given by

$$G_0(x - x') = \begin{cases} \tfrac{1}{2}(x - x') + B & (x - x') > 0 \\ -\tfrac{1}{2}(x - x') + B & (x - x') < 0 \end{cases}$$

where now the plane heat source has been translated to arbitrary location x'. Constant B can not be determined, but since all solutions to the Laplace equation may be written with an additive constant, there is no loss of generality by taking $B = 0$ to give

$$G_0(x - x') = -\tfrac{1}{2} |x - x'|$$

This is the plane source solution in an infinite body. Again it is not a GF because the homogeneous boundary condition at $x \to \infty$ cannot be satisfied.

The line source and the plane source are not GFs because of a problem with the boundary conditions. The auxiliary equation for the GF always has a general solution, but the homogeneous boundary conditions cannot always be satisfied. There are several other geometries for which this problem occurs and such geometries do not have a steady GF. For example, the X22 geometry has no steady GF and neither do finite geometries with specified heat flux on *all* of the boundaries (boundary conditions of type 2, also called Neumann boundary conditions). A list of one-dimensional geometries without a steady GF is given in Table 4.5.

A physical reason that some geometries do not have a steady GF function comes from the perspective of a GF as the response to a heat source. In steady heat transfer, any heat introduced inside the body must either flow out of the boundaries or flow off to infinity if the body is of infinite extent. If all the boundaries are insulated, there is nowhere for the heat to go and, consequently, there is no steady GF.

Steady *temperature* distributions can exist in bodies with no steady GF, but the steady GF method cannot be used to find the temperature. For example, the $X22$ geometry has a linear temperature distribution if the same amount of heat that flows into the body at $x = 0$ also flows out at $x = L$. In this simple case, the temperature distribution can be found by applying the *nonhomogeneous* boundary conditions to the general solution of the differential equation. The steady temperature can always be found with the transient GF solution equation (GFSE) in the limit as time becomes large ($t \rightarrow \infty$). Any questions on the existence of the steady-state temperature can be answered this way.

Barton (1989) uses a pseudo-GF to deal with those geometries that do not have steady GFs because of insulated boundaries. The pseudo-GF differs from the ordinary GF by an additive constant carefully chosen to satisfy the homogeneous boundary conditions. In physical terms, the additive constant cancels out the heat flow introduced by the heat source. A modified GFSE is then needed to calculate temperatures from the pseudo-GFs.

4.6.2 Method of Embedding

The point source solution may be distributed through three-dimensional space to generate the line and plane sources: this is called the method of embedding (Barton, 1989) because the lower dimensions can be calculated as though they are embedded in three-dimensional space. The GFSE is used to distribute point sources in the infinite body. This method shows the relationship between the point, line, and plane source solutions and the conditions under which a steady-state GF exists in the infinite body.

First the line source will be discussed. Consider a distribution of point sources of uniform strength g_0 along a line segment oriented in the z direction and located at $(x'$,

Table 4.5 One-dimensional geometries with no steady Green's functions

One-dimensional rectangular	Radial cylindrical	Radial spherical
$X00$	$R00$	$RS02$
$X10$	$R02$	$RS22$
$X20$	$R10$	
$X22$	$R20$	
$X30$	$R22$	
	$R30$	

Note: Homogeneous boundary conditions cannot be satisfied for these cases; however, nonhomogeneous steady solutions are possible.

y', $-L \le z' \le L$). The steady-state temperature is given by the energy generation term of the GFSE:

$$T(x, y, z) = \int_{-L}^{L} \frac{g_0}{k} G(x, y, z | x', y', z') \, dz'$$

$$= \int_{-L}^{L} \frac{g_0}{k} \frac{1}{4\pi} [(x - x')^2 + (y - y')^2 + (z - z')^2]^{-1/2} \, dz'$$

$$= \frac{g_0}{4\pi k} \ln \left[\frac{z + L + [(z + L)^2 + r^2]^{1/2}}{z - L + [(z - L)^2 + r^2]^{1/2}} \right] \qquad (4.124)$$

where $r^2 = (x - x')^2 + (y - y')^2$. The temperature is well behaved for finite values of L, but in the limit as the line segment becomes very long ($L \to \infty$), the temperature blows up like $\ln L$, demonstrating that there is no steady-state GF corresponding to the line source in an infinite body.

The plane heat source may be derived by distributing point sources uniformly over a circular disk and then letting the disk become very large. Let the circular disk be located in the plane $z = $ constant at location $z' = 0$, and let the radius of the disk be a. The temperature induced by a uniform distribution of unit-strength point sources is given by the GFSE for volume energy generation in cylindrical coordinates with the volume integral over the surface of the disk:

$$G(r - r', \phi - \phi', z - z') = \int_{r'=0}^{a} \int_{\phi=0}^{2\pi} \frac{1}{4\pi l} r' dr' d\phi \qquad (4.125)$$

where l is the distance between a point on the circular disk and the observation point (r, ϕ, z). $G(r - r', \phi - \phi', z - z')$ represents the GF, the response to the unit-strength heat source. Since we are interested in deriving the plane source solution, the result should depend only on coordinate z (distance perpendicular to the disk) and it should not depend on coordinates (r, ϕ). For simplicity locate the observation point at $r = 0$ as shown in Fig. 4.4 to give a simple form for vector magnitude l:

$$l = |\mathbf{r} - \mathbf{r}'| = [(r')^2 + (z - z')^2]^{1/2}$$

Substitute this expression for l into Eq. (4.125) and the integral may be carried out to give the GF

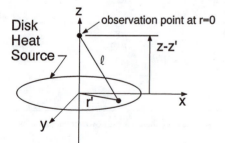

Figure 4.4 Disk heat source in the infinite body.

$$G(z - z') = \frac{1}{2} [a^2 + (z - z')^2]^{1/2} - \frac{1}{2} |z - z'| \qquad (4.126)$$

This is the GF at locations along the centerline of the disk heat source. As the disk becomes large ($a \gg z - z'$) the GF may be written approximately as

$$G(z - z') \approx \frac{1}{2} a - \frac{1}{2} |z - z'| \qquad (4.127)$$

The steady GF exists for any disk of finite size, but in the limit as $a \to \infty$, the GF blows up. That is, the GF does not exist for the plane heat source in the infinite body. Compare the GF, Eq. (4.127), to the plane source solution, Eq. (4.122). The functional form of the GF is identical to the plane heat source solution except for an additive term proportional to the size of the disk. This additive term shows that the plane heat source solution is not a GF.

The lesson here is that steady sources uniformly distributed over a *bounded region* of infinite three-dimensional space have a steady-state solution, and sources uniformly distributed over an *unbounded region* do not have a steady-state solution. This conclusion is limited to a *uniform* distribution of sources; in the next section sources are distributed nonuniformly (in source-sink pairs) to form steady solutions.

4.6.3 Method of Images

Infinite-body source solutions may be added together to generate GFs with the method of images. Some geometries do not have steady GF and the method of images does not apply, but for those geometries that do admit steady solutions, the method of images applies exactly as it does for the transient GF.

The steady GF generated by the method of images are constructed from source-sink pairs. A source introduces heat and a sink removes heat from the body. For a source-sink pair, the condition that $G \to 0$ far from the source will be satisfied by cancelation of the source and the sink. This is equivalent to introducing a net heat flow of zero into the infinite body.

Table 4.1, which lists method of images solutions for transient cases, may be applied to steady cases by replacing function $K(\cdot)$ by the appropriate steady-source solution. Steady cases $X11$ and $X12$ may be found in Table 4.1 by substituting the plane source solution for $K(\cdot)$ (however, better forms of one-dimensional steady GFs are listed in Appendix X). In two dimensions, cases $X10Y00$, $X11Y00$, and $X12Y00$ may be found by substituting the line source solution for $K(\cdot)$ in Table 4.1. In three dimensions, the solutions $X1JY00Z00$, where $J = 0, 1, 2$, may be found by substituting the point source solution.

For example, the steady GF for the $X10Y00$ geometry is given by combining a line source located at x' and a line sink located at $-x'$. From Table 4.1, the listing for the $X10$ geometry may be used to give

$$G_{X10Y00}(x, y|x', y') = -\frac{1}{2\pi} \ln \{[(x - x')^2 + (y - y')^2]^{1/2}\}$$

$$+ \frac{1}{2\pi} \ln \{[(x + x')^2 + (y - y')^2]^{1/2}\}$$

or,

$$G_{X10Y00}(x, y|x', y') = \frac{1}{4\pi} \ln \left[\frac{(x + x')^2 + (y - y')^2}{(x - x')^2 + (y - y')^2} \right]$$

This function satisfies the homogeneous boundary condition $G(x = 0) = 0$ because the source and sink are equidistant from every point on the plane $x = 0$. As $|x| \to \infty$, the homogeneous boundary condition $G(x \to \infty) = 0$ is also satisfied because the effect of the source and the sink cancel.

4.6.4 Limit Method

The steady GF can be calculated from the large-time transient GF by integrating over time and taking the limit as $t \to \infty$. This is the limit method, and in this book the steady GF is defined by the limit method in Eq. (3.89). Many steady GFs can be written down immediately in integral form with the limit method. Two examples of the limit method are discussed below.

Example 4.5: Point source in the infinite body. Find the steady point-source solution from the transient point source solution $G_{X00Y00Z00}$.

SOLUTION The point source will be located at the origin ($x' = 0, y' = 0, z' = 0$) for convenience. Later the source can be translated to any position. The limit method is given by the integral

$$G(x, y, z|0, 0, 0) = \lim_{t \to \infty} \alpha \int_{\tau=0}^{t} G_{X00Y00Z00}(x, y, z, t|0, 0, 0, \tau) \, d\tau \quad (4.128)$$

The product solution may be used for the transient $X00Y00Z00$ GF to give

$$G_{X00Y00Z00}(x, y, z, t|0, 0, 0, \tau) = G_{X00}(x, t|0, \tau) \, G_{Y00}(y, t|0, \tau) \, G_{Z00}(z, t|0, \tau)$$

where $G_{X00}(x, t|0, \tau) = [4\pi\alpha(t - \tau)]^{-1/2} \exp \left[\frac{-x^2}{4\alpha(t - \tau)} \right]$

$$G_{Y00}(y, t|0, \tau) = [4\pi\alpha(t - \tau)]^{-1/2} \exp \left[\frac{-y^2}{4\alpha(t - \tau)} \right]$$

$$G_{Z00}(z, t|0, \tau) = [4\pi\alpha(t - \tau)]^{-1/2} \exp \left[\frac{-z^2}{4\alpha(t - \tau)} \right]$$

Then Eq. (4.128) may be written

$$G(x, y, z|0, 0, 0) = \lim_{t \to \infty} \alpha \int_{\tau=0}^{t} [4\pi\alpha(t - \tau)]^{-3/2}$$

$$\times \exp \left[\frac{-x^2 + y^2 + z^2}{4\alpha(t - \tau)} \right] d\tau \quad (4.129)$$

Note that the product of the three one-dimensional GFs is also the same as the $RS00$ GF given in Appendix RS for the case $r' = 0$. The above integral may be evaluated to give

$$G(x, y, z|0, 0, 0) = \lim_{t \to \infty} \frac{1}{4\pi r} \text{erfc} \left[\frac{r}{(4\alpha t)^{1/2}} \right] = \frac{1}{4\pi |r|} \quad (4.130)$$

where $r^2 = (x^2 + y^2 + z^2)$. This is the steady point-source solution located at $r' = 0$, as discussed in Section 4.6.1.

Example 4.6: Parallelpiped with specified surface temperature—*X*11*Y*11*Z*11 case. Find the steady GF in the parallelpiped with temperature boundary conditions (type 1) on all six surfaces.

SOLUTION The parallelpiped body is shown in Fig. 4.5. The limit method integral for this case is given by

$$G(x, y, z|x', y', z') = \lim_{t \to \infty} \alpha \int_{\tau=0}^{t} G_{X11Y11Z11}(x, y, z, t|x', y', z', \tau) \, d\tau$$

The transient GF for the *X*11*Y*11*Z*11 geometry is given by the product of one-dimensional transient solutions: $G_{X11}G_{Y11}G_{Z11}$. The function G_{X11} is given in Appendix X:

$$G_{X11}(x, t|x', \tau) = \frac{2}{a} \sum_{m=1}^{\infty} e^{-m^2\pi^2\alpha(t-\tau)/a^2} \sin \frac{m\pi x}{a} \sin \frac{m\pi x'}{a} \quad (4.131)$$

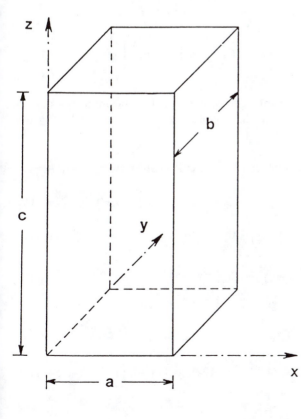

Figure 4.5 Parallelpiped geometry for Example 4.6.

where a is the length of the body in the x direction. The functions G_{Y11} and G_{Z11} are similar; for example, G_{Y11} is given by Eq. (4.131) with x and a replaced by y and b, respectively. Replace the transient GF into the integral to give

$$G(x, y, z|x', y', z') = \lim_{t \to \infty} \frac{8\alpha}{abc} \sum_{m=1}^{\infty} \sum_{n=1}^{\infty} \sum_{p=1}^{\infty} \int_{\tau=0}^{t}$$

$$\times \exp\left[-\alpha\pi^2(t - \tau) \left(\frac{m^2}{a^2} + \frac{n^2}{b^2} + \frac{p^2}{c^2} \right) \right] d\tau$$

$$\times \sin\left(m\pi \frac{x}{a} \right) \sin\left(m\pi \frac{x'}{a} \right) \sin\left(p\pi \frac{z}{c} \right)$$

$$\times \sin\left(p\pi \frac{z'}{c} \right) \sin\left(n\pi \frac{y}{b} \right) \sin\left(n\pi \frac{y'}{b} \right)$$

When the time integral is carried out and the limit taken, the steady GF becomes

$$G(x, y, z|x', y', z') = 8 \sum_{m=1}^{\infty} \sum_{n=1}^{\infty} \sum_{p=1}^{\infty} \sin\left(m\pi \frac{x}{a} \right)$$

$$\times \sin\left(m\pi \frac{x'}{a} \right) \sin\left(p\pi \frac{z}{c} \right) \sin\left(p\pi \frac{z'}{c} \right)$$

$$\times \sin\left(n\pi \frac{y}{b} \right) \sin\left(n\pi \frac{y'}{b} \right)$$

$$\times \left[abc\pi^2 \left(\frac{m^2}{a^2} + \frac{n^2}{b^2} + \frac{p^2}{c^2} \right) \right]^{-1} \tag{4.132}$$

Example 4.7: Two-dimensional slab with one side semi-infinite—X11Y20 case. Find the steady-state GF for the region $0 < x < a$, $y > 0$ with $G = 0$ at $x = 0$ and at $x = a$ and $\partial G/\partial y = 0$ at $y = 0$.

SOLUTION The limit method integral for this case is given (with $u = t - \tau$)

$$G(x, y|x', y') = \alpha \int_0^{\infty} G_{X11}(x, u|x') G_{Y20}(y, u|y') \, du \tag{4.133}$$

where $G_{X11}(x, u|x')$ is given by Eq. (4.131) with $du = t - \tau$ and $G_{Y20}(y, u|y')$ is [see Eq. (X20.1), Appendix X]

$$G_{Y20}(y, u|y') = (4\pi\alpha u)^{-1/2} \left(e^{-(y-y')^2/(4\alpha u)} + e^{-(y+y')^2/(4\alpha u)} \right) \tag{4.134}$$

Integrals of the form (see integral 12 in Table F.6, Appendix F)

$$\int_0^{\infty} u^{-1/2} e^{-a^2 u - b^2 u^{-1}} \, du = \frac{\pi^{1/2}}{a} e^{-2ab} \tag{4.135}$$

are needed. Then, using Eqs. (4.131), (4.134), and (4.135) in Eq. (4.133) gives

$$G(x, y|x', y') = \frac{1}{\pi} \sum_{m=1}^{\infty} \frac{1}{m} (e^{-m\pi(y-y')/a} + e^{-m\pi(y+y')/a})$$

$$\times \sin \frac{m\pi x}{a} \sin \frac{m\pi x'}{a} \qquad (4.136)$$

Observe for the point $y = y'$, $x = x'$ (with x not at 0 or a) that the value of G is unbounded, which is unlike the behavior of the one-dimensional GFs in the x coordinate. GFs in the cylindrical coordinate system also have this unbounded behavior for r and r' going to zero.

REFERENCES

Arpaci, V. S., 1966, *Conduction Heat Transfer*, Addison-Wesley, Reading, Mass.

Barton, G., 1989, *Elements of Green's Functions and Propagation*, Oxford University Press, London.

Beck, J. V., 1984, Green's Functions for Transient Heat Conduction Problems, *Int. J. Heat Mass Transfer*, vol. 27, no. 8, pp. 1235–1244.

Beyer, W. H., 1976, *CRC Standard Mathematical Tables*, 24th ed., CRC Press, Cleveland, Ohio.

Brebbia, C. A., and Walker, S., 1978, Introduction to Boundary Element Methods, in *Recent Advances in Boundary Element Methods*, ed. C. A. Brebbia, Pentech Press, Plymouth.

Carslaw, H. S., and Jaeger, J. C., 1959, *Conduction of Heat in Solids*, 2d ed., Oxford University Press, New York.

Luikov, A. V., 1968, *Analytical Heat Diffusion Theory*, Academic Press, New York.

Morse, P. M., and Feshbach, H., 1953, *Methods of Theoretical Physics*, McGraw-Hill, New York.

Ozisik, M. N., 1980, *Heat Conduction*, Wiley, New York.

PROBLEMS

Note: In many of the problems in this chapter the partial answers can be obtained by using the GFs tabulated in the appendixes. Integrals may be found in GF appendixes, in Appendix F, and at the end of Chapter 5. Unless otherwise requested, the reader should use these tables.

4.1 Using the method of images, find the GF for the region $0 < x < \infty$, $0 < y < \infty$, with the boundary conditions of $\partial G/\partial x = 0$ at $x = 0$ and $\partial G/\partial y = 0$ at $y = 0$. Also find the GF using the product of the appropriate GFs and relate the corresponding terms.

4.2 Using the method of images, find the GF for the region $0 < x < L$, $0 < y < \infty$, with the boundary conditions of $\partial G/\partial x = 0$ at $x = 0$ and L, and $\partial G/\partial y = 0$ at $y = 0$. Also find the GF using the product of the appropriate GFs and relate the corresponding terms.

4.3 Determine the temperature distribution for an instantaneous plane heat source at $x = x'$ and at time $t = \tau$ for the boundary conditions of $T = T_i$ at $x = 0$ and insulation at $x = L$ and an initial temperature of T_i. Give two mathematical forms of the solution.

4.4 Determine the GF for a line source at $x = x'$, $y = y'$ for the boundary condition of the third kind at $y = 0$ and for the region of $-\infty < x < \infty$, $y > 0$.

4.5 Using a computer, evaluate $LG_{X11}(x, t|x', \tau)$ at $x/L = x'/L = 0.5$ for times $\alpha(t - \tau)/L^2 = 0.025$, 0.1, 0.5, and 1.0. Use two different expressions, one from Table 4.1 and the other from Tables 4.2 and 4.3. Determine the number of terms required for each expression for the different dimensionless times for the errors to be less than 0.0001 in value. Compare the values with those obtained from $LG_{X00}(\cdot)$.

4.6 Evaluate $LG_{X22}(x, t|x', \tau)$ at $x/L = x'/L = 0.5$ for times $\alpha(t - \tau)/L^2 = 0.025$, 0.1, 0.5, and 1.0. Use two different expressions, one from Table 4.1 and the other from Tables 4.2 and 4.3. Determine the number of terms required for each expression for the different dimensionless times for the errors to be less than 0.0001. Compare the values with those obtained from $LG_{X00}(\cdot)$ and $LG_{X11}(\cdot)$.

4.7 Using expressions in Table 4.3, consider boundary conditions of the first and second kinds and also of the third kind for small values of B_1 and B_2 compared to 1 and both K_1 and K_2 not equal to zero (one K can be zero). Find an approximation of the first eigenvalue, β_1, using the approximate relation,

$$\cot x = \frac{1}{x} - \frac{x}{3}$$

in the eigencondition in Table 4.3. The use of this approximation yields a more accurate equation than a two-term approximation for tan x. Why?

4.8 Show that eigenvalues calculated using the eigencondition in Table 4.3 gives $\beta_{m+1} = \beta_m + \pi$ for large β_m values.

4.9 Give the expressions for the GFs for the cases represented by $X00$, $X00Y00$, and $X00Y00Z00$. What is the physical significance for each case?

4.10 What do

$$\int_a^b G_{X00}(x, t|x', \tau)\, dx' \qquad \int_a^b G_{X00Y00}(x, y, t|x', y', \tau)\, dx'$$

and $\displaystyle\int_a^b G_{X00Y00Z00}(x, y, z, t|x', y', z', \tau)\, dx'$

represent? What are mathematical expressions for each? (Perform the integration either explicitly or by using a table.)

4.11 An instantaneous volume source from $-a$ to a in an infinite body is to be approximated by a finite number of line sources. Show that the exact solution is

$$\frac{\text{erfc}\,[(x - a)^2/(4\alpha u)] - \text{erfc}\,[(x + a)^2/(4\alpha u)]}{2} \qquad \text{where } u = t - \tau$$

(The detailed derivation of this equation is not required if an appropriate integral in the book can be used.) This solution is to be approximated by a series of plane sources. Derive and evaluate the expressions for (a) a single source at $x = 0$, (b) three equally spaced, and (c) five equally spaced plane sources. Show that these approximations can be used to obtain

(a) $\text{erf}\,(z) \cong 2z/\pi^{1/2}$

(b) $\text{erf}\,(z) \cong (2z/\pi^{1/2})\,(1 + 2\,e^{-4z^2/9})/3$

(c) $\text{erf}\,(z) \cong (2z/\pi^{1/2})\,(1 + 2\,e^{-4z^2/25} + 2\,e^{-16z^2/25})/5$

Evaluate and compare these expressions with the exact values at $z = 0.05, 0.25, 1,$ and 2.

4.12 Obtain the Green's functions for the cases denoted $X23$ and $X13Z00$.

4.13 Obtain the GF for the cases denoted $R02Z20$ and $R01\Phi00Z10$.

4.14 Derive the steady-state GF for the $X11$ case by starting with

$$\frac{d^2 G}{dx^2} = -\delta(x - x')$$

Two regions are to be considered: $0 < x < x'$ and $x' < x < L$. Using the continuity condition at $x = x'$ and integrating the above equation to get

$$\left.\frac{dG}{dx}\right|_{x'+} - \left.\frac{dG}{dx}\right|_{x'-} = 1$$

provides two more conditions for obtaining the four constants of integration. Show that the solution can be written as given in Table X.1.

4.15 Derive the steady-state GF for the $X11$ case using the limit method and starting with Eq. (4.131) for $G_{X11}(\cdot)$.

The answer is

$$G_{X11}(x, x') = \frac{2a}{\pi^2} \sum_{m=1}^{\infty} \frac{1}{m^2} \sin\frac{m\pi x}{a} \sin\frac{m\pi x'}{a}$$

4.16 (a) Program on a computer the expression for $G_{X11}(\cdot)/a$ given in Problem 4.15 as a function of x/a, x'/a and M; M is the maximum number of terms used.

(b) Calculate using the computer program $G_{X11}(\cdot)/a$ as a function of the number of terms for $x/a =$

$x'/a = 1/2$. Also tabulate the errors by using the nonseries solution of Problem 4.14. How many terms are needed to obtain accuracy within 1%? By observing the dependence the error as a function of number of terms, how many terms would be needed to obtain 0.1%?

(c) Based on the observed characteristics of the error, suggest and use a procedure to correct the running value of $G_{X11}(\cdot)/a$ as a function of the number of terms in the summation. A possible expression is

$$S'_M = S_M + \frac{(S_M - S_{M-p})\,(M - p)}{p}$$

where $p = 2, 4, \ldots$, and S_M represents the summation after $m = M$ terms. Does it speed convergence?

4.17 Using the program developed in Problem 4.16, repeat parts b and c for the case of $x/a = 1/4$ and $x'/a = 1/2$.

4.18 Verify that in two-dimensional cylindrical coordinates the function

$$G_0(r|r') = -\frac{1}{2\pi}\ln|r - r'|$$

satisfies the differential equation $\nabla^2 G_0 = -\delta(r - r')$. Does G_0 satisfy the auxiliary problem for the two-dimensional cylindrical-coordinate GF?

4.19 Show by direct integration of the energy equation that the steady GF for the X23 geometry is given by

$$G(x|x') = \begin{cases} L\left(\dfrac{1 + B}{B} - \dfrac{x}{L}\right) & 0 \le x' \le x \\[2mm] L\left(\dfrac{1 + B}{B} - \dfrac{x'}{L}\right) & x \le x' \le L \end{cases}$$

where $B = hL/k$ is the Biot number.

4.20 Show that the temperature caused by a heat source distributed along a finite line segment of length $2L$ reduces to the line source solution plus another term for $r \ll L$. See Eq. (4.124). In the limit, as $L \to \infty$, is the temperature bounded?

4.21 (a) Use the limit method to solve for the steady-state GF for the problem denoted $X11Y10$. Use the $X11$ GF best for small times.

(b) Use the $X11$ GF best for large times.

4.22 Use the limit method to solve for the steady-state GF for the problem denoted $X11Y10Z12$. Use the $X11$ and $Z12$ GFs best for large times.

CHAPTER FIVE

TIME PARTITIONING, ACCELERATION OF SERIES, AND NUMERICAL INTEGRATION

5.1 INTRODUCTION

For heat conduction in finite bodies, expressions for temperature usually involve infinite series. Slow convergence of these infinite-series expressions can require that many terms be evaluated to obtain accurate numerical values. The alternative Green's function solution equation (AGFSE) provides one way to improve convergence for some nonhomogeneous problems. This difficulty can also be alleviated by time partitioning, acceleration of the series, and numerical integration.

The remainder of this section introduces some one-dimensional problems considered in this chapter. Section 5.2 discusses some convergence problems and provides motivation for the study of methods of improved evaluation of the Green's function (GF) solutions. Section 5.3 develops the time-partitioning concepts, and Section 5.4 discusses some important integrals. Section 5.5 provides several methods for acceleration of infinite series. Section 5.6 provides a computer program and background for the efficient numerical evaluations of integrals.

5.1.1 Problems Considered in this Chapter

In this chapter the method of time partitioning is introduced for one-dimensional problems described by XIJ ($I, J = 1, 2, 3$) with homogeneous boundary conditions. The same concepts apply for other coordinate systems, multiple dimensions, and nonhomogeneous boundary conditions. Two types of problems are considered: those containing a nonzero initial temperature distribution $F(x)$; and those containing an energy generation term $g(x, t)$.

The describing partial differential equation for the temperature in one-dimensional slab geometries is

$$\frac{\partial^2 T}{\partial x^2} + \frac{1}{k} g(x, t) = \frac{1}{\alpha} \frac{\partial T}{\partial t} \qquad 0 < x < L \qquad t > 0 \qquad (5.1)$$

119

Notice that this equation contains an energy equation term, and hence is nonhomogeneous. Homogeneous boundary conditions of the first, second, and third kinds can be written as

$$T = 0 \quad \text{or} \quad \frac{\partial T}{\partial x} = 0 \quad \text{or} \quad \frac{k\partial T}{\partial n} + hT = 0 \tag{5.2}$$

for $x = 0$ or L and $t > 0$. For the nonzero initial temperature distribution one writes

$$T(x, 0) = F(x) \quad 0 < x < L \tag{5.3}$$

Analogous to Eq. (1.3), the solution of the above problem using GFs for a finite body with homogeneous boundary conditions is

$$T(x, t) = \int_{x'=0}^{L} G(x, t|x', 0) \, F(x') \, dx'$$

$$+ \frac{\alpha}{k} \int_{\tau=0}^{t} \int_{x'=0}^{L} G(x, t|x', \tau) \, g(x', \tau) \, dx' \, d\tau \tag{5.4}$$

The GF $G(\cdot)$ is given in Chapter 4 for many cases, and more extensive tables of GFs for rectangular coordinates are given in Appendix X.

5.1.2 Two Basic Functions

The GF for one-dimensional slab bodies have the form of infinite series of basic functions. (For semi-infinite or infinite bodies, the GF is usually given as a *finite* sum of such functions.) There are two types of basic functions that occur in the expression for $G_{XIJ}(\cdot)$ for $I = 1, 2$ and $J = 0, 1,$ and 2. One is the fundamental heat conduction function, $K(z + x', t - \tau)$,

$$K(z + x', t - \tau) = [4\pi\alpha(t - \tau)]^{-1/2} \exp\left[-\frac{(z + x')^2}{4\alpha(t - \tau)}\right] \tag{5.5}$$

The variable z is $2nL + x$ or $2nL - x$. See Eq. (4.2). In this function the variables x' and τ occur in the same group as in the argument of $\exp(\cdot)$. This is a compact form, but integrations involving $K(\cdot)$ can be quite complicated and can be difficult to do analytically.

The other type of basic function involves the product of an exponential that is a function of only $t - \tau$ and two identical eigenfunctions, one a function of x and the other of x', and a norm N_m,

$$\exp\left[-\frac{\beta_m^2\alpha(t - \tau)}{L^2}\right] \frac{\mathbf{X}(\beta_m, x) \, \mathbf{X}(\beta_m, x')}{N_m} \tag{5.6}$$

The norm can be L, $L/2$, or a more complicated function. The eigenfunctions for the $X11$ and $X12$ geometries are

$$\mathbf{X}(\beta_m, x) = \sin\frac{\beta_m x}{L}$$

$$\text{and} \quad \mathbf{X}(\beta_m, x) = \cos \frac{\beta_m x}{L}$$

for the $X21$ and $X22$ geometries. The basic function given by Eq. (5.6) is more convenient for mathematical manipulation than $K(\cdot)$ given by Eq. (5.5), because the dependent variables x, x', and τ all occur in different terms of Eq. (5.6). Thus, an integral on one variable (x, x', or τ) acts only on one term and does not affect integration on the other two variables. Whenever practical, the product form given by Eq. (5.6) is preferred for this reason.

5.2 CONVERGENCE PROBLEMS

Unfortunately, the convenient form of the GF given by Eq. (5.6), also called the large-time form, cannot efficiently be used for small times. For small times, a finite body (such as a plate) behaves as if it were a semi-infinite body, since at small times each boundary condition affects only a small region near its boundary. Small times are defined by dimensionless time $\alpha t/L^2 \leq 0.06$, or $t \leq 0.06 L^2/\alpha$ seconds.* Under this circumstance, the large-time form of the solution requires many terms of the infinite series for the GF. This is inefficient and possibly inaccurate if too few terms are used. The other form of the GF, given by Eq. (5.5), also called the small-time form, can be evaluated accurately at small times with only a few terms of its infinite series; it tends to be more difficult to evaluate analytically, however.

There is an important case when the large-time GF also has convergence difficulties for large values of $\alpha t/L^2$. It occurs when $G(\cdot)$ is integrated over the dummy time variable τ. For example, let $g(x', \tau)$ in Eq. (5.4) be simply $g_0 \, \delta(x_0 - x')$. This is a continuous (that is, constant over time) source of heat of strength g_0 located at position x_0. Then, the second integral of Eq. (5.4) contains typical terms of

$$\frac{\alpha}{k} g_0 \sum_{m=1}^{\infty} \int_{\tau=0}^{t} \exp\left[-\frac{\beta_m^2 \alpha(t - \tau)}{L^2} \right] d\tau \, \frac{\mathbf{X}(\beta_m, x) \, \mathbf{X}(\beta_m, x_0)}{N_m} \tag{5.7}$$

[The integral over x' has been evaluated with the sifting property of the Dirac delta function, $\delta(x_0 - x')$.] Next, only the integral over τ is considered, but the upper limit is replaced by $t - \Delta t$, where Δt is discussed below. Then the τ integral can be expressed as

$$\int_{\tau=0}^{t-\Delta t} \exp\left[-\frac{\beta_m^2 \alpha(t - \tau)}{L^2} \right] d\tau = \frac{L^2}{\alpha \beta_m^2} (e^{-\beta_m^2 \alpha \Delta t/L^2} - e^{-\beta_m^2 \alpha t/L^2}) \tag{5.8}$$

If $\Delta t = 0$ in Eq. (5.8), then the term $\exp(-\beta_m^2 \alpha \Delta t/L^2)$ becomes unity, and the first term of the time integral becomes $L^2/(\alpha \beta_m^2)$. When this term is replaced back into the infinite sum in Eq. (5.7), for the $X21$ and $X22$ cases at $x = 0$, the resulting term is proportional to

---

*The number given here as 0.06 may vary between 0.025 and 0.25 depending on the circumstances; more important is the concept that a dimensionless number $\alpha t/(L \cdot L)$, sometimes called the Fourier number, defines the small-time regime.

$$\sum_{m=1}^{\infty} \frac{1}{\beta_m^2} \tag{5.9}$$

which is part of the expression for the temperature. In many cases β_m is approximately equal to m times π:

$$\beta_m \approx m\pi \tag{5.10}$$

for large values of m. For large m, the "tail" of the summation of Eq. (5.9) for $m = M$, $M + 1$, $M + 2$, etc., is approximately

$$\sum_{m=M}^{\infty} \frac{1}{\pi^2 m^2} \approx \frac{1}{\pi^2} \int_M^{\infty} \frac{1}{m^2} \, dm = \frac{1}{\pi^2 M} \tag{5.11}$$

Hence, the tail of the summation is proportional to $1/M$. This means that a very large number of terms in the series is needed if accurate temperature values are desired. For example, if M is equal to 100, the error in neglecting the tail is approximately $1/(100\pi^2) = 0.0010$; for $M = 1000$, the error is one-tenth as large, but there is 10 times as much computation. Note that

$$\sum_{m=1}^{\infty} \frac{1}{\pi^2 m^2} = \frac{1}{6} = 0.16667$$

and so using $M = 100$ would result in an error of $0.001/0.1667$ or a 0.6% error. One reason that analytical solutions are used is to obtain the "exact" solution which, in practice, usually means an error of 0.01% or less. In this example, accuracy of 0.01% would require the large number of over 6000 terms in the infinite series.

If the integral over $\tau = t - \Delta t$ to t can be found using the short-time GF, then instead of evaluating Eq. (5.9), it is only necessary to evaluate the sum

$$\sum_{m=1}^{M-1} \frac{1}{\beta_m^2} \exp\left[-\frac{\beta_m^2 \alpha \Delta t}{L^2} \right] \tag{5.12}$$

which requires many fewer terms for *nonzero* values of $\alpha \Delta t / L^2$. If the tail of Eq. (5.12) is calculated, and $\beta_m \approx m\pi$, the result is

$$\sum_{m=M}^{\infty} \frac{1}{m^2 \pi^2} \exp\left[-\frac{m^2 \pi^2 \, \alpha \Delta t}{L^2} \right] \approx \frac{1}{M \pi^{3/2}} \, \text{ierfc}\left[M\pi \left(\frac{\alpha \Delta t}{L^2} \right)^{1/2} \right] \tag{5.13}$$

which reduces to the Eq. (5.11) result for $\Delta t = 0$. For nonzero values of $\alpha \Delta t / L^2$, the right side of Eq. (5.13) decreases very rapidly as M increases. As an example, let $\alpha \Delta t / L^2$ be the small value of 0.025. Then, $\pi(\alpha \Delta t / L^2)^{1/2} = 0.497 \approx 0.5$, and then ierfc(0.5M) takes on the values 400E-7, 30E-7, and 0.9E-7, for $M = 4$, 5, and 6, respectively. Hence for $\alpha \Delta t / L^2 = 0.025$ and small values of M such as 4, the error by dropping the tail of the summation is negligible. (Larger $\alpha \Delta t / L^2$ values cause the right-hand side of Eq. (5.13) to decrease even more rapidly as M increases.) The contribution for $\tau = t - \Delta t$ to t in Eq. (5.8) is obtained using just a few terms of the small time GFs.

Consequently, partitioning the time integral in Eq. (5.4) has great potential to improve the computational efficiency of solutions obtained with the GF method, particularly for two- and three-dimensional problems; time partitioning is discussed in the following section. Another method of treating the problem of slow convergence is to use the concept of acceleration of series. Sometimes the analytical evaluation of integrals is so difficult that it is better to use numerical techniques; this is common in multidimensional problems.

5.3 TIME PARTITIONING

In the GF equation given by Eq. (5.4), the $G(\cdot)$ functions for finite bodies can be of the small-time form (see the $X11$, $X12$, $X21$, and $X22$ GFs listed in Table 4.1), or they can be of the large-time form (see Section 4.4). The small-time and large-time forms of the GF are each solutions to the heat conduction boundary value problem, given by Eqs. (5.1)–(5.3). These two solutions are mathematically equivalent, as required by the uniqueness property of solutions of linear boundary value problems (Carslaw and Jaeger, 1959, pp. 35–38). The numerical values are identical for the two solutions for the same conditions. The solution of a boundary value problem is unique, but the expansion of that solution in infinite series form may not be unique. That is, the small-time and large-time solutions are different infinite-series expansions of the same solution.

Many small-time GFs are derived from Laplace transform solutions of the heat conduction equation (refer to Section 4.3 for an example). For plates, small-time GFs take the form of an infinite series of fundamental heat conduction functions given by Eq. (5.5). See also Eq. (4.1) and the XIJ case of Table 4.1. For sufficiently small times, the value of a GF at any x is unaffected by the boundaries or at most by a single boundary. Hence, the GFs at sufficiently small t's can be described by the same GFs as for infinite or semi-infinite bodies. Consequently, the small time GFs can be represented by only the few terms which emphasize the effects of a single boundary.

In contrast, many large-time GF expressions are derived from the separation of variables method of solution of the heat conduction equation. For slab bodies, large-time GFs are composed of infinite series of basic functions given by Eq. (5.6). The large-time GFs incorporate the effect of the finite nature of the body and require only a few terms for sufficiently large times. The large time GFs contain eigenvalues that are based on the finite thickness of the plate. As a consequence, the small and large time GFs emphasize different aspects of the physical problem in a manner so that only a few terms, in their respective infinite series, are usually needed.

Time partitioning can speed evaluation of the infinite series expressions compared to using a single form of the series. To take advantage of the different convergence properties of the small-time and large-time solutions, Eq. (5.4) can be written as

$$
T(x, t) = \left[\int_{x'=0}^{L} G^S(x, t|x', 0)\, F(x')\, dx' \right]\Bigg|_{\alpha t/L^2 < \alpha t_1/L^2}
$$

$$
+ \left[\int_{x'=0}^{L} G^L(x, t|x', 0)\, F(x')\, dx' \right]\Bigg|_{\alpha t/L^2 > \alpha t_1/L^2}
$$

$$+ \frac{\alpha}{k} \int_{\tau=0}^{t-\Delta t} \int_{x'=0}^{L} G^L(x, t|x', \tau) \, g(x', \tau) \, dx' \, d\tau$$

$$+ \frac{\alpha}{k} \int_{\tau=t-\Delta t}^{t} \int_{x'=0}^{L} G^S(x, t|x', \tau) \, g(x', \tau) \, dx' \, d\tau \qquad (5.14)$$

where $G^S(\cdot)$ and $G^L(\cdot)$ correspond to small-time and large-time GFs, respectively. The dimensionless times $\alpha t_1/L^2$ and $\alpha \Delta t/L^2$ are small compared with unity.

The second integral in Eq. (5.14), which is the integral containing $G^L(\cdot)F(x')$, does not have the same convergence problems as the third integral in Eq. (5.14), which is the integral containing $G^L(\cdot)g(x', \tau)$. For this reason, in most cases $\alpha t_1/L^2$ can be taken to be zero. That is, time partitioning is usually not required for the initial condition term, and the first integral in Eq. (5.14) is not required. This is desirable for finite bodies, because $G^S(\cdot)$ is usually a more difficult function to analytically manipulate than $G^L(\cdot)$.

For the convergence reasons mentioned in the previous section, both $G^S(\cdot)$ and $G^L(\cdot)$ may be needed in the $g(x', \tau)$ integral. The value of $\alpha \Delta t/L^2$ for time partitioning is usually chosen to be between 0.025 and 0.25. The benefit of choosing a small value of $\alpha \Delta t/L^2$ is that only a few terms of the series for $G^S(\cdot)$ will be needed in the last integral of Eq. (5.14).

The integrals over τ in Eq. (5.14) deserve special comment because the large time GF $G^L(\cdot)$ is associated with the small τ's, and the small time GF $G^S(\cdot)$ is associated with the large τ's. At first glance this may seem contradictory. The contradiction is resolved by noting that the $G(x, t|x', \tau)$ functions are always a function of $t - \tau$; it is not t, but rather $\alpha(t - \tau)/L^2$, that should be "small" or "large." Note that $\tau \to t$ corresponds to $t - \tau \to 0$. This point can be illustrated by an alternative form of Eq. (5.14) which uses a new time variable. Let

$$u = t - \tau \qquad (5.15)$$

so that the GFs can in general be written

$$G^S(x, t|x', \tau) = G^S(x, x', u) \qquad (5.16a)$$

$$G^L(x, t|x', \tau) = G^L(x, x', u) \qquad (5.16b)$$

Then, with $\alpha t_1/L^2$ set equal to zero, Eq. (5.14) can be written as

$$T(x, t) = \int_{x'=0}^{L} G^L(x, x', t) \, F(x') \, dx'$$

$$+ \frac{\alpha}{k} \int_{u=0}^{\Delta t} \int_{x'=0}^{L} G^S(x, x', u) \, g(x', t - u) \, dx' \, du$$

$$+ \frac{\alpha}{k} \int_{u=\Delta t}^{t} \int_{x'=0}^{L} G^L(x, x', u) \, g(x', t - u) \, dx' \, du \qquad (5.17)$$

where the order of the last two integrals has been changed from that in Eq. (5.14).

Refer to Section 6.6 for complete examples in which time partitioning is applied to find expressions for the temperature in flat plates that are numerically efficient for all values of time.

5.4 INTEGRALS OCCURRING IN TEMPERATURE SOLUTION

The solution of the one-dimensional GF equation requires integration over x' and τ. Since there are only two types of basic functions (one for small values of $\alpha(t - \tau)/L^2$, and one for large values), many of the integrals required for the GF method can be tabulated in advance. The large-time GFs involve simple products of a function of τ multiplied by another function of x', so the integrations over τ and over x' can be performed independently. For small-time GFs, which involve $K(z \pm x', t - \tau)$, integration over τ is not independent of x', and the integrals can be more complicated than for the large time GFs. In the case of integrating $K(z \pm x', t - \tau)$ first over x', additional (and sometimes more complicated) functions are found; if the integration over τ is then needed, the set of integrals that must be solved is larger than for the large-time GFs.

Some integrals for the small-time and the large-time expressions for $G(\cdot)$ are given in this section. Tables 5.1–5.5 list integrals of the small-time G's, and Tables 5.6 and 5.7 list integrals of the large-time G's.

5.4.1 Integrals for Small-time Green's Functions

Integrals over the basic element of the small time GF $K(\cdot)$ are usually performed over x' first (that is, the τ integration is left to the end), if integrals are required over space

Table 5.1 Small-time function integrations over x' of $\int_a^b K(z - x', t - \tau)$
$$F(x')\,dx' = \int_a^b [4\pi\alpha(t - \tau)]^{-1/2} e^{-[(z-x')^2]/[4\alpha(t - \tau)]} F(x')\,dx'$$

$F(x')$	Integral
$\delta(x_0 - x')$	$K(z - x_0, t - \tau),\ a < x_0 < b;$ otherwise zero
1	$\dfrac{1}{2}\left(\text{erfc}\left\{\dfrac{z - b}{[4\alpha(t - \tau)]^{1/2}}\right\} - \text{erfc}\left\{\dfrac{z - a}{[4\alpha(t - \tau)]^{1/2}}\right\}\right)$
$\dfrac{x'}{L}$	$\dfrac{z}{2L}\left(\text{erfc}\left\{\dfrac{z - b}{[4\alpha(t - \tau)]^{1/2}}\right\} - \text{erfc}\left\{\dfrac{z - a}{[4\alpha(t - \tau)]^{1/2}}\right\}\right)$
	$\qquad + \dfrac{2\alpha(t - \tau)}{L}[-K(z - b, t - \tau) + K(z - a, t - \tau)]$
$\left(\dfrac{x'}{L}\right)^2$	$\left[\dfrac{1}{2}\left(\dfrac{z}{L}\right)^2 + \dfrac{\alpha(t - \tau)}{L^2}\right]\left(\text{erfc}\left\{\dfrac{z - b}{[4\alpha(t - \tau)]^{1/2}}\right\} - \text{erfc}\left\{\dfrac{z - a}{[4\alpha(t - \tau)]^{1/2}}\right\}\right)$
	$\qquad + \dfrac{2\alpha(t - \tau)}{L^2}[-(z + b)K(z - b, t - \tau) + (z + a)K(z - a, t - \tau)]$
$\exp\left(\dfrac{-Ux'}{2\alpha}\right)$	$\dfrac{1}{2}\exp\left[\dfrac{U^2(t - \tau)}{4\alpha} - \dfrac{Uz}{2\alpha}\right]\text{erf}\left\{\dfrac{x' - z}{[4\alpha(t - \tau)]^{1/2}} + \dfrac{U(t - \tau)^{1/2}}{2\alpha^{1/2}}\right\}\Bigg|_a^b$

Note: $\text{erfc}(0) = 1$, $\text{erfc}(\infty) = 0$, $\text{erfc}(-\infty) = 2$. See also Appendix E.

Table 5.2 Small-time function integrations over x' of $\int_a^b K(z + x', t - \tau)$ $F(x') \, dx' = \int_a^b K(-z - x', t - \tau) F(x') \, dx'$

$F(x')$	Integral
$\delta(x_0 - x')$	$K(z + x_0, t - \tau)$, $a < x_0 < b$; otherwise zero
1	$\dfrac{1}{2}\left(-\text{erfc}\left\{\dfrac{z + b}{[4\alpha(t - \tau)]^{1/2}}\right\} + \text{erfc}\left\{\dfrac{z + a}{[4\alpha(t - \tau)]^{1/2}}\right\}\right)$
$\dfrac{x'}{L}$	$\dfrac{z}{2L}\left(\text{erfc}\left\{\dfrac{z + b}{[4\alpha(t - \tau)]^{1/2}}\right\} - \text{erfc}\left\{\dfrac{z + a}{[4\alpha(t - \tau)]^{1/2}}\right\}\right)$
	$\qquad + \dfrac{2\alpha(t - \tau)}{L}\,[-K(z + b, t - \tau) + K(z + a, t - \tau)]$
$\left(\dfrac{x'}{L}\right)^2$	$\left[\dfrac{1}{2}\left(\dfrac{z}{L}\right)^2 + \dfrac{\alpha(t - \tau)}{L^2}\right]\left(-\text{erfc}\left\{\dfrac{z + b}{[4\alpha(t - \tau)]^{1/2}}\right\} + \text{erfc}\left\{\dfrac{z + a}{[4\alpha(t - \tau)]^{1/2}}\right\}\right)$
	$\qquad + \dfrac{2\alpha(t - \tau)}{L^2}[(z - b)K(z + b, t - \tau) - (z - a)K(z + a, t - \tau)]$

Table 5.3 Integral over x' for small-time typical term for XIJ, $I, J = 1, 2$
FIN $(i, z) = \int_{x'=0}^{L} [K(z - x', t) + (-1)^I K(z + x', t)] \, (x'/L)^i \, dx'$

0. $\text{FIN}\,(0, z) = \dfrac{1}{2}\left\{\text{erfc}\left[\dfrac{z - L}{(4\alpha t)^{1/2}}\right] - \text{erfc}\left[\dfrac{z}{(4\alpha t)^{1/2}}\right]\right\}$

$\qquad + (-1)^I \dfrac{1}{2}\left\{\text{erfc}\left[\dfrac{z}{(4\alpha t)^{1/2}}\right] - \text{erfc}\left[\dfrac{z + L}{(4\alpha t)^{1/2}}\right]\right\}$

1. $\text{FIN}\,(1, z) = \dfrac{z}{2L}\left\{\text{erfc}\left[\dfrac{z - L}{(4\alpha t)^{1/2}}\right] - \text{erfc}\left[\dfrac{z}{(4\alpha t)^{1/2}}\right]\right\} + 2\dfrac{\alpha t}{L}\,[K(z, t) - K(z - L, t)]$

$\qquad + (-1)^I \left(\dfrac{z}{2L}\left\{\text{erfc}\left[\dfrac{z}{(4\alpha t)^{1/2}}\right] - \text{erfc}\left[\dfrac{z + L}{(4\alpha t)^{1/2}}\right]\right\} + 2\dfrac{\alpha t}{L}\,[2K(z, t) - K(z + L, t)]\right)$

2. $\text{FIN}\,(2, z) = \left[\dfrac{1}{2}\left(\dfrac{z}{L}\right)^2 + \dfrac{\alpha t}{L^2}\right]\left\{\text{erfc}\left[\dfrac{z - L}{(4\alpha t)^{1/2}}\right] - \text{erfc}\left[\dfrac{z}{(4\alpha t)^{1/2}}\right]\right\}$

$\qquad + \dfrac{2\alpha t}{L^2}\,[z K(z, t) - (z + L) K(z - L, t)]$

$\qquad + (-1)^I \left(\left[\dfrac{1}{2}\left(\dfrac{z}{L}\right)^2 + \dfrac{\alpha t}{L^2}\right]\left\{\text{erfc}\left[\dfrac{z}{(4\alpha t)^{1/2}}\right] - \text{erfc}\left[\dfrac{z + L}{(4\alpha t)^{1/2}}\right]\right\}\right.$

$\qquad\qquad \left. + \left(\dfrac{2\alpha t}{L^2}\right)[(z - L) K(z + L, t) - z K(z, t)]\right)$

and τ. Moreover, the integral for the initial condition [see the first integral in Eq. (5.4), for example], is only over x'. For these reasons, the integrals over x' are considered first.

Tables 5.1 and 5.2 contain several integrals over x' of the functions $K(z - x', t - \tau)$ and $F(x')$ between the limits of a and b:

$$\int_a^b K(z - x', t - \tau) F(x') \, dx' = \int_a^b K(-z + x', t - \tau) F(x') \, dx' \quad (5.18)$$

Table 5.4 Integrals related to the short-time Green's functions

1. $\displaystyle\int_{t-\Delta t}^{t} K(z, t-\tau)\, d\tau = \left(\frac{\Delta t}{\alpha}\right)^{1/2} \text{ierfc}\left[\frac{|z|}{(4\alpha\Delta t)^{1/2}}\right]$

2. $\displaystyle\int_{t-\Delta t}^{t} \frac{\alpha(t-\tau)}{L^2} K(z, t-\tau)\, d\tau = \frac{|z|}{L^2}\Delta t \left[\left(\frac{4\alpha\Delta t}{z^2}\right)^{1/2} i^3\, \text{erfc}\frac{|z|}{(4\alpha\Delta t)^{1/2}} + i^2\, \text{erfc}\frac{|z|}{(4\alpha\Delta t)^{1/2}}\right]$

3. $\displaystyle\int_{t-\Delta t}^{t} \frac{1}{t-\tau} K(z, t-\tau)\, d\tau = \frac{1}{|z|}\, \text{erfc}\frac{|z|}{(4\alpha\Delta t)^{1/2}}$

4. $\displaystyle\int_{t-\Delta t}^{t} \text{erfc}\left\{\frac{z}{[4\alpha(t-\tau)]^{1/2}}\right\} d\tau = 4\Delta t\, i^2\, \text{erfc}\left[\frac{|z|}{(4\alpha\Delta t)^{1/2}}\right]$

5. $\displaystyle\int_{t-\Delta t}^{t} \frac{\alpha(t-\tau)}{L^2}\, \text{erfc}\left\{\frac{z}{[4\alpha(t-\tau)]^{1/2}}\right\} d\tau$

$$= \frac{\alpha}{L^2}(4\Delta t)^2\left\{\frac{1}{4}i^2\, \text{erfc}\left[\frac{|z|}{(4\alpha\Delta t)^{1/2}}\right] - i^4\, \text{erfc}\left[\frac{|z|}{(4\alpha\Delta t)^{1/2}}\right]\right\}$$

Table 5.5 Convolution integrals related to the short-time GFs

1. $\displaystyle\int_0^t \left(\frac{\tau}{t_0}\right)^{m/2} K(z, t-\tau)\, d\tau = \int_0^t \left(\frac{\tau}{t_0}\right)^{m/2} K(-z, t-\tau)\, d\tau$

$$= \frac{1}{2\alpha^{1/2}}\, \Gamma\left(\frac{m}{2}+1\right)\left(\frac{4t}{t_0}\right)^{(m+1)/2} t_0^{1/2} i^{m+1}\, \text{erfc}\left[\frac{|z|}{(4\alpha t)^{1/2}}\right], \quad m = -1, 0, 1, 2, \ldots$$

2. $\displaystyle\int_0^t \left(\frac{\tau}{t_0}\right)^{m/2} \frac{\alpha(t-\tau)}{L^2} K(z, t-\tau)\, d\tau = t_0 \frac{|z|}{L^2}\Gamma\left(\frac{m}{2}+1\right) 2^m \left(\frac{t}{t_0}\right)^{(m+2)/2}\left[\left(\frac{4\alpha t}{z^2}\right)^{1/2} i^{m+3}\, \text{erfc}\frac{|z|}{(4\alpha t)^{1/2}}\right.$

$$\left. + i^{m+2}\, \text{erfc}\frac{|z|}{(4\alpha t)^{1/2}}\right], \quad m = -2, -1, 0, 1, 2, \ldots$$

3. $\displaystyle\int_0^t \left(\frac{\tau}{t_0}\right)^{m/2} \frac{1}{t-\tau} K(z, t-\tau)\, d\tau = \frac{1}{|z|}\left(\frac{4t}{t_0}\right)^{m/2} i^m\, \text{erfc}\left[\frac{|z|}{(4\alpha t)^{1/2}}\right]\Gamma\left(\frac{m}{2}+1\right), \quad m = 0, 1, 2, \ldots$

4. $\displaystyle\int_0^t \left(\frac{\tau}{t_0}\right)^{m/2}\text{erfc}\left\{\frac{z}{[4\alpha(t-\tau)]^{1/2}}\right\} d\tau$

$$= t_0 \Gamma\left(\frac{m}{2}+1\right)\left(\frac{4t}{t_0}\right)^{(m+2)/2} i^{m+2}\, \text{erfc}\left[\frac{z}{(4\alpha t)^{1/2}}\right], \quad m = -1, 0, 1, 2, \ldots$$

5. $\displaystyle\int_0^t \left(\frac{\tau}{t_0}\right)^{m/2}\frac{\alpha(t-\tau)}{L^2}\, \text{erfc}\left\{\frac{z}{[4\alpha(t-\tau)]^{1/2}}\right\} d\tau = t_0\frac{\alpha t_0}{L^2}\left(\frac{4t}{t_0}\right)^{(m+4)/2}$

$$\times \left\{\frac{1}{4}\Gamma\left(\frac{m}{2}+1\right)i^{m+2}\, \text{erfc}\left[\frac{z}{(4\alpha t)^{1/2}}\right] - \Gamma\left(\frac{m}{2}+1\right)i^{m+4}\, \text{erfc}\left[\frac{z}{(4\alpha t)^{1/2}}\right]\right\}, \quad m = -1, 0, 1, 2, \ldots$$

6. $\displaystyle\int_0^t \tau^{m/2}\exp\left[\frac{hx}{k}+\frac{h^2}{k^2}\alpha(t-\tau)\right]\text{erfc}\left\{\frac{x}{[4\alpha(t-\tau)]^{1/2}}+\frac{h}{k}[\alpha(t-\tau)]^{1/2}\right\} d\tau$

$$= \frac{\Gamma\left(\frac{m}{2}+1\right)}{\alpha^{m/2+1}}\left(-\frac{k}{h}\right)^{m+2}\left\{\exp\left[\frac{hx}{k}+\frac{h^2}{k^2}\alpha t\right]\text{erfc}\left[\frac{x}{(4\alpha t)^{1/2}}+\frac{h}{k}(\alpha t)^{1/2}\right]\right.$$

$$\left. - \sum_{j=0}^{m+1}\left[-\frac{h}{k}(4\alpha t)^{1/2}\right]^j i^j\, \text{erfc}\left[\frac{x}{(4\alpha t)^{1/2}}\right]\right\}, \quad m = -1, 0, 1, 2, 3, \ldots$$

Table 5.6 Values of $LK(z, t)$ and $2^m\Gamma(m/2 + 1)i^m erfc[(z/L)/(4\alpha t/L^2)^{1/2}]$, $m = 0, 1, 2$, and 3 for $\alpha t/L^2 = 0.025$

z/L	$LK(z, t)$	erfc (*)	$\pi^{1/2}$ ierfc (*)	$4i^2$ erfc (*)	$6\pi^{1/2}i^3$ erfc (*)
0.00	0.178412E+01	0.100000E+01	0.100000E+01	0.100000E+01	0.100000E+01
0.10	0.161434E+01	0.654721E+00	0.537867E+00	0.462797E+00	0.408168E+00
0.20	0.119593E+01	0.371093E+00	0.254325E+00	0.189594E+00	0.148058E+00
0.30	0.725371E+00	0.179712E+00	0.104384E+00	0.679725E−01	0.472357E−01
0.40	0.360208E+00	0.736383E−01	0.367998E−01	0.211140E−01	0.131310E−01
0.50	0.146450E+00	0.253473E−01	0.110492E−01	0.563409E−02	0.315450E−02
0.60	0.487489E−01	0.279036E−02	0.280629E−02	0.128224E−02	0.650200E−03
0.70	0.132856E−01	0.174512E−02	0.599621E−03	0.247402E−03	0.114280E−03
0.80	0.296442E−02	0.346619E−03	0.107319E−03	0.402680E−04	0.170378E−04
0.90	0.541551E−03	0.569941E−04	0.160328E−04	0.550611E−05	0.214502E−05
1.00	0.809991E−04	0.774422E−05	0.199366E−05	0.630326E−06	0.227179E−06

$* = (z/L)/(0.025)^{1/2}$

This equation is valid because

$$K(z, t) = K(-z, t) \tag{5.19}$$

The $F(x')$ functions in Table 5.1 include the Dirac delta function $\delta(x - x')$ and also the functions $F(x') = 1$, x'/L, and $(x'/L)^2$. Table 5.2 gives similar integrals for the function $K(z + x', t - \tau)$.

Table 5.3 gives some integrals over x' from $x' = 0$ to L of the function

$$[K(z - x', t) + (-1)^I K(z + x', t)] \left(\frac{x'}{L}\right)^i$$

where $i = 0, 1, 2$ and $I = 1, 2$. These results apply to the $X11$, $X12$, $X21$, and $X22$ GFs given in Table 4.1.

The integrals in Tables 5.1 and 5.2 produce additional functions which have the forms

$$\frac{\alpha(t - \tau)}{L^2} K(z, t - \tau) \quad erfc\left\{\frac{z}{[4\alpha(t - \tau)]^{1/2}}\right\} \quad \frac{\alpha(t - \tau)}{L^2} erfc\left\{\frac{z}{[4\alpha(t - \tau)]^{1/2}}\right\} \tag{5.20}$$

Integrals of each of these functions over τ are given in Table 5.4. Also contained in Table 5.4 are integrals over τ of $K(w, t - \tau)$ and $K(w, t - \tau)/(t - \tau)$. These integrals have limits of τ from $t - \Delta t$ to t, to apply to time partitioning; if the desired range of τ is 0 to t, simply replace Δt by t.

Convolution integrals for some of the above functions, Eq. (5.20), are given in Table 5.5. A convolution integral involves integration of the product of two functions, one a function of $t - \tau$ and the other a function of τ; and, the integration range is from $\tau = 0$ to t. The integrals in Table 5.5 were obtained by using the well-known properties of the Laplace transform of a convolution integral.

All of the integrals in Table 5.5 involve the integrated complementary error

Table 5.7 Large-time function integration over x', $\int_a^b X(\beta_m x'/L) \, F(x') \, dx'$

No.	$F(x')$	$X(\beta_m x'/L)$	Integral
1	$L\delta(x_0 - x')$	$\sin \dfrac{\beta_m x'}{L}$	$L \sin \dfrac{\beta_m x_0}{L}$, $a < x_0 < b$
2	1	$\sin \dfrac{\beta_m x'}{L}$	$-\dfrac{L}{\beta_m} \cos u \Big|_{\beta_m a/L}^{\beta_m b/L}$
3	x'/L	$\sin \dfrac{\beta_m x'}{L}$	$\dfrac{L}{\beta_m^2} [\sin(u) - u \cos u] \Big|_{\beta_m a/L}^{\beta_m b/L}$
4	$(x'/L)^2$	$\sin \dfrac{\beta_m x'}{L}$	$\dfrac{2L}{\beta_m^3} \left[u \sin(u) + \left(1 - \dfrac{u^2}{2}\right) \cos u \right] \Big|_{\beta_m a/L}^{\beta_m b/L}$
5	$L\delta(x_0 - x')$	$\cos \dfrac{\beta_m x'}{L}$	$L \cos \dfrac{\beta_m x_0}{L}$, $a < x_0 < b$
6	1	$\cos \dfrac{\beta_m x'}{L}$	$\dfrac{L}{\beta_m} \sin u \Big|_{\beta_m a/L}^{\beta_m b/L}$
7	$\dfrac{x'}{L}$	$\cos \dfrac{\beta_m x'}{L}$	$\dfrac{L}{\beta_m^2} [\cos(u) + u \sin u] \Big|_{\beta_m a/L}^{\beta_m b/L}$
8	$\left(\dfrac{x'}{L}\right)^2$	$\cos \dfrac{\beta_m x'}{L}$	$\dfrac{2L}{\beta_m^3} \left[u \cos(u) + \left(\dfrac{u^2}{2} - 1\right) \sin u \right] \Big|_{\beta_m a/L}^{\beta_m b/L}$
9	$e^{-\gamma x'}$	$\cos \dfrac{\beta_m x'}{L}$	$\dfrac{e^{-\gamma L u/\beta_m}}{\gamma^2 + (\beta_m/L)^2} \left[\dfrac{\beta_m}{L} \sin(u) - \gamma \cos u\right] \Big|_{\beta_m a/L}^{\beta_m b/L}$
10	$\sin(\gamma x')$	$\cos \dfrac{\beta_m x'}{L}$	$\left[\dfrac{\cos[(C_m - 1)u]}{2(\beta_m/L)(C_m - 1)} \right.$ $\left. + \dfrac{\cos[(C_m + 1)u]}{2(\beta_m/L)(C_m + 1)} \right] \Big|_{\beta_m a/L}^{\beta_m b/L}$; $C_m = \gamma L/\beta_m$

function, $i^m \text{erfc}(u)$. Since infinite series of such terms often appear in the small-time GFs, it is instructive to note how the terms decrease with increasing values of the argument u. For this reason, it is helpful to plot the function

$$y = 2^m \, \Gamma\left(\frac{m}{2} + 1\right) i^m \text{erfc}(u) \qquad (5.21a)$$

$$i^m \text{erfc}(u) = \int_u^\infty i^{m-1} \text{erfc}(x) \, dx \qquad m = 1, 2, \text{ and } 3 \qquad i^0 \text{erfc}(u) \qquad (5.21b)$$

See Fig. 6.2 and also Appendix E for more information regarding $i^m \text{erfc}(u)$. The function $\Gamma(\cdot)$ is called the gamma function; for $m = 1, 2,$ and 3, $\Gamma(m/2 + 1)$ is equal to $\pi^{1/2}/2$, 1, and $3\pi^{1/2}/4$, respectively. For each m value from $m = 0, 1,$ and $2, \ldots$, the value of y, defined by Eq. (5.21a), is equal to unity for $u = 0$. More importantly, note that except for $u = 0$, each curve for a specified m value has a smaller magnitude at a given u than the u curve for a smaller m (≥ 0) value. This means that $y(u)$ decreases more rapidly with u for $m = 1, 2, \ldots$ than for the $m = 0$ curve:

$$y|_{m=0} = \text{erfc } (u) \tag{5.22}$$

Note that

$$\frac{\text{erfc } (|z|/(4\alpha t)^{1/2})}{L\, K(z, t)} = \frac{\text{erfc } [|z/L|/(4w)^{1/2}]}{(4\pi w)^{-1/2} \exp [-(z/L)^2/(4w)]} < 1 \tag{5.23}$$

for $z/L > 0$ and small values of w ($w \leq 0.08$) where $w = \alpha t/L^2$. In conclusion, when terms such as

$$2^m\, \Gamma\left(\frac{m}{2} + 1\right) i^m \text{ erfc } \left[\frac{|2nL + x|}{(4\alpha t)^{1/2}}\right] \tag{5.24}$$

have to be evaluated, the terms for increasing values of m will decrease more rapidly than for the fundamental heat conduction solution,

$$2^m\, \Gamma\left(\frac{m}{2} + 1\right) i^m \text{ erfc } \left[\frac{|2nL + x|}{(4\alpha t)^{1/2}}\right] < L\, K(2nL + x, t) \tag{5.25}$$

for values of $\alpha t/L^2 < 0.08$ and for $m = 0, 1, 2$, and so on.

To make the point even clearer and also to provide numerical values, Table 5.6 gives a tabulation of

$$L\, K(z, t) = \left(\frac{4\pi\alpha t}{L^2}\right)^{-1/2} \exp \left(\frac{-z^2}{4\alpha t}\right)$$

and also

$$2^m\, \Gamma\left(\frac{m}{2} + 1\right) i^m \text{ erfc } \left[\frac{z/L}{(4\alpha t/L^2)^{1/2}}\right]$$

for $m = 0, 1, 2, 3$; for $\alpha t/L^2 = 0.025$; and, for $z/L = 0, 0.1, 0.2, \ldots, 1.0$. As stated above, for a given z/L value greater than zero, the entries in each column after the second column in Table 5.6 are less than the previous values (except for the first row).

5.4.2 Integrals for Large-time Green's Functions

The integrals for the large-time GFs are easier to obtain than for the small-time GFs. The x' and τ integrations are independent because the terms have the form

$$\exp \left[-\frac{\beta_m^2 \alpha(t - \tau)}{L^2}\right] X(\beta_m, x)\, X(\beta_m, x') \tag{5.26}$$

Incidentally, not only can integrations be independently performed for τ and x', but they can also be performed independently for x (for finding the spatial average temperature). Furthermore, since the $X(\beta_m, x)$ and $X(\beta_m, x')$ functions have exactly the same form, expressions for integrations over x' can also be used for integrations over x. Hence, an integration over x' followed by integration over x is relatively simple. This is not true for the small-time GF, because first integrating over x' results in new functions that must then be integrated over x.

Table 5.8 lists some integrals for $X(\beta_m, x') = \sin(\beta_m x'/L)$. Similar results are obtained for $X(\beta_m, x') = \cos(\beta_m x'/L)$. Table 5.8 provides integrals of $\exp[-\beta_m^2 \alpha(t - \tau)/L^2]$ and $f(\tau)$,

$$\int_{\tau=0}^{t-\Delta t} \exp\left[-\frac{\beta_m^2 \alpha(t - \tau)}{L^2}\right] f(\tau)\, d\tau$$

where $f(\tau)$ is one of the functions $t_0 \delta(\tau)$, 1, τ/t_0, $(\tau/t_0)^{-1/2}$, $(\tau/t_0)^{1/2}$, or $\exp(-C\tau)$. For $f(\tau) = (\tau/t_0)^{-1/2}$ and $(\tau/t_0)^{1/2}$, the integral involves the Dawson integral,

$$\text{Dawson integral} = F(x) = e^{-x^2} \int_0^x e^{u^2}\, du \tag{5.27}$$

Tables of this function are available in Abramowitz and Stegun (1964).

5.5 ACCELERATION OF SERIES

One of the common outcomes of using the methods of separation of variables (and equivalently in many cases, the use of GFs) for the solution of partial differential equations is the answer in the form of infinite series. As discussed above, some of these series may converge slowly. Hence there is a need for speeding the convergence of these series. This section discusses several methods of accomplishing this. One method is called Aitken's δ^2 process. Another is for alternating series and is called Euler's transformation. The former can be used when the convergence is approximately geometric; it can be used if the series is or is not alternating (when the terms alternate

Table 5.8 Large-time function integrations over time,
$\int_0^{t-\Delta t} \exp[-\beta_m^2 \alpha(t - \tau)/L^2] f(\tau)\, d\tau$

$f(\tau)$	Integral [$F(\cdot)$ defined in Eq. (5.27)]
$t_0 \delta(\tau)$	$t_0 \exp(-\beta_m^2 t^+)$
1	$\dfrac{L^2}{\alpha \beta_m^2} [\exp(-\beta_m^2 \Delta t^+) - \exp(-\beta_m^2 t^+)]$
$\dfrac{\tau}{t_0}$	$t_0 \left(\dfrac{L^2}{\alpha t_0 \beta_m^2}\right)^2 \{[\beta_m^2(t^+ - \Delta t^+) - 1] \exp(-\beta_m^2 \Delta t^+) + \exp(-\beta_m^2 t^+)\}$
$\left(\dfrac{\tau}{t_0}\right)^{-1/2}$	$t_0 \left(\dfrac{L^2}{\alpha t_0 \beta_m^2}\right)^{1/2} \exp(-\beta_m^2 \Delta t^+)\, F[\beta_m(t^+ - \Delta t^+)^{1/2}]$
$\left(\dfrac{\tau}{t_0}\right)^{1/2}$	$t_0 \left(\dfrac{L^2}{\alpha t_0 \beta_m^2}\right)^{3/2} \exp(-\beta_m^2 \Delta t^+) \{\beta_m(t^+ - \Delta t^+) - F[\beta_m(t^+ - \Delta t^+)^{1/2}]\}$
$e^{-C\tau}$	$\dfrac{L}{\beta_m^2 \alpha - CL^2} \{\exp[-C(t - \Delta t) - \beta_m^2 \Delta t^+] - \exp(-\beta_m^2 t^+)\}$

$$t^+ = \frac{\alpha t}{L^2}, \quad \Delta t^+ = \frac{\alpha \Delta t}{L^2}$$

in signs). For this reason, Aitken's δ^2 process is discussed in greater detail. Other methods are also discussed.

Consider now the Aitken procedure for accelerating series. It is also called the Aitken-Shanks transformation. Suppose our series, which has an exact value of S is such that

$$S \approx S_n + \beta_1 \beta_2^n \tag{5.28}$$

where S_n is the truncated series up to the nth term, and β_1 and β_2 are constants to be determined. In other words, we suppose that the remainder in the infinite series, $S - S_n$, is fitted approximately by the simple expression $\beta_1 \beta_2^n$ for some constants β_1 and β_2. If the approximation is valid, S_n values at three different n's will yield three simultaneous equations for the three unknowns, S, β_1, and β_2. By using three successive values of n, an improved value of S_n, denoted S_n', can be derived:

$$S_n' \approx S_{n+1} - \frac{(S_{n+1} - S_n)^2}{S_{n+1} - 2S_n + S_{n-1}} \tag{5.29}$$

This expression should be used as written because there are equivalent algebraic forms that are much more susceptible to roundoff error. Equation (5.29) can also be used with $n + 1$ replaced by $n + p$ and $n - 1$ replaced by $n - p$, where p is any positive integer. Furthermore, it is possible to repeat the same process on the sequence of S_n' values obtained using Eq. (5.29). In some cases the process can be repeated several times further and improvements can be obtained.

The second method, Euler's transformation, is an important method for accelerating the convergence of series; it is particularly appropriate for alternating series. It can be used on some nonconvergent as well as convergent series. It is powerful but limited to alternating series. However, a transformation is possible to convert a series of positive terms to an alternating series. See Press et al. (1986).

A third method applies a correction of the known form of the remainder. This method has the problem of knowing the functional form of the remainder. It can also be used with the Aitken's δ^2 process to further accelerate the evaluation of the sum. See Example 5.2 for an illustration of its use.

The final method of accelerating series compares the present series with a closely related one with a known value. This is illustrated by Example 5.4.

Example 5.1 Use the Aitken δ^2 process three times in succession to evaluate the series

$$S = \sum_{i=1}^{\infty} \frac{1}{i^2} \tag{5.30}$$

SOLUTION Results of repeated use of Eq. (5.29) are given in Table 5.9. The exact value is $\pi^2/6 = 1.64493407$. The first column of Table 5.9 gives the number of terms and the second column gives the running sum for n terms. The third column is the first Aitken process values and the other columns are for second- and third-order Aitken processes. The acceleration of the series is not very effective. The

Table 5.9 Example 5.1 showing of acceleration of Σi^{-2} using Aitken's δ^2 process

n	S_n	S_n'	S_n''	S_n'''
1	1.000000			
2	1.250000	1.450000		
3	1.361111	1.503968	1.575465	
4	1.423611	1.534722	1.590296	1.618209
5	1.463611	1.554520	1.599981	1.622752
6	1.491389	1.568312	1.606776	1.626025
7	1.511797	1.578464	1.611798	1.628473
8	1.527422	1.586246	1.615658	1.630368
9	1.539768	1.592399	1.618716	1.631876
10	1.549768	1.597387	1.621197	1.633103
20	1.596163	1.620553	1.632749	1.638846
40	1.620244	1.632590	1.638762	1.641849
100	1.634984	1.639959	1.642447	1.643696
200	1.639947	1.642440	1.643687	1.643881
500	1.642936	1.643935	1.644434	1.644433
1000	1.643935	1.644434	1.644669	1.644674

third Aitken process is as accurate at $n = 10$ as the original sum after $n = 40$. After 1000 terms, the Aitken values are accurate to four significant figures while the direct evaluation of the sum is accurate only to three.

Example 5.2 Use the linear correction for the remainder as implied by Eq. (5.11) and then use Aitken's δ^2 process for the summation given in Example 5.1.

SOLUTION The summation S_n is as found in Example 5.1 and shown in column two of Table 5.10. The correction based on Eq. (5.11) is

$$S_{L,n} = S_n + \frac{1}{n}$$

and is given in the third column of Table 5.10; these values are actually much better than those given in Table 5.9, which is for repeated use of Aitken's process on S_n. Aitken's process can also be used to improve the $S_{L,n}$ values, since the value for $n = 250$ for the second process is equivalent to the $n = 750$ value of the linear correction. By $n = 250$, the combination of the linear correction and Aitken's process yields values accurate to six significant figures, which is clearly much better than given in Table 5.9.

Example 5.3 Use the Aitken δ^2 process to evaluate the series

$$S = \sum_{i=1}^{\infty} (-1)^{i+1} \frac{1}{i^2} \tag{5.31}$$

Use up to the third order.

Table 5.10 Example 5.2 showing of acceleration of Σi^{-2} using Aitken's δ^2 process upon linear corrected sum

n	S_n	$S_{L,n}$	$S'_{L,n}$	$S''_{L,n}$
1	1.000000	2.000000		
2	1.250000	2.250000	2.097826	
3	1.361111	1.861111	1.718835	1.678945
4	1.423611	1.756944	1.682744	1.657809
5	1.463611	1.713611	1.667997	1.652724
10	1.549768	1.660879	1.650257	1.646713
20	1.596163	1.648795	1.646221	1.645363
40	1.620244	1.645885	1.645251	1.645040
80	1.632512	1.645170	1.645013	1.644960
100	1.634984	1.645085	1.644984	1.644951
150	1.638290	1.645001	1.644956	1.644942
250	1.640942	1.644958	1.644942	1.644937
500	1.642936	1.644940	1.644936	1.644933
750	1.643602	1.644937	1.644935	1.644935

Table 5.11 Example 5.3 showing acceleration of sum of $(-1)^{n+1}1/n^2$ using Aitken's δ^2 process

n	S_n	S'_n	S''_n	S'''_n
1	1.000000			
2	0.750000	0.826923		
3	0.861111	0.821111	0.822537	
4	0.798611	0.823001	0.822447	0.822468
5	0.838611	0.822218	0.822474	0.822467
6	0.810833	0.822598	0.822464	0.822467
7	0.831241	0.822392	0.822468	0.822467
8	0.815616	0.822513	0.822466	0.822467
9	0.827962	0.822437	0.822467	0.822467
10	0.817962	0.822487	0.822467	0.822467
20	0.821279	0.822468	0.822467	0.822467
30	0.821930	0.822467	0.822467	0.822467

SOLUTION Results of repeated using of Eq. (5.29) are given in Table 5.11. The exact value is $\pi^2/12 = 0.82246703$. By the tenth term in the summation, the third Aitken's δ^2 process is giving six correct significant figures, while the direct .summation takes about 1000 terms to reach this accuracy. Evidently alternating series can converge much more rapidly than the nonalternating series; this conclusion applies both for the direct evaluation and the use of the Aitken's δ^2 process.

Example 5.4 Use the known value of $\Sigma \, i^{-2}$ to evaluate

$$S = \sum_{i=1}^{\infty} \frac{1}{i^2 + 10} \qquad (5.32)$$

SOLUTION The summation minus the known summation can be written as

$$S - \frac{\pi^2}{6} = \sum_{i=1}^{\infty} \frac{1}{i^2 + 10} - \sum_{i=1}^{\infty} \frac{1}{i^2} = \sum_{i=1}^{\infty} \frac{-10}{(i^2 + 10)i^2}$$

This series converges much more rapidly than the original one. After the value, 0.11982, is found, the series S is given by

$$S = \frac{\pi^2}{6} - 0.11982 = 1.52511$$

About 50 terms are needed to obtain this value, but as few as 18 terms will give about five significant figures; this is much faster than direct integration.

5.6 NUMERICAL INTEGRATION

The emphasis in this book is on obtaining analytical expressions for various transient heat conduction problems through the use of the GF method. This almost always involves integrals over time and frequently over space as well. These integrals can be performed analytically in many, but not all cases. Numerical integration, however, can always be performed. With numerical integration, some of the generality of the analytical solutions is lost, but solutions can be obtained that might not be possible with exact integration. For engineering problems, numerical values are usually the final desired results, rather than infinite series expressions; in such cases, numerical evaluation of integrals may be equivalent to direct integration.

In this section, numerical integration over time is illustrated and a similar technique can be used over space. There are many integration schemes that are possible, such as trapezoidal and Simpson's rules. These methods involve evaluation of the integrand at the extreme values of the argument τ, such as at $\tau = t - \Delta t$ and t. In some cases the GF method integrand cannot be evaluated (it goes to infinity) as τ goes to t. To avoid this difficulty, techniques such as the rectangular rule, Newton-Cotes open-ended formulas, and Gaussian quadrature could be used. (See Abramowitz and Stegun, 1964, p. 887.) This difficulty can be avoided, however, by considering $t - \tau$ values slightly larger than zero and by considering several small regions. The method suggested by Press et al. (1986) will be given; this method builds on the extended trapezoidal rule.

The method given below has a number of advantages. One of these is that the method can be efficiently used to obtain values of any desired accuracy. Another is that it can be programmed readily. Such a computer program is suggested. It is built upon the concepts of Press et al. (1986).

The algorithm in the program is based on the extended trapezoidal rule,

$$\int_a^b f(\tau) \, d\tau = \Delta t \left[\frac{1}{2} f_1 + f_2 + f_3 + \cdots + f_{N-1} + \frac{1}{2} f_N \right]$$
$$+ O\left[\frac{(b - a)^3 f''}{N^2} \right] \tag{5.33}$$

where Δt is equal to

$$\Delta t = \frac{b - a}{N - 1} \qquad (5.34)$$

and N is an integer and $f_i = f[a + (i - 1)\Delta t]$. There are several reasons why Eq. (5.33) is a powerful building block. One is that, for a fixed function $f(t)$ to be integrated between the fixed limits a and b, one can double the number of intervals in the extended trapezoidal rule without losing the benefit of previous work. Another is that the integral value can be improved by using a combination of the values for N and $2N$ terms, denoted S_n and S_{2N}, by using

$$S = \frac{4}{3} S_{2N} - \frac{1}{3} S_N, \qquad (5.35)$$

which yields values of the integral S that are order N^{-4}, which is very good. Furthermore, it can be used to write a program to provide specified accuracy. Below is some FORTRAN coding for accomplishing the above objectives. For more details, see Press et al. (1986).

Example 5.5 The fundamental heat conduction solution, $K(z, t - \tau)$ [see Eq. (1.4)], is to be integrated over τ from $\tau = 0$ to 1 for the values $z = 0$, 0.01, 0.1, 1, 2, and 3. The thermal diffusivity α is set equal to 1.0.

SOLUTION This problem was solved using the extended trapezoidal rule and its improvements discussed above. The FORTRAN coding to accomplish this is given in Table 5.12. FUNC(Z) contains the integrand of $(4\pi u)^{-1/2} \exp(-z^2/4u)$ where u is $t - \tau$. Because $(4\pi u)^{-1/2}$ is not defined at $u = 0$, the integration cannot start at $u = 0$. Instead the integration is started at the small value of 2×10^{-8}. For smaller values than that (or some other small value), there are at least two possibilities. One is that there is negligible contribution, as for $z = 0.01$ and larger in Table 5.13. The other is that the integrand at $z = 0$ reduces to the simple expression, $(4\pi u)^{-1/2}$, which can be integrated analytically to $(t/\pi)^{1/2}$. Hence, $(2\text{E-}8/\pi)^{1/2} = 0.000080$ should be added to the $z = 0$ column of Table 5.13, resulting in the $z = 0$ value of 0.56423, which is higher than the exact value of 0.56419. Because of the rapid change in the integrand near $u = 0$ for the $z = 0$ case (somewhat larger u values for $z > 0$), it may be necessary to obtain the required accuracy to break the integration interval into parts. This can be determined by subdividing an interval into two parts and noting if the same result is obtained by the whole interval and the sum of the values for the two parts. The exact values for this example are given by ierfc $[z/(4\alpha t)^{1/2}]$, where α and t are both 1.

Example 5.6 Evaluate the expression

$$T = \int_0^t \{1 - u^{1/2}[\pi^{-1/2} - \text{ierfc}\,(u^{-1/2})]^2\}\,du$$

Table 5.12 **Program for evaluating an integral from *a* to *b*. The evaluated function is the fundamental heat conduction solution, $K(x, x', t - \tau)$**

```
      PROGRAM TRAPIN
C     A AND B ARE INPUT QUANTITIES:file name is INT-TR.FOR
C     VARIOUS PARTS OF PROGRAM ARE ADAPTED FROM "NUMERICAL RECIPES"
C     BY W H PRESS, B P FLANNERY, S A TEUKOLSKY AND VETTERLING
C     CAMBRIDGE UNIVERSITY PRESS, NY, 1986
      A = 0.001
      B = 1.0
      CALL QSIMP(A,B,S)
      WRITE(*,*)' VALUE OF THE INTEGRAL'
      WRITE(*,5)S
   20 CONTINUE
    5 FORMAT(1X,4F13.6)
      END
      SUBROUTINE QSIMP(A,B,S)
C     RETURNS S AS THE INTEGRAL OF THE FUNCTION FROM A TO B.
C     PARAMETER EPS IS SET TO RETURN THE DESIRED ACCURACY.
C     DOUBLE PRECISION IN ENTIRE PROGRAM MAY BE NEEDED IF EPS IS
C     MUCH SMALLER
C     KMAX IS THE MAXIMUM ALLOWED NUMBER OF ITERATIONS
      EPS = 1.0E - 5
      KMAX = 17
      OST = - 1.0E25
      OS = - 1.0E25
      K = 1
C     BELOW WRITE CAN BE OMITTED IF DESIRED
      WRITE(*,*)'      K          S'
   50 CALL TRAPZD(A,B,ST,K)
C     BELOW STATEMENT IMPROVES THE ACCURACY BY COMBINING
C     S FOR 2N AND S FOR N STEPS
      S = (4.0*ST - OST)/3.0
      IF(ABS(S - OS) .LT. EPS*ABS(OS)) RETURN
      OS = S
      OST = ST
C     BELOW WRITE CAN BE OMITTED, IF DESIRED
      WRITE(*,6)K,S
      K = K + 1
  100 IF(K .LT. KMAX)GOTO 50
      PAUSE 'TOO MANY STEPS'
    6 FORMAT(I6,F10.5)
      END
      FUNCTION FUNC(Z)
C     USER SUPPLIED INTEGRAND
      PI = 4.0*ATAN(1.0)
      X = 0.0
      EE = SQRT(4.0*PI*Z)
      EE2 = EXP( - X*X/(4.0*Z))
      FUNC = EE2/EE
C     END USER SUPPLIED INTEGRAND
      RETURN
      END
      SUBROUTINE TRAPZD(A,B,S,N)
C     COMPUTES THE N'TH STAGE OF REFINEMENT OF AN EXTENDED TRAPEZOIDAL
```

(Continued)

Table 5.12 Program for evaluating an integral from a to b. The evaluated function is the fundamental heat conduction solution, $K(x, x', t - \tau)$ (Continued)

```
C     RULE.   S SHOULD NOT BE MODIFIED BETWEEN CALLS. S IS OUTPUT
      IF(N .EQ. 1) THEN
          S = 0.5*(B – A)*(FUNC(A) + FUNC(B))
          IT = 1
      ELSE
          TNM = IT
          DEL = (B – A)/TNM
          X = A + 0.5*DEL
          SUM = 0.0
          DO 100 J = 1,IT
              SUM = SUM + FUNC(X)
              X = X + DEL
100       CONTINUE
          S = 0.5*(S + (B – A)*SUM/TNM)
          IT = 2*IT
      ENDIF
      RETURN
      END
```

Table 5.13 Results for Example 5.6 using extended trapezoidal program for integration of fundamental heat conduction solution from $t = 0$ to 1 with $\alpha = 1$ and for various z values

Range of t	Values of the integral					
	$z = 0.0$	$z = 0.01$	$z = 0.1$	$z = 1.0$	$z = 2.0$	$z = 3.0$
2E-8, 2E-6	0.00072	0.0	0.0	0.0	0.0	0.0
2E-6, 2E-4	0.00718	0.00396	0.0	0.0	0.0	0.0
2E-4, 0.02	0.07183	0.07094	0.03956	0.0	0.0	0.0
0.02, 1.0	0.48442	0.48433	0.47660	0.19964	0.05025	0.008623
Total	0.56415	0.55923	0.51616	0.19964	0.05025	0.008623
Exact	0.56419	0.55920	0.51560	0.19964	0.05025	0.008623

This is the average temperature over a square region on a semi-infinite body; the square region is heated by a constant heat flux.

SOLUTION This is a very difficult expression to evaluate analytically, but it can be evaluated readily by using numerical methods. It can be done by using the same FORTRAN program given in Table 5.12 but the function subroutine replaced by

```
      FUNCTION FUNC(Z)
C     USER SUPPLIED INTEGRAND
      PI = 4.0*ATAN(1.0)
      X = 0.0
      EE = SQRT(PI*Z)
      ARG = 1.0D + 0/SQRT(Z)
```

```
      CALL ERFCP(ARG,PERF,PERFC,PIERFC)
      FUNC=(1.0-SQRT(Z)*((1.0/SQRT(PI))-PIERFC))**2
      FUNC=FUNC/EE
C     END USER SUPPLIED INTEGRAND
      RETURN
      END
```

and where the ERFCP subroutine is found in Appendix E. The resulting value for the limits of 0.0001 and 1.0 is 0.610132. Using the approximation of ierfc $(z) = -z + \pi^{-1/2}$ for very small z values results in the integral value from 0 to E-4 of

$$\pi^{-1/2}\left[2u^{1/2} - 2u + \frac{2}{3}u^{3/2}\right]\Bigg|_{0.0001} = 0.01117$$

Then the integral becomes 0.62130, which compares very well with 0.621320 found another way.

REFERENCES

Abramowitz, M., and Stegun, I. A., eds., 1964, *Handbook of Mathematical Functions with Formulas, Graphs, and Mathematical Tables*, National Bureau of Standards, Series 55, U.S. Government Printing Office, Washington, D.C.

Carslaw, H. S., and Jaeger, J. C., 1959, *Conduction of Heat in Solids*, 2d ed., Oxford, London.

Press, W. H., Flannery, B. P., Teukolsky, S. A., and Vetterling, W. T., 1986, *Numerical Recipes, The Art of Scientific Computing*, Cambridge, New York.

PROBLEMS

5.1 Write a computer program to evaluate the sum given by Eq. (5.11) with $M = 1$. Evaluate the sum for the maximum number of terms equal to 100, 1000, 10,000, 100,000, and 1,000,000. Is it necessary to use double precision (that is about 15 significant figures rather than about 8)? Investigate the errors and verify that the errors are as indicated in Eq. (5.11).

5.2 Verify integral 2 of Table F.3, Appendix F, by using integration by parts.

5.3 Write computer programs for the $X22$ GF given by Eqs. (X22.1) and (X22.3) and verify that they give the same values for various x/L values such as 0, 0.5, and 1 and $\alpha(t - \tau)/L^2 = 0.01, 0.1, 1$, and 4. What conclusions can you draw regarding the number of terms?

5.4 Write a computer program to evaluate erf (x), erfc (x), ierfc (x) and i^2 erfc (x) and find values for steps of 0.1 for x values from -1.0 to 3.0. One of the subroutines in Appendix E can be used to evaluate erf (x).

5.5 Problem $X22B10T0$ using the long time GF has the solution given by Eq. (6.71). Also solve this problem using the time-partitioning concept.

5.6 Evaluate the sum

$$S = \sum_{i=1}^{\infty} \frac{i^2 + B}{i^2 + B^2 + B}\frac{1}{i^2}$$

by direct summation for $B = 1$.

5.7 Evaluate the sum in Problem 5.6 by using the Aitken δ^2 process.

5.8 Evaluate the sum in Problem 5.6 taking advantage of its known remainder for large values of i.

5.9 Evaluate the sum in Problem 5.6 by using the known value of $\Sigma\, i^{-2}$.

5.10 Evaluate the sum

$$S = \sum_{i=2}^{\infty} \frac{1}{i^2 - 1}$$

The value is known to be 0.75. Find an efficient way to evaluate it.

5.11 Evaluate the sum

$$S = \sum_{i=1}^{\infty} \frac{1}{i^4}$$

The value is known to be $\pi^4/90$. Find an efficient way to evaluate it.

5.12 Evaluate the sum at $x = 0.5$ given by sum 2 of Appendix SE. Find an efficient way to evaluate it and compare your value with the exact value.

5.13 Evaluate the double summation of Eq. (6.128) for $a = q_0 = b = 1$, $x = a_1 = 0.1$, and $y = 0$. Let t go to infinity. Select an efficient method of evaluation.

5.14 Evaluate the double summation of Eq. (6.128) for $a = q_0 = b = 1$, $x = a_1 = 0.1$, and $y = 0.5$. Let $\alpha t/a^2$ be equal to 0.005 and 0.01. What does your result say regarding time partitioning?

5.13 Evaluate numerically Eq. (6.142) for $q_0 = k = \alpha = x = y = t = 1$.

5.14 Evaluate numerically Eq. (6.142) for $q_0 = k = \alpha = x = y = 1$ and $t = 0.5$.

5.15 Evaluate numerically Eq. (6.151) for $p = X = 1$.

CHAPTER SIX

RECTANGULAR COORDINATES

6.1 INTRODUCTION

This chapter is concerned with heat conduction in bodies described by rectangular coordinates. Complete examples are included that demonstrate strategies for evaluating the integrals in the Green's function solution equation (GFSE).

An important feature of the Green's function (GF) solution method is the ability to simply write down the temperature in integral form. Once the problem is properly defined, one can jump to the solution and gain insight into the problem. For example, one can also immediately write down the alternative Green's function solution, and then the better form of the solution can be selected for evaluation. The student can concentrate on translating a physical heat transfer situation into a boundary value problem without getting lost in the details of the solution. There is a sense of accomplishment associated with jumping to the solution that can be a valuable part of the learning process. After the integral form is written down, the integrals can be examined. If they are familiar, the solution can be completed easily. If the integrals are unfamiliar they may be available in integral tables, approximate forms may be substituted, or finally numerical integration will always yield an answer.

The examples in this chapter are concerned with only one nonhomogeneous term at a time. The nonhomogeneous term may be the initial condition, the volume energy generation, or the boundary condition. Practical situations often involve two or more nonhomogeneous terms, but because the GF solution equation is the sum of the contributions from the various nonhomogeneous terms, the temperature resulting from initial conditions, boundary conditions, and volume energy generation can simply be added together for the complete solution.

One-dimensional geometries are emphasized in this chapter and the one-dimensional GFSE is given in Section 6.2. Semi-infinite bodies are discussed in Section 6.3. Flat plates are discussed in Section 6.4 through 6.6. Time partitioning is demonstrated for flat plates in Section 6.6. Some two-dimensional cases are discussed in Sections 6.7 and 6.8, and some steady-state cases are discussed in Section 6.9.

6.2 ONE-DIMENSIONAL GREEN'S FUNCTIONS SOLUTION EQUATION

The heat conduction equation for homogeneous one-dimensional bodies in the rectangular coordinate system is

$$\frac{\partial^2 T}{\partial x^2} + \frac{1}{k} g(x, t) = \frac{1}{\alpha} \frac{\partial T}{\partial t} \tag{6.1}$$

with initial condition

$$T(x, 0) = F(x) \tag{6.2}$$

and with boundary conditions

$$k_i \left. \frac{\partial T}{\partial n_i} \right|_{x_i} + h_i T|_{x_i} = f_i(t) - (\rho c b)_i \left. \frac{\partial T}{\partial t} \right|_{x_i} \tag{6.3}$$

where n_i is an outward normal from the body at the boundary, and x_i represents the two boundaries ($i = 1, 2$). Equation (6.3) represents five different kinds of boundary conditions by the choice of k_i, h_i, f_i, and b_i. These boundary conditions are discussed in detail in Chapter 2.

The solution of the temperature problem given in Eqs. (6.1)–(6.3) is given by the GFSE for one-dimensional rectangular coordinates (refer to Section 3.2 for a derivation)

$$T(x, t) = \int_{x'=0}^{L} G(x, t|x', 0) F(x') \, dx' \qquad \text{(for the initial condition)}$$

$$+ \alpha \sum_{i=1}^{s} \left[\frac{(\rho c b)_i}{k_i} G(x, t|x', 0) F(x') \right]_{x'=x_i} \qquad \begin{array}{l} \text{(for boundary conditions} \\ \text{of the fourth and fifth} \\ \text{kinds only)} \end{array}$$

$$+ \int_{\tau=0}^{t} \int_{x'=0}^{L} \frac{\alpha}{k} G(x, t|x', \tau) g(x', \tau) \, dx' \, d\tau \qquad \begin{array}{l} \text{(for volume} \\ \text{energy generation)} \end{array}$$

$$+ \alpha \int_{\tau=0}^{t} d\tau \sum_{i=1}^{2} \left[\frac{f_i(\tau)}{k_i} G(x, t|x_i, \tau) \right] \qquad \begin{array}{l} \text{(for boundary conditions} \\ \text{of the second through} \\ \text{fifth kinds)} \end{array}$$

$$- \alpha \int_{\tau=0}^{t} d\tau \sum_{i=1}^{2} \left[f_i(\tau) \left. \frac{\partial G}{\partial n_i'} \right|_{x'=x_i} \right] \qquad \begin{array}{l} \text{(for boundary conditions} \\ \text{of the first kind only)} \end{array} \tag{6.4}$$

where $G(x, t|x', \tau)$ is the GF. For each different set of boundary conditions there is a different GF that must be used in the GFSE.

6.3 SEMI-INFINITE ONE-DIMENSIONAL BODIES

In this section, the cases under consideration are semi-infinite bodies denoted by $X I0$, $I = 1, 2, 3, 4, 5$. The GFs for infinite and semi-infinite bodies are listed in Table

Table 6.1 GF for infinite and semi-infinite bodies

$$G_{XI0}(x, t|x', \tau) = [4\pi\alpha(t - \tau)]^{-1/2} \left\{ exp \frac{-(x - x')^2}{4\alpha(t - \tau)} + M \, exp \frac{-(x + x')^2}{4\alpha(t - \tau)} \right\}$$

$$- \frac{1}{L} \left\{ M \, D_1 \, L \, ER(x + x', t - \tau, D_1) \right.$$

$$+ \frac{E_1}{(1 - 4 \, B_1 C_1)^{1/2}} [S_2 \, ER(x + x', t - \tau, S_2)$$

$$\left. - S_1 \, ER(x + x', t - \tau, S_1)] \right\}$$

	M	$D_1 L$	E_1
$X00$	0	0	0
$X10$	-1	0	0
$X20$	1	0	0
$X30$	1	B_1	0
$X40$	-1	$(C_1)^{-1}$	0
$X50$	-1	0	1

where $ER(x, t, D) = exp \, (Dx + D^2\alpha t) \, erfc \left[\frac{x}{(4\alpha t)^{1/2}} + D(\alpha t)^{1/2} \right]$

$$S_1 = \frac{L}{2C_1} [1 - (1 - 4 \, B_1 \, C_1)^{1/2}], \, C_1 < \frac{1}{4} B_1$$

$$S_2 = \frac{L}{2C_1} [1 + (1 - 4 \, B_1 \, C_1)^{1/2}]$$

$$B_1 = \frac{hL}{k}$$

$$C_1 = \frac{\rho cb}{\rho cL}$$

and, L is a reference length that cancels out.

6.1. A complete listing of rectangular-coordinate GFs, including certain derivatives, integrals, and approximations is given in Appendix X.

For the semi-infinite cases, the GFs have only one form, do not involve infinite series, and are mathematically well behaved everywhere except at the point $x - x' = 0$ and $t - \tau = 0$, where every GF approaches a Dirac delta function. The temperatures calculated by integrating these GFs are mathematically well behaved for any location x and for $t > 0$.

6.3.1 Initial Conditions

For the case of *spatially uniform* initial conditions in semi-infinite bodies, the appropriate integrals in the GFSE, Eq. (6.4), are known in closed form. The resulting temperature expressions for homogeneous boundary conditions are listed in Table 6.2

Table 6.2 Temperatures in a semi-infinite body for uniform initial temperature of T_i for cases $XI0B0T1$, $I = 1, 2, 3$, and cases $XI0B0T10$, and $XI0B0T01$ for $I = 4, 5$

$$T(x, t) = T_i \left(1 - \frac{1}{2} (1 - M) \text{ erfc} \left[\frac{x}{(4\alpha t)^{1/2}} \right] \right.$$

$$- M(D_0 - F_0) ER(x, t, D_1) + \frac{1}{R} \left\{ \left[E_0 - \frac{F_1}{2} (1 - R) \right] ER(x, t, S_1) \right.$$

$$\left. \left. - \left[E_0 - \frac{F_1}{2} (1 + R) \right] ER(x, t, S_2) \right\} \right), \quad R = [1 - 4 B_1 C_1]^{1/2}$$

Number	M	D_0	$D_1 L$	E_0	F_0	F_1
$X10B0T1$	-1	0	0	0	0	0
$X20B0T1$	1	0	0	0	0	0
$X30B0T1$	-1	1	B_1	0	0	0
$X40B0T10$	1	1	C_1^{-1}	0	1	0
$X40B0T01$	1	1	C_1^{-1}	0	0	0
$X50B0T10$	-1	0	0	0	0	1
$X50B0T01$	-1	0	0	1	0	0

Note: ER($\cdot$) is given in Table 6.1.

in compact form. The integrals that were used to create Table 6.2 are listed in the integral tables in Appendix F.

In Table 6.2, two cases involve a surface film of high conductivity, numbered $X40$ and $X50$. The notation for the initial condition $T01$ in Table 6.2 refers to a zero initial temperature in the film and a uniform initial temperature in the body. Conversely the notation $T10$ refers to a uniform initial temperature in the film and a zero initial temperature in the body. If both the film and the body have the same uniform initial temperature, the problem can always be formulated with no contribution from the initial condition by defining a new temperature variable, $T - T_0$, where T_0 is the initial temperature.

Semi-infinite bodies with *spatially varying* initial conditions are now considered. Consider the initial temperature distribution of

$$F(x') = \begin{cases} T_0 & a < x' < b \\ 0 & \text{otherwise} \end{cases} \quad (6.5)$$

The GF for boundary conditions of type 1 or 2 can be written in terms of the fundamental heat conduction solution $K(\cdot)$,

$$G_{XI0}(x, t|, x', 0) = K(x - x', t) + (-1)^I K(x + x', t) \quad (6.6)$$

where $I = 1$ or 2. The solution for the temperature is obtained by substituting Eq. (6.6) in Eq. (6.4), the GFSE, to give (case $XI0B0T5$; $I = 1$ or 2):

$$T(x, t) = \frac{T_0}{2} \left(\text{erfc} \left[\frac{-x + a}{(4\alpha t)^{1/2}} \right] - \text{erfc} \left[\frac{-x + b}{(4\alpha t)^{1/2}} \right] \right.$$

$$\left. + (-1)^I \left\{ \text{erfc} \left[\frac{x + a}{(4\alpha t)^{1/2}} \right] - \text{erfc} \left[\frac{x + b}{(4\alpha t)^{1/2}} \right] \right\} \right) \tag{6.7}$$

Next consider the initial temperature distribution of a linear function of x' over part of the body,

$$F(x') = \begin{cases} T_0 \dfrac{x'}{L} & \text{for } a < x' < b \\ 0 & \text{elsewhere} \end{cases} \tag{6.8}$$

The length L can have any desired significance; it is only present to make Eq. (6.8) dimensionally consistent. The integrals in the GF equation can then be evaluated using Tables 5.1 and 5.2 with z replaced by x', and $t - \tau$ replaced by t. The solution is (case *XI0B0T2*; $I = 1$ or 2; use Table 5.3)

$$T(x, t) = T_0 \left[\frac{x}{2L} \left(\text{erfc} \left[\frac{x - b}{(4\alpha t)^{1/2}} \right] - \text{erfc} \left[\frac{x - a}{(4\alpha t)^{1/2}} \right] \right. \right.$$

$$\left. + (-1)^I \left\{ \text{erfc} \left[\frac{x + b}{(4\alpha t)^{1/2}} \right] - \text{erfc} \left[\frac{x + a}{(4\alpha t)^{1/2}} \right] \right\} \right)$$

$$+ \frac{2\alpha t}{L} \{ K(x - a, t) - K(x - b, t)$$

$$+ (-1)^I [K(x + a, t) - K(x + b, t)] \} \Bigg] \tag{6.9}$$

For boundary condition of the first kind ($I = 1$), consider the special case of $a = 0$ and $b \to \infty$, which is for a linear initial temperature $F(x') = T_0 x'/L$ over the entire body, $0 \leq x \leq \infty$. In this case the temperature given by Eq. (6.9) reduces to

$$T(x, t) = T_0 \frac{x}{L} \tag{6.10}$$

This is a time-independent solution for the case denoted *X10B0T2*. For the boundary condition of the second kind ($I = 2$) with $a = 0$ and $b \to \infty$, Eq. (6.9) gives (case *X20B0T2*)

$$T(x, t) = T_0 \left\{ \frac{x}{L} \text{erfc} \left[\frac{x}{(4\alpha t)^{1/2}} \right] + \frac{4\alpha t}{L} K(x - b, t) \right\} \tag{6.11}$$

This solution is always transient and never reaches a steady state. The transient deviation from the initial straight-line temperature distribution begins at $x = 0$ and spreads to larger x values as time increases.

6.3.2 Boundary Conditions

Temperature expressions resulting from time-invariant boundary conditions are listed in Table 6.3 for five kinds of boundary conditions. These temperature expressions were found by evaluating the integrals in the GFSE. Two mathematical functions that appear in Table 6.3 are erfc and ierfc, which are the complementary error function and the integral of the complementary error function, respectively. Table 6.4 lists some numerical values of $\mathrm{erf}[(4z)^{-1/2}]$, $\mathrm{erfc}[(4z)^{-1/2}]$ and $\mathrm{ierfc}[(4z)^{-1/2}]$; the function z could be dimensionless time $\alpha t/x^2$.

The following examples demonstrate the use of Table 6.1 to find the GF and demonstrate strategies for finding the integrals that occur for various nonhomogeneous boundary conditions.

Example 6.1: Semi-infinite body with specified surface temperature—X10B-T0 case. Find the temperature distribution in the semi-infinite body with specified surface temperature $f(t)$ and with zero initial condition. The volume energy generation is zero.

Table 6.3 Temperatures for semi-infinite bodies for constant source term at $x = 0$; case $X I0B1T0$, $I = 1, 2, 3$, and $X I0B1T00$ for $I = 4$ and 5

$$
T(x, t) = H_0 (1 + M) \left(\frac{t}{k\rho c}\right)^{1/2} \mathrm{ierfc} \left[\frac{x}{(4\alpha t)^{1/2}}\right]
$$

$$
+ \left(K_0 - \frac{MD_0}{kD_1} + \frac{E_1}{h}\right) \mathrm{erfc} \left[\frac{x}{(4\alpha t)^{1/2}}\right]
$$

$$
+ \frac{MD_0}{kD_1} ER\,(x, t, D_1) + \frac{E_1}{kR} \left[\frac{1}{S_2'} ER\,(x, t, S_2') - \frac{1}{S_1'} ER\,(x, t, S_1')\right]
$$

Number	M	H_0	K_0	D_0	D_1	E_1
$X10B1T0$	-1	0	T_0	0	0	0
$X20B1T0$	1	q_0	0	0	0	0
$X30B1T0$	-1	0	0	$q_0 + hT_\infty$	h/k	0
$X40B1T00$	1	q_0	0	q_0	$\rho c/(\rho cb)_1$	1
$X50B1T00$	-1	$hT_\infty + q_0$	0	0	0	$hT_\infty + q_0$

$$
S_1' \equiv \frac{\rho c}{2(\rho cb)_1} [1 - R] \qquad S_2' \equiv \frac{\rho c}{2(\rho cb)_1} [1 + R]
$$

$$
R \equiv \left[1 - 4\frac{h}{k}\frac{(\rho cb)_1}{\rho c}\right]^{1/2}
$$

$$
ER\,(x, t, D) = \exp\,[Dx + D^2\alpha t]\,\mathrm{erfc} \left[\frac{x}{(4\alpha t)^{1/2}} + D(\alpha t)^{1/2}\right]
$$

Table 6.4 Table of solutions for semi-infinite bodies

z	erf $[(4z)^{1/2}]$	erfc $[(4z)^{-1/2}]$	$(4z)^{1/2}$ ierfc $[(4z)^{-1/2}]$
0.01	1.000000	0.000000	0.000000
0.02	0.999999	0.000001	0.000000
0.03	0.999955	0.000045	0.000002
0.04	0.999593	0.000407	0.000029
0.05	0.998435	0.001565	0.000135
0.06	0.996108	0.003892	0.000393
0.07	0.992474	0.007526	0.000867
0.08	0.987581	0.012419	0.001603
0.09	0.981578	0.018422	0.002625
0.10	0.974653	0.025347	0.003943
0.20	0.886154	0.113846	0.030732
0.30	0.803294	0.196706	0.071893
0.40	0.736448	0.263552	0.118437
0.50	0.682689	0.317311	0.166631
0.60	0.638690	0.361310	0.214891
0.70	0.601975	0.398025	0.262515
0.80	0.570805	0.429195	0.309190
0.90	0.543943	0.456057	0.354791
1.00	0.520500	0.479500	0.399282
2.00	0.382925	0.617075	0.791186
3.00	0.316909	0.683091	1.115053
4.00	0.276326	0.723674	1.396355
5.00	0.248170	0.751830	1.648248
6.00	0.227170	0.772830	1.878325
7.00	0.210732	0.789268	2.091402
8.00	0.197413	0.802587	2.290758
9.00	0.186336	0.813664	2.478736
10.00	0.176937	0.823063	2.657085
20.00	0.125633	0.874367	4.109212
30.00	0.102721	0.897279	5.231819
40.00	0.089021	0.910979	6.181053
50.00	0.079656	0.920344	7.018707
60.00	0.072736	0.927264	7.776780
70.00	0.067353	0.932647	8.474394
80.00	0.063013	0.936987	9.124053
90.00	0.059416	0.940584	9.734466
100.00	0.056372	0.943628	10.311989
200.00	0.039878	0.960122	14.977634
300.00	0.032564	0.967436	18.560385
400.00	0.028204	0.971796	21.581687
500.00	0.025227	0.974773	24.243940
600.00	0.023030	0.976970	26.651048
700.00	0.021322	0.978678	28.864768
800.00	0.019945	0.980055	30.925355
900.00	0.018805	0.981195	32.860778
1000.00	0.017840	0.982160	34.691403
2000.00	0.012615	0.987385	49.468958
3000.00	0.010300	0.989700	60.809023
4000.00	0.008920	0.991080	70.369425
5000.00	0.007979	0.992021	78.792445

(Continued)

Table 6.4 Table of solutions for semi-infinite bodies *(Continued)*

6000.00	0.007284	0.992716	86.407516
7000.00	0.006743	0.993257	93.410346
8000.00	0.006308	0.993692	99.928455
9000.00	0.005947	0.994053	106.050420
10000.00	0.005642	0.994358	111.840738

SOLUTION This is the $X10B\text{-}T0$ geometry shown in Fig. 6.1. The GFSE gives the temperature as

$$T(x, t) = \alpha \int_{\tau=0}^{t} f(\tau) \left. \frac{\partial G_{X10}}{\partial x'} \right|_{x'=0} d\tau \tag{6.12}$$

The GF G_{X10} is found from Table 6.1 by choosing $M = -1$, $D_1 = 0$, and $E_1 = 0$:

$$G_{X10}(x, t|x', \tau) = \frac{1}{[4\pi\alpha(t - \tau)]^{1/2}} \left\{ \exp\left[\frac{-(x - x')^2}{4\alpha(t - \tau)} \right] - \exp\left[\frac{-(x + x')^2}{4\alpha(t - \tau)} \right] \right\} \tag{6.13}$$

The derivative of the GF with respect to x' is required here in the form $\partial/\partial x' = -\partial/\partial n_i$ at $x' = 0$. The derivative of G_{X10} is given in Appendix X as

$$\left. \frac{\partial G_{X10}}{\partial x'} \right|_{x'=0} = \frac{x}{(4\pi)^{1/2}[\alpha(t - \tau)]^{3/2}} \exp\left[\frac{-x^2}{4\alpha(t - \tau)} \right]$$

$$= \frac{x}{\alpha(t - \tau)} K(x, t - \tau) \tag{6.14}$$

where $K(\cdot)$ is the fundamental heat conduction solution. The temperature solution can then be written as

$$T(x, t) = \alpha \int_{\tau=0}^{t} f(\tau) \frac{x}{\alpha(t - \tau)} K(x, t - \tau) d\tau \tag{6.15}$$

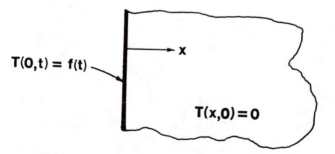

Figure 6.1 Geometry for semi-infinite body in Example 6.1.

  a. Case X10B1T0. For the case where the boundary temperature is constant, $f(t) = T_0$, the integral in Eq. (6.15) is given in Table 5.4 as integral 3

$$T(x, t) = T_0 \text{ erfc} \left[\frac{x}{(4\alpha t)^{1/2}} \right] \qquad (6.16)$$

or

$$\frac{T(x, t)}{T_0} = 1 - \text{erf} \left[\frac{x}{(4\alpha t)^{1/2}} \right] \qquad (6.17)$$

  Compare this solution to case *X10B0T1* listed in Table 6.2 which is the temperature caused by a zero boundary temperature and uniform initial condition. The two solutions differ by a constant and a change of sign. The heat conduction equation is linear, so that multiplying a solution by (-1) gives another solution; and, adding a constant to a solution gives another solution.

  b. Case X10B3T0. In the case where the boundary temperature is a polynomial in $t^{n/2}$, such as

$$f(t) = a_{-1} t^{-1/2} + a_0 + a_1 t^{1/2} + a_2 t + a_3 t^{3/2} + \cdots \qquad (6.18)$$

then the integral in Eq. (6.15) can be written as the sum of the effects of each term in the polynomial. For the general term of such a polynomial, let $f(\tau) = T_0(\tau/t_0)^{n/2}$, where T_0 has units of temperature and t_0 is some reference time (t_0 could be 1 s). Then the integral in Eq. (6.15) may be written

$$T(x, t) = \alpha \int_{\tau=0}^{t} T_0 \left(\frac{\tau}{t_0} \right)^{n/2} \frac{x}{\alpha(t - \tau)} K(x, t - \tau) \, d\tau \qquad (6.19)$$

This integral is listed in Table 5.5 and the temperature resulting from the applied surface temperature $T_0(t/t_0)^{n/2}$ may be written

$$T(x, t) = T_0 \Gamma \left(1 + \frac{n}{2} \right) \left(4 \frac{t}{t_0} \right)^{n/2} i^n \text{erfc} \left[\frac{x}{(4\alpha t)^{1/2}} \right] \qquad (6.20)$$

where $n = -1, 0, 1, \ldots$, and so on. The gamma function $\Gamma(1 + n/2)$ takes the values $\pi^{1/2}$, 1, $\pi^{1/2}/2$, and 1 for $n = -1, 0, 1$, and 2, respectively. The function $i^n \text{erfc}(\cdot)$ is the repeated integral of the error function plotted in Fig. 6.2. Some numerical values of $i^1 \text{erfc}(z)$ are listed in Table 6.4. The $i^n \text{erfc}(z)$ function is related to $\text{erfc}(z)$ by

$$i^0 \text{erfc}(z) = \text{erfc}(z) \qquad (6.21a)$$

$$i^1 \text{erfc}(z) = \frac{1}{\pi^{1/2}} e^{-z^2} - z \, \text{erfc}(z) \qquad (6.21b)$$

$$2n \, i^n \text{erfc}(z) = i^{n-2} \text{erfc}(z) - 2z \, i^{n-1} \text{erfc}(z) \qquad (6.21c)$$

Refer to Appendix E for other properties of the error function.

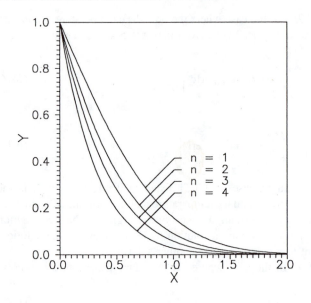

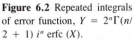

Figure 6.2 Repeated integrals of error function, $Y = 2^n \Gamma(n/2 + 1) i^n \text{erfc}(X)$.

In the case where the surface temperature $f(t)$ is periodic in time, the Laplace transform technique can be used on the integral in Eq. (6.15) to good advantage. Refer to Carslaw and Jaeger (1959), pages 399–402, for a general discussion. If the periodic function is a sine or cosine function, then the steady periodic part of the solution is given by Carslaw and Jaeger (1959), pages 64–70.

Example 6.2: Semi-infinite body with specified surface heat flux—*X20B-T0* case. Find the temperature in the semi-infinite body that has a heat flux boundary condition and zero initial condition.

SOLUTION This is the *X20B-T0* geometry shown in Fig. 6.3. The GFSE for the temperature takes the form

$$T(x, t) = \alpha \int_{\tau=0}^{t} \frac{f(\tau)}{k} G_{X20}(x, t|0, \tau) \, d\tau \qquad (6.22)$$

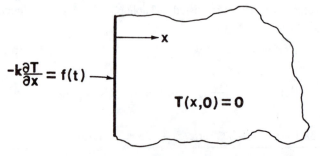

Figure 6.3 Geometry for semi-infinite body in Example 6.2.

The heat flux at the boundary is $f(t)$ with units of W/m^2. Note that the $X20$ GF is evaluated at the surface $x' = 0$. The $X20$ GF given in Table 6.1 is the sum of two fundamental heat conduction solutions, so the temperature can be written as

$$T(x, t) = \alpha \int_{\tau=0}^{t} \frac{f(\tau)}{k} [K(x - 0, t - \tau) + K(x + 0, t - \tau)] \, d\tau$$

$$= 2\alpha \int_{\tau=0}^{t} \frac{f(\tau)}{k} K(x, t - \tau) \, d\tau \tag{6.23}$$

a. Case X20B1T0. In the case where $f(t) = q_0$, a constant heat flux, the integral in Eq. (6.23) is given in Table 5.4 with Δt replaced by t, and the temperature is given by

$$T(x, t) = \frac{q_0}{k} (4\alpha t)^{1/2} \text{ ierfc} \left[\frac{x}{(4\alpha t)^{1/2}} \right] \tag{6.24}$$

This expression is also listed in Table 6.3. The temperature is plotted in Fig. 6.4 in terms of $(T - T_0)/(q_0 a/k)$, where a is the reference length. Sometimes the quantity $(4\alpha t)^{1/2}$ is used as a reference length. The quantity $q_0 (4\alpha t/\pi)^{1/2}/k$ is the surface temperature on the semi-infinite body resulting from the heat flux q_0 (ierfc$(0) = 1/\sqrt{\pi}$). Equation (6.24) can also be obtained from Eq. (6.20) for $n = 1$; that is, a surface temperature proportional to $t^{1/2}$ produces a steady surface heat flux.

b. Case X20B3T0. For the case where the surface heat flux is $f(t) = q_0(t/t_0)^{n/2}$, for $n = -1, 0, 1$, and so on, the integral in Eq. (6.23) is given in Table 5.5. After some simplification, the temperature is given by

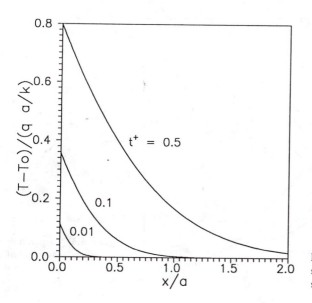

Figure 6.4 Temperature in semi-infinite body with constant heat flux at surface.

$$T(x, t) = \frac{q_0}{k} (4\alpha t)^{1/2} \left(\frac{t}{t_0}\right)^{n/2} \Gamma\left(1 + \frac{n}{2}\right) 2^n$$

$$i^{n+1} \operatorname{erfc}\left[\frac{x}{(4\alpha t)^{1/2}}\right] \qquad n = -1, 0, 1, \ldots \qquad (6.25)$$

Note that for $n = 0$, this solution reduces to the constant heat flux case.

Example 6.3: Semi-infinite body with convection—X30B1T0 case. Find the temperature in a semi-infinite body due the sudden application of the convection boundary condition where T_∞ is constant. The convection boundary condition is

$$-k \left.\frac{\partial T}{\partial x}\right|_{x=0} + hT\big|_{x=0} = hT_\infty \qquad (6.26)$$

where h is a constant, and in general the fluid temperature T_∞ is a function of time.

SOLUTION This is the X30B1T0 case. The temperature is plotted in Fig. 6.5 and the geometries shown in Fig. 6.6. The temperature solution is given by the GF equation as

$$T(x, t) = \alpha \int_{\tau=0}^{t} \frac{hT_\infty}{k} G_{X30}(x, t|0, \tau) \, d\tau \qquad (6.27)$$

Note that x' is evaluated at the surface, $x' = 0$. The function G_{X30} is listed in Table 6.1, and Eq. (6.27) becomes

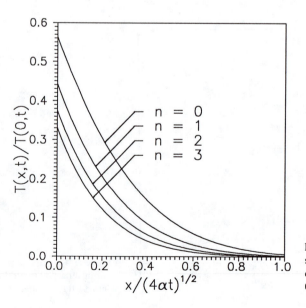

Figure 6.5 Temperature in semi-infinite body with surface convection for $h(\alpha t)^{1/2}/k = 0.1$, 0.5, 2.0, 00.

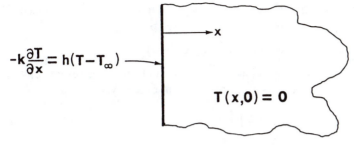

Figure 6.6 Geometry for the semi-infinite body in Example 6.3.

$$T(x, t) = \alpha \int_{\tau=0}^{t} \frac{hT_\infty}{k} \left(\frac{2}{[4\pi\alpha(t - \tau)]^{1/2}} \exp\left[\frac{-x^2}{4\alpha(t - \tau)} \right] \right.$$

$$- \frac{h}{k} \exp\left[\frac{hx}{k} + \frac{h^2}{k^2} \alpha(t - \tau) \right] \text{erfc} \left\{ \frac{x}{2[\alpha(t - \tau)]^{1/2}} \right.$$

$$\left. \left. + \frac{h}{k} [\alpha(t - \tau)]^{1/2} \right\} \right) d\tau \tag{6.28}$$

This contains a difficult integral if $T_\infty = T_\infty(t)$. Note that if the temperature were evaluated at $x = 0$, the integral would be less difficult. Usually, the surface temperature resulting from a boundary condition is much easier to find than the temperature everywhere inside the body.

For the case where T_∞ is time invariant, the integral for any value of x given in Table 5.5 is used to obtain

$$T(x, t) = T_\infty \left\{ \text{erfc}\left[\frac{x}{2(\alpha t)^{1/2}} \right] - \exp\left(\frac{hx}{k} + \alpha t \frac{h^2}{k^2} \right) \right.$$

$$\left. \times \text{erfc}\left[\frac{x}{2(\alpha t)^{1/2}} + \frac{h}{k} (\alpha t)^{1/2} \right] \right\} \tag{6.29}$$

6.3.3 Volume Energy Generation

Next consider the temperature in a semi-infinite body caused by volume energy generation. The boundary conditions and the initial condition are homogeneous. The temperature is given by the GFSE,

$$T(x, t) = \frac{\alpha}{k} \int_{\tau=0}^{t} \int_{x'=0}^{\infty} G(x, t|x', \tau) g(x', \tau) \, dx' \, d\tau \tag{6.30}$$

This expression is more complicated than the temperature resulting from a boundary condition because there are two integrals to evaluate.

Consider the case when the volume energy generation $g(x', \tau)$ is either independent of τ or a product of a function of x' and a function of τ,

$$g(x', \tau) = g_x(x') \, g_t(\tau) \tag{6.31}$$

Then the integrations over x' previously discussed can be used. For example, suppose the volume energy generation is given by one term of a polynomial in time:

$$g(x', \tau) = g_0 \left(\frac{\tau}{t_0}\right)^{n/2} \qquad n = -1, 0, 1, 2, \ldots \tag{6.32}$$

where g_0 is a constant with units of W/m^3. That is, $g(x', \tau)$ is independent of x' and is proportional to $\tau^{n/2}$. The time t_0 is any convenient value and could be one unit, such as 1 s.

The solution for the temperature when $g(\cdot)$ is given by Eq. (6.32) can be found for boundary conditions of the first and second kinds using the GF given by Eq. (6.6). The integration over the body (x' in this case) is usually considered first, and the integrals required are listed in Tables 5.1 and 5.2 [for $F(x') = 1$, $a = 0$, and $b \to \infty$]. Integration of Eq. (6.30) over x' yields

$$T(x, t) = \frac{\alpha}{2k} \int_{\tau=0}^{t} g_0 \left(\frac{\tau}{t_0}\right)^{n/2} \left(2 - \mathrm{erfc}\left\{\frac{x}{[4\alpha(t - \tau)]^{1/2}}\right\}\right.$$

$$\left. + (-1)^I \, \mathrm{erfc}\left\{\frac{x}{[4\alpha(t - \tau)]^{1/2}}\right\}\right) d\tau \tag{6.33}$$

The remaining integral on τ in Eq. (6.33) is listed in Table 5.5 to give for $I = 1$ (boundary condition of the first kind)

$$T(x, t) = \frac{g_0 \alpha t}{k} \left\{\frac{1}{n/2 + 1} \left(\frac{t}{t_0}\right)^{n/2}\right.$$

$$\left. - 4 \, \Gamma\left(\frac{n}{2} + 1\right) \left(\frac{t}{t_0}\right)^{n/2} i^{n+2} \, \mathrm{erfc}\left[\frac{x}{(4\alpha t)^{1/2}}\right]\right\} \tag{6.34a}$$

for $n = -1, 0, 1, 2$, etc. This is case $X10B0T0G t3$ and the temperature is plotted in Fig. 6.7.

In the case $I = 2$ for the boundary condition of the second kind, Eq. (6.33) gives,

$$T(x, t) = \frac{g_0 \alpha t}{k} \frac{1}{n/2 + 1} \left(\frac{t}{t_0}\right)^{n/2} \tag{6.34b}$$

This is case $X20B0T0G t3$, the temperature in a semi-infinite body with spatially uniform heat generation and an insulated boundary. The temperature does not depend on position because there is no heat flow in the body; the temperature increases everywhere at the same rate. The same result could have been obtained by a simple lumped capacitance description that is appropriate when the temperature is spatially uniform: $\rho c \, \partial T / \partial t = g(t)$.

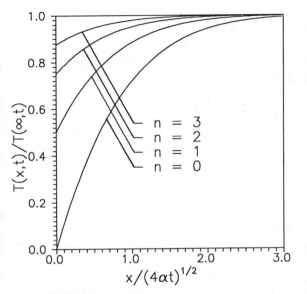

Figure 6.7 Temperature in the semi-infinite body with energy generation $g \simeq g_0 t^{n/2}$.

6.4 FLAT PLATES: SMALL-TIME GREEN'S FUNCTIONS

The cases discussed in this section are one-dimensional flat plates, denoted XIJ, I, $J = 1, 2, 3, 4, 5$. The small-time GF for these cases are infinite-series expressions or approximate truncated infinite series. The small-time GFs are listed in Appendix X. In general, for $\alpha(t - \tau)/L^2 < 0.05$, only three terms of the expressions for the small-time GF are needed for accuracy to four decimal places. Many of these expressions were derived from a Laplace transform solution of the auxiliary equation for the GF.

6.4.1 Initial Conditions

Consider the small-time solutions of the following equation:

$$\frac{\partial^2 T}{\partial x^2} = \frac{1}{\alpha} \frac{\partial T}{\partial t} \qquad 0 < x < L \tag{6.35}$$

The boundary conditions are homogeneous and the initial temperature distribution is

$$T(x, 0) = F(x) \tag{6.36}$$

The solution using GFs is

$$T(x, t) = \int_{x'=0}^{L} G(x, t|x', 0) F(x') \, dx' \tag{6.37}$$

Consider one term in a quadratic initial temperature distribution,

$$F(x') = T_0 \left(\frac{x'}{L}\right)^i \qquad 0 < x' < L \qquad i = 0, 1, 2 \tag{6.38}$$

For boundary of the first and second kinds, introduce the expressions for G_{XIJ} and $F(x')$ in Eq. (6.37) and the temperature is given by

$$T(x, t) = T_0 \sum_{n=-\infty}^{\infty} (-1)^{I+J} \text{FIN}(i, 2nL + x) \tag{6.39}$$

where I and J describe the boundary conditions types at $x = 0$ and $x = L$, respectively, and where

$$\text{FIN}(i, z) = \int_{x'=0}^{L} [K(z - x', t) + (-1)^I K(z + x', t)] \left(\frac{x'}{L}\right)^i dx' \tag{6.40}$$

This is the $XIJB00T(i + 1)$ case, where $i = 0, 1, 2; I = 1$ or 2; and, $J = 1$ or 2. The function $\text{FIN}(i, z)$ stands for F initial and i denotes x'/L to the ith power. Refer to Table 5.3 for closed form expressions for the function $\text{FIN}(i, z)$.

The solution given by Eq. (6.39) is valid for all $t > 0$, but for small time only a few terms of the series are needed. As $\alpha t/L^2$ increases, the number of significant terms in the infinite series increases.

Example 6.4: Slab with zero-temperature boundaries—X11B00T1 case. Find the temperature in a slab body with zero temperature boundary conditions and with a spatially uniform initial condition $F(x) = T_0$.

SOLUTION This is the $X11B00T1$ case and the solution is given by Eq. (6.39) where $i = 0$ and $I = J = 1$. This case involves the function $\text{FIN}(0, 2nL + x)$ (see Table 5.3):

$$\text{FIN}(i = 0, 2nL + x) = \frac{1}{2} \left\{ \text{erfc}\left[\frac{(2n-1)L + x}{(4\alpha t)^{1/2}}\right] - 2\,\text{erfc}\left[\frac{2nL + x}{(4\alpha t)^{1/2}}\right] \right.$$
$$\left. + \text{erfc}\left[\frac{(2n+1)L + x}{(4\alpha t)^{1/2}}\right] \right\} \tag{6.41}$$

Some values of these arguments are given in Table 6.5 for various values of n. The major contributions to the temperature given by Eq. (6.39) for small times come from the smaller values of $|n|$ such as 0 and 1. For $n = 0$ the function $\text{FIN}(0, 2nL + x)$ gives

$$\text{FIN}(0, x) = \frac{1}{2} \left\{ 2 - \text{erfc}\left[\frac{L - x}{(4\alpha t)^{1/2}}\right] - 2\,\text{erfc}\left[\frac{x}{(4\alpha t)^{1/2}}\right] \right.$$
$$\left. + 2\,\text{erfc}\left[\frac{L + x}{(4\alpha t)^{1/2}}\right] \right\} \tag{6.42a}$$

Table 6.5

n	$(2n - 1)L + x$	$2nL + x$	$(2n + 1)L + x$
-3	$-7L + x$	$-6L + x$	$-5L + x$
-2	$-5L + x$	$-4L + x$	$-3L + x$
-1	$-3L + x$	$-2L + x$	$-L + x$
0	$-L + x$	x	$L + x$
1	$L + x$	$2L + x$	$3L + x$
2	$3L + x$	$4L + x$	$5L + x$
3	$5L + x$	$6L + x$	$7L + x$

For $n = -1$ the function $\text{FIN}(0, 2nL + x)$ gives

$$\text{FIN}(0, -2L + x) = \frac{1}{2}\left\{ - \text{erfc}\left[\frac{3L - x}{(4\alpha t)^{1/2}}\right] - 2\,\text{erfc}\left[\frac{2L - x}{(4\alpha t)^{1/2}}\right]\right.$$

$$\left. - \text{erfc}\left[\frac{L - x}{(4\alpha t)^{1/2}}\right]\right\} \tag{6.42b}$$

Note 1. The identity $\text{erfc}(-u) = 2 - \text{erfc}(u)$ has been used in Eq. (6.42) to put positive arguments in each of the terms $\text{erfc}(\cdot)$. The quantity $(x - L)$ is zero or negative since $0 \leq x \leq L$. Recall that $\text{erfc}(u \to +\infty) = 0$ but that $\text{erfc}(u \to -\infty) = 2$, so that positive arguments ensures that each of the $\text{erfc}(\cdot)$ terms will converge to zero as $|n| \to \infty$.

Note 2. The identity $\text{erfc}(-u) = 2 - \text{erfc}(u)$ applied to the $\text{FIN}(0, 2nL + x)$ term for $n = -1$ produced three constant terms that canceled to zero. This cancelation occurs for every $n < 0$ and it has an important numerical consequence. As you add more terms to the infinite series for the temperature to improve the accuracy, it is important to find a value for each $\text{FIN}(0, 2nL + x)$ as a unit and then add that value to the temperature. This will avoid excessive loss of significant digits resulting from subtracting numbers that are very close in value.

For small values of $\alpha t/L^2$, the dominant terms in Eq. (6.39) for the temperature in the $X11B00T1$ case are given by the largest terms from Eqs. (6.42a) and (6.42b) multiplied by the initial temperature T_0:

$$T(x, t) \approx T_0\left\{ 1 - \text{erfc}\left[\frac{x}{(4\alpha t)^{1/2}}\right] - \text{erfc}\left[\frac{L - x}{(4\alpha t)^{1/2}}\right]\right.$$

$$\left. - \text{erfc}\left[\frac{L + x}{(4\alpha t)^{1/2}}\right] - \text{erfc}\left[\frac{2L - x}{(4\alpha t)^{1/2}}\right]\right\} \tag{6.43}$$

Near the boundary $x = 0$ and for $\alpha t/L^2 < 0.025$, the quantity $\text{erfc}[L/(4\alpha t)^{1/2}]$ is less than 0.0001 and the temperature is given approximately by the first two terms of Eq. (6.43):

$$T(x, t) \approx T_0 \left\{ 1 - \text{erfc} \left[\frac{x}{(4\alpha t)^{1/2}} \right] \right\}$$

$$= T_0 \, \text{erf} \left[\frac{x}{(4\alpha t)^{1/2}} \right] \quad \begin{matrix} x \ll L \\ \alpha t/L^2 \text{ small} \end{matrix} \tag{6.44a}$$

This result is identical to the semi-infinite case $X10B0T1$. That is, near the boundary at small time, the temperature in a flat plate is given by the semi-infinite case with the same boundary condition. For $\alpha t/L^2 < 0.025$ and for $x \approx L$, the dominant terms are

$$T(x, t) \approx T_0 \left\{ 1 - \text{erfc} \left[\frac{L - x}{(4\alpha t)^{1/2}} \right] \right\}$$

$$= T_0 \, \text{erf} \left[\frac{L - x}{(4\alpha t)^{1/2}} \right] \quad \begin{matrix} L \cdot x \ll L \\ \alpha t/L^2 \text{ small} \end{matrix} \tag{6.44b}$$

6.4.2 Volume Energy Generation

Early-time solutions of

$$\frac{\partial^2 T}{\partial x^2} + \frac{1}{k} g(x, t) = \frac{1}{\alpha} \frac{\partial T}{\partial t} \quad 0 < x < L \tag{6.45}$$

are discussed in this section for homogeneous boundary conditions and zero initial condition. Actually the solutions are valid for all times but they are computationally efficient for early times.

The solution for the temperature using GFs is

$$T(x, t) = \frac{\alpha}{k} \int_{\tau=0}^{t} \int_{x'=0}^{L} G(x, t|x', \tau) \, g(x', \tau) \, dx' \, d\tau \tag{6.46}$$

The discussion will be limited to cases for which the volume energy generation $g(x', \tau)$ is the product of a function of x' and a function of τ,

$$g(x', \tau) = g_x(x') \, g_t(\tau) \tag{6.47}$$

The integration over x' in Eq. (6.46) is similar to that for the nonzero initial temperature distribution, Eq. (6.37). Integrals over time τ are given in Tables 5.4 and 5.5.

As an example, the case where $g(x', \tau) = g_0 L \, \delta(x' - x_0) g_t(\tau)$ is examined. This is a plane heat source located at x_0 with a time-variable source strength given by $g_t(\tau)$. Using Eq. (6.46) and the small-time GF for geometry XIJ given by Table 4.1 gives

$$T(x, t) = \frac{\alpha}{k} \int_{\tau=0}^{t} \int_{x'=0}^{L} G(x, t|x', \tau) \, g_0 L \, \delta(x' - x_0) g_t(\tau) \, dx' \, d\tau$$

$$= \frac{\alpha}{k} g_0 L \int_{\tau=0}^{t} G(x, t|x_0, \tau) \, g_t(\tau) \, d\tau$$

$$= \frac{\alpha}{k} g_0 L \sum_{n=-\infty}^{\infty} \int_{\tau=0}^{t} (-1)^{(l+J)} [K(2nL + x - x_0, t - \tau)$$

$$+ (-1)^l K(2nL + x + x_0, t - \tau)] g_t(\tau) d\tau \qquad (6.48)$$

where $I = 1$ or 2 and $J = 1$ or 2 determines the type of boundary conditions.

Suppose the time variation of the plane source strength $g_t(\tau)$ is given by

$$g_t(\tau) = \left(\frac{\tau}{t_0}\right)^{m/2} \qquad m = -1, 0, 1, \ldots \qquad (6.49)$$

where t_0 is some convenient positive time value. Then the time integral in Eq. (6.48) is given in Table 5.5:

$$T(x, t) = \frac{1}{2}\left(\frac{\alpha t_0}{L^2}\right)^{1/2} \frac{g_0 L^2}{k} \Gamma\left(\frac{m}{2} + 1\right) \left(\frac{4t}{t_0}\right)^{(m+1)/2}$$

$$\times \sum_{n=-\infty}^{\infty} (-1)^{-(l+J)n} \left\{ i^{m+1} \operatorname{erfc}\left[\frac{|2nL + x - x_0|}{(4\alpha t)^{1/2}}\right]\right.$$

$$+ (-1)^l i^{m+1} \operatorname{erfc}\left[\frac{|2nL + x + x_0|}{(4\alpha t)^{1/2}}\right]\right\} \qquad (6.50)$$

This solution applies to geometries described by the number $X1JB00T0Gx7t3$ for I, J = 1, 2. The plane source at x_0 can vary with time as given by Eq. (6.49) with $m = -1, 0, 1, 2,$ and so on. A particularly important value of m is $m = 0$, which gives the temperature resulting from a continuous constant plane source; for $m = 0$, the t_0 values cancel in Eq. (6.50).

One possible location for the plane source is at $x_0 = 0$. For this location and case XIJ with $I = 1$ (that is, geometries X11 and X12), $T(x, t)$ is equal to zero,

$$T(x, t) = 0 \qquad \text{for all } x \text{ and } t \qquad (6.51)$$

while for cases XIJ with $I = 2$ (that is, geometries X21 and X22), Eq. (6.50) gives

$$T(x, t) = \left(\frac{\alpha t_0}{L^2}\right)^{1/2} \frac{g_0 L^2}{k} \Gamma\left(\frac{m}{2} + 1\right) \left(\frac{4t}{t_0}\right)^{(m+1)/2}$$

$$\times \sum_{n=-\infty}^{\infty} (-1)^{Jn} \left\{ i^{m+1} \operatorname{erfc}\left[\frac{|2nL + x|}{(4\alpha t)^{1/2}}\right]\right\} \qquad (6.52a)$$

By isolating the $n = 0$ term, the temperature can be written as a sum over $n = 1$ to ∞:

$$T(x, t) = \left(\frac{\alpha t_0}{L^2}\right)^{1/2} \frac{g_0 L^2}{k} \Gamma\left(\frac{m}{2} + 1\right) \left(\frac{4t}{t_0}\right)^{(m+1)/2} \left(i^{m+1} \operatorname{erfc}\left[\frac{x}{(4\alpha t)^{1/2}}\right]\right.$$

$$+ \sum_{n=1}^{\infty} (-1)^{Jn} \left\{ i^{m+1} \operatorname{erfc}\left[\frac{|2nL + x|}{(4\alpha t)^{1/2}}\right]\right.$$

$$+ i^{m+1} \operatorname{erfc}\left[\frac{|2nL - x|}{(4\alpha t)^{1/2}}\right]\right\}\right) \tag{6.52b}$$

For small values of $\alpha t/L^2$ (such as $\alpha t/L^2 < 0.1$), only a few terms of the summation are needed.

Equation (6.52) was derived as the temperature resulting from the space- and time-varying volume energy source

$$g(x, t) = g_0 L \, \delta(x = 0) \left(\frac{t}{t_0}\right)^{m/2} \tag{6.53a}$$

which is a plane heat source located at $x = 0$, but this plane source produces an effect identical to a prescribed heat flux at $x = 0$ given by

$$-k \frac{\partial T}{\partial x}\bigg|_{x=0} = q_0 \left(\frac{t}{t_0}\right)^{m/2} \tag{6.53b}$$

Therefore, g_0 and q_0 in Eq. (6.53a, b) are related by

$$q_0 = g_0 L \tag{6.53c}$$

where q_0 has units of W/m^2 and g_0 has units of W/m^3. Equation (6.52) has been described as the temperature for the case X2JB00T0Gx7t3 (plane heat source at $x = 0$), but because the plane heat source at $x = 0$ is equivalent to a prescribed heat flux at $x = 0$, the description X2JB30T0 also applies to Eq. (6.52).

6.5 FLAT PLATES: LARGE-TIME GREEN'S FUNCTIONS

Large-time GFs are usually derived from a separation of variables solution of the energy equation. The separation of variables technique is discussed in Chapter 4. The large-time GFs for slab bodies have the general form

$$G(x, t|x', \tau) = \frac{X_0(x)}{N_0} + \sum_{m=1}^{\infty} \exp\left[-\frac{\beta_m^2 \alpha(t - \tau)}{L^2}\right] \frac{X_m(x) \, X_m(x')}{N_m} \tag{6.54}$$

where the eigenfunctions, $X_0(x)$ and $X_m(x)$, and the norms N_0 and N_m are given in Tables 4.2 and 4.3. Each GF also has associated eigenvalues β_m. For cases involving only boundary conditions of kinds 1 or 2, the eigenvalues are given in Table 4.2. For cases with boundary conditions of types 3, 4, or 5, the eigenvalues must be found numerically as roots of the characteristic equation listed in Table 4.3. A complete list of large-time GFs with derivatives and useful approximations is given in Appendix X.

6.5.1 Initial Conditions

The temperature in a body resulting from a nonzero initial temperature distribution is discussed in this section. As an example, consider the initial temperature distribution given by

$$F(x') = \begin{cases} T_0 & a < x' < b \\ 0 & \text{otherwise} \end{cases} \tag{6.55}$$

For the specific case of a body with zero temperature at boundary $x = 0$ and with one of two possible boundary conditions at $x = L$ described by number $X1J$ where $J = 1$ or 2, the temperature distribution is found from the initial-temperature term of the GFSE with the GF given by Eq. (6.54):

$$T(x, t) = 2T_0 \sum_{m=1}^{\infty} e^{-\beta_m^2 \alpha t/L^2} \sin\left(\frac{\beta_m x}{L}\right) \frac{\cos(\beta_m a/L) - \cos(\beta_m b/L)}{\beta_m} \tag{6.56}$$

The eigenvalues β_m depend on whether $J = 1$ or 2. The number for this case is $X1JB00T5$ for $J = 1$ or 2. The presence of the term $\exp(-\beta_m \alpha t/L^2)$ multiplying by all the other terms in Eq. (6.56) causes rapid numerical convergence of the series for large dimensionless time $\alpha t/L^2 > 0.025$.

In evaluating temperature from an infinite series, a central issue is the number of terms of the series that must be evaluated for acceptable accuracy. Clearly an infinite number of terms cannot be evaluated and the series must be truncated. A reasonable criterion for truncating a series whose convergence is controlled by the exponential factor $\exp(-\beta_m \alpha t/L^2)$ is that the exponential factor should be less than 0.001 in the last (smallest) term retained. Then the *argument* of the exponential factor in the last term that is retained can be determined from

$$\exp\frac{-\beta_m \alpha t}{L^2} < 0.001 \quad \text{or} \quad \frac{-\beta_m \alpha t}{L^2} < \ln 0.001$$

$$= -6.9 \quad \text{or} \quad \frac{\beta_m \alpha t}{L^2} > 6.9 \tag{6.57}$$

For the $X11$ case ($J = 1$) the eigenvalues are $\beta_m = m\pi$, and the above convergence criterion can be solved for m where m is the number of the last term retained in the series:

$$m \le \left(\frac{6.9}{\pi^2} \frac{L^2}{\alpha t}\right)^{1/2} \tag{6.58}$$

The number of terms needed in the truncated series depends on dimensionless time $\alpha t/L^2$. For $\alpha t/L^2 = 0.01$, $m \le 8$ which means that only eight terms of the infinite series are sufficient to make the exponential factor smaller than 0.001. For $\alpha t/L^2 = 0.025$, only five terms of the series are sufficient. For "large" values of $\alpha t/L^2$, such as 0.17 or larger, only one term of the series (the $m = 1$ term) is sufficient for the $X12$ case. For the $X11$ case, if $\alpha t/L^2$ is larger than 0.31, then one term of the series is sufficient to make the exponential factor less than 0.001.

Next, consider the uniform initial temperature

$$F(x') = T_0 \quad 0 < x < L$$

applied to a body with homogeneous boundary conditions of the first kind at both $x = 0$ and L. This is case $X11B00T1$ and the temperature is given by Eq. (6.56) with

$a = 0$ and $b = L$:

$$T(x, t) = T_0 \frac{4}{\pi} \sum_{m=1,3,\ldots}^{\infty} e^{-m^2\pi^2\alpha t/L^2} \sin\left(m\pi \frac{x}{L}\right) \frac{1}{m} \qquad (6.59a)$$

The eigenvalues are $\beta_m = m\pi$. The first two terms of this infinite series can be used to approximate the temperature for $\alpha t/L^2$ not too small:

$$T(x, t) \approx T_0 \frac{4}{\pi} \left[e^{-\pi^2\alpha t/L^2} \sin\left(\pi \frac{x}{L}\right) + \frac{1}{3} e^{-9\pi^2\alpha t/L^2} \sin\left(3\pi \frac{x}{L}\right) \right] \qquad (6.59b)$$

Equation (6.59b) gives satisfactory accuracy for $\alpha t/L^2 \geq 0.025$. The related small-time expression, Eq. (6.43), is accurate for $\alpha t/L^2 < 0.025$. The least accurate range for Eq. (6.59b) is in its lower limit ($\alpha t/L^2 \approx 0.025$) and the least accurate range for Eq. (6.43) (the small-time form of the same problem) is near its upper limit ($\alpha t/L^2 \approx 0.025$). Hence it is instructive to evaluate both expressions for the temperature when they are least accurate at the middle of the body: at $x = L/2$ and at $\alpha t/L^2 = 0.025$. Equation (6.43) evaluated at $x = L/2$ is

$$T\left(\frac{L}{2}, t\right) \approx T_0 \left\{ 1 - 2 \, \text{erfc}\left[\frac{1}{4(\alpha t/L^2)^{1/2}} \right] + 2 \, \text{erfc}\left[\frac{3}{4(\alpha t/L^2)^{1/2}} \right] \right\} \qquad (6.60a)$$

which has the numerical components of

$$T\left(\frac{L}{2}, 0.025\right) \approx T_0[1 - 2(0.0253473) + 2(0.197\text{E-10})] = 0.949305 \, T_0 \qquad (6.60b)$$

The components of Eq. (6.59b) at $x = L/2$ and $\alpha t/L^2 = 0.025$ are

$$T\left(\frac{L}{2}, 0.025\right) = T_0 \frac{4}{\pi} [0.7813437 + \frac{1}{3}(0.108537)(-1)] = 0.94877 \, T_0 \qquad (6.60c)$$

The expression given by Eq. (6.60b) is slightly more accurate, but both expressions are less than 0.1% in error. Again only two terms are needed for each temperature expression near $\alpha t/L^2 = 0.025$.

6.5.2 Plane Heat Source

Consider a plane heat source located at x_0 described by

$$g(x', \tau) = g_0 L \, \delta(x_0 - x') \, g_t(\tau) \qquad (6.61)$$

Then, using this expression for $g(x', \tau)$ in the GFSE gives

$$T(x, t) = \frac{\alpha}{k} \int_{\tau=0}^{t} G(x, t|x_0, \tau) g_0 L \, g_t(\tau) \, d\tau \qquad (6.62)$$

Here the integral on x' has been evaluated using the sifting property of the Dirac delta function. For $G(\cdot)$ given by the large-time GF from Eq. (6.54), Eq. (6.62) becomes

$$T(x, t) = \frac{\alpha}{k} g_0 L \int_{\tau=0}^{t} \sum_{m=1}^{\infty} e^{-\beta_m^2 \alpha(t-\tau)/L^2} \frac{X(\beta_m x/L) \, X(\beta_m x_0/L)}{N_m} g_t(\tau) \, d\tau$$

$$+ \frac{\alpha}{k} g_0 L \int_{\tau=0}^{t} \frac{X_0}{N_0} g_t(\tau) \, d\tau \tag{6.63}$$

The temperature caused by a number of time-varying plane sources can be investigated with different functions $g_t(\tau)$ in Eq. (6.63). One of the simplest is for $g_t(\tau) = 1$, a constant for which the time integral in Eq. (6.63) may be evaluated as

$$T(x, t) = \frac{g_0 L^3}{k} \sum_{m=1}^{\infty} (1 - e^{-\beta_m^2 \alpha t/L^2}) \frac{X(\beta_m x/L) \, X(\beta_m x_0/L)}{N_m \beta_m^2}$$

$$+ \delta_{2I} \delta_{2J} \frac{g_0 L^2}{k} \frac{\alpha t}{L^2} \tag{6.64a}$$

This solution is denoted $XIJB00T0Gx7t1$ where I and J can be 1 or 2. The symbol δ_{IJ} is called the Kronecker delta and is defined to be

$$\delta_{IJ} = \begin{cases} 1 & \text{for } I = J \\ 0 & \text{for } I \neq J \end{cases} \tag{6.64b}$$

Do not confuse δ_{IJ} with the Dirac delta function $\delta(\cdot)$ defined in Chapter 1. In Eq. (6.63) there is a contribution for the $\delta_{2I}\delta_{2J}$ term only for $I = J = 2$. The term associated with $\delta_{2I}\delta_{2J}$ comes from the $m = 0$ term of the summation for $G(\cdot)$ which must be treated in a special manner when $I = J = 2$ because in this case $\beta_{m=0} = 0$ is an eigenvalue. There are two parts in this solution: a steady-state part, and a transient part. The steady-state part of Eq. (6.64a) can be written as

$$T(x) = \frac{g_0 L^3}{k} \sum_{m=1}^{\infty} \frac{X(\beta_m x/L) \, X(\beta_m x_0/L)}{N_m \beta_m^2} \tag{6.65}$$

for the $X11$, $X12$, and $X21$ cases. The $X22$ case does not in general have a steady-state part. The series given by Eq. (6.65) for the steady-state part converges very slowly. This slow convergence can be avoided because a simple linear function for the steady-state solution for the $X11$, $X12$, and $X21$ cases may be found with steady-state GFs (refer to Section 6.9). The steady-state solution for the $X11$ case is

$$T(x) = \begin{cases} \dfrac{g_0 x(L - x_0)}{k} & 0 \leq x \leq x_0 \\[3mm] \dfrac{g_0 x_0(L - x)}{k} & x_0 \leq x \leq L \end{cases} \tag{6.66}$$

Equation (6.66) is the steady-state GF multiplied by the source strength. The solution for the $X12$ case is

$$T(x) = \begin{cases} \dfrac{g_0 x L}{k} & 0 \le x \le x_0 \\[3mm] \dfrac{g_0 x_0 L}{k} & x_0 \le x \le L \end{cases} \tag{6.67}$$

Algebraic expressions such as Eqs. (6.66) and (6.67) are clearly much easier to evaluate than the infinite-series expression Eq. (6.65). Furthermore, the simple linear dependence on x can be seen in these equations, while it is not apparent in Eq. (6.65). When it is convenient to do so, the separate solution for the steady state should be obtained.

Next, two specific temperature expressions are given that are drawn from the general expressions discussed above. The solution of the $X11B00T0Gx7t1$ problem is obtained from Eq. (6.64), Tables 4.2 and 4.3, and Eq. (6.66),

$$T(x, t) = \frac{g_0 L^2}{k} \frac{x}{L} \left(1 - \frac{x_0}{L} \right)$$

$$- \frac{2g_0 L^2}{k} \sum_{m=1}^{\infty} e^{-m^2 \pi^2 \alpha t / L^2} \frac{\sin (m\pi x/L) \sin (m\pi x_0/L)}{m^2 \pi^2} \tag{6.68}$$

for $0 \le x \le x_0$. For $x_0 \le x \le L$, the same expression applies where the x and x_0 symbols are interchanged.

For the case of $T = 0$ at $x = 0$ and $\partial T/\partial x = 0$ at $x = L$ (i.e., $X12B00T0Gx7t1$), the solution is

$$T(x, t) = \frac{g_0 L^2}{k} \min \left(\frac{x}{L}, \frac{x_0}{L} \right)$$

$$- \frac{2g_0 L^2}{k} \sum_{m=1}^{\infty} e^{-\beta_m^2 \alpha t / L^2} \frac{\sin (\beta_m x/L) \sin (\beta_m x_0/L)}{\beta_m^2} \tag{6.69}$$

where $\min (x/L, x_0/L)$ means the minimum values of the choice between x/L and x_0/L, and where $\beta_m = (m - \frac{1}{2})\pi$.

The two specific temperature expressions given above as Eqs. (6.68) and (6.69) are relatively efficient expressions for computation for $\alpha t/L^2 > 0.025$. These equations are valid for values of $\alpha t/L^2$ that are even smaller, but more computationally efficient solutions for small times can be obtained by using the small time GFs for flat plates.

Approximate solutions at small times can also be obtained from the GFs for *semi-infinite* bodies. For example, for x/L and x_0/L both less than 0.5, the temperature solution of the problems $X11B00T0Gx7t1$ and $X12B00T0Gx7t1$ can be approximated at small times by the $X10B0T0Gx7t1$ problem. In other words, for sufficiently small times, the temperature distribution is affected most by the nearest boundary. This is the nature of diffusion—the influence of any transient driving term is localized in space at early time.

Example 6.5: Slab with one side heated, one side insulated—$X22B10T0$ case. Consider the flat plate insulated on one side and heated by a steady heat flux on the other side. Find the temperature using the standard and alternative

GFSEs. The boundary value problem is given by

$$\frac{\partial^2 T}{\partial x^2} = \frac{1}{\alpha}\frac{\partial T}{\partial t} \tag{6.70a}$$

$$-k\frac{\partial T}{\partial x}\bigg|_{x=0} = q_0 \qquad \frac{\partial T}{\partial x}\bigg|_{x=L} = 0 \tag{6.70b}$$

$$T(x, 0) = 0 \tag{6.70c}$$

a. Standard solution. The standard GF solution is given by Eq. (6.4) where the only nonhomogeneous term (the only driving term) is the heat flux at $x = 0$. This is the $X22B10T0$ case. Then using Eq. (6.4) and the $X22$ GF from Appendix X gives

$$T(x, t) = \frac{\alpha}{k}\int_{\tau=0}^{t} q_0 G_{X22}(x, t|0, \tau)\,d\tau$$

$$= \frac{\alpha q_0}{k}\int_{\tau=0}^{t}\frac{1}{L}\left(1 + 2\sum_{m=1}^{\infty}e^{-m^2\pi^2\alpha(t-\tau)/L^2}\cos\left(m\pi\frac{x}{L}\right)\right)d\tau$$

$$= \frac{q_0 L}{k}\left[\frac{\alpha t}{L^2} + \frac{2}{\pi^2}\sum_{m=1}^{\infty}\frac{1}{m^2}\cos\left(m\pi\frac{x}{L}\right)\left(1 - e^{-m^2\pi^2\alpha t/L^2}\right)\right] \tag{6.71}$$

This expression has three main parts. The first part is proportional to time and thus increases without limit over time. The last part contains an exponential factor that decays with time. The middle part that does not depend on time is given by

$$\frac{q_0 L}{k}\frac{2}{\pi^2}\sum_{m=1}^{\infty}\frac{1}{m^2}\cos\left(m\pi\frac{x}{L}\right) \tag{6.72}$$

This part of the temperature expression converges very slowly, that is, many terms of the infinite series must be evaluated for accurate numerical values, particularly for small values of x/L.

Next, another temperature expression with better convergence properties will be found with the alternative GFSE equation.

b. Alternative solution. The alternative GFSE (AGFSE) involves a known solution T^* that satisfies the boundary conditions but does not need to satisfy the initial condition. Since Eq. (6.71) contains a term proportional to time that dominates the temperature for large times, the T^* solution should display that behavior. The T^* solution for this problem is

$$T^*(x, t) = f(x) + \frac{q_0 L}{k}\frac{\alpha t}{L^2} \tag{6.73}$$

where $f(x)$ must be chosen to satisfy the boundary conditions. Substitute T^* into the energy equation:

$$\frac{\partial^2 T^*}{\partial x^2} = \frac{1}{\alpha}\frac{\partial T^*}{\partial t} \quad \text{or} \quad \frac{d^2 f}{dx^2} = \frac{q_0}{kL} \tag{6.74}$$

Solve Eq. (6.74) for $f(x)$ (by integrating twice) and then substitute $f(x)$ back into Eq. (6.73) to give

$$T^*(x, t) = \frac{q_0}{kL} \frac{x^2}{2} + C_1 x + C_2 + \frac{q_0 L}{k} \frac{\alpha t}{L^2} \tag{6.75}$$

Using the boundary conditions at $x = 0$ and L given by Eq. (6.70b, c) allows C_1 to be found as

$$C_1 = -\frac{q_0}{k} \tag{6.76}$$

Since C_2 cannot be found using these boundary conditions, it is set equal to zero. The choice of C_2 is arbitrary because both boundary conditions for T^* are gradient conditions and a constant can be subtracted from T^* without changing the properties of the solution. Then T^* is given by

$$T^*(x, t) = \frac{q_0 L}{k} \left[\frac{1}{2} \left(\frac{x}{L} \right)^2 - \frac{x}{L} + \frac{\alpha t}{L^2} \right] \tag{6.77}$$

Now T^* will be used in the alternative GFSE. The only nonzero integral in the alternative GFSE, Eq. (3.66), is the one corresponding to the initial condition. (Why does the last integral drop out in this case?) The alternative GFSE gives

$$T(x, t) = T^*(x, t) + \int_{x'=0}^{L} G_{X22}(x, t|x', 0) \left[-T^*(x', 0) \right] dx'$$

$$= T^*(x, t) - \frac{q_0}{k} \int_{x'=0}^{L} \left[1 + 2 \sum_{m=1}^{\infty} e^{-m^2 \pi^2 \alpha t/L^2} \right.$$

$$\left. \times \cos \left(m\pi \frac{x'}{L} \right) \cos \left(m\pi \frac{x}{L} \right) \right] \left[\frac{1}{2} \left(\frac{x'}{L} \right)^2 - \frac{x'}{L} \right] dx'$$

$$= \frac{q_0 L}{k} \left[\frac{\alpha t}{L^2} + \frac{1}{2} \left(\frac{x}{L} \right)^2 - \frac{x}{L} + \frac{1}{3} - \frac{2}{\pi^2} \sum_{m=1}^{\infty} \frac{1}{m^2} \right.$$

$$\left. \times \cos \left(m\pi \frac{x}{L} \right) e^{-m^2 \pi^2 \alpha t/L^2} \right\} \tag{6.78}$$

Equation (6.78) is valid for any time value but it has good convergence properties for $\alpha t/L^2 > 0.025$. For $\alpha t/L^2 < 0.025$ the temperature may be found approximately from the semi-infinite body solution with the same boundary heat flux (the X20B1T0 case) for five-digit numerical accuracy near $x = 0$ and with lesser accuracy near $x = L$.

It is interesting to equate the two expressions for the temperature found from the GFSE and AGFSE, Eqs. (6.71) and (6.78). Setting them equal and canceling identical terms leaves the equality

$$\frac{2}{\pi^2} \sum_{m=1}^{\infty} \frac{1}{m^2} \cos \left(m\pi \frac{x}{L} \right) = \frac{1}{2} \left(\frac{x}{L} \right)^2 - \frac{x}{L} + \frac{1}{3} \tag{6.79}$$

In effect, we have found the exact value of the infinite sum, and we can next investigate its convergence properties. Consider first the $x = 0$ location and thus the sum

$$\frac{2}{\pi^2} \sum_{m=1}^{\infty} \frac{1}{m^2} = \frac{2}{\pi^2} \left(\frac{1}{1} + \frac{1}{4} + \frac{1}{9} + \frac{1}{16} + \cdots \right) = \frac{1}{3} \qquad (6.80)$$

This sum converges but does so only after many terms. For the first four terms shown, the sum is 0.28848 and the exact value is $\frac{1}{3}$, an error of about 13%. The tail of the infinite series in Eq. (6.80) can be approximated by

$$\frac{2}{\pi^2} \sum_{m=M}^{\infty} \frac{1}{m^2} \approx \frac{2}{\pi^2} \int_{M}^{\infty} \frac{1}{m^2} \, dm = \frac{2}{\pi^2} \frac{1}{M} \qquad (6.81)$$

which shows that the series indeed converges slowly. For example, to reduce the error in half requires doubling the number of terms; also to make the error less than 0.01% requires more than 6000 terms.

Consider now the behavior of the series given by the left-hand side of Eq. (6.79) at $x = L$ which behaves much better than at $x = 0$. Since $\cos(m\pi) = (-1)^m$, two successive terms in the summation can be written as

$$\frac{2}{\pi^2} \left[\frac{1}{m^2} - \frac{1}{(m+1)^2} \right] = \frac{2}{\pi^2} \left[\frac{2m+1}{m^2(m+1)^2} \right] \qquad (6.82)$$

when m is an even number. For large m, this expression is proportional to m^{-3}, which converges much more rapidly, since

$$\frac{4}{\pi^2} \sum_{m=M, M+2, \ldots}^{\infty} \frac{1}{m^3} = \frac{4}{\pi^2} \sum_{n=M/2, M/2+1, \ldots}^{\infty} \frac{1}{(2n)^3}$$

$$\approx \frac{4}{\pi^2} \int_{n=M/2}^{\infty} \frac{1}{(2n)^3} \, dn = \frac{1}{\pi^2 M^2} \qquad (6.83)$$

Note that doubling M reduces the error in the tail by a factor of 4. Even more important, note that for an error of less than 0.01%, only 78 terms are needed; that is, the maximum m need be only 78 at $x = L$ compared to the maximum m of 6000 for $x = 0$ if the left-hand side of Eq. (6.79) is used. In any case, it is more convenient to use the exact value of the infinite sum given by Eq. (6.79) than to evaluate the sum approximately.

6.6 FLAT PLATES: TIME PARTITIONING

The infinite series for the GF expressions for flat plates are mathematically convergent (except at $x - x' = 0$ and $t = \tau = 0$ where the GF is singular). However there is no guarantee how many terms of the series are needed to find accurate numerical values for the GF. If the GF is integrated over time to find the temperature, the resulting infinite series may converge even more slowly than the GF and it may be difficult to calculate accurate temperature values for certain values of dimensionless time. In this section, the method of time partitioning is demonstrated as a means to calculate accurate temperature values at any dimensionless time.

Time partitioning was introduced in Section 5.3. In this method, the time integral from the GFSE is split or partitioned into two regions and both large-time and small-time GF are used to find the temperature. The use of the words ''small'' and ''large'' times can be misleading because the time scale for GFs is $\alpha(t - \tau)/L^2$, while the time scale for temperature is $\alpha t/L^2$, and we are concerned with large and small values of both time scales.

If one is interested in temperatures at small times, only the small-time GF is needed. However, if one is interested in temperatures at large times, both small-time and large-time GF are needed in the time partitioning method. The explanation is related to the different time scales for temperature and GFs. The GF (or its derivative) is integrated over time on variable τ on the interval $0 \leq \tau \leq t$ to find the temperature at dimensionless time $\alpha t/L^2$. This is equivalent to integration on variable $(t - \tau)$ on interval $t \geq (t - \tau) \geq 0$, so that there is an important integration over $(t - \tau)$ near zero (GF time scale $\alpha(t - \tau)/L^2$ is small) even if only large dimensionless times for temperature (time scale $\alpha t/L^2$) are of interest.

Let Δt be the partition time, and let the time integrals in the GFSE be partitioned at $\tau = t - \Delta t$. Then the time-partitioned form of the GFSE may be written as [see Eq. (5.14)],

$$
\begin{aligned}
T(x, t) = {} & \int_{\tau=0}^{t-\Delta t} \int_{x'=0}^{L} \frac{\alpha}{k} G^L(x, t|x', \tau) \, g(x', \tau) \, dx' \, d\tau && \text{(for volume}\\
&&& \text{energy generation)}\\[4pt]
& + \int_{\tau=t-\Delta t}^{t} \int_{x'=0}^{L} \frac{\alpha}{k} G^S(x, t|x', \tau) \, g(x', \tau) \, dx' \, d\tau\\[4pt]
& + \alpha \int_{\tau=0}^{t-\Delta t} d\tau \sum_{i=1}^{2} \left[\frac{f_i(\tau)}{k_i} G^L(x, t|x_i, \tau) \right] && \text{(for boundary conditions}\\
&&& \text{of the second through}\\
&&& \text{fifth kinds)}\\[4pt]
& + \alpha \int_{\tau=t-\Delta t}^{t} d\tau \sum_{i=1}^{2} \left[\frac{f_i(\tau)}{k_i} G^S(x, t|x_i, \tau) \right]\\[4pt]
& - \alpha \int_{\tau=0}^{t-\Delta t} d\tau \sum_{i=1}^{2} \left[f_i(\tau) \frac{\partial G^L}{\partial n_i'} \bigg|_{x'=x_i} \right] && \text{(for boundary conditions}\\
&&& \text{of the first kind only)}\\[4pt]
& - \alpha \int_{\tau=t-\Delta t}^{t} d\tau \sum_{i=1}^{2} \left[f_i(\tau) \frac{\partial G^S}{\partial n_i'} \bigg|_{x'=x_i} \right]
\end{aligned}
$$

$$(6.84)$$

(Terms that do not contain time integrals have been omitted for clarity.) Note that $G^S(\cdot)$ in Eq. (6.84) is associated with small values of $(t - \tau)$, and $G^L(\cdot)$ is associated with large values of $(t - \tau)$. A typical value for $\alpha(\Delta t)/L^2$ is about 0.025, for which the expressions for $G^S(\cdot)$ require only three terms of the infinite series for accuracy to four decimal places.

6.6.1 Boundary Conditions

Two examples are presented to demonstrate time partitioning in flat plates with non-homogeneous boundary conditions.

Example 6.6: Slab with one side heated, one side at fixed temperature—
X21B10T0 case. Find the temperature in a flat plate suddenly heated by a constant
heat flux q_0 at $x = 0$ and with a fixed zero temperature at $x = L$. The initial
temperature is zero and there is no volume heat generation. Use time partitioning
to find expressions that are numerically efficient for all values of time.

SOLUTION This is the $X21$ geometry shown in Fig. 6.8. The temperature for small
values of time ($\alpha t/L^2 < 0.025$) is most efficiently found from the GF equation
involving the small-time GF:

$$T(x, t) = \alpha \int_{\tau=0}^{t} \frac{q_0}{k} G_{X21}^S(x, t|0, \tau) \, d\tau \qquad \text{for } \frac{\alpha t}{L^2} < 0.025 \qquad (6.85)$$

Note that the GF in Eq. (6.85) is evaluated at $x' = 0$ where the heat flux is
located. The GF $G_{X21}^S(\cdot)$ is given in Appendix X as an infinite series. Substituting
$G_{X21}^S(\cdot)$ into Eq. (6.85) gives

$$T(x, t) = \frac{\alpha q_0}{k} \int_{\tau=0}^{t} \frac{2}{[4\pi\alpha(t - \tau)]^{1/2}} \sum_{n=-\infty}^{\infty} (-1)^n$$

$$\times \exp\left[\frac{-(2nL + x)^2}{4\alpha(t - \tau)}\right] d\tau \qquad (6.86)$$

This integral can be stated in terms of the fundamental heat conduction solution,
$K(w_n, t - \tau)$, as

$$T(x, t) = 2\frac{\alpha q_0}{k} \sum_{n=-\infty}^{\infty} (-1)^n \int_{\tau=0}^{t} K(w_n, t - \tau) \, d\tau \qquad (6.87)$$

where $w_n = 2nL + x$. The integral in Eq. (6.87) is then given in Table 5.5 as

$$T(x, t) = 2\frac{q_0 L}{k} \sum_{n=-\infty}^{\infty} (-1)^n \left(\frac{\alpha t}{L^2}\right)^{1/2} \text{ierfc}\left[\frac{|2n + x/L|}{2(\alpha t/L^2)^{1/2}}\right] \qquad (6.88)$$

This expression is numerically efficient for $\alpha t/L^2 < 0.025$, for which only three
terms of the series ($n = 0, 1, -1$) are sufficient to give a temperature that is
exact to over 13 digits. (This can be shown by evaluating the "tail" of the series,
$n = \pm 2, \pm 3$, etc.)

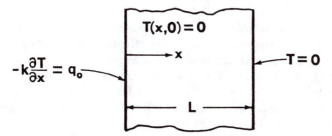

Figure 6.8 Geometry for slab body with one side heated and one side at a fixed temperature.

For larger values of time, the time-partitioning form of the GF equation, Eq. (6.84), gives

$$T(x, t) = \alpha \int_{\tau=0}^{t-\Delta t} d\tau \frac{f_i(\tau)}{k_i} G^L(x, t|0, \tau)$$

$$+ \alpha \int_{\tau=t-\Delta t}^{t} d\tau \frac{f_i(\tau)}{k_i} G^S(x, t|0, \tau) \qquad (6.89)$$

for $t > \Delta t$.

The two integrals in Eq. (6.89) will next be examined one at a time. The first integral in Eq. (6.89) involves the large-time GF, $G_{X21}^L(\cdot)$, which is given in Tables 4.2 and 4.3:

$$\alpha \int_{\tau=0}^{t-\Delta t} d\tau \frac{f_i(\tau)}{k_i} G^L(x, t|0, \tau) = \alpha \int_{\tau=0}^{t-\Delta t} d\tau \frac{q_0}{k} \frac{2}{L}$$

$$\times \sum_{m=1}^{\infty} e^{-\beta_m^2 \alpha(t-\tau)/L^2} \cos \frac{\beta_m x}{L} \qquad (6.90)$$

where $\beta_m = \pi(m - \frac{1}{2})$. This integral can be evaluated by a simple change of variable. Let $\lambda = t - \tau$ and $d\tau = -d\lambda$. Then, Eq. (6.90) becomes

$$\alpha \int_{\tau=0}^{t-\Delta t} d\tau \frac{f_i(\tau)}{k_i} G^L(x, t|0, \tau) = 2 \frac{q_0 L}{k} \sum_{m=1}^{\infty} \cos \left(\frac{\beta_m x}{L} \right)$$

$$\times \frac{1}{\beta_m^2} (e^{-\beta_m^2 \alpha \Delta t/L^2} - e^{-\beta_m^2 \alpha t/L^2}) \qquad (6.91)$$

for $t > \Delta t$. This expression is non-negative and it rapidly converges because the arguments of the exponential factors are never too small. The numerical convergence is controlled by the exponential factors, not by the $1/\beta_m^2$ term.

The second integral in Eq. (6.89) involves the small-time GF and it is similar to the integral evaluated in Eq. (6.88). Finally, the temperature given by Eq. (6.89) for $t > \Delta t$ is given by

$$T(x, t) = 2 \frac{q_0 L}{k} \sum_{n=-1}^{1} (-1)^n \left(\frac{\alpha \Delta t}{L^2} \right)^{1/2} \text{ierfc} \left[\frac{|2n + x/L|}{2(\alpha \Delta t/L^2)^{1/2}} \right]$$

$$+ 2 \frac{q_0 L}{k} \sum_{m=1}^{\infty} \cos \left(\frac{\beta_m x}{L} \right) \frac{1}{\beta_m^2} (e^{-\beta_m^2 \alpha \Delta t/L^2} - e^{-\beta_m^2 \alpha t/L^2}) \qquad (6.92)$$

where $\beta_m = \pi(m - \frac{1}{2})$ and $\alpha \Delta t/L^2 = 0.025$. Note the term Δt that appears in the error function term. This expression is very efficient numerically, as only three terms of each infinite series are needed for accuracy to four decimal places for $\alpha t/L^2 \approx 0.025$, and the accuracy improves as $\alpha t/L^2$ increases. More terms of the series will give more accuracy. The accuracy of Eq. (6.92) is determined by comparing it with the classic Fourier series solution of the same problem [given by Eq. (6.92) in the limit as $\Delta t \rightarrow 0$]. Many, many terms of the classic Fourier series solution are required for accurate numerical values.

Alternative solution. The alternative GF solution method can be applied to this problem in place of time partitioning to improve the numerical convergence of the solution at large times. The alternative GF solution method is introduced in Section 3.4.

With the alternative GF method the solution is sought in the form $T(x, t) = T^* + T'(x, t)$ where the function T^* satisfies the boundary conditions but does not necessarily satisfy the initial conditions. In the present case, T^* reduces to the steady-state temperature distribution in the body:

$$T^*(x) = \frac{q_0 L}{k} \left(1 - \frac{x}{L} \right) \tag{6.93}$$

The alternative GF solution equation can then be used to find

$$T(x, t) = T^*(x) + \int_{x'=0}^{L} G_{X21}^{L}(x, t | x', 0) [-T^*(x')] \, dx' \tag{6.94}$$

or

$$T(x, t) = \frac{q_0 L}{k} \left(1 - \frac{x}{L} \right) - 2 \frac{q_0 L}{k} \sum_{m=1}^{\infty} e^{-\beta_m^2 \alpha t / L^2} \frac{\cos (\beta_m x/L)}{\beta_m^2} \tag{6.95}$$

This expression has better convergence properties than the classic Fourier series solution at large times $(\alpha t/L^2 > 0.025)$.

Example 6.7: Slab with convection on both sides. A large flat plate of thickness $2L$, initially at temperature T_0, is quenched in a large tank of fluid at temperature T_∞. The heat transfer coefficient for the quenching process is h, a constant. Find the temperature distribution $T(x, t)$ using time partitioning.

SOLUTION The geometry of the quenching problem is shown in Fig. 6.9a. This problem is modeled as the $X32$ geometry shown in Fig. 6.9b. The centerline of the plate is a plane of symmetry, which is modeled as an insulated boundary. The initial condition is made homogeneous by defining a new variable $T = T - T_0$, and the fluid temperature becomes $(T_\infty - T_0)$. This is the $X32B10T0$ case.

The GF solution using time partitioning is given by Eq. (6.84),

$$T(x, t) - T_0 = \alpha \int_{\tau=0}^{t-\Delta t} d\tau \, \frac{h(T_\infty - T_0)}{k} G_{X32}^{L}(x, t | 0, \tau)$$

$$+ \alpha \int_{\tau=t-\Delta t}^{t} d\tau \, \frac{h(T_\infty - T_0)}{k} G_{X32}^{S}(x, t | 0, \tau) \tag{6.96}$$

This equation will be analyzed as

$$T(x, t) - T_0 = T^L(x, \Delta t) + T^S(x, t, \Delta t)$$

where T^L and T^S are the first and second integrals in Eq. (6.96), respectively.

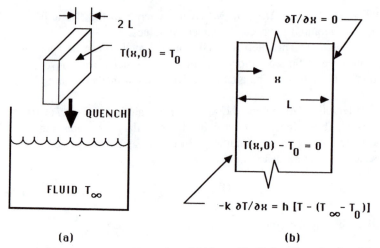

(a) **(b)**

Figure 6.9 (a) Quenching of large plate of thickness $2L$. (b) One-dimensional model using $X32$ geometry on $0 \le x \le L$.

Consider the first integral in Eq. (6.96), named $T^L(x, t, \Delta t)$. The large-time $X32$ GF is listed in Tables 4.2 and 4.3, and the result is

$$T^L(x, t, \Delta t) = \alpha \int_{\tau=0}^{t-\Delta t} d\tau \, \frac{h(T_\infty - T_0)}{k} \frac{2}{L} \sum_{m=1}^{\infty} e^{-\beta_m^2 \alpha(t-\tau)/L^2}$$

$$\times \frac{\beta_m^2 + B^2}{\beta_m^2 + B^2 + B} \cos\left[\beta_m\left(1 - \frac{x}{L}\right)\right] \cos \beta_m \quad (6.97a)$$

where

$$B = \frac{hL}{k} \quad \text{and} \quad \beta_m \tan \beta_m = B \quad (6.97b)$$

The integral in Eq. (6.97a) operates only on the exponential term, and the result is given by

$$T^L(x, t, \Delta t) = \frac{2h(T_\infty - T_0)L}{k} \sum_{m=1}^{\infty} \frac{1}{\beta_m^2} \left(e^{-\beta_m^2 \alpha \Delta t/L^2} - e^{-\beta_m^2 \alpha t/L^2}\right)$$

$$\times \frac{\beta_m^2 + B^2}{\beta_m^2 + B^2 + B} \cos\left[\beta_m\left(1 - \frac{x}{L}\right)\right] \cos \beta_m \quad (6.98)$$

Only a few terms of the series are required for good numerical results.

Consider the second integral in Eq. (6.96), with G_{X32}^S taken from Appendix X:

$$T^S(x, \Delta t) = \alpha \int_{\tau=0}^{t} \frac{h(T_\infty - T_0)}{k} \left(\frac{1}{[4\pi\alpha(t - \tau)]^{1/2}}\right.$$

$$\times \left\{ 2 \exp\left[\frac{-x^2}{4\alpha(t-\tau)}\right] + \exp\left[\frac{-(2L-x)^2}{4\alpha(t-\tau)}\right] \right\}$$

$$-\frac{h}{k} \exp\left[\frac{hx}{k} + \frac{h^2}{k^2}\alpha(t-\tau)\right]$$

$$\times \text{erfc}\left\{\frac{x}{2[\alpha(t-\tau)]^{1/2}} + \frac{h}{k}[\alpha(t-\tau)]^{1/2}\right\}\Bigg) \, d\tau \qquad (6.99)$$

Note that G^S_{X32} contains some terms identical to G_{X30} discussed in Example 6.3. At very small times, the $X32$ geometry at $x = 0$ acts like the semi-infinite $X30$ geometry. In general, slab-body small-time GFs contain terms that are the same as the semi-infinite GFs, so that insight gained from semi-infinite problems applies to slab-body problems. In the present example, the integral in Eq. (6.99) may be solved with the help of Table 5.5 to give

$$T^S(x, \Delta t) = (T_\infty - T_0)\left\{\frac{h}{k}(\alpha\Delta t)^{1/2}\,\text{ierfc}\left[\frac{|2L-x|}{(4\alpha\Delta t)^{1/2}}\right]\right.$$

$$+\,\text{erfc}\left[\frac{x}{2(\alpha\Delta t)^{1/2}}\right] - \exp\left[\frac{hx}{k} + \alpha\Delta t\,\frac{h^2}{k^2}\right]$$

$$\times\,\text{erfc}\left[\frac{x}{2(\alpha\Delta t)^{1/2}} + \frac{h}{k}(\alpha\Delta t)^{1/2}\right]\Bigg\} \qquad (6.100)$$

Finally, the temperature over the whole range of time is given by

$$T(x, t) - T_0 = \begin{cases} T^S(x, t) & \text{for } t < \Delta t \\ T^S(x, \Delta t) + T^L(x, t, \Delta t) & \text{for } t > \Delta t \end{cases} \qquad (6.101)$$

where the partition time is given by $\alpha\Delta t/L^2 = 0.025$, T^L is given by Eq. (6.98) and T^S is given by Eq. (6.100). Note that for $t > \Delta t$, the expression for $T^S(x, \Delta t)$ does not depend on the time t. The quantity Δt is a constant, so the spatial distribution $T^S(x, \Delta t)$ may be calculated once and stored for evaluating the temperature at different values of $t > \Delta t$.

The eigenvalues β_m must be found to get numerical values for $T^L(x, t, \Delta t)$, and formulas for approximating β_m are listed in Appendix X. Note that the $X23$ GF could be used instead of the $X32$ GF to carry out this example, but the $x = 0$ and $x = L$ locations would have to be reversed and the $X23$ GF would be evaluated at $x = L$ where the convection occurs; that is, $G_{X23}(x, t|x' = L, \tau)$ would be replaced into Eq. (6.96).

6.6.2 Volume Energy Generation

The time partitioning procedure is now illustrated for finite one-dimensional cases having boundary conditions of the first and second kinds and with spatially uniform volume energy generation over a portion of the body. That is,

$$g(x', \tau) = \begin{cases} g_0L\, g_t(\tau) & 0 < x' < a \\ 0 & a < x' < L \end{cases} \tag{6.102}$$

The source strength is spatially uniform from $x' = 0$ to a and is zero otherwise. The solution is found using time partitioning for $t > \Delta t$ where Δt is the partition time. For $t < \Delta t$ the early time solution is used. For the initial temperature $F(x)$ being zero, Eq. (6.84) gives for $t > \Delta t$,

$$T(x, t > \Delta t) = \frac{\alpha}{k} \int_{\tau=0}^{t-\Delta t} \int_{x'=0}^{a} G^L(x, t|x', \tau)\, g_0L\, g_t(\tau)\, dx'\, d\tau$$

$$+ \frac{\alpha}{k} \int_{\tau=t-\Delta t}^{t} \int_{x'=0}^{a} G^S(x, t|x', \tau)\, g_0L\, g_t(\tau)\, dx'\, d\tau \tag{6.103}$$

Now, as usual, consider the integrals over x' first. The integral of $G^L(\cdot)$ over x' can be written as

$$\int_{x'=0}^{a} G^L(x, t|x', \tau)\, dx' = \sum_{m=0}^{\infty} \exp\left[-\frac{\beta_m^2 \alpha(t - \tau)}{L^2}\right]$$

$$\times \frac{\mathbf{X}(\beta_m x/L)\,\mathbf{IX}(\beta_m a/L)}{N_m} \tag{6.104}$$

where $\mathbf{IX}(\cdot)$ is defined to be

$$\mathbf{IX}\left(\frac{\beta_m a}{L}\right) = \int_0^a \mathbf{X}\left(\frac{\beta_m x}{L}\right) dx' \tag{6.105}$$

The $\mathbf{X}(\beta_m x/L)$ functions are eigenfunctions for $G^L(\cdot)$ listed in Table 4.2 and they are either sin $(\cdot)$ or cos $(\cdot)$ depending on the boundary conditions. The integral of $G^S(\cdot)$ over x' is found for the general case for XIJ, $I, J = 1, 2$, by using Tables 4.1, 5.1, and 5.2:

$$\int_{x'=0}^{a} G^S(x, t|x', \tau)\, dx'$$

$$= \frac{1}{2} \sum_{n=-\infty}^{\infty} (-1)^{(I+J)n} \left(\text{erfc}\left\{\frac{2nL + x - a}{[4\alpha(t - \tau)]^{1/2}}\right\} \right.$$

$$- \text{erfc}\left\{\frac{2nL + x}{[4\alpha(t - \tau)]^{1/2}}\right\} + (-1)^I \text{erfc}\left\{\frac{2nL + x}{[4\alpha(t - \tau)]^{1/2}}\right\}$$

$$\left. - (-1)^I \text{erfc}\left\{\frac{2nL + x + a}{[4\alpha(t - \tau)]^{1/2}}\right\} \right) \tag{6.106}$$

For the small time of $\alpha\Delta t/L^2 = 0.025$ and $n \geq 0$, Table 5.6 shows that there will be a significant contribution to the summation in Eq. (6.106) only when the numerator of the arguments of erfc$(\cdot)$ is less than L. For the negative values of n, more care is required, as pointed out in connection with Table 6.5. Recall the identity

$$\text{erfc}(-z) = 2 - \text{erfc}(z) \tag{6.107}$$

The $n = -1$ term inside the summation of Eq. (6.106) may be written with the above identity to give

$$(n = -1 \text{ term}) = (-1)^{I-J}\left[2 - \text{erfc}\left\{ \frac{2L - x + a}{[4\alpha(t - \tau)]^{1/2}} \right\} \right.$$

$$- \left(2 - \text{erfc}\left\{ \frac{2L - x}{[4\alpha(t - \tau)]^{1/2}} \right\} \right)$$

$$+ (-1)^I\left(2 - \text{erfc}\left\{ \frac{2L - x}{[4\alpha(t - \tau)]^{1/2}} \right\} \right)$$

$$\left. - (-1)^I\left(2 - \text{erfc}\left\{ \frac{2L - x - a}{[4\alpha(t - \tau)]^{1/2}} \right\} \right) \right] \qquad (6.108a)$$

As there a significant contribution (for $\alpha \Delta t / L^2 < 0.025$) only for the arguments of the numerators being greater than L, Eq. (6.108a) may be written approximately as

$$(n = -1 \text{ term}) \approx (-1)^J \text{erfc}\left\{ \frac{2L - x - a}{[4\alpha(t - \tau)]^{1/2}} \right\} \qquad (6.108b)$$

Finally, using the appropriate $n = 0$ terms and Eq. (6.108b), Eq. (6.106) gives

$$\int_{x'=0}^{a} G^S(x, t|x', \tau)\, dx' \approx \frac{1}{2}\left(\text{erfc}\left\{ \frac{x - a}{[4\alpha(t - \tau)]^{1/2}} \right\} \right.$$

$$- [1 - (-1)^I)] \text{erfc}\left\{ \frac{x}{[4\alpha(t - \tau)]^{1/2}} \right\}$$

$$\left. + (-1)^J \text{erfc}\left\{ \frac{2L - x - a}{[4\alpha(t - \tau)]^{1/2}} \right\} \right) \qquad (6.109)$$

For the special case of $a = L$ and for the $X22$ geometry (i.e., $I = J = 2$), Eq. (6.109) becomes

$$\int_{x'=0}^{L} G^S_{X22}(x, t|x', \tau)\, dx' = 1 \qquad (6.110)$$

This partial result for the $X22$ case with spatially uniform volume energy generation over the entire body means that there is no dependence on x or on $t - \tau$ in the integration of $G^S(\cdot)$ over $x' = 0$ to L. The body is acting as if it is "lumped." Notice that this is not true for the $X11$, $X12$, and $X21$ cases. Also, this is a partial result because in general there is an integration over τ yet to be performed in Eq. (6.103).

The time integration of Eq. (6.103) is now considered. This, in turn, requires a choice of the form of $g_t(\tau)$. Two cases are considered here:

$$g_t(\tau) = 1 \qquad \text{and} \qquad g_t(\tau) = \frac{\tau}{t_0} \qquad (6.111\text{a, b})$$

For the first of these, $g_t(\tau) = 1$, integration over τ in Eq. (6.103) yields, for $t > \Delta t$,

$$T(x, t) = \frac{g_0 L^2}{k} \left(L \sum_{m=1}^{\infty} \left(e^{-\beta_m^2 \alpha \Delta t / L^2} - e^{-\beta_m^2 \alpha t / L^2} \right) \frac{X(\beta_m x/L) \mathbf{I} X(\beta_m a/L)}{N_m} \right.$$

$$+ \delta_{2I} \delta_{2J} \frac{\alpha(t - \Delta t)}{L^2} + 2 \frac{\alpha \Delta t}{L^2} \left\{ i^2 \, \mathrm{erfc} \left[\frac{x - a}{(4\alpha \Delta t)^{1/2}} \right] \right.$$

$$- [1 - (-1)^I] \, i^2 \, \mathrm{erfc} \left[\frac{x}{(4\alpha \Delta t)^{1/2}} \right]$$

$$\left. \left. + (-1)^J i^2 \, \mathrm{erfc} \left[\frac{2L - x - a}{(4\alpha \Delta t)^{1/2}} \right] \right\} \right) \tag{6.112}$$

where Tables 5.8 and 5.4 have been used. The cases covered by Eq. (6.112) are denoted *XIJB00T0Gx5t1* for *I*, *J* = 1, 2. Equation (6.112) can also be broken into steady state and transient parts.

Now consider the linear time variation of the volume energy generation, $g_t(\tau) = \tau/t_0$. The solution for $t > \Delta t$ is

$$T(x, t) = \frac{g_0 L^2}{k} \left[\frac{L^2}{\alpha t_0} L \sum_{m=1}^{\infty} \left\{ [\beta_m^2(t^+ - \Delta t^+) - 1] e^{-\beta_m^2 \Delta t^+} - e^{-\beta_m^2 t^+} \right\} \right.$$

$$\times \frac{X(\beta_m x/L) X(\beta_m a/L)}{N_m \beta_m^4} + \delta_{2I} \delta_{2J} \frac{1}{2} \frac{t - \Delta t}{t_0} \frac{\alpha(t - \Delta t)}{L^2}$$

$$+ \frac{2\alpha \Delta t}{L^2} \left(\frac{t - \Delta t}{t_0} i^2 \, \mathrm{erfc} \left[\frac{x - a}{(4\alpha \Delta t)^{1/2}} \right] \right.$$

$$+ \frac{4\Delta t}{t_0} i^4 \, \mathrm{erfc} \left[\frac{x - a}{(4\alpha \Delta t)^{1/2}} \right]$$

$$- [1 - (-1)^I] \left\{ \frac{t - \Delta t}{t_0} i^2 \, \mathrm{erfc} \left[\frac{x}{(4\alpha \Delta t)^{1/2}} \right] \right.$$

$$+ \frac{4\Delta t}{t_0} i^4 \, \mathrm{erfc} \left[\frac{x - a}{(4\alpha \Delta t)^{1/2}} \right] \right\}$$

$$+ (-1)^J \left\{ \frac{t - \Delta t}{t_0} i^2 \, \mathrm{erfc} \left[\frac{2L - x - a}{(4\alpha \Delta t)^{1/2}} \right] \right.$$

$$\left. \left. \left. + \frac{4\Delta t}{t_0} i^4 \, \mathrm{erfc} \left[\frac{2L - x - a}{(4\alpha \Delta t)^{1/2}} \right] \right\} \right) \right] \tag{6.113}$$

where $t^+ = \alpha t/L^2$ and $\Delta t^+ = \alpha \Delta t/L^2$. Equation (6.113) has the notation of *XIJB00T0Gx5t2*, with *I*, *J* = 1, 2. Notice that for the case *I* = *J* = 2, (X22), the Kronecker delta terms gives $\delta_{2I} \delta_{2J} = 1$ and the temperature increases like t^2; hence, there is no steady-state portion in Eq. (6.113). The standard separation of variables procedure does not work for this problem because the source term is not a constant.

This GF method works, however, and it is not difficult to evaluate Eq. (6.113) for $\alpha \Delta t / L^2 = 0.025$. For even smaller times, Eq. (6.113) can be used with Δt replaced by t to give the small-time solution, which means that only the terms containing $i^4 \mathrm{erfc}(\cdot)$ are present.

A comment is in order regarding the solution given by Eq. (6.113) which is for $g_t(\tau) = \tau/t_0$. The time integration in Eq. (6.103) for the small-time GF is integrated over τ from $\tau = t - \Delta t$ to t. The integrand contains terms of the form of $\mathrm{erfc}\{z[4\alpha (t - \tau)]^{1/2}\}$, as given in Eq. (6.113). The interesting point is that although the magnitude of τ is varying with t (from $t - \Delta t$ to t), the small time integration result given in Eq. (6.113) depends in a very simple way on t. For example, the $i^2 \mathrm{erfc}(\cdot)$ and $i^4 \mathrm{erfc}(\cdot)$ functions are independent of t and depend only on Δt. The t terms enter only in a simple multiplicative manner as a coefficient for the $i^2 \mathrm{erf}(\cdot)$ terms. As a consequence, the $i^2 \mathrm{erfc}(\cdot)$ and $i^4 \mathrm{erfc}(\cdot)$ terms need to be evaluated just once for a given Δt value, and they do not have to be reevaluated as t increases.

The same comment can be made regarding the small-time terms in Eq. (6.112), but the independence of $i^2 \mathrm{erfc}(\cdot)$ with respect to t is not so surprising because $g_t(\tau) = 1$ in this equation (independent of t).

6.7 TWO-DIMENSIONAL RECTANGULAR BODIES

Transient temperatures in two-dimensional rectangular bodies are discussed in this section. The GF for two-dimensional cases can be found by multiplying one-dimensional GF together for boundary conditions of type 0, 1, 2, and 3. Thus for many cases the temperature solution can be written down immediately in integral form.

Multidimension cases are often more difficult than one-dimensional cases because the integrals in the GFSE are more difficult. Often the spatial integrals can be evaluated, but sometimes the time integral cannot be evaluated in closed form. In this event, numerical methods may be required to get accurate numbers for the temperature.

Some two-dimension rectangle cases are solved in the literature. Ozisik (1980) has some examples with boundary conditions of type 1. Zang (1987, 1988) has used time partitioning with three time regions to take into account the different dimensionless time scales for rectangles with unequal sides. In this section two examples are discussed for boundary conditions of type 1 and 2.

Example 6.8: Rectangular body with several different boundary conditions— **X21B10Y21B01 case.** Consider a rectangle with zero initial temperature, with one side uniformly heated, one side at a fixed temperature, T_0, one side at a fixed temperature of zero, and one side insulated. Find the temperature by using large time GFs.

SOLUTION This is the $X21B10Y21B01$ case and the geometry is shown in Fig. 6.10. The boundary value problem is given by

$$\frac{\partial^2 T}{\partial x^2} + \frac{\partial^2 T}{\partial y^2} = \frac{1}{\alpha} \frac{\partial T}{\partial t} \tag{6.114}$$

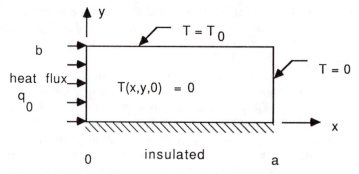

Figure 6.10 Geometry for rectangular body in Example 6.8.

$$T(x, y, 0) = 0 \tag{6.115a}$$

$$-k \frac{\partial T}{\partial x}\bigg|_{x=0} = q_0 = \text{constant} \tag{6.115b}$$

$$T(a, y, t) = 0 \tag{6.115c}$$

$$\frac{\partial T}{\partial x}\bigg|_{y=0} = 0 \tag{6.115d}$$

$$T(x, b, t) = T_0 \tag{6.115e}$$

The integral expression for the temperature can be written down immediately from the GFSE. There are two terms to account for the heating at $x = 0$ and the nonzero temperature at $y = b$:

$$T(x, y, t) = \alpha \int_{\tau=0}^{t} d\tau \int_{y'=0}^{b} \frac{q_0}{k} G_{X21Y21}(x, y, t|0, y', \tau)\, dy'$$

$$- \alpha \int_{\tau=0}^{t} d\tau \int_{x'=0}^{a} T_0 \frac{\partial G_{X21Y21}}{\partial y'}\bigg|_{y'=0} dx' \tag{6.116}$$

The GF is formed by multiplying two one-dimensional GFs together. That is,

$$G_{X21Y21}(x, t|x', \tau) = G_{X21}(x, t|x', \tau)\, G_{Y21}(y, t|y', \tau) \tag{6.117}$$

where G_{X21} and G_{Y21} can be readily obtained from Appendix X as

$$G_{X21}(x, t|x', \tau) = \frac{2}{a} \sum_{m=1}^{\infty} e^{-\beta_m^2 \alpha(t-\tau)/a^2} \cos\left(\frac{\beta_m x}{a}\right) \cos\frac{\beta_m x'}{a} \tag{6.118a}$$

$$G_{Y21}(y, t|y', \tau) = \frac{2}{b} \sum_{n=1}^{\infty} e^{-\beta_n^2 \alpha(t-\tau)/b^2} \cos\left(\frac{\beta_n y}{b}\right) \cos\frac{\beta_n y'}{b} \tag{6.118b}$$

where

$$\beta_m = \pi(m - \tfrac{1}{2}) \qquad \beta_n = \pi(n - \tfrac{1}{2}) \tag{6.119}$$

The spatial integrals in Eq. (6.116) operate only on the cosine terms. The time integral can be carried out independently on the product of the exponentials:

$$\int_{\tau=0}^{t} e^{-\beta_m^2 \alpha(t-\tau)/a^2} \, e^{-\beta_n^2 \alpha(t-\tau)/b^2} \, d\tau = \frac{1}{\alpha C} (1 - e^{-\alpha t C}) \qquad (6.120)$$

where $C = (\beta_m/a)^2 + (\beta_n/b)^2$. Then the temperature is given by Eq. (6.116) with Eq. (6.117) and (6.118) (Beck, 1984),

$$T(x, y, t) = 4 \sum_{m=1}^{\infty} \sum_{n=1}^{\infty} (1 - e^{-\alpha t C}) \cos\left(\frac{\beta_m x}{a}\right) \cos\left(\frac{\beta_n y}{b}\right)$$

$$\times (-1)^n \left\{ \frac{q_0 a}{k} \frac{1}{\beta_n[\beta_m^2 + \beta_n^2(a/b)^2]} \right.$$

$$\left. + T_0 \frac{\beta_n(-1)^m}{\beta_m[\beta_n^2 + \beta_m^2(b/a)^2]} \right\} \qquad (6.121)$$

where $C = (\beta_m/a)^2 + (\beta_n/b)^2$. There are two difficulties with this solution. First, this is the large-time solution suitable only for $\alpha t/b^2$ and $\alpha t/a^2$ large (greater than 0.05, say). Second, the most difficult part of the solution to evaluate directly is the steady-state part for $T_0 \neq 0$ and $q_0 = 0$:

$$T(x, y) = 4T_0 \sum_{m=1}^{\infty} \sum_{n=1}^{\infty} \cos\left(\frac{\beta_m x}{a}\right) \cos\left(\frac{\beta_n y}{b}\right) \frac{\beta_n(-1)^{m+n}}{\beta_m[\beta_n^2 + \beta_m^2(b/a)^2]} \qquad (6.122)$$

This part of the solution converges slowly because for $m \gg 1$ and $n \gg 1$, the series converges something like $n(-1)^{m+n}/(mn^2 + m^3)$ which is painfully close to the slowly converging series $1/n^2$. Time partitioning to improve convergence of this expression is discussed in Zang (1987). Other methods are given in Chapter 5.

Example 6.9: Rectangular body heated over part of one face. Consider a rectangle heated over part of one face. The other faces are held at a fixed temperature of zero and the initial temperature is also zero. The geometry is shown in Figure 6.11. The boundary value problem is given by

$$\frac{\partial^2 T}{\partial x^2} + \frac{\partial^2 T}{\partial y^2} = \frac{1}{\alpha} \frac{\partial T}{\partial t} \qquad 0 < x < a \qquad 0 < y < b \qquad t > 0 \qquad (6.123a)$$

$$T(0, y, t) = T(a, y, t) = T(x, b, t) = T(x, y, 0) = 0 \qquad (6.123b)$$

$$-k \frac{\partial T}{\partial y}\bigg|_{y=0} = \begin{cases} q_0 & 0 < x < a_1 \\ 0 & a_1 < x < a \end{cases} \qquad (6.123c)$$

(a) Solve the problem using the large-time GFs.

(b) Solve the problem using small-time GFs and retain only the terms needed for small times near $x = a_1$, and near $y = 0$.

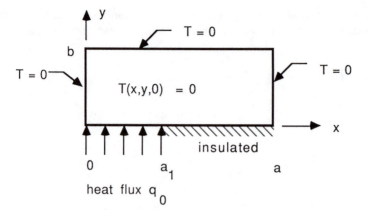

Figure 6.11 Geometry for rectangular body heated over part of one face.

SOLUTION The number for this case is $X11B00Y21B(x5)0T0$. The GFSE for this problem is

$$T(x, y, t) = \frac{\alpha q_0}{k} \int_{x'=0}^{a_1} \int_{\tau=0}^{t} G_{X11}(x, t|x', \tau) G_{Y21}(y, t|0, \tau)\, dx'd\tau \quad (6.124)$$

a. Large-time solution. The large-time forms of the GFs are

$$G_{X11}^L(x, t|x', \tau) = \frac{2}{a} \sum_{m=1}^{\infty} e^{-m^2\pi^2\alpha(t-\tau)/a^2} \sin\left(m\pi\frac{x}{a}\right) \sin\left(m\pi\frac{x'}{a}\right) \quad (6.125a)$$

$$G_{Y21}^L(y, t|y', \tau) = \frac{2}{b} \sum_{n=1}^{\infty} e^{-\beta_n^2\alpha(t-\tau)/b^2} \cos\left(\beta_n\frac{y}{b}\right) \cos\left(\beta_n\frac{y'}{b}\right) \quad (6.125b)$$

where $\beta_n = \pi(n - \frac{1}{2})$. Solving the problem using the GFSE and $G^L(\cdot)$ requires the integrals

$$\int_{x'=0}^{a_1} \sin\left(m\pi\frac{x'}{a}\right) dx' = \frac{a}{m\pi}\left[1 - \cos\left(m\pi\frac{a_1}{a}\right)\right] \quad (6.126a)$$

$$\int_{\tau=0}^{t} e^{-(\cdot)\alpha(t-\tau)}\, d\tau = \frac{1}{(\cdot)\alpha}(1 - e^{-(\cdot)\alpha t}) \quad (6.126b)$$

where $(\cdot)$ is equal to

$$(\cdot) = \left(\frac{m^2\pi^2}{a^2} + \frac{\beta_n^2}{b^2}\right) \quad (6.127)$$

Using these integrals in Eq. (6.124) gives

$$T(x, y, t) = \frac{4q_0 a^2}{\pi k b} \sum_{m=1}^{\infty} \sum_{n=1}^{\infty} (1 - e^{-(\cdot)\alpha t}) \frac{1 - \cos(m\pi a_1/a)}{m[m^2\pi^2 + (a^2/b^2)\beta_n^2]}$$

$$\times \sin\left(m\pi\frac{x}{a}\right) \cos\left(\beta_n\frac{y}{b}\right) \quad (6.128)$$

There are two parts to this solution: steady-state and transient. The steady-state part converges something like $1/m^3$, which is faster than the steady state in the previous example but which may still require many terms of the series for accurate evaluation.

b. *Small-time solution.* At early times, any temperature changes occur near the heated boundary $y = 0$, and elsewhere, the temperature remains zero. The small time GFs useful for the early time solution are given in Appendix X in the form of infinite series. Near the point $x = a_1$ and $y = 0$, however, just the dominant terms of the series may be used. An equivalent point of view at early time is to replace the rectangle by the quarter-infinite body described by number $X10B0Y20(x5)T0$. The appropriate GFs are

$$G_{X11}^{S}(x, t|x', \tau) \simeq G_{X10}(x, t|x', \tau)$$

$$= \frac{1}{[4\pi\alpha(t - \tau)]^{1/2}} \left(e^{-(x-x')^2/[4\alpha(t-\tau)]}\right.$$

$$\left.- e^{-(x+x')^2/[4\alpha(t-\tau)]}\right) \qquad (6.129)$$

$$G_{Y21}^{S}(y, t|0, \tau) \simeq G_{Y20}(y, t|0, \tau)$$

$$= \frac{2}{[4\pi\alpha(t - \tau)]^{1/2}} e^{-y^2/[4\alpha(t-\tau)]} \qquad (6.130)$$

Next, replace these GFs into the temperature expression given by Eq. (6.124). The integral over x' should be familiar; by focusing on the area of interest near $x = a_1$ the integral over x' may be written as

$$\int_{x'=0}^{a_1} G_{X10}^{S}(x, t|x', \tau) \, dx'$$

$$= \frac{1}{2} \left(\text{erfc} \left\{\frac{x - a_1}{[4\alpha(t - \tau)]^{1/2}}\right\} - 2 \text{ erfc} \left\{\frac{x}{[4\alpha(t - \tau)]^{1/2}}\right\}\right.$$

$$\left.+ \text{erfc} \left\{\frac{x + a_1}{[4\alpha(t - \tau)]^{1/2}}\right\}\right)$$

$$\simeq \frac{1}{2} \text{erfc} \left\{\frac{x - a_1}{[4\alpha(t - \tau)]^{1/2}}\right\} \qquad (6.131)$$

for x near a_1. Then the solution for small y values and for x near a_1 becomes

$$T(x, y, t) = \frac{\alpha q_0}{k} \int_{\tau=0}^{t} \frac{1}{2} \text{erfc} \left\{\frac{x - a}{[4\alpha(t - \tau)]^{1/2}}\right\}$$

$$\times \frac{1}{[\pi\alpha(t - \tau)]^{1/2}} e^{-y^2/[4\alpha(t-\tau)]} \, d\tau \qquad (6.132)$$

This is a difficult integral and it will be evaluated below with an approximate integrand. This integral is evaluated exactly in Section 6.8 in the form of an infinite series.

The integral in Eq. (6.132) may be evaluated in closed form if an approximation for the complementary error function is used. The erfc(z) function for "small" values of z can be approximated by

$$
\text{erfc}(z) = \begin{cases} 1 - Az & -A^{-1} < z < A^{-1} \\ 0 & z > A^{-1} \\ 2 & z < -A^{-1} \end{cases} \qquad (6.133)
$$

where $A = 2/\pi^{1/2}$. Using this approximation then gives, for the temperature for small y and near $x = a_1$,

$$
T(x, y, t) = T_0 + \frac{\alpha q_0}{k} \int_{u=u_m}^{t} \frac{1}{2} \left[1 - A \frac{x - a_1}{(4\alpha u)^{1/2}} \right]
$$

$$
\times \frac{1}{(\pi \alpha u)^{1/2}} e^{-y^2/[4\alpha u]} du \qquad t > u_m \qquad (6.134)
$$

where $u_m = A^2(x - a_1)^2/\alpha$. Note that the region of $u = 0$ to u_m (which corresponds to $\tau = t$ to $t - u_m$) has no contribution to the temperature using the above approximation for erfc(z). This equation also implies that the region under which the approximation for erfc(z) is useful is given by coordinate x in the range

$$
a_1 - A^{-1}(4\alpha t)^{1/2} < x < a_1 + A^{-1}(4\alpha t)^{1/2} \qquad (6.135)
$$

This equation defines what the phrase "x near a_1 at early time" means in describing the range of application of Eq. (6.134).

For smaller x values (but not near $x = 0$), the temperature distribution is given by

$$
T(x, y, t) = \frac{\alpha q_0}{k} \int_{u=0} \frac{1}{(\pi \alpha u)^{1/2}} e^{-y^2/4\alpha u} du
$$

$$
= 2q_0 \left(\frac{t}{k\rho c} \right)^{1/2} \text{ierfc} \left[\frac{y}{(4\alpha t)^{1/2}} \right] \qquad (6.136)
$$

This result is exactly the same as for a semi-infinite body that is uniformly heated over its entire surface.

For x values larger than a_1, the surface at $y = 0$ is insulated. Sufficiently far from a_1 indicated by

$$
x > a_1 + A^{-1}(4\alpha u)^{1/2} \qquad (6.137)
$$

the temperature near the surface $y = 0$ is simply zero.

Surface temperature. The temperature on the heated surface can be found directly by substituting $y = 0$ into the temperature expression at any point in the derivation. Often the surface temperature is easier to find than interior temperatures. The surface temperature is given by

$$T(x, 0, t) \simeq \frac{\alpha q_0}{k} \int_{u_m}^{t} \frac{1}{2} \left[1 - \frac{x - a_1}{(4\alpha u)^{1/2}} \right] \frac{1}{(\pi \alpha u)^{1/2}} \, du$$

$$= \frac{\alpha q_0}{k} \left[\frac{1}{(\pi \alpha)^{1/2}} (t^{1/2} - u_m^{1/2}) - A \frac{x - a_1}{2\pi^{1/2} \alpha} \ln \left(\frac{t}{u_m} \right) \right]$$

$$= q_0 \left[\left(\frac{t}{\pi k \rho c} \right)^{1/2} - \frac{A(x - a_1)}{k \sqrt{\pi}} \right]$$

$$- \frac{q_0(x - a_1)}{k} \frac{A}{2\sqrt{\pi}} \ln \frac{\alpha t}{A^2(x - a_1)^2} \tag{6.138}$$

for $- A^{-1}(4\alpha t)^{1/2} < x - a_1 < A^{-1}(4\alpha t)^{1/2}$ and $x \neq a_1$.

For larger values of x in the range $a_1 < x < a$, such that

$$x > a_1 + A^{-1}(4\alpha t)^{1/2}$$

the surface temperature is simply zero, and for smaller values of x in the range $0 < x < a_1$ such that

$$x < a_1 - A^{-1}(4\alpha t)^{1/2} \tag{6.139a}$$

the surface temperature is given by

$$T(x, 0, t) = 2 q_0 \left(\frac{t}{\pi k \rho c} \right)^{1/2} \tag{6.139b}$$

which is the same as the surface temperature for a uniformly heated semi-infinite body.

6.8 TWO-DIMENSIONAL SEMI-INFINITE BODIES

The temperature in a semi-infinite body heated over half of the surface and insulated over the other half is treated in this section. This is a basic solution of two-dimensional heat conduction because it serves as a building block for other solutions and it is a kernel function for the unsteady surface element method discussed in Chapter 12.

The temperature is presented first in integral form, and then two series expressions for the integral are presented to evaluate the temperature efficiently at any location in the body and at any value of time.

6.8.1 Integral Expression for the Temperature

The geometry for the semi-infinite body heated over the half-plane is shown in Fig. 6.12. The initial temperature is zero and the spatially uniform heat flux q_0 begins at time zero. This is the $X00Y20B5T0$ case. The temperature is given by the GF equation in the form

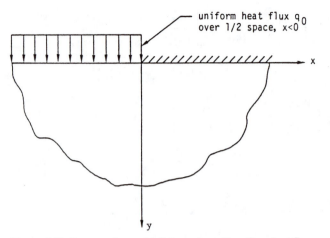

Figure 6.12 Geometry for semi-infinite region with uniform heat flux q_0 over half-space $\infty < x < 0$ and $y = 0$.

$$T(x, y, t) = \frac{\alpha q_0}{k} \int_{\tau=0}^{t} \int_{x'=-\infty}^{0} G_{X00Y20}(x, y, t|x', 0, \tau)\, dx'\, d\tau \quad (6.140)$$

where q_0 is a constant. Note that the GF is evaluated at the surface $y' = 0$, and that the integral over surface extends over only the heated half plane $-\infty < x' < 0$. The GF is given by a product solution of two familiar one-dimensional GFs, $G_{X00Y20} = G_{X00}\, G_{Y20}$.

The integral on x' in Eq. (6.140) falls only on G_{X00}, and this integral should be familiar, so Eq. (6.140) can be written

$$T(x, y, t) = \frac{1}{2} \frac{\alpha q_0}{k} \int_{\tau=0}^{t} G_{Y20}(y, t|0, \tau)\, \text{erfc} \left\{ \frac{x}{[4\alpha(t - \tau)]^{1/2}} \right\} d\tau \quad (6.141)$$

The function G_{Y20} is listed in Appendix X (replace x by y wherever it appears in the listing for G_{X20}), and the general expression for the temperature can be written

$$T(x, y, t) = \frac{1}{2} \frac{\alpha q_0}{k} \left(\frac{1}{\pi} \right)^{1/2} \int_{\tau=0}^{t} \frac{2}{[4\alpha(t - \tau)]^{1/2}}$$

$$\times \exp \left[\frac{-y^2}{4\alpha(t - \tau)} \right] \text{erfc} \left\{ \frac{x}{[4\alpha(t - \tau)]^{1/2}} \right\} d\tau \quad (6.142)$$

This expression is valid for all location in the body ($-\infty < x < \infty$, $y \geq 0$) and for any time $t \geq 0$.

6.8.2 Special Cases

The time integral in Eq. (6.142) can be evaluated in closed form in two special cases.

Surface temperature. For the special case of $y = 0$, the temperature on the surface is given by (Carslaw and Jaeger, 1959, p. 264)

$$T(x, 0, t) = \frac{q_0}{k} \left(\frac{\alpha t}{\pi}\right)^{1/2} \left\{ \text{erfc} \left[\frac{x}{2(\alpha t)^{1/2}}\right] - \frac{x}{2(\pi \alpha t)^{1/2}} E_1\left(\frac{x^2}{4\alpha t}\right) \right\} \quad (6.143)$$

The function $E_1(\cdot)$ is the exponential integral, defined by

$$E_1(z) = \int_z^\infty \frac{e^{-u}}{u} \, du \quad (6.144)$$

It is tabulated in Abramowitz and Stegun (1964) and it is available in computer libraries. See also Appendix F in the back of this book.

Centerline temperature. For the special case of $x = 0$, the temperature at the centerline is given by

$$T(0, y, t) = \frac{q_0}{k} (\alpha t)^{1/2} \text{ierfc} \left[\frac{y}{2(\alpha t)^{1/2}}\right] \quad (6.145)$$

which is exactly one-half of the solution for a semi-infinite body heated over the entire $y = 0$ surface.

6.8.3 Series Expression for the Temperature

The time integral for the temperature, Eq. (6.142), is evaluated in this section with series expressions. To begin, the time integral is written with a change of variables using

$$u = \frac{y}{2[\alpha(t - \tau)]^{1/2}} \quad (6.146)$$

and Eq. (6.142) can be written as

$$T(x, y, t) = \frac{q_0 y}{2k\pi^{1/2}} \int_{y/(4\alpha t)^{1/2}}^\infty \frac{du}{u^2} e^{-u^2} \text{erfc} \left(\frac{xu}{y}\right) \quad (6.147)$$

Further, a set of dimensionless variables will be used to present the temperature results:

$$X = \frac{x}{2(\alpha t)^{1/2}} \qquad Y = \frac{y}{2(\alpha t)^{1/2}} \quad (6.148a, b)$$

$$p = \frac{y}{x} = \frac{Y}{X} \qquad \Theta = \frac{T}{(q_0/k)(\alpha t/\pi)^{1/2}} \quad (6.148c, d)$$

Notice that the variable p is independent of time. With these new variables, Eq. (6.147) can be written

$$\Theta(p, Y) = Y \int_Y^\infty \frac{du}{u^2} e^{-u^2} \text{erfc} \left(\frac{u}{p} \right) \tag{6.149}$$

The number of independent variables has been reduced from three (x, y, t) in Eq. (6.147) to two dimensionless variables (p, Y) in Eq. (6.149). The time dependence of the temperature has been absorbed into the coordinates and into the dimensionless temperature by normalizing them by the "length" $\sqrt{(\alpha t)}$.

This type of coordinate transformation is called a similarity transformation, and the variables are called similarity variables. Heat conduction problems can be solved this way where the solution depends on a penetration depth $\sqrt{(\alpha t)}$, usually because the geometry has no intrinsic length scale. Certain fluid flow problems may also be solved with similarity transformations. Equation (6.149) can be integrated by parts to give (Litkouhi, 1982)

$$\Theta(X, Y) = \pi^{1/2} \text{ierfc} (Y) - e^{-Y^2} \text{erf} (X) - \frac{X}{\pi^{1/2}} E_1(X^2 + Y^2)$$

$$+ 2pY \int_X^\infty e^{-p^2 u^2} \text{erfc} (u) \, du \tag{6.150}$$

Here u is a dummy variable.

The integral in the last term of Eq. (6.150) can be represented by a function H defined as

$$H(X, Y) = \frac{2p}{\pi^{1/2}} \int_X^\infty e^{-p^2 u^2} \text{erf} (u) \, du \tag{6.151}$$

Recall that $p = Y/X$. Then the general temperature solution for a constant heat flux over the half plane can be written

$$\Theta(X, Y) = \pi^{1/2} \text{ierfc} (Y) - e^{-Y^2} \text{erf} (X)$$

$$- \frac{X}{\pi^{1/2}} E_1(X^2 + Y^2) + \pi^{1/2} Y H(X, Y) \tag{6.152}$$

The general solution given by Eq. (6.152) is valid for all times and any location in the body. However, Eq. (6.152) is recommended only for $X > 0$. For $X < 0$, a complementary expression is recommended:

$$\Theta(X < 0, Y) = Y \pi^{1/2} \text{ierfc} (Y) - \Theta(X > 0, Y) \tag{6.153}$$

where the first term on the right-hand side of Eq. (6.153) is the solution to the same problem if the entire surface was heated by a constant heat flux.

The function $H(X, Y) = H(X, p)$ can be represented in a series form for the three different regions indicated in Fig. 6.13.

Region $|p| > 1$. The region $|p| > 1$ represents the region closest to the surface of the semi-infinite body. In this region, $H(X, p)$ is given by

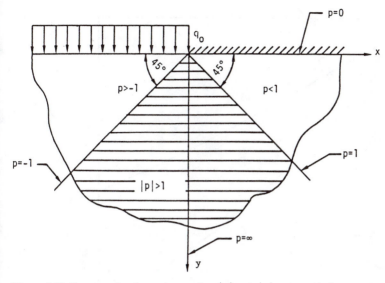

Figure 6.13 Geometry showing various regions $|p| < 1$, $|p| = 1$, and $|P| > 1$.

$$H(X, Y) = H(X, p) = \frac{2}{\pi} \sum_{n=0}^{\infty} \frac{(-1)^n \, \Gamma(n + 1, p^2 X^2)}{p^{2n+1}(2n + 1) \, n!} \tag{6.154}$$

where the truncated exponential function which is defined in Abramowitz and Stegun (1964)

$$\Gamma(n, u) = \int_u^{\infty} e^{-t} \, t^{n-1} \, dt \tag{6.155}$$

Region $|p| < 1$. For the region $|p| < 1$, the expression (6.154) cannot be used for $H(X, Y)$ since the term p^{2n+1} appearing in the denominator causes the summation to diverge. In this case the following expression is provided:

$$H(X, p) = 1 - \text{erf}(X) \, \text{erf}(pX) - \frac{2}{\pi} \sum_{n=0}^{\infty} \frac{(-1)^n p^{2n+1} \Gamma(n + 1, X^2)}{(2n + 1) \, n!} \tag{6.156}$$

Region $|p| = 1$. On the line $|p| = 1$, it can be shown that $H(X, p)$ is given by

$$H(X, 1) = -H(X, -1) = \frac{1 - \text{erf}^2(X)}{2} \tag{6.157}$$

Next some numerical results are presented. Figure 6.14 is a plot of function $H(X, p)$ versus X as calculated from the series expressions. [Numerical results for $H(X, p)$ to six decimal places are tabulated in Litkouhi, 1982.] Dimensionless temperature in the semi-infinite body is plotted versus X in Fig. 6.15. Recall that $\Theta(X, Y)$ is normalized by the time, so time does not explicitly appear in the figure.

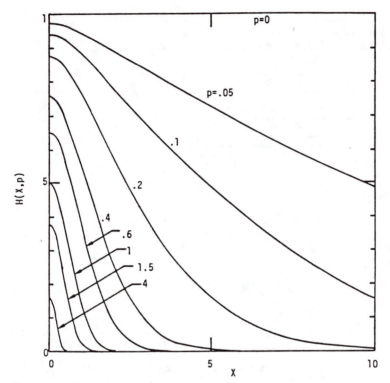

Figure 6.14 Function $H(X, p)$ versus X for different values of p.

6.8.4 Application to the Strip Heat Source

Other boundary conditions can be obtained by using the half-plane solution and superposition, and Fig. 6.16 shows several geometries that are possible. One case of interest is the semi-infinite body heated by a constant heat flux over an infinite strip of width $2a$ and insulated elsewhere as shown in Fig. 6.16a. This solution can be found from the superposition of two half-plane solutions: one half-plane is located at $x - a = 0$ with a positive heat flux, and the other half-plane is located at $x + a = 0$ with a negative heat flux. The resulting temperature is given by (Litkouhi, 1982)

$$
\Theta(x^+, y^+, t^+) = e^{-(y^{+2}/4t^+)} \left\{ - \operatorname{erf} \left[\frac{x^+ - 1}{(4t^+)^{1/2}} \right] + \operatorname{erf} \left[\frac{x^+ + 1}{(4t^+)^{1/2}} \right] \right\}
$$

$$
- \left(\frac{x^+ - 1}{(4\pi t^+)^{1/2}} \right) E_1 \left[\frac{(x^+ - 1)^2 + (y^+)^2}{4t^+} \right]
$$

$$
+ \left[\frac{x^+ + 1}{(4\pi t^+)^{1/2}} \right] E_1 \left[\frac{(x^+ + 1)^2 + (y^+)^2}{4t^+} \right]
$$

$$
+ \frac{\pi^{1/2} y^+}{(4t^+)^{1/2}} H \left[\frac{x^+ - 1}{(4t^+)^{1/2}}, p \right] - H \left[\frac{x^+ + 1}{(4t^+)^{1/2}}, p \right] \tag{6.158}
$$

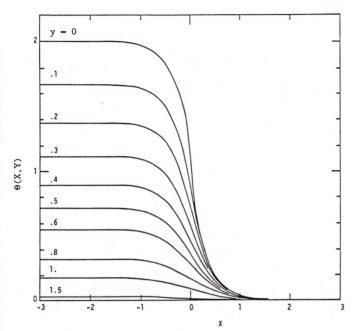

Figure 6.15 Dimensionless temperature $\Theta(X, Y)$ versus X for different values of Y in semi-infinite body with uniform heat flux over half-space $x < 0$ and $y = 0$.

where now the coordinates are normalized by a, the characteristic length:

$$x^+ = \frac{x}{a} \qquad y^+ = \frac{y}{a} \tag{6.159a, b}$$

$$t+ = \frac{\alpha t}{a^2} \qquad p = \frac{y}{x} = \frac{y^+}{x^+} \tag{6.159c, d}$$

and $\Theta = T/[(q_0/k)(\alpha t/\pi)^{1/2}]$ as before. Note that the definition of parameter p has not changed from when it was introduced in Eq. (6.148).

Surface temperature. For the special case of $y^+ = 0$, the surface of the semi-infinite body, the temperature due to the heated strip is (Carslaw and Jaeger, 1959)

$$\Theta(x^+, 0, t^+) = \operatorname{erf}\left[\frac{x^+ + 1}{(4t^+)^{1/2}}\right] - \operatorname{erf}\left[\frac{x^+ - 1}{(4t^+)^{1/2}}\right]$$

$$+ \left[\frac{x^+ + 1}{(4\pi t^+)^{1/2}}\right] E_1\left[\frac{(x^+ + 1)^2}{4t^+}\right]$$

$$- \left[\frac{x^+ - 1}{(4\pi t^+)^{1/2}}\right] E_1\left[\frac{(x^+ - 1)^2}{4t^+}\right] \tag{6.160}$$

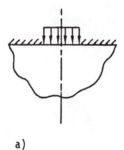

a)

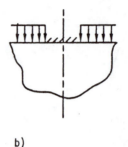

b)

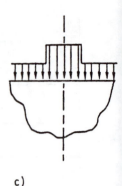

c)

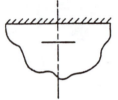

d)

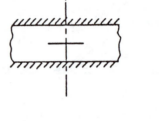

e)

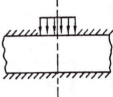

f)

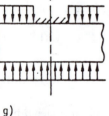

g)

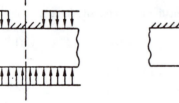

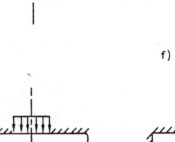

—T=0

h)

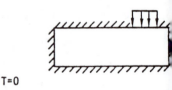

i)

Figure 6.16 Various possible cases that can be treated using solution given in Fig. 6.15 as a building block.

6.8.5 Discussion

Round-off error. The expressions for the temperature in the strip heater case are recommended only for $x^+ > 0$ due to the possibility of computer round-off error. The geometry is symmetric about the x axis so that the temperature for $x^+ < 0$ can easily be found from $T^+(x^+ < 0, y^+, t^+) = T^+(x^+ > 0, y^+, t^+)$.

Round-off error comes from subtracting two numbers that are close in value. For the strip heater problem, round-off error can come from the two superposed half-plane solutions. The temperature due to the heated strip can be written as

$$\Theta_{strip}(x^+) = \Theta_{half\text{-}plane}(x^+ - 1) - \Theta_{half\text{-}plane}(x^+ + 1) \qquad (6.161)$$

(For the moment, the dependence on y^+ and t^+ has been left out.) Now, the physics of the heat transfer problem requires that sufficiently far from the heated strip, the temperature must approach zero. As $x^+ \to +\infty$, the two superposed solutions each approach zero (within the computation limits of the computer) because the half-plane solution is heated on the *left half* of the plane. There is no round-off error associated with the temperature at $x^+ > 0$. As $x^+ \to -\infty$, however, the two half-plane solutions are evaluated near their heated regions and the half-plane temperatures can be very large (especially near the surface $y^+ = 0$); the strip-heater temperature is near zero due to cancelation of the nearly equal half-plane temperatures. This process of canceling when $x^+ < 0$ can be demonstrated by a numerical example.

Suppose the temperature is evaluated directly at $x^+ = -3$, $y^+ = 0$, and $t^+ = 0.5$ for the heated strip located over ($-1 \le x^+ \le 1$). The numerical value will be calculated with seven-digit accuracy using floating-point notation appropriate for a computer. Using Eq. (6.161) with $x^+ = -3$,

$$\Theta_{strip}(x^+ = -3) = \Theta_{half\text{-}plane}(-3 - 1) - \Theta_{half\text{-}plane}(-3 + 1)$$

$$= 0.2000000E+01 - 0.1999587E+01$$

$$= 0.413E\text{-}03$$

Note that the two half-plane temperatures are nearly equal, so the subtraction problem has reduced the accuracy from seven digits to three digits. Loss of accuracy is only part of the error, however, because a computer working with seven-digit accuracy will usually give the answer in seven digits, such as 0.4132662E-03 where the last four digits of the mantissa are computer-generated gibberish (round-off error). Most computers won't tell you when this type of error occurs. Again, for the strip heater problem, this type of error can be avoided by evaluating the temperature only at $x^+ > 0$ and using symmetry to find the temperature at $x^+ < 0$.

Lack of a steady state. The heated half-plane temperature, $T(x, y, t)$, has no steady state. As $t \to \infty$, the temperature increases without limit. In Eq. (6.152), this dependence on time is hidden by the normalized temperature Θ which results in a dimensionless temperature expression that does not explicitly depend on time; however, the actual temperature in degrees Kelvin represented by Eq. (6.152) and Figure 6.15 does depend on time and there is no steady state.

It is not always clear if a semi-infinite body with heat flux boundary conditions has a steady-state temperature. In general, a semi-infinite body will have a steady-state temperature if a *finite* amount of heat (joules) is added to the body. There are at least three ways that a finite amount of heat can be added to a semi-infinite body: through a heated region that is finite in spatial extent, through a short duration of heating, or through a net zero heat flow into a body (sources and sinks of heat that balance out). For example, the heated strip solution discussed in this section is infinite in extent in the z direction and an infinite amount of heat enters the body per unit time; consequently, there is no steady state. As a counter example, a semi-infinite body heated over its surface for a short period and insulated thereafter always has a steady-state temperature of zero if you wait long enough after the heating has ended; in the limit of an infinitesimally short heating period, the temperature is similar to the GF G_{X20}, which goes to zero as $t - \tau$ goes to infinity.

6.9 STEADY STATE

Steady-state solutions have already been touched on in connection with the alternative GF solution method in Examples 6.5 and 6.6. In this section, three examples of steady heat conduction in rectangular coordinates are presented, one each in one, two, and three dimensions. For one-dimensional steady cases in rectangular coordinates the GFs are listed in Appendix X, Table X.1. For two- and three-dimensional cases, the steady GF must be found on a case-by-case basis.

Example 6.10: Slab body with steady internal energy generation—X21B00G-case. In a one-dimension plate with an insulated surface at $x = 0$ and a fixed temperature $T = 0$ at $x = L$, find the steady temperature caused by internal energy generation.

SOLUTION The solution is given by the energy generation term of the steady GFSE:

$$T(x) = \frac{1}{k} \int_{x'=0}^{L} g(x') \, G_{X21}(x|x') \, dx' \qquad (6.162)$$

The GF is given in Appendix X, Table X.1, as

$$G_{X21}(x|x') = \begin{cases} L - x & 0 \le x' \le x \\ L - x' & x \le x' \le L \end{cases} \qquad (6.163)$$

The GF is piecewise continuous, so the integral in Eq. (6.162) must be evaluated in two parts:

$$T(x) = \frac{1}{k} (L - x) \int_{x'=0}^{x} g(x') \, dx' + \frac{1}{k} \int_{x'=x}^{L} g(x') \, (L - x') \, dx' \qquad (6.164)$$

Over the region $0 \le x' \le x$, this GF is constant with respect to variable (x') to satisfy the insulated boundary condition at $x = 0$.

    a. X21B00G1 case. For uniform internal energy generation, $g(x') = g_0$, the above integrals give

$$T(x) = \frac{1}{2} \left[1 - \left(\frac{x}{L} \right)^2 \right] \frac{g \, L^2}{k} \qquad (6.165)$$

    b. X21B00G4 case. For heating caused by radiation (nuclear or electromagnetic), the internal energy generation is largest at the surface $x = L$ and is attenuated exponentially inside the body. This behavior may be described by $g(x) = g_0 \, e^{-b(1-x/L)}$ and substituted into Eq. (6.164) to give

$$T(x) = \left[\frac{1}{b^2} - \frac{1}{b^2} e^{-b(1-x/L)} - \frac{1}{b} \left(1 - \frac{x}{L} \right) e^{-b} \right] \frac{g L^2}{k} \qquad (6.166)$$

Example 6.11: Two-dimensional slab heated over a small region. Find the steady temperature in a two-dimensional slab caused by a uniform heat flux q_0 over a small region $-a \leq x' \leq a$ and insulated elsewhere on one side of the slab, and a zero temperature on the other side. The region is very large in the x direction and has thickness L in the y direction. This geometry is related to the study of surface-mounted heated films.

SOLUTION This is the $X00Y21$ geometry, and the temperature distribution in the body is driven by heating on the surface $y = 0$. The geometry is shown in Fig. 6.16h. The steady temperature is given by the surface heating term of the GFSE:

$$T(x, y) = \int_{x'=-a}^{a} \frac{q_0}{k} G_{X00Y21}(x, y | x', y = 0) \, dx' \qquad (6.167)$$

The steady GF may be found from the method of limits and by the product of two one-dimensional transient GFs:

$$G_{X00Y21}(x, y | x', y') = \lim_{t \to \infty} \alpha \int_{\tau=0}^{t} G_{X00}(x, t | x', \tau) G_{Y21}(y, t | y', \tau) \, d\tau \qquad (6.168)$$

The transient GFs are available in Appendix X, and Eq. (6.168) may be written

$$G_{X00Y21}(x, y | x', y') = \lim_{t \to \infty} \alpha \int_{\tau=0}^{t} d\tau \, [4\pi\alpha(t - \tau)]^{-1/2}$$

$$\times \exp\left[-\frac{(x - x')^2}{4\alpha(t - \tau)} \right] \frac{2}{L} \sum_{m=1}^{\infty}$$

$$\times \exp\left[-\frac{\pi^2(m - 1/2)^2 \alpha(t - \tau)}{L^2} \right]$$

$$\times \cos\left[\frac{\pi(m - 1/2)y'}{L} \right] \cos\left[\frac{\pi(m - 1/2)y}{L} \right] \qquad (6.169)$$

Note that the large-time form of the function G_{Y21} is used. The time integral in the above equation involves the error function and is given in integral 12 in Table F.6, Appendix F. After the limit is taken, the result is

$$G_{X00Y21}(x, y | x', y') = -\sum_{m=1}^{\infty} \frac{2 L}{|x - x'|}$$

$$\times \exp\left[\frac{-\pi(m - 1/2) |x - x'|}{L} \right]$$

$$\times \cos\left[\frac{\pi(m - 1/2)y'}{L} \right] \cos \frac{\pi(m - 1/2)y}{L} \qquad (6.170)$$

The absolute value $|x - x'|$ is introduced by the time integral, and it reflects the symmetry of the GF about $(x - x') = 0$ and it also guarantees that the exponential term dies away as $|x - x'|$ increases.

Now that the GF has been found, the temperature caused by heating the body over a small region may be found from Eq. (6.167):

$$T(x, y) = -\frac{q_0}{k} \int_{x'=-a}^{a} \sum_{m-1}^{\infty} \frac{dx'}{\pi(m - 1/2)}$$

$$\times \exp\left[\frac{-\pi(m - 1/2)\,|x - x'|}{L}\right]$$

$$\times \cos\left[\frac{\pi(m - 1/2)y}{L}\right] \tag{6.171}$$

The absolute value must be treated carefully by examining $(x - x') > 0$ separately from $(x - x') < 0$. The result is two expressions for the temperature depending on the region:

For $|x| > a$,

$$T(x, y) = \frac{2q_0 L}{k} \sum_{m=1}^{\infty} \cos\left[\frac{\pi(m - 1/2)y}{L}\right] \frac{1}{\pi^2(m - 1/2)^2}$$

$$\times \left\{ \exp\left[-\pi(m - 1/2)\frac{|x| - 9}{L}\right] - \exp\left[-\pi(m - 1/2)\frac{|x| + 9}{L}\right] \right\}$$

and for $|x| < a$,

$$T(x, y) = \frac{2q_0 L}{k} \sum_{m=1}^{\infty} \cos\left[\frac{\pi(m - 1/2)y}{L}\right]$$

$$\times \frac{1}{\pi^2(m - 1/2)^2} \left\{ 2 - \exp\left[-\pi(m - 1/2)\frac{x + 9}{L}\right] \right.$$

$$\left. - \exp\left[-\pi(m - 1/2)\frac{x - 9}{L}\right] \right\} \tag{6.172}$$

The convergence of the infinite series for the temperature is controlled by how rapidly $\exp(z) \to 0$ as the argument becomes large with increasing m. The series converges rapidly for $|x| >> a$, but the series solution is difficult to evaluate at the edge of the heated region $x = a$ or $x = -a$ for $a/L << 1$.

Another form of the steady solution for this geometry can be found with the method-of-images form of the GF, given by Table 4.1 with the line source solution substituted for $K(\cdot)$ in the X21 case:

$$G_{X00Y21}(x, y|x', y') = -\sum_{n=-\infty}^{\infty} (-1)^n \frac{1}{4\pi} \{\ln [(2nL + y - y')^2$$

$$+ (y - y')^2] + \ln [(2nL + x + x')^2$$

$$+ (x - x')^2]\} \tag{6.173}$$

This form of the GF may be substituted into the temperature expression Eq. (6.167) to produce a temperature series that is well behaved at $|x| = a$. The convergence of this series expression for the GF is controlled by the cancellation of terms with alternating signs.

Example 6.12: Parallelpiped with specified surface temperature—X11Z11Y11 case. Find the steady temperature in the parallelpiped with five faces at zero temperature and one face (at $x = 0$) maintained at temperature T_0.

SOLUTION The GF for this geometry was treated in Example 4.6 and the parallelpiped body is shown in Fig. 4.5. The GF for this case is given by

$$G(x, y, z|x', y', z')$$

$$= 8 \sum_{m=1}^{\infty} \sum_{n=1}^{\infty} \sum_{p=1}^{\infty} \sin\left(m\pi \frac{x}{a}\right) \sin\left(m\pi \frac{x'}{a}\right) \sin\left(p\pi \frac{z}{c}\right)$$

$$\times \sin\left(p\pi \frac{z'}{c}\right) \sin\left(n\pi \frac{y}{b}\right) \sin\left(n\pi \frac{y'}{b}\right)$$

$$\times \left[abc\pi^2 \left(\frac{m^2}{a^2} + \frac{n^2}{b^2} + \frac{p^2}{c^2}\right)\right]^{-1} \tag{6.174}$$

The temperature for this case is given by the boundary term of the steady GFSE, Eq. (3.33),

$$T(x, y, z) = -\int_{y'=0}^{b} dy' \int_{z'=0}^{c} dz' \, T_0 \left.\frac{\partial G}{\partial n'}\right|_{x'=0} \tag{6.175}$$

where the surface integral is carried out over the $x = 0$ face of the parallelpiped. The required derivative and integrals are elementary, and the temperature is

$$T(x, y, z) = 8T_0 \sum_{m=1}^{\infty} \sum_{n=1}^{\infty} \sum_{p=1}^{\infty} [1 - (-1)^p][1 - (-1)^n]$$

$$\times \sin\left(m\pi \frac{x}{a}\right) \sin\left(p\pi \frac{z}{c}\right) \sin\left(n\pi \frac{y}{b}\right)$$

$$\times \left[a^2 np\pi^3 \left(\frac{m^2}{a^2} + \frac{n^2}{b^2} + \frac{p^2}{c^2}\right)\right]^{-1} \tag{6.176}$$

REFERENCES

Abramowitz, M., and Stegun, I. A., 1964, *Handbook of Mathematical Functions*, National Bureau of Standards, Applied Mathematics Series 55, U.S. Government Printing Office, Washington, D.C.

Beck, J. V., 1984, Green's Function Solution for Transient Heat Conduction Problems, *Int. J. Heat Mass Transfer*, vol. 27, pp. 1135–1244.

Carslaw, H. S., and Jaeger, J. C., 1959, *Conduction of Heat in Solids*, 2d ed., Oxford University Press, New York.

Litkouhi, B., 1982, Surface Element Method in Transient Heat Conduction Problems, Ph.D. thesis, Michigan State University Mechanical Engineering Department, East Lansing, Mich.

Ozisik, M. N., 1980, *Heat Conduction*, Wiley, New York.

Zang, P. H., 1987, Symbolic, Algebraic, and Numeric Solutions to Heat Conduction Problems using Green's Functions, Ph.D. thesis, Michigan State University Mechanical Engineering Department, East Lansing, Mich.

Zang, P. H., 1988, CANSS—A Prototype Program for Exact Symbolic Solutions to Heat Transfer Problems, HTD Vol. 105, Symbolic Computations in Fluid Mechanics and Heat Transfer, presented at the ASME Winter Annual Meeting, Chicago, Ill., Dec. 1988.

PROBLEMS

6.1 Use the standard Green's function solution equation (GFSE) to obtain the temperature distribution for the problem

$$\frac{\partial^2 T}{\partial x^2} + \frac{g(x)}{k} = \frac{1}{\alpha}\frac{\partial T}{\partial t} \qquad 0 < x < L \qquad t > 0$$

$$T(0, t) = 0 \qquad T(L, t) = 0 \qquad T(x, 0) = 0$$

where $g(x) = g_0 = $ constant for $0 < x < L_1 < L$
$\qquad\qquad = 0$ otherwise

Use the large time GF.

6.2 Solve Problem 6.1 using the AGFSE.

6.3 Solve, using the GFSE, the problem

$$\frac{\partial^2 T}{\partial x^2} = \frac{1}{\alpha}\frac{\partial T}{\partial t} \qquad 0 < x < L \qquad t > 0$$

$$-k\left.\frac{\partial T}{\partial x}\right|_{x=0} = q_0 \qquad T(L, t) = T_0 \qquad T(x, 0) = T_0$$

Use the large time GF.

6.4 Solve Problem 6.3 using the AGFSE.

6.5 Solve Problem 6.3 using the GFSE with the small time GF.

6.6 Consider the following one-dimensional problem.

$$\frac{\partial^2 T}{\partial x^2} = \frac{1}{\alpha}\frac{\partial T}{\partial t} \qquad -k\left.\frac{\partial T}{\partial x}\right|_{x=0} = q_0 \qquad -k\left.\frac{\partial T}{\partial x}\right|_{x=L} = h(T|_{x=L} - T_\infty)$$

and initial condition $T(x, 0) = 0$.

(a) Using $T(x, t) = T^*(x) + T_1(x, t)$, write down an alternative boundary value problem for $T_1(x, t)$, where $T^*(x)$ is the solution to the following steady problem.

$$\frac{\partial^2 T^*}{\partial x^2} = 0$$

$$-k\left.\frac{\partial T^*}{\partial x}\right|_0 = q_0$$

$$-k\left.\frac{\partial T^*}{\partial x}\right|_{x=L} = h(T^*|_{x=L} - T_\infty)$$

(b) Carry out the transient solution using the *large-time* form of the GF and the AGFSE. Write your answer in terms of dimensionless parameters hL/x, x/L, and dimensionless temperature $(T - T_\infty)/(q_0L/k)$.

6.7 Exponential heating is sometimes used to model runaway heating of nuclear fuel rods. Write down the integral form of the temperature for the following problem with exponential heating and convection cooling. Assume that the GF has the name $G_{X33}(x, t|x', \tau)$. Do not evaluate the GF. Do not evaluate the integrals.

$$\frac{\partial^2 T}{\partial x^2} + \frac{1}{k} g(t) = \frac{1}{\alpha} \frac{\partial T}{\partial t}$$

$$T(x, 0) = 0$$

$$-k \left. \frac{\partial T}{\partial n} \right|_{x_i} = h(T|_{x_i} - T_\infty), \, i = 1, 2.$$

$$g(t) = g_0 e^{at}$$

6.8 Find the temperature in the semi-infinite body with uniform heating at the boundary with the Laplace transform method:

$$\frac{\partial^2 T}{\partial x^2} = \frac{1}{\alpha} \frac{\partial T}{\partial t}$$

$$-k \left. \frac{\partial T}{\partial x} \right|_{x=0} = q_0$$

$$T(x, 0) = 0$$

Hint: Take the derivative of the energy equation to find a boundary value problem on $q(x, t)$.

6.9 Find the prescribed surface temp, $f(t)$, such that when applied to the semi-infinite solid with zero initial conditions ($X10B$-$T0$ case) gives a constant heat flux at the surface. Check the result by differentiation; that is, show

$$-k \left. \frac{\partial T}{\partial x} \right|_{x=0} = \text{constant}$$

6.10 Find the small-time form of the temperature for a one-dimensional slab geometry with one surface heated with a constant heat flux, one surface insulated, and zero initial conditions ($X22B10T0$ case). Compare your result to the (semi-infinite) $X20B1T0$ solution listed in Table 6.3, and comment on the differences.

6.11 Find the temperature in a semi-infinite body resulting from the following surface temperature.

$$F(t) = \begin{cases} T_1 & 0 < t \leq t_1 \\ T_2 & t > t_1 \end{cases}$$

The initial temperature is zero. What is the number of this case?

6.12 Find the temperature in a semi-infinite body heated at the surface by a square pulse of heat:

$$q(t) = \begin{cases} q_0 & 0 < t < t_1 \\ 0 & t > t_1 \end{cases}$$

Find the steady-state temperature as $t \to \infty$.

6.13 Suppose the surface temperature on a semi-infinite solid due to surface heating is given by

$$T(t) - T_0 = a\sqrt{\frac{t}{t_0}} + b\left(\frac{t}{t_0}\right)$$

where a, b, and t_0 are constants. Find the surface heat flux that caused the temperature to rise.

6.14 Find the prescribed surface temperature, $f(t)$, such that when applied to the semi-infinite solid with zero initial condition ($X10B$-$T0$ case), the surface heat flux is given by

$$-k \left. \frac{\partial T}{\partial x} \right|_{x=0} = q_0 \left(\frac{t}{t_0}\right)^{n/2} \quad \text{for } n = 0, 1, 2, \ldots$$

6.15 Write down the GF solution equation for the following two-dimensional case. Do not derive the GF; do not solve the integrals. However, use the correct form of dv and ds_i. Use the name $G_{X22Y11}(x, y, t|x', y', \tau)$ in your expression.

$$\frac{\partial^2 T}{\partial x^2} + \frac{\partial^2 T}{\partial y^2} + \frac{g_0}{k} = \frac{1}{\alpha}\frac{\partial T}{\partial t} \qquad g_0 \text{ is constant}$$

$$T(x, y, 0) = ax + by + c$$

$$\frac{\partial T}{\partial x}(x = 0, y, t) = 0$$

$$\frac{\partial T}{\partial x}(x = L_x, y, t) = 0$$

$$T(x, y = 0, t) = T_0$$

$$T(x, y = L_y, t) = 0$$

6.16 Consider the surface temperature on a semi-infinite body heated over two infinite strips of equal size, case $X00T0Y20B(x5)$.

(a) Find the surface temperature resulting from the following anti-symmetric heat flux distribution:

$$q(x, t) = \begin{cases} 0 & t = 0 \\ -q_0 & t > 0 \quad -b < x < -a \\ +q_0 & t > 0 \quad a < x < b \end{cases}$$

(b) Where does the maximum temperature occur?

(c) Plot the steady state surface temperature $T(x, y = 0)$.

6.17 Find the surface temperature $T(x, y, t)$ on a semi-infinite body heated by a line source located at $x = 0$, $y = 0$. The surface heating is given by:

$$q(x, t) = \begin{cases} 0 & t = 0 \\ q_0\delta(x) & t > 0 \end{cases}$$

6.18 Find the steady-state temperature at $y = 0$ for the $X00T0Y21B(x5)0$ case (strip heat source) as follows:

(a) First find the integral form of the transient temperature with the large-time form of the GF. The boundary conditions are the following:

$$T(x, y = D, t) = 0$$

$$-k\frac{\partial T(x, 0, t)}{\partial y} = \begin{cases} q_0 & -a < x < a \\ 0 & \text{elsewhere} \end{cases}$$

$$T(x \rightarrow -\infty, y, t) = 0$$

$$T(x \rightarrow +\infty, y, t) = 0$$

$$T(x, y, 0) = 0$$

(b) Evaluate the temperature at $y = 0$ and evaluate the integrals.

(c) Suggest one method to improve the convergence of the series expression.

6.19 (a) Find an integral expression for a semi-infinite body heated at the surface over a rectangular area (three-dimensional problem). This is the $X00Y00Z20B(x5y5)T0$ case. The surface heating is given by

$$-k\frac{\partial T(x, y, z = 0, t)}{\partial y} = \begin{cases} q_0 & -a < y < a \\ & -b < x < b \quad t > 0 \\ 0 & \text{elsewhere on surface} \end{cases}$$

Initially the temperature is zero.

(b) Find the average temperature on the rectangle in the form of an integral on τ (evaluate spatial integrals).

6.20 A rectangular parallel piped $0 \le x \le L$, $0 \le y \le L$ is initially at temperature zero. Surfaces $y = 0$, $y = L$, and $x = 0$ are insulated. Surface $x = 0$ is heated by constant heat flux q_0 over $0 < y < L/2$ and is insulated over $L/2 < y < L$. Find the temperature at location $x = 0$, $y = 0$ as a function of time.

(a) Using large-time GFs.

(b) Using time-partitioning at $\alpha\Delta t/L^2 = 0.005$.

6.21 Consider a semi-infinite solid with a thin, high conductivity film at $-\delta \le x \le 0$. Let $x > 0$ be the semi-infinite body. Find the temperature at $x = 0$ for the following heating condition:

$$-k \frac{\partial T}{\partial x} (0, t) = q_0$$

$$T(x, 0) = 0$$

This is case $X40B1T0$. Compare your answer to the $X20B1T0$ case. What is a dimensionless parameter that describes the added effect of the thin surface film on the heated semi-infinite body at early times after heating begins?

6.22 Consider the same geometry as in Problem 6.21, but now the surface of the thin film is suddenly heated by a convection process. The initial temperature is zero. Find the transient temperature and compare it to the $X30B1T0$ case.

6.23 Solve, using the GFSE, the problem

$$\frac{\partial^2 T}{\partial x^2} + \frac{g_0 e^{-x/x_0}}{k} = \frac{1}{\alpha} \frac{\partial T}{\partial t} + \frac{1}{\alpha} U_0 \frac{\partial T}{\partial x} \qquad 0 < x < \infty$$

$$T(0, t) = T_0, \, T(x, 0) = 0$$

The quantities, g_0, U_0, and x_0, are constants. What is the number of this case?

6.24 Solve the following problem of two-dimensional heat flow.

$$\frac{\partial^2 T}{\partial x^2} + \frac{\partial^2 T}{\partial y^2} = \frac{1}{\alpha} \frac{\partial T}{\partial t}$$

$$T(0, y, t) = T(a, y, t) = T(x, b, t) = T_0, \, T(x, y, 0) = T_0$$

$$T(x, 0, t) = \begin{cases} T_1 \ne T_0 & \text{for } 0 < x < a_1 < a \\ T_0 & \text{for } a_1 < x < a \end{cases}$$

(a) Use the GFSE with large-time GFs.
(b) Use the GFSE with small-time GFs.

CHAPTER SEVEN

RADIAL HEAT FLOW IN CYLINDRICAL COORDINATES

7.1 INTRODUCTION

Green's functions (GFs) for radial flow of heat in the cylindrical coordinate system (r, ϕ, z) are discussed in this chapter. For radial flow of heat, the temperature depends on position r and time t, and the heat conduction equation has the form

$$\frac{1}{r} \frac{\partial}{\partial r} \left[r \frac{\partial T}{\partial r} \right] + \frac{1}{k} g(r, t) = \frac{1}{\alpha} \frac{\partial T}{\partial t} \qquad (7.1)$$

That is, the temperature does not depend on ϕ or z.

In Section 7.2, the GF equation for radial heat flow is discussed. Section 7.3 covers the infinite body with radial heat flow. Sections 7.4–7.6 cover cylinders, and Section 7.7 covers the infinite body with a cylindrical hole.

7.2 GREEN'S FUNCTION SOLUTION EQUATION FOR RADIAL HEAT FLOW

Radial heat flow occurs when the boundary conditions, initial conditions, and energy generation do not depend on angle ϕ or axial position z, so that $T = T(r, t)$. The radial GF equation is given by

$$T(r, t) = \int_{r'} G(r, t | r', 0) \, F(r') \, 2\pi r' \, dr'$$

$$+ \alpha \sum_{i=1}^{s} \frac{(\rho c b)_i}{k_i}$$

$$\times G(r, t|r_i, 0) F(r_i) 2\pi r_i \qquad \begin{array}{l} \text{(for boundary conditions} \\ \text{of fourth and fifth} \\ \text{kinds only)} \end{array}$$

$$+ \int_{\tau=0}^{t} \int_{r'} \frac{\alpha}{k} G(r, t|r', \tau) g(r', \tau) 2\pi r' \, dr' \, d\tau$$

$$+ \alpha \int_{\tau=0}^{t} \sum_{i=1}^{s} \frac{f_i(r_i, t)}{k_i} \qquad \begin{array}{l} \text{(for boundary conditions} \\ \text{of the second} \\ \text{through fifth kinds)} \end{array}$$

$$\times G(r, t|r_i, \tau) 2\pi r_i \, d\tau$$

$$- \alpha \int_{\tau=0}^{t} \sum_{j=1}^{s} f_j(r_j, t)$$

$$\times \left. \frac{\partial G}{\partial n'_j} \right|_{r'=r_j} 2\pi r_j \, d\tau \qquad \begin{array}{l} \text{(for boundary condition} \\ \text{of the first kind only)} \end{array}$$

$$\tag{7.2}$$

Note that $dv' = 2\pi r' \, dr'$, and the integrals over boundary surface s_i have been replaced by $2\pi r_i$, the area per unit length. Equation (7.2) may be applied to bodies with boundary conditions of type 0 through 5. However, the radial heat flow GF actually listed in this book (Appendix R) are denoted $G_{RIJ}(\cdot)$, where $I, J = 0, 1, 2$, and 3.

7.3 INFINITE BODY

7.3.1 The $R00$ Green's Function

The GF for the radial flow of heat in the infinite body is denoted $G_{R00}(r, t|r', \tau)$. This GF can be interpreted as the response to a cylindrical surface heat source located at radius r' (refer to Fig. 7.1), and it is given by

$$G_{R00}(r, t|r', \tau) = \frac{1}{4\pi\alpha(t - \tau)} \exp\left[\frac{-(r^2 + r'^2)}{4\alpha(t - \tau)}\right] I_0\left[\frac{rr'}{2\alpha(t - \tau)}\right] \tag{7.3}$$

for $0 \leq r \leq \infty$ and $0 \leq r' \leq \infty$. The function $I_0(\cdot)$ is the modified Bessel function of the first kind of order zero [$I_0(0) = 1$ and $I_0(z \to \infty) \to \infty$]. Refer to Appendix B for more information on the Bessel functions. The units of $G_{R00}(\cdot)$ are m^{-2}. Note that the reciprocity relation holds for this GF because r and r' can be reversed and the function is unchanged.

In the special case where $r' = 0$, the cylindrical source that generates the function $G_{R00}(\cdot)$ collapses into a line source located at $r' = 0$, given by

$$G_{R00}(r, t|0, \tau) = \frac{1}{4\pi\alpha(t - \tau)} \exp\left[\frac{-r^2}{4\alpha(t - \tau)}\right] \tag{7.4a}$$

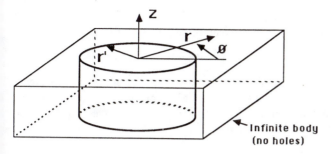

Figure 7.1 Cylindrical surface heat source located at r'.

Recall that a line source can also be represented by the product of two plane sources, and that $r' = 0$ corresponds to the point $x' = 0$, $y' = 0$. Thus, the identity is

$$G_{R00}(r, t|0, \tau) = G_{X00}(x, t|0, \tau)\, G_{Y00}(y, t|0, \tau) \qquad (7.4b)$$

This product solution also demonstrates that the units of $G_{R00}(\cdot)$ are $\mathrm{m^{-1} m^{-1}} = \mathrm{m^{-2}}$.

7.3.2 Derivation of the R00 Green's Function

There are several ways to derive the R00 GF from first principles (Ozisik, 1980, p. 107; Carslaw and Jaeger, 1959, p. 259). The following derivation involves an infinite body with heat generation in rectangular coordinates. The GFSE in two-dimensional rectangular coordinates is given by

$$T(x, y, t) = \frac{\alpha}{k} \int_{\tau=0}^{t} \int_{x'} \int_{y'} G_{X00Y00}(x, y, t|x', y', \tau)\, g(x', y', \tau)\, dy'\, dx'\, d\tau \qquad (7.5)$$

The appropriate heat generation term is an instantaneous cylindrical surface heat source shown in Fig. 7.1 and given by

$$g(x', y', \tau) = g_0 \frac{\delta(\tau - \tau_0)\, \delta(r' - r_0)}{2\pi r_0} \qquad (7.6)$$

Parameter r_0 is the radius of the cylindrical surface heat source that introduces heat at time τ_0 and g_0 (J/m) is the strength of the heat source per unit length of the cylindrical surface. [The strength of the heat source *per unit area* is $g_0/(2\pi r_0)$].

The appropriate two-dimensional GF in rectangular coordinates is given by

$$G_{X00Y00}(x, y, t|x', y', \tau) = \frac{1}{4\pi\alpha(t - \tau)} \exp\left[-\frac{(x - x')^2 + (y - y')^2}{4\alpha(t - \tau)}\right] \qquad (7.7)$$

Recall that G_{X00Y00} represents the response to an instantaneous line heat source located at (x', y', τ), and that $G_{X00Y00} = G_{X00}\, G_{Y00}$.

To evaluate the temperature in Eq. (7.5), the integral over the infinite body must be changed to cylindrical coordinates. First, the distance between points (x, y) and (x', y') that appears in the expression for G_{X00Y00} must be converted to cylindrical coordinates. If the cylindrical coordinates of points (x, y) and (x', y') are (r, ϕ) and (r', ϕ'), respectively, then the distance between these points is given by

$$R^2 = (x - x')^2 + (y - y')^2 = r^2 + (r')^2 - 2rr' \cos (\phi - \phi') \quad (7.8)$$

Second, the spatial integrals in Eq. (7.5) that extend over the entire (x', y') plane, where $dA = dx' \, dy'$ must be converted to equivalent integrals in the $(r' \phi')$ coordinate system over $0 \le r' < \infty$ and $0 \le \phi' \le 2\pi$ with $dA = r' \, dr' \, d\phi'$. The Eq. (7.5) can be combined with Eq. (7.6)–(7.8) to give

$$T(r, \phi, t) = \frac{\alpha}{k} \int_{\tau=0}^{t} \int_{r'=0}^{\infty} \int_{\phi'=0}^{2\pi} \left\{ \frac{1}{4\pi\alpha(t - \tau)} \right.$$

$$\times \exp \left[- \frac{r^2 + (r')^2 - 2rr' \cos (\phi - \phi')}{4\alpha(t - \tau)} \right]$$

$$\times g_0 \left. \frac{\delta(\tau - \tau_0) \, \delta(r' - r_0)}{2\pi r_0} \right\} d\tau \, r' \, dr' \, d\phi' \quad (7.9)$$

The integrals over r' and τ can be evaluated easily with the sifting property of the Dirac delta functions:

$$T(r, \phi, t) = \frac{\alpha}{k} \int_{\phi'=0}^{2\pi} d\phi' \left\{ \frac{r_0}{4\pi\alpha(t - \tau_0)} \right.$$

$$\times \exp \left[- \frac{r^2 + r_0^2 - 2rr_0 \cos (\phi - \phi')}{4\alpha(t - \tau_0)} \right] \left. \frac{g_0}{2\pi r_0} \right\}$$

$$= \frac{\alpha}{k} \frac{g_0/(2\pi)}{4\pi\alpha(t - \tau_0)} \exp \left[\frac{- (r^2 + r_0^2)}{4\alpha(t - \tau_0)} \right]$$

$$\times \int_{\phi'=0}^{2\pi} \exp \left[\frac{rr_0 \cos (\phi - \phi')}{2\alpha(t - \tau_0)} \right] d\phi' \quad (7.10)$$

The final integral on ϕ' is given by Watson (1944). The GF is given by the temperature divided by the source strength, or

$$G_{R00}(r, t | r_0, \tau_0) = \frac{T(r, \phi, t)}{\alpha g_0/k}$$

$$= \frac{1}{4\pi\alpha(t - \tau_0)} \exp \left[\frac{- (r^2 + r_0^2)}{4\alpha(t - \tau_0)} \right] I_0 \left[\frac{rr_0}{2\alpha(t - \tau_0)} \right] \quad (7.11)$$

Note that the result does not depend on angle ϕ. Finally, the GF is usually written with the heat source located at (r', τ) instead of at (r_0, τ_0), to give the same result as in Eq. (7.3).

7.3.3 Approximations for the *R00* Green's Function

The *R00* GF usually must be integrated to find the temperature, but it is not an easy function to integrate. Most integrals of function G_{R00} must be evaluated numerically unless a simple approximate expression can be found. A few approximate expressions for G_{R00} are listed in Appendix R, and Table 7.1 is a reference list of these approximations. These approximations are composed of exponentials and powers and they are generally easier to manipulate than the exact expression for G_{R00}. Table 7.1 lists the region of application, the maximum error, and the location in Appendix R of several approximate expressions for G_{R00}.

7.3.4 Temperatures from Initial Conditions

The temperature in an infinite body resulting from a nonuniform initial condition is given by the Green's function solution equation (GFSE) as

$$T(r, t) = \int_{r'} G_{R00}(r, t | r', 0) \, F(r') \, 2\pi r' \, dr' \qquad (7.12)$$

In this section, the above integral is discussed for the specific case of a uniform initial temperature near the origin and zero temperature elsewhere:

$$F(r') = \begin{cases} T_0 & 0 \le r' \le a \\ 0 & r' > a \end{cases} \qquad (7.13)$$

This is the $R00T5$ case. The transient temperature is given by

$$T(r, t) = T_0 \int_{r'=0}^{a} G_{R00}(r, t | r', 0) \, 2\pi r' \, dr'$$

$$= \frac{T_0}{4\pi\alpha t} \int_{r'=0}^{a} \exp\left[\frac{-(r^2 + r'^2)}{4\alpha t}\right] I_0\left[\frac{rr'}{2\alpha t}\right] 2\pi r' \, dr' \qquad (7.14)$$

Note that the integral is written over $0 \le r' \le a$ because $F(r')$ is zero elsewhere. In general, this integral must be evaluated numerically, and some numerical values of this integral are listed in Table R00.1 in Appendix R.

Over the region $0 \le r < a/2$, the temperature given by Eq. (7.14) remains within 0.03% of T_0 for small values of the time parameter ($\alpha t/a^2 < 0.01$). That is, $T(r = a/2, \alpha t/a^2 = 0.01) = 0.9997 \, T_0$.

Several approximate closed form expressions are also available for the integral given by Eq. (7.14), and these expressions are listed in Appendix R. For example,

Table 7.1 Approximate expressions for $G_{R00}(r, t | r', \tau)$ listed in Appendix R

Range of application	Error (%)	Equation number in Appendix R
$\dfrac{\alpha(t - \tau)}{rr'} < 0.25$	0.016	R00.4
$\dfrac{\alpha(t - \tau)}{rr'} > 0.33$	-0.012	R00.5
$\dfrac{\alpha(t - \tau)}{r^2}$ large and $\dfrac{\alpha(t - \tau)}{(r')^2}$ large		R00.6

for $\alpha t/a^2 < 0.25$ and at $r/a = 1.0$, the temperature resulting from initial temperature T_0 over $0 \leq r' \leq a$ is given approximately by Eq. (R00.9), Appendix R:

$$\frac{T(r, t)}{T_0} = \frac{1}{2}\left[1 - \left(\frac{u}{\pi}\right)^{1/2} - \frac{1}{4\sqrt{\pi}} u^{3/2} \right] \quad \text{where } u = \frac{\alpha t}{a^2} \quad (7.15)$$

Equation (7.15) and several other approximate expressions for the integral given by Eq. (7.14) are summarized in Table 7.2 with their region of application, maximum error, and location in Appendix R. Some of the expressions referenced in Table 7.2 have been found by integration of the expressions for G_{R00} referenced in Table 7.1.

In the special case $r = 0$, the temperature in the infinite body in Eq. (7.14) may be found in closed form. This temperature is given by Eq. (7.14) evaluated at $r = 0$:

$$T(r = 0, t) = \frac{T_0}{4\pi\alpha t} \int_{r'=0}^{a} \exp\left[\frac{-r'^2}{4\alpha t}\right] 2\pi r' \, dr' \quad (7.16a)$$

Note that $I_0(0) = 1$. This integral can be evaluated by a change of variables to $z = r'/(4\alpha t)^{1/2}$ to give

$$T(0, t) = 2T_0 \int_{z=0}^{a/(4\alpha t)^{1/2}} e^{-z^2} z \, dz = T_0 \left(1 - \exp\frac{-a^2}{4\alpha t}\right) \quad (7.16b)$$

This expression is exact for all t. Thus, the temperature at $r = 0$ decays with time as $(1 - e^{-1/4u})$, where $u = \alpha t/a^2$, the time parameter.

7.4 SEPARATION OF VARIABLES FOR RADIAL HEAT FLOW

The following derivation is an example of the separation of variables method that shows how the Bessel functions arise for cylindrical geometries. For the geometries RIJ, $I = 0, 1, 2, 3$, and $J = 1, 2, 3$, the large-time GFs can be derived by this method. For a complete discussion of the separation of variables method for cylinders, refer to Ozisik (1980, Chapter 3). It is important to note that Ozisik's notation for GFs in cylindrical and spherical coordinates differs from this book by a factor of (2π); that is, G (Ozisik, 1980)$/2\pi = G$ (this volume).

In this section the separation of variables technique will be demonstrated with the $R01$ GF (solid cylinder with temperature boundary conditions), but the method also

Table 7.2 Approximate closed-form expressions for $\int_{r'=0}^{a} G_{R00}(r, t|r', \tau) \, 2\pi r' \, dr'$

Range of application	Error (%)	Equation number in Appendix R
$u < 0.1$, $r/a \geq 1$		R00.7
$u < 0.25$, $r/a = 1$	1.3	R00.9
$u < 0.01$, $0.5 < r/a < 1$	0.03	R00.10
$u \geq 0.25$, $(r/a)^2/(4u)$ small	-0.016	R00.11

Note: (1) $u = \alpha(t - \tau)/a^2$. (2) As $a \to \infty$, the integral approaches the value 1.0.

applies to hollow cylinders. Consider the following initial-value problem for a solid cylinder:

$$\frac{1}{r}\frac{\partial}{\partial r}\left[r\frac{\partial T}{\partial r}\right] = \frac{1}{\alpha}\frac{\partial T}{\partial t} \tag{7.17}$$

$$T(b, t) = 0 \tag{7.18}$$

$$T(0, t) < M \tag{7.19}$$

$$T(r, 0) = F(r) \tag{7.20}$$

The initial condition is an arbitrary function of position. There is no energy generation and the boundary condition at $r = b$ is homogeneous. An equivalent boundary condition at $r = 0$ is that the temperature is symmetric, $\partial T/\partial r = 0$. The same solution can be derived with either condition.

The separation of variables technique produces a series solution of the form

$$T(r, t) = \sum_{n=1}^{\infty} T_n(r, t) \tag{7.21}$$

where $T_n(r, t)$ satisfies the differential equation and the boundary conditions. Individually the $T_n(r, t)$ solutions do not satisfy the initial condition given by Eq. (7.20), and the series form is used precisely to satisfy the initial condition. The issue of convergence raised by the infinite series in Eq. (7.21) is an important one, but for the purpose of this book, the solution converges for heat conduction problems and the results are physically meaningful.

The separation of variables method assumes that the solutions $T_n(r, t)$ have the form

$$T_n(r, t) = \mathbf{R}(r)\,\theta(t) \tag{7.22}$$

That is, the dependence on r and t has been separated into a product of a function of position and a function of time. The function T_n must satisfy the differential equation

$$\frac{\partial^2 T_n}{\partial r^2} + \frac{1}{r}\frac{\partial T_n}{\partial r} = \frac{1}{\alpha}\frac{\partial T_n}{\partial t} \tag{7.23}$$

Substitute Eq. (7.22) in Eq. (7.23) to give, after some rearrangement,

$$\frac{1}{\mathbf{R}}\left[\frac{\partial^2 \mathbf{R}}{\partial r^2} + \frac{1}{r}\frac{\partial \mathbf{R}}{\partial r}\right] = \frac{1}{\alpha\theta}\frac{\partial \theta}{\partial t} = -\lambda^2 \tag{7.24}$$

The negative constant $-\lambda^2$ is introduced because (a) a function of r set equal to a function of t must both be equal to a constant function, and (b) the negative value is required to give physically meaningful results for $\theta(t)$. Equation (7.24) represents two ordinary differential equations. The equation for $\mathbf{R}$ is

$$\frac{d^2\mathbf{R}}{dr^2} + \frac{1}{r}\frac{d\mathbf{R}}{dr} + \lambda^2\mathbf{R} = 0 \tag{7.25a}$$

This is the Bessel equation of order zero, and the elementary solutions are

$$R(r) = A J_0(\lambda r) + B Y_0(\lambda r) \qquad (7.25b)$$

where $J_0(\cdot)$ and $Y_0(\cdot)$ are Bessel functions of order zero and A and B are constants. A graph of these functions is shown in Fig. 7.2.

The differential equation for $\theta(t)$ is

$$\frac{d\theta}{dt} + \lambda^2 \alpha \theta = 0 \qquad (7.26a)$$

and the elementary solution is

$$\theta(t) = C e^{-\lambda^2 \alpha t} \qquad (7.26b)$$

where C is a constant. Thus, the solution $T_n(r, t)$ is given by Eq. (7.23), with Eqs. (7.25b) and (7.26b):

$$T_n(r, t) = e^{-\lambda_n^2 \alpha t} [A_n J_0(\lambda_n r) + B_n Y_0(\lambda_n r)] \qquad (7.27)$$

Here new names have been given to the constants λ_n, A_n, and B_n, which must be determined from the boundary conditions and the initial condition for each geometry. Up to this point the analysis applies to both solid and hollow cylinders.

Next, the boundary conditions are applied to the general solution given by Eq. (7.27), so the following analysis applies only to the $R01$ geometry. At $r = 0$, the natural boundary condition given by Eq. (7.19) yields

$$\lim_{r \to 0} e^{-\lambda_n^2 \alpha t} [A_n J_0(\lambda_n r) + B_n Y_0(\lambda_n r)] \neq \infty \qquad (7.28)$$

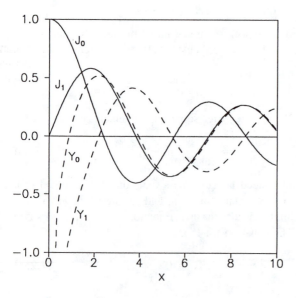

Figure 7.2 Bessel functions $J_0(x)$, $Y_0(x)$, $J_1(x)$, $Y_1(x)$.

In the limit as $r \to 0$, the function $J_0(\lambda_n r)$ goes to one, but the function $Y_0(\lambda_n r)$ becomes infinite. The term containing $Y_0(\lambda_n r)$ does not belong in the solution, and Eq. (7.28) can be satisfied only by

$$B_n = 0 \tag{7.29}$$

Next the temperature boundary condition at $r = b$ given by Eq. (7.18) is applied to the general solution to give

$$T_n(b, t) = 0 = e^{-\lambda_n^2 \alpha t} A_n J_0(\lambda_n b) \quad \text{or} \quad 0 = A_n J_0(\lambda_n b) \tag{7.30}$$

The exponential is never zero, so it may be canceled out. The constant A_n cannot be zero or the entire solution will be identically zero, a trivial result. Equation (7.30) is satisfied by choosing

$$J_0(\beta_n) = 0 \tag{7.31}$$

where $\beta_n = \lambda_n b$ are the dimensionless eigenvalues for $n = 1, 2,$ and so on. There are an infinite number of eigenvalues that are distinct for each cylinder geometry. The first few eigenvalues are listed in Appendix B for the cylinder cases $R01$, $R02$, $R03$, $R11$, $R12$, and $R22$.

Next, the initial condition must be satisfied. So far, the complete solution has the form

$$T(r, t) = \sum_{n=1}^{\infty} e^{-\beta_n^2 \alpha t / b^2} A_n J_0\left(\frac{\beta_n r}{b}\right) \tag{7.32}$$

The initial condition requires that

$$T(r, 0) = F(r) = \sum_{n=1}^{\infty} A_n J_0\left(\frac{\beta_n r}{b}\right) \tag{7.33}$$

The initial condition can be satisfied if an arbitrary function $F(r)$ can be expressed as an infinite series of Bessel functions. In Chapter 4, expansions of arbitrary functions in terms of Fourier sine and cosine series arose from one-dimensional plate cases. Fourier series are a special case of the general theory of orthogonal functions (Wylie and Barrett, 1982). Bessel functions are simply another class of functions for which infinite-series expansions are possible, and the infinite series expansion is needed to satisfy the initial condition.

The orthogonality condition for $J_0(\cdot)$ on $0 \le r \le b$ is (Appendix B)

$$\int_0^b J_0\left(\frac{\beta_m r}{b}\right) J_0\left(\frac{\beta_n r}{b}\right) 2\pi r \, dr = \begin{cases} 0 & m \ne n \\ \pi b^2 J_1^2(\beta_n) & m = n \end{cases} \tag{7.34a}$$

To apply the orthogonality condition to find A_n, multiply Eq. (7.33) by $J_0(\beta_m r/b)$ and integrate over the volume of the cylinder ($0 \le r \le b$):

$$\int_0^b J_0\left(\frac{\beta_m r}{b}\right) F(r) \, 2\pi r \, dr = \int_0^b \sum_{n=1}^{\infty} A_n J_0\left(\frac{\beta_n r}{b}\right) J_0\left(\frac{\beta_m r}{b}\right) 2\pi r \, dr \tag{7.34b}$$

The orthogonality condition applied to the right-hand side of Eq. (7.34b) gives exactly one nonzero term from the infinite series, at $m = n$. Solving for A_n gives

$$A_n = \frac{1}{\pi b^2 J_1^2(\beta_n)} \int_0^b J_0\left(\frac{\beta_n r'}{b}\right) F(r') \, 2\pi r' \, dr' \tag{7.35}$$

Note that the subscript n is really a dummy subscript, and any letter could be substituted. Also, the variable of integration has been written as r', as it too is a dummy variable.

Next, replace A_n into the solution given by Eq. (7.32) to give the particular solution to the initial-value problem (case R01B0T-),

$$T(r, t) = \sum_{n=1}^{\infty} e^{-\beta_n^2 \alpha t/b^2} \int_{r'=0}^b \frac{J_0(\beta_n r'/b) \, J_0(\beta_n r/b)}{\pi b^2 J_1^2(\beta_n)} F(r') \, 2\pi r' \, dr' \tag{7.36}$$

After some rearrangement, this solution can be written

$$T(r, t) = \int_{r'=0}^b F(r') \left[\frac{1}{\pi b^2} \sum_{n=1}^{\infty} e^{-\beta_n^2 \alpha t/b^2} \frac{J_0(\beta_n r'/b) \, J_0(\beta_n r/b)}{J_1^2(\beta_n)} \right] 2\pi r' \, dr' \tag{7.37}$$

This is the separation of variables result for the temperature resulting from an arbitrary initial condition on the R01 geometry.

Finally, the GF can be deduced from the separation of variables solution by also solving the initial value problem [Eqs. (7.17)–(7.20)] with the GFSE to give

$$T(r, t) = \int_{r'=0}^b F(r') \, [G_{R01}(r, t|r', 0)] \, 2\pi r' \, dr' \tag{7.38}$$

Equations (7.37 and (7.38) are solutions to the same boundary value problem, and since a boundary value problem has only one unique solution, the expression in brackets in Eq. (7.37) must be identically $G_{R01}(r, t|r', 0)$, the GF evaluated at $\tau = 0$:

$$G_{R01}(r, t|r', 0) = \frac{1}{\pi b^2} \sum_{n=1}^{\infty} e^{-\beta_n^2 \alpha t/b^2} \frac{J_0(\beta_n r'/b) \, J_0(\beta_n r/b)}{J_1^2(\beta_n)} \tag{7.39}$$

The last step in finding the GF from the separation of variables solution is to replace $(t - 0)$ in Eq. (7.39) by $(t - \tau)$. Recall that the time dependence of all GFs is in the form $(t - \tau)$. Then,

$$G_{R01}(r, t|r', \tau) = \frac{1}{\pi b^2} \sum_{n=1}^{\infty} e^{-\beta_n^2 \alpha(t-\tau)/b^2} \frac{J_0(\beta_n r'/b) \, J_0(\beta_n r/b)}{J_1^2(\beta_n)} \tag{7.40}$$

This GF is also listed in Appendix R.

This method for finding the large-time GFs can be used on all of the solid cylinder and hollow cylinder cases, denoted G_{RIJ} for which $I = 0, 1, 2, 3$ and $J = 1, 2, 3$. It is not necessary to derive these GFs, however, since they are listed in Appendix R.

7.5 SOLID CYLINDERS

Some worked examples are next discussed for the temperature in solid cylinders. Time partitioning is introduced on a case-by-case basis because the choice of an appropriate small-time GF depends on time, on geometry, and on location in the cylinder.

7.5.1 Initial Conditions

Example 7.1: Solid cylinder with zero surface temperature—*R01B0T-* case. Find the temperature in a solid cylinder, $0 \le r \le b$, with initial temperature $F(r)$ and a boundary temperature fixed at $T = 0$.

SOLUTION This is the *R01B0T-* case and it was examined in Section 7.4. The temperature is given by Eq. (7.37), where the expression in brackets is the GF $G_{R01}(r, t|r', 0)$. The eigenvalues β_n are defined by the eigencondition $J_0(\beta_n) = 0$, and the first 10 values of β_n are listed in Appendix B. The integral on r' acts on just a portion of Eq. (7.37):

$$T(r, t) = \frac{1}{\pi b^2} \sum_{n=1}^{\infty} e^{-\beta_n^2 \alpha t/b^2} \frac{J_0(\beta_n r/b)}{J_1^2(\beta_n)} \int_{r'=0}^{b} F(r') J_0\left(\frac{\beta_n r'}{b}\right) 2\pi r' \, dr' \quad (7.41)$$

a. Case R01B0T1. If the initial temperature is uniform, $F(r) = T_0$, then the above integral is given by

$$\int_{r'=0}^{b} J_0\left(\frac{\beta_n r'}{b}\right) 2\pi r' \, dr' = \frac{2\pi b^2 J_1(\beta_n)}{\beta_n} - 0 \quad (7.42)$$

and the temperature resulting from initial temperature T_0 becomes (case *R01B0T1*)

$$T(r, t) = 2T_0 \sum_{n=1}^{\infty} e^{-\beta_n^2 \alpha t/b^2} \frac{J_0(\beta_n r/b)}{\beta_n J_1(\beta_n)} \quad (7.43)$$

For $\alpha t/b^2$ small, the temperature near the center of the cylinder (at $r = 0$) remains at T_0, because the effect of the surface temperature has not yet penetrated to the center of the cylinder.

b. Case R01B0T5. For the initial condition

$$F(r) = \begin{cases} T_0 & 0 \le r \le a \\ 0 & a < r \le b \end{cases}$$

the temperature is given by

$$T(r, t) = 2 \, T_0 \frac{a}{b} \sum_{n=1}^{\infty} e^{-\beta_n^2 \alpha t/b^2} \frac{J_1(\beta_n a/b) J_0(\beta_n r/b)}{\beta_n J_1^2(\beta_n)} \quad (7.44)$$

This solution converges efficiently for large values of time. Small time expressions for the temperature for the case when $a/b \ne 1$ can be found by approximating the cylinder as an infinite body. Initially, heat diffuses outward from the point $r = a$, and it takes a little time before the diffusion is influenced by the zero-temperature boundary at $r = b$. During this small time period, the temperature distribution is identical to that in an infinite body with the same initial condition. Thus, the appropriate early-time GF is G_{R00}, and the expressions referenced in Table 7.2 (integral of G_{R00}) may be used to find the temperature at small times. The criterion for small time is $\alpha t/(b - a)^2$ small (<0.01) because it is the distance between the initial temperature region and the boundary, $(b - a)$, that determines the time span of infinite-body behavior.

Example 7.2: Solid cylinder with surface convection—R03B0T1 case. Find the transient temperature in a cylinder initially at temperature T_0 that is suddenly quenched in a large tank of fluid at temperature T_∞ with heat transfer coefficient h.

SOLUTION The boundary and initial conditions are given by

$$T \text{ is finite as } r \to 0$$

$$-k\,\frac{\partial T(b,\,t)}{\partial r} = h(T(b,\,t) - T_\infty)$$

$$T(r,\,0) = T_0 \tag{7.45}$$

This boundary value problem has two nonhomogeneous conditions resulting from the two temperatures T_0 and T_∞. Two integrals from the GFSE are needed to directly describe this problem; however, one nonhomogeneous condition can be removed by defining a new variable $(T - T_\infty)$. The new boundary and initial conditions are given by

$$(T - T_\infty) \text{ is finite as } r \to 0$$

$$-k\,\frac{\partial (T - T_\infty)}{\partial r}\bigg|_{r=b} - h(T\big|_{r=b} - T_\infty) = 0$$

$$T(r,\,0) - T_\infty = (T_0 - T_\infty) \tag{7.46}$$

Note that the boundary condition at $r = b$ is now homogeneous in terms of variable $(T - T_\infty)$. Variable $T - T_0$ could have been chosen, but it would result in a form of the solution less well suited to numerical evaluation at small values of dimensionless time. The temperature in the cylinder is now given by the initial condition term of the GFSE:

$$T(r,\,t) - T_\infty = \int_{r'=0}^{b} (T_0 - T_\infty)\,G_{R03}(r,\,t|b,\,0)\,2\pi r'\,dr' \tag{7.47}$$

Using the $R03$ GF listed in Appendix R gives

$$T(r,\,t) - T_\infty = (T_0 - T_\infty)\int_{r'=0}^{b} \sum_{n=1}^{\infty} e^{-\beta_n^2 \alpha t/b^2}$$

$$\times\; \frac{\beta_n^2 J_0(\beta_n r/b)J_0(\beta_n r'/b)}{\pi b^2(B^2 + \beta_n^2)J_0^2(\beta_n)}\,2\pi r'\,dr' \tag{7.48}$$

where $B = hb/k$ (the Biot number) and eigenvalues β_n are the roots of

$$-\beta_n J_1(\beta_n) + B\,J_0(\beta_n) = 0 \tag{7.49}$$

Values of β_n for several values of B are given in Carslaw and Jaeger (1959).

The integral on r' in Equation (7.48) was given earlier in Example 7.1, so the temperature in the cylinder is given by

$$T(r, t) - T_\infty = 2(T_0 - T_\infty) \sum_{n=1}^{\infty} e^{-\beta_n^2 \alpha t/b^2} \frac{\beta_n J_1(\beta_n) J_0(\beta_n r/b)}{(B^2 + \beta_n^2) J_0^2(\beta_n)} \qquad (7.50)$$

7.5.2 Boundary Conditions

Example 7.3: Solid cylinder with elevated surface temperature—$R01B1T0$ case. Find the temperature in a solid cylinder, $0 \le r \le b$, that has zero initial condition and has temperature T_0 suddenly applied at boundary $r = b$.

DIRECT SOLUTION The temperature resulting from a boundary temperature is given by the last term of Eq. (7.2) with $r_j = b$:

$$T(r, t) = -\alpha \int_{\tau=0}^{t} T_0 \left. \frac{\partial G}{\partial n'} \right|_{r'=b} 2\pi b \, d\tau \qquad (7.51)$$

The required $R01$ GF and its derivative $\partial G_{R01}/\partial n'$ is given in Appendix R, so the integral in Eq. (7.51) is given by

$$\alpha T_0 \int_{\tau=0}^{t} \frac{2}{b^2} \sum_{n=1}^{\infty} e^{-\beta_n^2 \alpha(t-\tau)/b^2} \frac{\beta_n J_0(\beta_n r/b)}{J_1(\beta_n)} \, d\tau \qquad (7.52)$$

and the eigenvalues are given by $J_0(\beta_m) = 0$. The integral on τ is easily evaluated to give

$$T(r, t) = 2T_0 \sum_{n=1}^{\infty} (1 - e^{-\beta_n^2 \alpha t/b^2}) \frac{J_0(\beta_n r/b)}{\beta_n J_1(\beta_n)} \qquad (7.53)$$

This solution suffers from poor numerical convergence which can be made clear by writing the solution as the sum of two series,

$$T(r, t) = 2T_0 \sum_{n=1}^{\infty} \frac{J_0(\beta_n r/b)}{\beta_n J_1(\beta_n)} - 2T_0 \sum_{n=1}^{\infty} e^{-\beta_n^2 \alpha t/b^2} \frac{J_0(\beta_n r/b)}{\beta_n J_1(\beta_n)} \qquad (7.54)$$

The first series converges slowly and it does not depend on the dimensionless time. Time partitioning could be used to find a temperature expression that converges more efficiently, but in this case there is a simple alternative solution.

ALTERNATIVE SOLUTION The alternative solution method discussed in Chapter 3 is useful for improving the numerical convergence of the temperature driven by nonhomogeneous boundary conditions. The alternative solution method is very simple when the steady-state solution is a simple function. In this case, the steady-state solution is simply $T(r, t \to \infty) = T_0$. Let the known solution be $T^*(r, t) = T_0$, and let the unknown temperature be given by $T(r, t) = T^*(r, t) + T'(r, t)$. Temperature T^* is of course also a solution to the transient energy equation. A new solution is now sought for the temperature $T'(r, t) = T(r, t) - T^*(r, t)$ subject to the following boundary value problem:

$$\frac{\partial^2 T'}{\partial r^2} + \frac{1}{r}\frac{\partial T'}{\partial r} = \frac{1}{\alpha}\frac{\partial T'}{\partial t}$$

$$T'(b, t) = T(b, t) - T^*(b, t) = T_0 - T_0 = 0 \qquad (7.55)$$

$$T'(r, 0) = T(r, 0) - T^*(r, 0) = 0 - T_0$$

Then the alternative solution is given by Eq. (3.66):

$$T(r, t) = T_0 + \int_{r'=0}^{b} (-T_0)$$

$$\times \left[\sum_{n=1}^{\infty} e^{-\alpha\beta_n^2 t/b^2} \frac{J_0(\beta_n r'/b)\, J_0(\beta_n r/b)}{\pi b^2 J_1^2(\beta_n)} \right] 2\pi r'\, dr' \qquad (7.56)$$

Effectively, the boundary heating problem has been transformed into an initial heating problem, and this integral has been solved previously as in Example 7.1:

$$T(r, t) = T_0 - T_0 \sum_{n=1}^{\infty} e^{-\alpha\beta_n^2 t/b^2} \frac{J_0(\beta_n r/b)}{\beta_n J_1(\beta_n)} \qquad (7.57)$$

This expression converges better than Eq. (7.54) for all values of $\alpha t/b^2$.

Example 7.4: Solid cylinder with heating at the surface—R02B1T0 case. A cylinder whose initial temperature is zero is heated by a suddenly applied surface heat flux q_0. Find (a) the surface temperature on the cylinder at early time, and (b) the spatial average temperature in the cylinder at any time.

SOLUTION The boundary conditions are given by

$$\frac{\partial T(0, t)}{\partial r} = 0 \qquad \text{(symmetry condition)}$$
$$\qquad (7.58)$$
$$-k\frac{\partial T(b, t)}{\partial r} = q_0 \qquad T(r, 0) = 0$$

　　a. Surface temperature at early time. The surface temperature is given by the GFSE evaluated at $r = b$:

$$T(b, t) = \alpha \int_{\tau=0}^{t} \frac{q_0}{k} G_{R02}(b, t\,|\,b, \tau)\, 2\pi b\, d\tau \qquad (7.59)$$

This is a case where time partitioning can improve the numerical convergence of the temperature expression. For small times, only the small-time form of the GF is needed to find the temperature. For the G_{R02} small-time form, Eq. (R02.5) from Appendix R is appropriate:

$$G_{R02}(b, t\,|\,b, \tau) \approx \frac{1}{2\pi b^2}\left[\frac{1}{\sqrt{\pi}}(u)^{-1/2} + \frac{1}{2} + \frac{3}{4\sqrt{\pi}}(u)^{1/2} + \frac{3}{8}u \right] \qquad (7.60)$$

where $u = \alpha(t - \tau)/b^2 < 0.1$. Then the surface temperature can be written in terms of the integral given by Eq. (7.59) with a change of variable to $u = \alpha(t - \tau)/b^2$:

$$T(b, t) = \int_{u=0}^{\alpha t/b^2} \frac{q_0 b}{k} \left[\frac{1}{\sqrt{\pi}} (u)^{-1/2} + \frac{1}{2} + \frac{3}{4\sqrt{\pi}} (u)^{1/2} + \frac{3}{8} u \right] du \quad (7.61)$$

or, for $\alpha t/b^2 < 0.1$,

$$T(b, t) = \frac{q_0 b}{k} \left[\frac{2}{\sqrt{\pi}} \left(\frac{\alpha t}{b^2} \right)^{1/2} + \frac{1}{2} \left(\frac{\alpha t}{b^2} \right) \right.$$

$$\left. + \frac{1}{2\sqrt{\pi}} \left(\frac{\alpha t}{b^2} \right)^{3/2} + \frac{3}{16} \left(\frac{\alpha t}{b^2} \right)^2 \right] \quad (7.62)$$

In the above expression, the first term inside the brackets

$$\frac{2}{\sqrt{\pi}} \left(\frac{\alpha t}{b^2} \right)^{1/2}$$

is the same as the temperature on a plane wall. From this perspective, the next term ($\alpha t/2b^2$) is the first correction term for the curvature of the cylinder wall (Beck et al., 1985).

    b. Spatial average temperature. The spatial average temperature in the cylinder may be found from an overall energy balance on the cylinder

$$q_{\text{in}} = q_{\text{storage}} \quad \text{or} \quad q_0 = \rho c \pi b^2 \frac{\partial T_{\text{av}}}{\partial t} \quad (7.63)$$

This may be integrated from the initial temperature of zero to find

$$T_{\text{av}}(t) = \frac{q_0 t}{\rho c \pi b^2} \quad (7.64)$$

Note that the spatial average temperature increases linearly with time. The same behavior occurs in a body with uniform energy generation if the boundary is insulated (*R02B0T0G1*); in both cases there is heating specified and no heat loss.

7.5.3 Volume Energy Generation

Example 7.5: Solid cylinder with uniform energy generation—*R01B0T0G1* case. A cylinder is initially at zero temperature and the boundary at $r = b$ is maintained at $T = 0$. Find the temperature in the cylinder resulting from a uniform internal energy generation g_0 (W/m^3).

SOLUTION The temperature is given by the GFSE:

$$T(r, t) = \int_{\tau=0}^{t} \int_{r'=0}^{b} \frac{\alpha}{k} g_0 G_{R01}(r, t|r', \tau) 2\pi r' \, dr' \, d\tau \quad (7.65)$$

Using the large-time form of G_{R01} from Appendix R, the temperature is given by

$$T(r, t) = 2 \frac{g_0 b^2}{k} \sum_{n=1}^{\infty} (1 - e^{-\beta_n^2 \alpha t/b^2}) \frac{J_0(\beta_n r/b)}{\beta_n J_1(\beta_n)} \qquad (7.66)$$

where the eigenvalues are roots of the equation $J_0(\beta_n) = 0$. This solution is a candidate for time partitioning. At early time, the interior of the cylinder will behave like an infinite body and the zero-temperature boundary at $r = b$ will have only a local influence. Refer to Appendix R for a suitable small-time form of G_{R01}.

At large time values the convergence of Eq. (7.66) is controlled by a term that does not depend on time. This is the steady solution, and it can be found independently (by solving the steady boundary value problem) to give a better expression for numerical evaluation. Equation (7.66) can be written

$$T(r, t) = \frac{g_0 b^2}{4k} \left[1 - \left(\frac{r}{b} \right)^2 \right] - 2 \frac{g_0 b^2}{k} \sum_{n=1}^{\infty} e^{-\beta_n^2 \alpha t/b^2} \frac{J_0(\beta_n r/b)}{\beta_n J_1(\beta_n)} \qquad (7.67)$$

The same result can also be found from the alternate GFSE.

Example 7.6: Solid cylinder with nonuniform energy generation—*R02B0T0Gr5* case. Consider the solid cylinder $0 \leq r \leq b$ initially at zero temperature with an insulated boundary. The cylinder is heated by volume energy generation

$$g(r', \tau) = \begin{cases} 0 & 0 \leq r \leq a \\ g_0 & a \leq r \leq b \end{cases} \qquad (7.68)$$

where g_0 (W/m³) is the energy generation rate. The energy generation is zero deep inside the cylinder and it has the value g_0 near the surface of the cylinder. Refer to Fig. 7.3. This geometry approximately describes microwave heating of food or nuclear radiation heating of reactor control rods (approximately, because actual radiation heating is attenuated inside the body). Find the temperature after a long period.

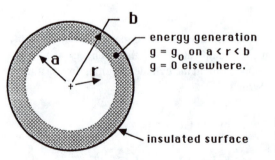

energy generation
g = g₀ on a < r < b
g = 0 elsewhere.

insulated surface

Figure 7.3 Solid cylinder with nonuniform energy generation and insulated surface for Example 7.6.

SOLUTION This is the $R02$ geometry with energy generation. The GF equation for this case is given by the second term of Eq. (7.2), and G_{R02} is listed in Appendix R. The temperature in the cylinder is

$$
T(r, t) = \frac{\alpha}{k} \int_{\tau=0}^{t} \int_{r'=a}^{b} \frac{g_0}{\pi b^2} \left[1 + \sum_{n=1}^{\infty} e^{-\beta_n^2 \alpha(t-\tau)/b^2} \right.
$$

$$
\left. \times \frac{J_0(\beta_n r/b) J_0(\beta_n r'/b)}{J_0^2(\beta_n)} \right] 2\pi r' \, dr' \, d\tau \tag{7.69}
$$

where the eigenvalues β_n are defined by $J_1(\beta_n) = 0$. Note that the integral on r' is evaluated over (a, b) as a result of the piecewise constant energy generation function. The integrals can be evaluated to give

$$
T(r, t) = \frac{g_0 b^2}{k} \left[1 - \left(\frac{a}{b} \right)^2 \right] \frac{\alpha t}{b^2} + 2 \frac{g_0 ab}{k} \sum_{n=1}^{\infty} e^{-\beta_n^2 \alpha t/b^2}
$$

$$
\times \frac{J_0(\beta_n r/b) J_1(\beta_n a/b)}{\beta_n^3 J_0^2(\beta_n)} - 2 \frac{g_0 ab}{k} \sum_{n=1}^{\infty} \frac{J_0(\beta_n r/b) J_1(\beta_n a/b)}{\beta_n^3 J_0^2(\beta_n)} \tag{7.70}
$$

The above solution contains three terms, one of which does not depend on time. At $t = 0$, the temperature is zero as required because the second and third terms cancel out, not because each term is zero.

The first term of the above solution is the spatial average temperature in the cylinder, $T_{av}(t)$, which can be found by integrating Eq. (7.70) over the volume of the cylinder:

$$
T_{av}(t) \equiv \frac{1}{\pi b^2} \int_{r=0}^{b} T(r, t) \, 2\pi r \, dr = \frac{g_0 b^2}{k} \left[1 - \left(\frac{a}{b} \right)^2 \right] \frac{\alpha t}{b^2} \tag{7.71}
$$

The average temperature increases with time because the heat that is added has no place to go (the boundary is insulated). There is no steady-state temperature. The spatial average temperature can also be found from an overall energy balance $q_{in} = q_{storage}$.

In the limit as $a/b \to 1$, the cylinder will be heated uniformly over its volume. In this case the temperature given by Eq. (7.70) reduces to the spatial average temperature given by Eq. (7.71). No heat can escape at the boundary, and every point in the cylinder is heated equally.

The second term of the solution given by Eq. (7.70) is a transient term

$$
2 \frac{g_0 ab}{k} \sum_{n=1}^{\infty} e^{-\beta_n^2 \alpha t/b^2} \frac{J_0(\beta_n r/b) J_1(\beta_n a/b)}{\beta_n^3 J_0^2(\beta_n)} \tag{7.72}
$$

This transient term decreases exponentially to zero over time. After the transient term becomes zero, the spatial distribution of the temperature stops changing. The temperature is said to be quasi-steady, because although the shape of the temperature distribution is fixed, the magnitude of the temperature distribution increases linearly with time according to the average temperature term.

The third term of the solution given by Eq. (7.70) is the quasi-steady temperature distribution which does not depend on time,

$$- 2 \frac{g_0 ab}{k} \sum_{n=1}^{\infty} \frac{J_0(\beta_n r/b) J_1(\beta_n a/b)}{\beta_n^3 J_0^2(\beta_n)} \tag{7.73}$$

For $t \gg 1$, this term describes the shape of the temperature distribution in the form of deviations from the average temperature given by Eq. (7.71). That is, the temperature is above average near $r = b$ (the heated region), and the temperature is below average near $r = 0$ (the unheated region).

7.6 HOLLOW CYLINDERS

Two examples are given for the temperature in hollow cylinders. Compared to solid cylinders, hollow cylinders have one more physical boundary and consequently the GF and the eigenconditions are more complex; however, all of the analytical techniques for solid cylinders also apply to hollow cylinders. Another approach for cylinders is the Galerkin-based GF developed in Chapters 10 and 11. For hollow cylinders, it may be possible to obtain numerical results more easily from the Galerkin-based GFs than from the analytical GFs discussed in this section.

Example 7.7: Hollow cylinder with zero surface temperature—R11B00T1 case. Consider the hollow cylinder $a \leq r \leq b$ with uniform initial temperature T_0. Find the temperature for $t > 0$ for the boundaries fixed at zero temperature.

SOLUTION The temperature due to an initial condition is given by the GFSE:

$$T(r, t) = \int_{r'=a}^{b} G_{R11}(r, t|r', 0) T_0 \, 2\pi r' \, dr' \tag{7.74}$$

Note that the integral is evaluated over the hollow cylinder, $a \leq r' \leq b$. The large-time GF is listed in Appendix R to give

$$T(r, t) = T_0 \int_{r'=a}^{b} \frac{\pi}{4a^2} \sum_{m=1}^{\infty} e^{-\beta_m^2 \alpha t/a^2}$$

$$\times \frac{\beta_m^2 J_0^2(\beta_m)}{J_0^2(\beta_m) - J_0^2(\beta_m b/a)} \, \mathbf{R}(r)\mathbf{R}(r') \, 2\pi r' \, dr' \tag{7.75}$$

where
$$\mathbf{R}(r) = J_0\left(\frac{\beta_m r}{a}\right) Y_0\left(\frac{\beta_m b}{a}\right) - J_0\left(\frac{\beta_m b}{a}\right) Y_0\left(\frac{\beta_m r}{a}\right) \tag{7.76}$$

and where the eigenvalues β_m satisfy

$$J_0(\beta_m) Y_0\left(\frac{\beta_m b}{a}\right) - J_0\left(\frac{\beta_m b}{a}\right) Y_0(\beta_m) = 0 \tag{7.77}$$

The first five eigenvalues are listed in Appendix B for various values of b/a (various cylinder geometries). The integral on r' operates only on $J_0(\beta_m r'/a)$ and $Y_0(\beta_m r'/a)$, and the integral can be carried out with the first integrals from Tables 7.3 and 7.4 to give (see also Appendix B)

$$T(r, t) = T_0 \frac{\pi^2}{2a} \sum_{m=1}^{\infty} e^{-\beta_m^2 \alpha t/a^2} \frac{\beta_m J_0^2(\beta_m)}{J_0^2(\beta_m) - J_0^2(\beta_m b/a)} \mathbf{R}(r)$$

$$\times \left[bJ_1\left(\beta_m \frac{b}{a}\right)Y_0\left(\beta_m \frac{b}{a}\right) - aJ_1(\beta_m)Y_0\left(\beta_m \frac{b}{a}\right) \right.$$

$$\left. - bJ_0\left(\beta_m \frac{b}{a}\right)Y_1\left(\beta_m \frac{b}{a}\right) + aJ_0\left(\beta_m \frac{b}{a}\right)Y_1(\beta_m) \right] \qquad (7.78)$$

This is the large-time form of the temperature in the hollow cylinder where $\mathbf{R}(r)$ is given by Eq. (7.76). There are four Bessel functions involved: J_0, Y_0, J_1, and

Table 7.3 Bessel function integral over r' of $\int J_0\left(\beta_m \dfrac{r'}{b}\right) F(r')\, r'\, dr'$

$F(r')$	Integral $\left(\text{let } x = \beta_m \dfrac{r'}{b}\right)$
1	$\left(\dfrac{b}{\beta_m}\right)^2 xJ_1(x)$
$\dfrac{r'}{b}$	$\dfrac{b^2}{\beta_m^3} [x^2 J_1(x) + xJ_0(x) - \int J_0(x)dx]$
$\left(\dfrac{r'}{b}\right)^2$	$\dfrac{b^2}{\beta_m^4} [(x^3 - 4x) J_1(x) + 2x^2 J_0(x)]$
$\ln \dfrac{r'}{b}$	$\dfrac{b^2}{\beta_m^2} \left[J_0(x) + x \ln\left(\dfrac{r'}{b}\right) J_1(x) \right]$

Table 7.4 Bessel function integral over r' of $\int Y_0\left(\beta_m \dfrac{r'}{b}\right) F(r')\, r'\, dr'$

$F(r')$	Integral $\left(\text{let } x = \beta_m \dfrac{r'}{b}\right)$
1	$\left(\dfrac{b}{\beta_m}\right)^2 xY_1(x)$
$\dfrac{r'}{b}$	$\dfrac{b^2}{\beta_m^3} [x^2Y_1(x) + xY_0(x) - \int Y_0(x)dx]$
$\left(\dfrac{r'}{b}\right)^2$	$\dfrac{b^2}{\beta_m^4} [(x^3 - 4x)Y_1(x) + 2x^2Y_0(x)]$
$\ln \dfrac{r'}{b}$	$\dfrac{b^2}{\beta_m^2} \left[Y_0(x) + x \ln\left(\dfrac{r'}{b}\right) Y_1(x) \right]$

Y_1. For large values of $\alpha t/a^2$ only a few terms of the series are needed for accurate numerical values.

Example 7.8: Hollow cylinder insulated inside—*R21B00T* case. Consider the hollow cylinder $a \leq r \leq b$ has a steady temperature distribution T_i due to steady heating at the boundary $r = a$ and a zero temperature $r = b$. That is,

$$-k \frac{\partial T_i(a)}{\partial r} = q_0 \tag{7.79}$$

$$T_i(b) = 0 \tag{7.80}$$

Suppose that for $t > 0$, the heat flux at $r = a$ suddenly becomes zero (the boundary becomes insulated). Find (a) the initial temperature distribution, and (b) the transient temperature due to the change in the heating at the boundary $r = a$.

SOLUTION

a. Initial temperature. The initial temperature may be found from the steady GFSE, Eq. (3.94), in radial cylindrical coordinates:

$$T_i(r) = \frac{q_0}{k} G(r|r' = a) \, 2\pi a \tag{7.81}$$

The steady GF is given by Table R.1 in Appendix R:

$$G_{R21}(r|r') = \begin{cases} \dfrac{1}{2\pi} \ln \left(\dfrac{b}{r'}\right) & r < r' \\[3mm] \dfrac{1}{2\pi} \ln \dfrac{b}{r} & r > r' \end{cases} \tag{7.82}$$

When G_{R21} is substituted into Eq. (7.81), the steady temperature is given by

$$T_i(r) = \frac{q_0 a}{k} \ln \frac{b}{r} = \frac{q_0 a}{k} \left(\ln \frac{b}{a} - \ln \frac{r}{a}\right) \tag{7.83}$$

This form of the steady temperature is convenient for part (b) discussed below. The steady temperature may also be found by direct integration of the steady energy equation

$$\frac{1}{r} \left[\frac{d}{dr} \left(r \frac{dT_i}{dr}\right)\right] = 0 \tag{7.84}$$

with boundary conditions given by Eqs. (7.79) and (7.80).

b. Transient temperature. The transient temperature is given by the initial condition form of the GF equation, Eq. (7.2), because the boundary conditions are homogeneous:

$$T(r, t) = \int_{r' = a}^{b} G_{R21}(r, t|r', 0) \, T_i(r') \, 2\pi r' \, dr' \tag{7.85}$$

The large-time form of the GF is given in Appendix R, and the initial temperature is given by Eq. (7.84) to give,

$$
T(r, t) = \int_{r'=a}^{b} \frac{\pi}{4a^2} \sum_{m=1}^{\infty} e^{-\beta_m^2 \alpha t/a^2} \frac{\beta_m^2 J_1^2(\beta_m)}{J_1^2(\beta_m) - J_0^2(\beta_m b/a)} R(r)R(r')
$$

$$
\times \frac{q_0 a}{k} \left(\ln \frac{a}{b} - \ln \frac{r'}{a} \right) 2\pi r' \, dr' \tag{7.86}
$$

where

$$
R(r) = J_0\left(\frac{\beta_m r}{a}\right) Y_0\left(\frac{\beta_m b}{a}\right) - J_0\left(\frac{\beta_m b}{a}\right) Y_0\left(\frac{\beta_m r}{a}\right) \tag{7.87}
$$

and where the eigenvalues β_m satisfy

$$
J_1(\beta_m) Y_0\left(\frac{\beta_m b}{a}\right) - J_0\left(\frac{\beta_m b}{a}\right) Y_1(\beta_m) = 0 \tag{7.88}
$$

The integral in Eq. (7.86) contains two basic forms:

$$
\int W_0\left(\frac{\beta_m r'}{a}\right) r' \, dr' \quad \text{and} \quad \int W_0\left(\frac{\beta_m r'}{a}\right) \left(\ln \frac{r'}{a}\right) r' \, dr'
$$

where $W_0(\cdot)$ is either J_0 or Y_0. These integrals are listed in Table 7.3 and Table 7.4 and can also be written as

$$
\int W_0\left(\beta_m \frac{r'}{a}\right) r' \, dr' = \frac{a^2}{\beta_m^2} \left(\beta_m \frac{r'}{a}\right) W_1\left(\beta_m \frac{r'}{a}\right) \tag{7.89}
$$

$$
\int W_0\left(\beta_m \frac{r'}{a}\right) \left(\ln \frac{r'}{a}\right) r' \, dr' = \frac{a^2}{\beta_m^2}
$$

$$
\times \left[W_0\left(\beta_m \frac{r'}{a}\right) + \beta_m \frac{r'}{a} \left(\ln \frac{r'}{a}\right) \right.
$$

$$
\left. \times W_1\left(\beta_m \frac{r'}{a}\right) \right] \tag{7.90}
$$

where W_1 is either J_1 or Y_1. After some simplification involving the eigencondition Eq. (7.88), to cancel some terms, Eq. (7.86) may be written

$$
T(r, t) = \pi^2 \frac{q_0 b}{k} \sum_{m=1}^{\infty} e^{-\beta_m^2 \alpha t/a^2} \frac{\beta_m J_1^2(\beta_m)}{J_1^2(\beta_m) - J_0^2(\beta_m b/a)} R(r)
$$

$$
\times \ln \frac{a}{b} \left[J_1\left(\beta_m \frac{b}{a}\right) Y_0\left(\beta_m \frac{b}{a}\right) - J_0\left(\beta_m \frac{b}{a}\right) Y_1\left(\beta_m \frac{b}{a}\right) \right] \tag{7.91}
$$

where $R(r)$ is given by Eq. (7.87). This expression involves four Bessel functions, J_0, Y_0, J_1, and Y_1. Only a few terms of the series are needed for $\alpha t/a^2$ large. For $\alpha t/a^2$ very small, the analysis can be repeated with an approximate small-time GF, such as $(1/2\pi a)G_{X20}$ for $r \approx a$ (refer to Example 7.3).

7.7 INFINITE BODY WITH A CIRCULAR HOLE

The radial heat flow in an infinite body containing a circular hole is discussed in this section. Some of the applications for this heat transfer geometry are buried pipes, oil wells, and a heated wire in a quiescent fluid at early time. The GFs for cases numbered $R10$, $R20$, and $R30$ are available in Appendix R. Do not confuse these numbers with the solid cylinder numbers $R01$, $R02$, and $R03$.

The GFs discussed below are derived from Laplace transformation methods (Carslaw and Jaeger, 1959, p. 334). The GFs for the infinite body with a hole involve integrals over a continuous range of eigenvalues instead of a series over discrete eigenvalues. Although the GFs are more complex, they are used to find the temperature as any other cylindrical GF.

Example 7.9: Infinite body with a circular hole and specified surface temperature—$R10B1T0$ case. An infinite body bounded internally by the circular hole $r = a$ has an initial temperature of zero. At $t > 0$ the surface $r = a$ has a fixed temperature T_0. Find the temperature in the body for $t > 0$.

SOLUTION The GF equation for radial flow of heat, Eq. (7.2), applies to this case as

$$T(r, t) = -\alpha \int_{\tau=0}^{t} T_0 \frac{\partial G_{R10}}{\partial n'} (r, t|a, \tau) \, 2\pi a \, d\tau \qquad (7.92)$$

The derivative $\partial G_{R10}/\partial n'$ is given in Appendix R as

$$-\frac{\partial G_{R10}}{\partial n'}\bigg|_{r'=a} = -\frac{1}{\pi^2 a^3} \int_{\beta=0}^{\infty} e^{-\beta^2 \alpha(t-\tau)/a^2}$$

$$\times \frac{\beta \left[J_0\left(\beta \frac{r}{a}\right) Y_0(\beta) - Y_0\left(\beta \frac{r}{a}\right) J_0(\beta) \right]}{J_0^2(\beta) + Y_0^2(\beta)} \, d\beta \qquad (7.93)$$

Then, replace Eq. (7.93) into Eq. (7.92) to find the temperature:

$$T(r, t) = T_0 \frac{2\alpha}{\pi a^2} \int_{\tau=0}^{t} \int_{\beta=0}^{\infty} e^{-\beta^2 \alpha(t-\tau)/a^2}$$

$$\times \frac{\beta \left[J_0\left(\beta \frac{r}{a}\right) Y_0(\beta) - Y_0\left(\beta \frac{r}{a}\right) J_0(\beta) \right]}{J_0^2(\beta) + Y_0^2(\beta)} \, d\beta \, d\tau \qquad (7.94)$$

The time integral may be evaluated with to give

$$T(r, t) = T_0 \frac{2}{\pi} \int_{\beta=0}^{\infty} [1 - e^{-\beta^2 \alpha t/a^2}]$$

$$\times \frac{\left[J_0\left(\beta \frac{r}{a}\right) Y_0(\beta) - Y_0\left(\beta \frac{r}{a}\right) J_0(\beta) \right]}{\beta[J_0^2(\beta) + Y_0^2(\beta)]} \, d\beta \qquad (7.95)$$

The integral on β must be evaluated numerically, but the temperature is bounded by T_0 at $r = a$ and the temperature decays to zero as $r \to \infty$ (as $\alpha t/r^2 \to 0$). At steady state defined by $\alpha t/r^2 \to \infty$, the temperature approaches T_0 everywhere. A plot of the temperature given by Eq. (7.95) is given by Carslaw and Jaeger (1959, p. 337). An approximate small-time form of this solution is also listed by Carslaw and Jaeger (1959) on p. 336; another approximate small-time solution is $G_{R10} \approx 1/(2\pi a) \, G_{X10}$ (refer to Example 7.3).

Example 7.10: Infinite body with a circular hole and specified surface heat flux—R20B-T0 case. An infinite body bounded internally by the circular hole $r = a$ has a zero initial temperature. At $t \geq 0$ the surface $r = a$ sees an instantaneous pulse of heat given by $q'\delta(t)$, where $\delta(t)$ is the Dirac delta function and q' has units of J/m². Find the surface temperature $T(a, t)$ due to this heat pulse.

SOLUTION The temperature is given by the GF equation (7.2), for a boundary condition of the second kind:

$$T(r, t) = \alpha \int_{\tau=0}^{t} \frac{q'\delta(\tau)}{k} G_{R20}(r, t|a, \tau) \, 2\pi a \, d\tau \qquad (7.96)$$

To evaluate the temperature at $r = a$ apply the sifting property of the Dirac delta function to the time integral to give

$$T(a, t) = 2\pi\alpha \frac{q'a}{k} G_{R20}(a, t|a, 0) = 2\pi \frac{q'a}{\rho c} G_{R20}(a, t|a, 0) \qquad (7.97)$$

Note that $q'a/(\rho c)$ has units of Km² and that G_{R20} has units of m⁻², as expected. The GF G_{R20} is listed in Appendix R as

$$G_{R20}(a, t|a, 0) = \frac{2}{\pi^2 a^2} \int_{\beta=0}^{\infty} \frac{e^{-\beta \alpha t/a^2}}{\beta[J_1^2(\beta) - Y_1^2(\beta)]} \, d\beta \qquad (7.98)$$

In general, this integral must be evaluated numerically. However, several approximate expressions for $G_{R20}(a, t|a, \tau)$ are listed in Appendix R. For small values of time $\alpha t/a^2$, the surface temperature is approximately

$$T(a, t) \approx \frac{q'}{a\rho c} [(\pi t^+)^{-1/2} - 0.5 + 0.413434 \, (t^+)^{1/2} - 0.299877 \, t^+$$

$$+ \, 0.154483 \, (t^+)^{3/2} - 0.045263 \, (t^+)^2 + 0.005484 \, (t^+)^{5/2}] \quad (7.99)$$

where $t^+ = \alpha t/a^2 < 6$. The first term inside the square brackets in Eq. (7.99) is the same as the temperature in a plane wall, and the second term is the first correction for the curvature of the cylindrical hole. For large values of time $\alpha t/a^2 > 6$, the surface temperature is approximately

$$T(a, t) \approx \frac{q'}{a\rho c} \frac{1}{t^+} \left\{ 1 - \frac{1}{2t^+} L \left[1 + \frac{3}{4t^+} (1 - L) \right] \right.$$

$$\left. - (\pi^2 + 4) \frac{C}{(4t^+)^2} \right\} \quad (7.100)$$

where $L = \ln (4t^+) - \gamma$
$\gamma = 0.57722$ (Euler's constant)
$C = 0.5$

Finally, some numerical values of $G_{R20}(a, t|a, \tau)$ are listed in Table GR20 in Appendix R.

REFERENCES

Beck, J. V., Keltner, N. R., and Schisler, I. P., 1985, Influence Functions for the Unsteady Surface Element Method, *AIAA J.*, vol. 23, pp. 1978–1982.

Carslaw, H. S., and Jaeger, J. C., 1959, *Conduction of Heat in Solids*, 2d ed., Oxford University Press, New York.

Ozisik, M. N., 1980, *Heat Conduction*, Wiley, New York.

Watson, G. N., 1944, *A Treatise on the Theory of Bessel Functions*, 2d ed., Cambridge University Press.

Wylie, C. R., and Barrett, L. C., 1982, *Advanced Engineering Mathematics*, McGraw-Hill, New York.

PROBLEMS

7.1 Derive the relation of $\alpha(t - \tau)/r^2 = 0.25$ for the time of maximum $G_{R00}(r, t|0, \tau)$ for a given r (not equal to zero).

7.2 Plot $b^2 G_{R00}(r, t|0, \tau)$ versus $\alpha(t - \tau)/b^2$ for $\alpha(t - \tau)/b^2$ values from 0.1 to 2 for $r/b = 0$ and 1.

7.3 Derive the dimensionless distance for the values of the GF to drop to 1% of the $r = 0$ value for a given value of $\alpha(t - \tau)$. (Answer: $r^2/\alpha(t - \tau) = 18.42$ or $\alpha(t - \tau)/r^2 = 0.054$.)

7.4 Plot $b^2 G_{R00}(r, t|0, \tau)$ versus r/b for $\alpha t/b^2 = 0.01$, 1, and 10. (Three separate plots are to be done.)

7.5 Derive the first few terms of Eq. (R00.4).

7.6 Under what conditions does $G_{X00}(r, t|r', \tau)$ approximate $G_{R00}(r, t|r', \tau) 2\pi r'$?

7.7 Derive Eq. (R00.5).

7.8 Derive the approximate expression below for $G_{R00}(r, t|r', \tau)$ by approximating the circular source by four line sources.

$$G_{R00}(r, t|r', \tau) \approx \frac{1}{16\pi\alpha(t - \tau)} \left\{ \exp \left[-\frac{(r - r')^2}{4\alpha(t - \tau)} \right] + \exp \left[-\frac{(r - r')^2}{4\alpha(t - \tau)} \right] + 2 \exp \left[-\frac{r^2 + r'^2}{4\alpha(t - \tau)} \right] \right\}$$

Show that the ratio of this approximate expression to the exact one is

$$\frac{G_{R00,\ app}}{G_{R00,\ exact}} = \frac{\cosh\{rr'/[2\alpha(t - \tau)]\} + 1}{2I_0\{rr'/[2\alpha(t - \tau)]\}}$$

Calculate values of this ratio, showing that the errors are less than 0.5% for $\alpha(t - \tau)/rr'$ greater than 0.5.

7.9 Compare the numerical values of $G_{R00}(a, t|0, \tau)$ with the average GF over r and r' from the center to $r = a$, which is denoted $\overline{G}_{R00}(t, \tau)$, Eq. (R00.15). Plot the values from $\alpha(t - \tau)/a^2 = 0$ to 2.

7.10 A line source is frequently used to measure the thermal conductivity. It is made of a thin wire which has an electric current flowing through it. The temperature of the wire is measured and its asymptotic response is used to measure k. Derive an expression using GFs for the temperature distribution in an infinite solid with a line source. The initial temperature is zero. The source is to simulate a wire of radius a and volume energy generation of g_0 in W/m³.

7.11 Find an expression for the temperature at $r = 0$ in infinite body with the following initial temperature distribution:

$$F(r) = \begin{cases} 0 & r' < a \\ T_0 & a \le r' \le b \\ 0 & r' > b \end{cases}$$

7.12 Find an expression for the temperature everywhere in an infinite body with the following initial temperature:

$$F(r') = \begin{cases} 0 & r' < a \\ T_0 \left(\frac{r_0}{r'}\right) & a < r' < b \ (r_0 \text{ is a constant}) \\ 0 & r' > b \end{cases}$$

7.13 Using the series definition of the Bessel function $J_\nu(z)$:

$$J_\nu(z) = \left(\frac{1}{2} z\right)^\nu \sum_{k=0}^\infty (-1)^k \frac{(z/2)^{2k}}{k! \Gamma(\nu + k + 1)}$$

show that

$$\frac{d}{dz}[J_0(z)] = -J_1(z).$$

7.14 Find a small-time temperature for the $R01B0T0G1$ case (Example 7.5).

7.15 Find the small-time temperature for the case $R03B0T1$ which represents quenching of a hot cylinder in a cold fluid.

7.16 Find the small-time temperature for the case $R02B0T0Gr5$ (Example 7.6).

7.17 In hot-wire anemometry, a heated wire is cooled by a fluid flow. An important issue is the time constant of the wire, which is the time for the heated wire to come to steady state. This problem is a simple model of the time constant of the wire alone (without supports).

(a) Find the *steady* temperature in a solid cylinder that is heated uniformly, $g(x, t) = g_0$, and cooled by convection, case $R03B0G1$.

(b) Find the spatial average, time-varying temperature in the wire heated uniformly by energy generation g_0 (case $R03B0G1T0$).

(c) As an estimate of the time constant, find the time it takes for the *average* temperature in the wire to reach 90% of the steady temperature.

7.18 Consider the $R21B10T0$ case:

(a) Write down the GF solution equation for this case.

(b) Find the transient temperature in *integral form* using the GF from Appendix R. *Hint*: Use $R(r)$ $R(r')$ as a shorthand notation for the eigenfunctions, and refer to Eq. (7.87).

(c) Carry out the integral on τ to find the temperature in closed form.

7.19 A steel rod 25 mm in diameter is heated to a temperature of 1000°C, then quenched in a liquid bath. The temperature of the bath remains constant and equal to 50°C. If the heat transfer coefficient is 10,000 W/m² K, calculate the time required for the center temperature to reach 500°F. What is the surface temperature at the calculated time? The thermophysical properties are $k = 32$ W/m K, $c_p = 700$ J/kg K, and $\rho = 7800$ kg/m³.

7.20 An electrical cable has a 1-cm diameter copper wire ($k = 400$ W/m K) and 0.5-cm thick electrical insulation ($k = 5$ W/m K) carries electricity. The current is 300 A and resistance is 0.006 ohm/m. When the ambient temperature is 25°C, use a steady-state solution to calculate the surface heat transfer coefficient if the wire temperature is not to exceed 100°C. For the same heat transfer coefficient, calculate the temperature variations as a function of time at the center of wire. The line frequency is 60 cycles per second.

CHAPTER EIGHT

OTHER GEOMETRIES IN
CYLINDRICAL COORDINATES

8.1 INTRODUCTION

In the previous chapter, infinite bodies and infinite cylinders were considered with temperature variation limited to the r coordinate in space: $T = T(r, t)$. In the present chapter, bodies are considered with temperature variation in any of the spatial coordinates (r, ϕ, z). Refer to the cylindrical coordinate system shown in Fig. 8.1. The heat conduction equation in cylindrical coordinates has the form:

$$\frac{1}{r} \frac{\partial}{\partial r} \left[r \frac{\partial T}{\partial r} \right] + \frac{1}{r^2} \frac{\partial^2 T}{\partial \phi^2} + \frac{\partial^2 T}{\partial z^2} + \frac{1}{k} g(r, \phi, z, t) = \frac{1}{\alpha} \frac{\partial T}{\partial t} \tag{8.1}$$

The GF analysis in this chapter applies to infinite bodies, thin shells, cylinders, and wedges. These are bodies for which the surfaces are described by a constant value of a coordinate (orthogonal bodies). Nonorthogonal bodies of finite extent can be treated with the Galerkin-based GF method discussed in Chapters 10 and 11. Some of the actual geometries discussed in this chapter are thin shells (Section 8.3), cylinders (Section 8.4), bodies heated by a disk heat source (Section 8.5) and wedges (Section 8.6). Steady solutions are discussed in Section 8.7.

8.2 LIMITING CASES

The ability to analyze multidimensional heat transfer geometries is an important feature of the GF method. Such geometries can be so challenging that the analysis of a limiting case, a simple approximation to the geometry of interest, is often an important step in the solution. Limiting cases can improve one's insight and contribute to a better understanding of the whole problem.

One-dimensional limiting cases can be important for checking the analysis of two- or three-dimensional geometry and for checking the numerical results. Under the

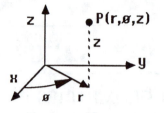

Figure 8.1 Cylindrical coordinate system.

limiting conditions, the multidimension expression for the temperature should reduce to the limiting-case expression and the multidimension computer program should give numerical values that agree with the limiting case. Numerical values for simple one-dimensional cases are sometimes tabulated in books such as this one, whereas numerical values for two-dimensional cases are rarely available. Comparison with more than one limiting case should be used whenever possible.

8.2.1 Fourier Number

All of the limiting cases discussed in this chapter depend upon a Fourier number $\alpha t/L^2$, where L is a characteristic length, t is the characteristic time, and α is the thermal diffusivity. The trick is to use the significant characteristic length. The characteristic length can depend on (1) time (early, middle, or late); (2) body shape (slab, cylinder, etc.); (3) location of the driving force for the transient heat transfer (at the surface or internally); or (4) location of the temperature of interest.

For example, in a solid cylinder heated at the boundary (case $R01B1T0$), the Fourier number is $\alpha t/b^2$, where b is cylinder radius. For sufficiently small values of this Fourier number, the temperature near $r = b$ is given approximately by the semi-infinite case $X10$. The $X10$ geometry is a limiting case for small time because the surface heating penetrates the cylinder so slightly that the curvature of the cylinder may be neglected.

For energy generation inside a body and for small dimensionless times, the characteristic length depends on the heating location. For example, in a cylinder heated by a cylindrical-surface heat source, $R01B0T0Gr7$, the significant Fourier number is $\alpha t/(b - r_0)^2$ where b is the cylinder radius, r_0 is the location of the cylindrical-surface heat source, and $r_0/b < 0.5$ (this last condition ensures that the boundary is far enough from the heat source). Then, for $\alpha t/(b - r_0)^2$ sufficiently small, the temperature is given by an infinite region ($R00$) heated by a cylindrical-surface heat source. The characteristic length is $b - r_0$, the distance from the heating location to the boundary.

8.2.2 Aspect Ratio

Limiting cases may be found by changing the aspect ratio of the body, the ratio of the width to the length of the body. For solid cylinders, the aspect ratio is b/L, where b is the cylinder radius and L is the cylinder length. In a solid cylinder there are two limiting cases based on variations of the aspect ratio. First, consider the cylinder with aspect ratio $b/L > 5$ shown in Fig. 8.2a. This cylinder is more like a flat disk, and

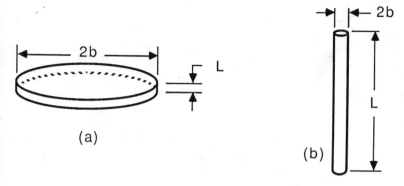

Figure 8.2 (*a*) Cylinder with aspect ratio $b/L > 5$. (*b*) Cylinder with aspect ratio $b/L < 1/10$.

depending how it is heated, the limiting case may be the one-dimensional slab of thickness L. Second, for the solid cylinder with aspect ratio $b/L < 1/10$ shown in Fig. 8.2*b*, the limiting case is the infinite cylinder of radius b for which $T = T(r, t)$.

Three-dimensional bodies may have two aspect ratios. For hollow cylinders, an additional aspect ratio is δ/b, where δ is the thickness of the cylinder wall.

8.2.3 Nonuniform Heating

When a body is heated nonuniformly over position or over time, the limiting case of uniform heating is useful for checking purposes. The uniformly heated cases are generally easier to analyze. For example, a cylinder heated over part of its surface is shown in Fig. 8.3, and it is described by the number $R01B(z5)Z11B11T0$. The limiting case of a uniformly heated cylinder, number $R01B1Z11B1T0$, is particularly easy to find by multiplying two one-dimension temperature solutions; refer to Example 8.2.

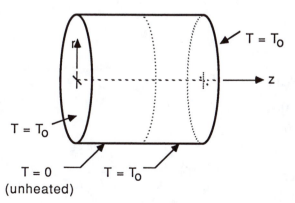

Figure 8.3 Cylinder with specified temperature over part of its surface.

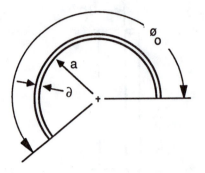

Figure 8.4 Thin shell.

8.3 THIN SHELLS, $T = T(\phi, t)$

Thin shells are bodies for which the coordinates (ϕ, t) completely describe the heat transfer. These are closely related to the one-dimensional rectangular cases, although there are some differences. A thin shell has radius a, thickness δ, and angle ϕ_0. Refer to Fig. 8.4. The shell is thin if the temperature at $r = a$ and at $r = a + \delta$ are approximately equal. If not, then the variable r must be included in the analysis, and the body must be analyzed with the variables (r, ϕ, t).

Example 8.1: Thin shell heated over half of its surface—$\Phi00T0G5$ case. Consider a thin-walled tube that is initially at temperature T_0, and is suddenly heated uniformly over half of its surface as shown in Fig. 8.5. The inside of the tube is cooled by forced convection according to $q_0 = h(T(\phi, t) - T_0)$, where h is the heat transfer coefficient (W/m^2 K), and T_0 is the local bulk temperature of the flowing fluid. Find (a) the temperature in the tube wall for large values of time, and (b) the steady-state temperature of the tube wall.

SOLUTION The energy equation for the thin-walled tube is given by

$$\frac{1}{a^2}\frac{\partial T^2}{\partial \phi^2} + \frac{1}{k}g(\phi, t) - m^2(T - T_0) = \frac{1}{\alpha}\frac{\partial T}{\partial t} \tag{8.2a}$$

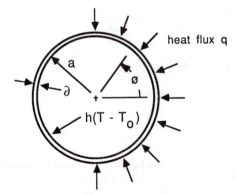

Figure 8.5 Thin-walled tube heated over half of outside surface and cooled by convection on inside surface.

The initial condition is

$$T(\phi, t = 0) = T_0 \tag{8.2b}$$

The boundary conditions take the form of periodic conditions on ϕ:

$$T(\phi, t) = T(\phi + 2n\pi, t)$$

$$\frac{\partial T}{\partial \phi}(\phi, t) = \frac{\partial T}{\partial \phi}(\phi + 2n\pi, t)$$

This is the $\Phi 00$ geometry, with volume energy generation and with a fin term. The volume energy generation term, $g(\phi, t)/k$, is used to simulate the heat flux boundary condition outside the tube. Ordinarily, a specified heat flux is introduced by $q_0 = -k\,\partial T/\partial r$ at the outer boundary of a circular tube. In the thin-wall case, the assumption is that $\partial T/\partial r \approx 0$, which means that there is no resistance to heat transfer in the r direction, and the thickness of the tube wall serves only to store heat according to its mass. This is called a lumped-capacity model of the actual tube wall. The surface heat flux q_0 may be simulated by volume energy generation through the tube wall according to

$$q_0 \times \text{Surface area} = g_0 \times \text{volume} \quad \text{or} \quad q_0 \times 2\pi a = g_0 \times 2\pi a\delta$$

The energy generation term is

$$g(\phi, t) = \begin{cases} \dfrac{q_0}{\delta} & \text{for } \dfrac{-\pi}{2} < \phi < \dfrac{\pi}{2} \quad \text{and for } t > 0 \\ 0 & \text{otherwise} \end{cases} \tag{8.3}$$

The fin term, $m^2(T - T_0)$, describes the cooling effect of the fluid flow inside the tube. The constant m^2 has units of meters^{-2} and it is given by $m^2 = h/(k\delta)$ where h is the heat transfer coefficient due to fluid flow inside the tube, k is the thermal conductivity of the tube wall, and δ is the thickness of the tube wall.

The GF solution involves a variable transformation to eliminate the fin term. Let a new temperature variable be given by

$$W(\phi, t) = [T(\phi, t) - T_0]\, e^{m^2\alpha t} \tag{8.4}$$

Refer to Section 3.5 for a complete discussion of this procedure. Then, Equation (8.2a) can be written with the new temperature variable as

$$\frac{1}{a^2}\frac{\partial^2 W}{\partial \phi^2} + \frac{1}{k} g(\phi, t)\, e^{m^2\alpha t} = \frac{1}{\alpha}\frac{\partial W}{\partial t} \tag{8.5a}$$

Note that the fin term is gone but the energy generation term is more complicated. The initial condition is homogeneous,

$$W(\phi, t = 0) = 0 \tag{8.5b}$$

and the boundary conditions on ϕ are unchanged.

The GF solution to this transformed equation is given by a single integral for the volume energy generation:

$$W(\phi, t) = \frac{\alpha}{k} \int_{\tau=0}^{t} \int_{\phi'=-\pi/2}^{\phi'=\pi/2} \frac{q_0}{\delta} e^{m^2\alpha\tau} G_{\Phi00}(\phi, t|\phi', \tau) \, \delta a d\phi' \, d\tau \quad (8.6)$$

The boundary conditions do not appear in the GFSE because there is no physical boundary; the periodic conditions are satisfied by the $\Phi00$ GF. The integral on ϕ' has limits of $-\pi/2 < \phi' < \pi/2$ because q_0, the external heat flux, is nonzero only over this interval.

The $G_{\Phi00}$ GF is listed in Appendix Φ, and it involves an infinite sum of cosine terms:

$$G_{\Phi00}(\phi, t|\phi', \tau) = \frac{1}{2\pi\delta a} + \frac{1}{\pi\delta a} \sum_{n=1}^{\infty} e^{-n^2\alpha(t-\tau)/a^2} \cos[n(\phi - \phi')] \quad (8.7)$$

There is no "small time" form of $G_{\Phi00}$, where "small" is defined by dimensionless time of order one, $\alpha t/\delta^2 = O(1)$, because the thin-wall approximation is not consistent with transient diffusion over distances on the order of δ. An examination of the temperature in the tube at small time would require a two-dimensional analysis using the $G_{R32\Phi00}$ GF (thick-walled tube).

The integrals in Eq. (8.7) may be evaluated in any order, but it is generally better to carry out the spatial integrals first. The integral on ϕ' operates only on the cosine term of $G_{\phi00}$ to give

$$W(\phi, t) = \frac{\alpha q_0}{k\delta} \left[\int_{\tau=0}^{t} \frac{1}{2} e^{m^2\alpha\tau} \, d\tau \right.$$

$$\left. + \sum_{m=1}^{\infty} \frac{2}{n\pi} \cos(n\phi) \, e^{-n^2\alpha t/a^2} \int_{\tau=0}^{t} e^{(m^2\alpha + n^2\alpha/a^2)\tau} \, d\tau \right] \quad (8.8)$$

In evaluating the integral on τ, do not confuse the term $e^{m^2\alpha\tau}$, which contains the constant fin parameter m^2, with the term $e^{-n^2\alpha(t-\tau)/a^2}$, which comes from the infinite series with index n.

The integral on τ gives, after some rearrangement,

$$W(\phi, t) = \frac{q_0 a}{k} \frac{a}{\delta} \left[\frac{1}{2m^2a^2} (e^{m^2\alpha t} - 1) \right.$$

$$\left. + \frac{2}{\pi} \sum_{m=1}^{\infty} \frac{\cos(n\phi)}{n(m^2a^2 + n^2)} (e^{m^2\alpha t} - e^{-n^2\alpha t/a^2}) \right] \quad (8.9)$$

Finally, to find the temperature in the original problem, convert back according to the transformation $T - T_0 = W(\phi, t) \, e^{-m^2\alpha t}$:

$$T(\phi, t) - T_0 = \frac{q_0 a}{k} \frac{a}{\delta} \left[\frac{1}{2m^2a^2} (1 - e^{-m^2\alpha t}) \right.$$

$$\left. + \frac{2}{\pi} \sum_{n=1}^{\infty} \frac{\cos(n\phi)}{n(m^2a^2 + n^2)} (1 - e^{-(m^2a^2 + n^2)\alpha t/a^2}) \right] \quad (8.10)$$

The steady-state temperature is given by the limit as $t \to \infty$, or,

$$T(\phi, t) - T_0 = \frac{q_0 a}{k} \frac{a}{\delta} \left[\frac{1}{2m^2 a^2} + \frac{2}{\pi} \sum_{n=1}^{\infty} \frac{\cos(n\phi)}{n(m^2 a^2 + n^2)} \right] \qquad (8.11)$$

8.4 CYLINDERS FOR WHICH $T = T(r, z, t)$

In this section cylinders are discussed for which the temperature depends on coordinates r and z. The GFs for these cases can be constructed by multiplying two one-dimensional GFs. That is,

$$G_{RZ} = (G_R)(G_Z) \qquad (8.12)$$

The boundary conditions of types 0, 1, 2, and 3, may be treated.

Two examples are given to illustrate the method. All of the analytical techniques previously introduced for one-dimension cases apply to these two-dimension cases.

Example 8.2: Finite cylinder with specified surface temperature—R01Z11 geometry. A finite cylinder of length L and radius b has a uniform initial temperature T_0. For $t > 0$, the entire surface of the cylinder is suddenly set to temperature T_1. Find the temperature in the cylinder for large times.

SOLUTION The cylinder is shown in Fig. 8.6. A detailed statement of the boundary and initial conditions of this example are

$$T(r = b, z, t) = T_1 \qquad (8.13a)$$

$$T(r, z = 0, t) = T_1 \qquad (8.13b)$$

$$T(r, z = L, t) = T_1 \qquad (8.13c)$$

$$T(r, z, t = 0) = T_0 \qquad (8.13d)$$

The heat conduction numbering system for this case is $R01B1T0Z11B11$.

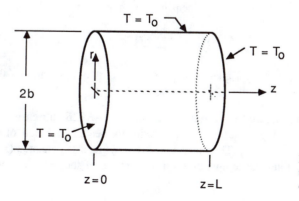

T = T₀

T = T₀

2b

T = T₀

z = 0

z = L

Figure 8.6 Solid cylinder with temperature boundary conditions, case $R01B1T1Z11B11$.

The best solution for this problem using GFs is to cast the problem as an initial-temperature case, by defining a new temperature $T = T - T_1$. Then, the boundary temperature is zero, and the initial temperature is $T_0 - T_1$. This is equivalent to the alternative GF solution method with $T^* = T_1$. The solution is given by

$$T(r, z, t) = \int_0^b 2\pi r' dr' \int_0^L dz' (T_0 - T_1) G_{R01Z11}(r, z, t|r', z', 0) \quad (8.14)$$

The GF G_{R01Z11} is found from a product solution as shown in Eq. (8.12), and substituted into the spatial integral in Eq. (8.14) to give

$$T(r, z, t) = (T_0 - T_1) \left[\int_0^b G_{R01}(r, t|r', 0) 2\pi r' dr' \right]$$

$$\times \left[\int_0^L G_{Z11}(z, t|z', 0) dz' \right] \quad (8.15)$$

The product of the two integrals in Eq. (8.15) can be interpreted as the product of two dimensionless temperatures, one for an infinite cylinder $R01B0T1$, and one for an infinite slab $X11B0T1$. A product solution for temperature is possible only for certain initial conditions, but a product solution for the GF is always possible for coordinates r and z. Refer to Section 4.5 for a discussion of this point.

Function G_{R01} comes from Appendix R and function G_{Z11} comes from Appendix X (with x and x' replaced by z and z'). The result for the temperature is

$$T(r, z, t) = (T_0 - T_1) \left[2 \sum_{m=1}^{\infty} e^{-\beta_m^2 \alpha t/b^2} \frac{J_0(\beta_m r/b)}{\beta_m J_1^2(\beta_m)} \right]$$

$$\times \left\{ 2 \sum_{n=1}^{\infty} e^{-n^2\pi^2\alpha t/L^2} \sin\left(n\pi \frac{z}{L}\right) \frac{[1 - (-1)^n]}{n\pi} \right\} \quad (8.16)$$

where the eigenvalues β_m are the roots of $J_0(\beta_m) = 0$; some values for β_m are given in Appendix B. The expression $[1 - (-1)^n]$ comes from evaluating the integral of the sine to give $\cos(0) - \cos(n\pi)$. This expression gives zero for all the odd terms in the second summation ($n = 1, 3, 5, \ldots$), and the summation on n could be rewritten in a form to represent just the nonzero terms:

$$\sum_{n=1}^{\infty} f(n) [1 - (-1)^n] = 2 \sum_{k=0}^{\infty} f(2k + 1) \quad (8.17)$$

Although the solution given by Equation (8.16) contains two summations, the summations converge very rapidly for large times $\alpha t/b^2 \gg 1$.

Example 8.3: Cylinder heated over half of its surface—R02BZ00 case. An infinite cylinder initially at zero temperature is suddenly heated over half of its surface ($z < 0$) with heat flux q_0. The other half of the cylinder is insulated ($z > 0$). Refer to Fig. 8.7. Find the temperature on the surface of the cylinder soon after the heating begins.

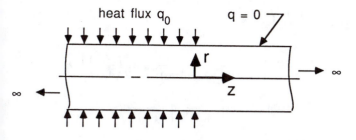

Figure 8.7 Infinite cylinder heated over half of the surface and insulated elsewhere, case R02B5T0Z00.

SOLUTION This is the R02Z00 geometry with heating at the surface and with no initial temperature or internal heat generation. The maximum temperature occurs on the surface of the cylinder and the surface temperature is given by the surface heating term of the GFSE:

$$T(b, z, t \mid b, z', \tau)$$

$$= \frac{q_0 \alpha}{k} \int_{\tau=0}^{t} \int_{z'=-\infty}^{0} G_{R20Z00}(b, z, t \mid b, z', \tau) \, 2\pi b \, dz' \, d\tau \quad (8.18)$$

The GF is evaluated at $r' = b$ where the heating occurs, and the temperature is evaluated at $r = b$. Note that the spatial integral over the surface involves differential area $ds = 2\pi b \, dz'$ and the integral limits are $-\infty < z' < 0$. The GF is given by the product $(G_{R02})(G_{Z00})$. To find the temperature soon after heating begins only the small-time forms of the GFs are needed. An approximate form of G_{R02} for small times evaluated at the surface $r = b$ is given in Appendix R:

$$G_{R02}(b, t \mid b', \tau) \approx \frac{1}{2\pi b^2} \left\{ \frac{b}{[\pi \alpha(t - \tau)]^{1/2}} \right.$$

$$\left. + \frac{1}{2} + \frac{3}{4\sqrt{\pi}} \left[\frac{\alpha(t - \tau)}{b^2} \right]^{1/2} \right\} \quad (8.19)$$

The GF G_{Z00} is given in Appendix X:

$$G_{Z00}(z, t \mid z', \tau) = \frac{1}{[4\pi\alpha(t - \tau)]^{1/2}} \exp \left[\frac{-(z - z')^2}{4\alpha(t - \tau)} \right] \quad (8.20)$$

It is generally better to evaluate spatial integrals first, and the integral on z' in Eq. (8.18) may be written

$$T(b, z, t) = \frac{q_0 \alpha}{k} \frac{1}{4\pi b} \int_{\tau=0}^{t} d\tau \left\{ \frac{b}{[\pi\alpha(t - \tau)]^{1/2}} + \frac{1}{2} \right.$$

$$\left. + \frac{3}{4\sqrt{\pi}} \left[\frac{\alpha(t - \tau)}{b^2} \right]^{1/2} \right\} \text{erfc} \left\{ \frac{z}{[4\alpha(t - \tau)]^{1/2}} \right\} \quad (8.21)$$

The integral on τ may be evaluated in three terms and the final result for the surface temperature may be written (see integral Table 5.5)

$$T(b, z, t) = \frac{q_0 b}{k} \frac{1}{4\pi} \left(2 (t^+)^{1/2} \right.$$

$$\times \left\{ \text{erfc} \left[\frac{z}{(4\alpha t)^{1/2}} \right] - \frac{z}{(4\pi\alpha t)^{1/2}} E_1 \left(\frac{z^2}{4\alpha t} \right) \right\}$$

$$+ 2 (t^+) \, i^2 \, \text{erfc} \left[\frac{z}{(4\alpha t)^{1/2}} \right] + \frac{12}{\sqrt{\pi}} (t^+)^2$$

$$\left. \times \left\{ i^2 \, \text{erfc} \left[\frac{z}{(4\alpha t)^{1/2}} \right] + i^4 \, \text{erfc} \left[\frac{z}{(4\alpha t)^{1/2}} \right] \right\} \right) \qquad (8.22)$$

where $t^+ = \alpha t / b^2$.

Discussion. One part of the given temperature expression is multiplied by $(t^+)^{1/2}$. This part is identical to the temperature in a semi-infinite plane body heated over half of its surface which was studied in Section 6.8. For early times, the surface of the cylinder displays behavior similar to a plane body. From this perspective the other terms in the temperature expression Eq. (8.22) are corrections to account for the curvature in the surface of the cylinder.

The temperature expression contains factors like $\sqrt{t^+}$, t^+, and $(t^+)^2$, which indicate that the temperature increases over time without limit; there is no steady-state solution since all the heat that enters the cylinder remains in the cylinder. The surface temperature is the largest temperature on the cylinder at any given time.

On the heated region of the cylinder ($z < 0$) and far away from the point $z = 0$, the temperature is described by one-dimensional radial heat conduction, $T = T(r, t)$. Here "far" is determined by $z^2/\alpha t \gg 1$, because the correct Fourier number along the z axis is $\alpha t/z^2$. On the nonheated end of the cylinder and for $z^2/\alpha t \gg 1$, the temperature is identically zero.

8.5 DISK HEAT SOURCE ON A SEMI-INFINITE BODY

In this section, the cylindrical GFs are applied to a semi-infinite body heated at the surface by a disk heat source. Over the disk heat source, the heat flux is constant with position and with time, while outside the disk, the surface is insulated. This case is a basic building block in transient heat conduction and in the surface element method discussed in Chapter 12. Applications of the disk heat source solution include constriction resistance, the intrinsic thermocouple, and laser heating of a flat surface.

The GF solution yields an exact solution in the form of an integral with limits of zero and infinity, and the integrand involves error functions and Bessel functions. The integral is difficult to evaluate numerically because the domain is infinite and because of the sinusoidal behavior of the Bessel functions. Though this integral represents a solution valid for any position (r, z), accurate numerical values are difficult to obtain directly except along the centerline ($r = 0$).

The purpose of this section is to present the exact solution with the GF method, to present closed-form expressions for some special cases, and to present series expressions for the surface temperature that are accurate and easy to evaluate numerically. Expressions for interior temperatures ($z > 0$) are given by Beck (1980, 1981).

8.5.1 Integral Expression for the Temperature

The geometry for the disk heat source problem is shown in Fig. 8.8. The surface of the semi-infinite body is insulated except for the disk $0 < r < a$. The initial temperature is zero. A mathematical statement of the energy equation and boundary conditions is given by

$$\frac{1}{r}\frac{\partial}{\partial r}\left(r\frac{\partial T}{\partial r}\right) + \frac{\partial^2 T}{\partial z^2} = \frac{1}{\alpha}\frac{\partial T}{\partial t} \tag{8.23}$$

$$-k\frac{\partial T(r, 0\ t)}{\partial z} = \begin{cases} q_0 & \text{for } 0 < r < a \\ 0 & \text{for } r > a \end{cases} \tag{8.24a}$$

$$T(r, z, t) \to 0 \quad \text{for } r \to \infty \quad \text{and } z \to \infty \tag{8.24b}$$

$$T(r, z, 0) = 0 \tag{8.24c}$$

The GFSE is

$$T(r, z, t) = \frac{\alpha}{k}\int_{\tau=0}^{t}\int_{r'=0}^{a} q_0\, G_{R00Z20}(r, z, t|r', 0, \tau)\, 2\pi r'\, dr'd\tau \tag{8.25}$$

The integral over the surface involves the area element $dA = 2\pi r'\, dr'$, and the GF is evaluated at the surface $z' = 0$.

The GF is given by the multiplication of two one-dimension functions, $G_{R00Z20} = (G_{R00})(G_{Z20})$, where

$$G_{Z20}(z, t|z' = 0, \tau) = \frac{2}{[4\pi\alpha(t - \tau)]^{1/2}}\exp\left[\frac{-z^2}{4\alpha(t - \tau)}\right] \tag{8.26}$$

from Appendix X, and where

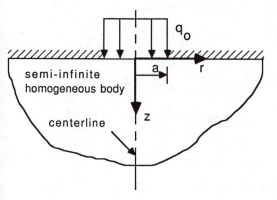

Figure 8.8 Semi-infinite body heated over a disk-shaped region centered at $r = 0$ and $z = 0$ and insulated elsewhere at $z = 0$.

$$G_{R00}(r, t|r', \tau) = \frac{1}{2\pi a^2} \int_{\beta=0}^{\infty} e^{-\beta^2 \alpha(t-\tau)/a^2} \beta J_0\left(\frac{\beta r}{a}\right) J_0\left(\frac{\beta r'}{a}\right) d\beta \qquad (8.27)$$

from Appendix R. Note that β is a dimensionless eigenvalue, and that G_{R00} has units of meters^{-2}. This form of G_{R00} is difficult to evaluate for reasons discussed above; but, because the integrand depends on r' and τ in *separate terms*, the GF equation (8.25) can be integrated separately over r' and τ leaving only the integral on β. The other form of G_{R00} listed in Appendix R contains the term $I_0[(rr')/(2\alpha(t - \tau)]$, and the integrals on r' and τ cannot be evaluated separately.

The GF G_{R00} and G_{Z20} can now be substituted into the expression for the temperature, and the integrals on r' and τ can be evaluated. The integral on r' acts only on the term $r' J_0(\beta r'/a)$, so the integral is given by (Appendix B)

$$\int_{r'=0}^{a} J_0\left[\frac{\beta r'}{a}\right] 2\pi r' \, dr' = \frac{2\pi a^2}{\beta} J_1(\beta) \qquad (8.28)$$

Combine the integral on r' with the temperature expression, Eq. (8.25), to get

$$T(r, z, t) = \frac{\alpha}{k} \int_{\tau=0}^{t} q_0 \frac{1}{[\pi\alpha(t - \tau)]^{1/2}} \exp\left[\frac{-z^2}{4\alpha(t - \tau)}\right]$$
$$\times \int_{\beta=0}^{\infty} e^{-\beta^2\alpha(t-\tau)/a^2} J_0\left(\frac{\beta r}{a}\right) J_1(\beta) \, d\beta \, d\tau \qquad (8.29)$$

The integral on τ can now be identified as (Appendix F, Table F.6),

$$\int_{\tau=0}^{t} \frac{1}{[\alpha(t - \tau)]^{1/2}} \exp\left[-\frac{\beta^2\alpha(t - \tau)}{a^2} - \frac{z^2}{4\alpha(t - \tau)}\right] d\tau$$
$$= \frac{2\pi^{1/2}}{2\beta\alpha}\left(e^{-\beta z/a}\left\{1 + \text{erf}\left[\frac{\beta(\alpha t)^{1/2}}{a} - \frac{z}{2(\alpha t)^{1/2}}\right]\right\}\right.$$
$$\left. - e^{\beta z/a} \text{erfc}\left[\frac{\beta(\alpha t)^{1/2}}{a} + \frac{z}{2(\alpha t)^{1/2}}\right]\right) \qquad (8.30)$$

Then, Eq. (8.29) can be written as

$$T(r, z, t) = \frac{1}{2}\frac{q_0 a}{k} \int_{\beta=0}^{\infty} J_0\left(\frac{\beta r}{a}\right) J_1(\beta) \left(e^{-\beta z/a}\left\{1 + \text{erf}\left[\frac{\beta(\alpha t)^{1/2}}{a}\right.\right.\right.$$
$$\left.\left.\left. - \frac{z}{2(\alpha t)^{1/2}}\right]\right\} - e^{\beta z/a} \text{erfc}\left[\frac{\beta(\alpha t)^{1/2}}{a} + \frac{z}{2(\alpha t)^{1/2}}\right]\right) \frac{d\beta}{\beta} \qquad (8.31)$$

This is the exact solution in integral form, valid for all values $r > 0$, $z > 0$, $t > 0$ (Carslaw and Jaeger, 1959).

At the surface where $z = 0$ the exact solution reduces to

$$\frac{T(r, 0, t)}{q_0 a/k} = \int_{\beta=0}^{\infty} \text{erf}\left[\frac{\beta(\alpha t)^{1/2}}{a}\right] J_0\left(\frac{\beta r}{a}\right) J_1(\beta) \frac{d\beta}{\beta} \qquad (8.32)$$

This expression still contains the difficult integral on β.

8.5.2 Closed-Form Expressions for the Temperature

In general, the infinite integral on β in Eq. (8.31) cannot be evaluated in closed form. In restricted cases, however, convenient temperature expressions may be found.

Steady-state temperature. Thomas (1957) derived an exact steady solution for the surface temperature ($z = 0$) in terms of known functions. The steady surface temperature for $0 < r < a$ is given by

$$\frac{T(r, 0, \infty)}{q_0 a/k} = \frac{2}{\pi} E\left(\frac{r}{a}\right) \tag{8.33}$$

and, for $r > a$,

$$\frac{T(r, 0, \infty)}{q_0 a/k} = \frac{2r}{\pi}\left[E\left(\frac{a}{r}\right) - (1 - r^{-2}) K\left(\frac{a}{r}\right)\right] \tag{8.34a}$$

The functions $K(\cdot)$ and $E(\cdot)$ are the complete elliptic integrals of the first and second kinds;

$$K(\epsilon) = \int_0^{\pi/2} (1 - \epsilon^2 \sin^2 \theta)^{-1/2} d\theta \tag{8.34b}$$

$$E(\epsilon) = \int_0^{\pi/2} (1 - \epsilon^2 \sin^2 \theta)^{1/2} d\theta \tag{8.34c}$$

These functions are tabulated in Abramowitz and Stegun (1964) and are available in computer libraries.

At $r \approx 1$, $[T(r, 0, \infty) - T_0]/(q_0 a/k) = 2/\pi$. For large values of r/a, Eq. (8.34a) can be approximated by

$$\frac{T(r, 0, \infty)}{q_0 a/k} = \frac{1}{2r}\left[1 + \frac{1}{2(2r/a)^2} + \frac{1}{2^2(2r/a)^4} + \cdots\right] \tag{8.35}$$

The leading term of Eq. (8.35) is proportional to $1/(r)$, which is the same as a steady point heat source on the surface.

Centerline temperature. At the centerline of the body at $r = 0$, the exact solution is given by (Carslaw and Jaeger, 1959)

$$\frac{T(0, z, t)}{q_0 a/k} = 2\frac{\alpha t}{a^2}\left\{\text{ierfc}\left[\frac{z}{2(\alpha t)^{1/2}}\right] - \text{ierfc}\left[\frac{(z^2 + a^2)^{1/2}}{2(\alpha t)^{1/2}}\right]\right\} \tag{8.36}$$

where ierfc ($\cdot$) is the integral of the complementary error function (see Appendix E).

The centerline temperature in Eq. (8.36) can be derived using Eq. (8.25) and using the following form of the R00 GF listed in Appendix R:

$$G_{R00}(r, t|r', \tau) = [4\pi\alpha(t - \tau)]^{-1} \exp\left[\frac{-(r^2 + r'^2)}{4\alpha(t - \tau)}\right] I_0\left[\frac{rr'}{2\alpha(t - \tau)}\right]$$

At $r = 0$, the modified Bessel function drops out ($I_0(0) = 1$), and the integrals on r' and τ in Eq. (8.25) produce function ierfc($\cdot$).

Surface temperature far from the disk source. The surface temperature far from the disk heat source behaves as if the heat is introduced by a point source and the temperature is given approximately by

$$\frac{T(r, 0, t)}{q_0 a/k} = \frac{1}{2(r/a)} \operatorname{erfc}\left[\frac{r}{2(\alpha t)^{1/2}}\right] \tag{8.37}$$

In the limit as $t \to \infty$, the steady-state surface temperature goes like $(1/r)$. At steady-state, Eq. (8.37) gives for $r/a = 8$ the value of 0.0625 while the exact value is 0.062623 which is 0.2% higher. For larger r/a, the error in using Eq. (8.37) is less, but the percent error for a given r/a tends to become larger as $\alpha t/a^2$ is reduced.

8.5.3 Series Expression for the Surface Temperature at Large Times

By using the relation erfc $(\cdot) = 1 - \operatorname{erf}(\cdot)$, Eq. (8.32) for the surface temperature is given by

$$\frac{T(r, 0, t)}{q_0 a/k} = \int_{\beta=0}^{\infty} J_0\left(\frac{\beta r}{a}\right) J_1(\beta) \frac{d\beta}{\beta}$$

$$- \int_{\beta=0}^{\infty} \operatorname{erfc}\left[\frac{\beta(\alpha t)^{1/2}}{a}\right] J_0\left(\frac{\beta r}{a}\right) J_1(\beta) \frac{d\beta}{\beta} \tag{8.38}$$

Notice that the first integral is a steady-state term and the second integral goes to zero as $t \to \infty$. Hence, the first integral is equal to the steady-state temperature given either by Eq. (8.33) or (8.34) depending on the range of r.

Consider now the second integral in Eq. (8.38). Using the dimensionless variables $r^+ = r/a$ and $t^+ = \alpha t/a^2$, an exact series expression for this integral is given by (Beck, 1981),

$$I_2 = -\frac{1}{2\sqrt{\pi} \, t^+} \sum_{k=1}^{\infty} \frac{(-1)^k}{C_{k-1} (t^+)^{k-1}} \sum_{j=1}^{k} \frac{k - j + 1}{k} U_{kj}^2 \tag{8.39}$$

where

$$C_k = 4^k (2k + 1)[(k + 1)!] \tag{8.40}$$

$$U_{k1} = 1 \tag{8.41a}$$

$$U_{kj} = U_{k,j-1} \frac{(k - j + 2)r^+}{(j - 1)} \qquad k = 1, 2, \ldots \qquad j = 2, 3, \ldots, k \tag{8.41b}$$

where Eq. (8.41b) is a recursion relation.

In summary for $0 < r^+ < 1$, a series expression for T at $z = 0$ is

$$\frac{T(r^+, 0, t)}{q_0 a/k} = \frac{2}{\pi} E(r^+) - I_2(r^+, t^+) \tag{8.42a}$$

where the fundamental dependence of I_2 is noted. For $r^+ > 1$, the temperature is given by

$$\frac{T(r^+, 0, t)}{q_0 a/k} = \frac{2r^+}{\pi} \left[E\left(\frac{1}{r^+}\right) - [1 - (r^+)^{-2}] K\left(\frac{1}{r^+}\right) \right] - I_2(r^+, t^+) \quad (8.42b)$$

The function $I_2(r^+, t^+)$ is calculated using Eqs. (8.39)–(8.41). These exact expressions are very efficient for 'large' times because the infinite summation in I_2 can be approximated with just a few terms.

In order to display clearly the nature of the summation in I_2, several terms are now given.

$$T(r^+, 0, t^+) = T(r^+, 0, \infty) - \frac{1}{2\sqrt{\pi t^+}} \left\{ 1 - \frac{1 + 2(r^+)^2}{24t^+} + \frac{1}{480(t^+)^2} \right.$$

$$\times [1 + 6(r^+)^2 + 3(r^+)^4] - \frac{1}{10752(t^+)^3} [1 + 12(r^+)^2$$

$$\left. + 18(r^+)^4 + 4(r^+)^6] + \cdots \right\} \quad (8.43)$$

Note that the denominators 24, 480, etc., are the C_k values given by Eq. (8.40). The number of terms required in the series for I_2 increases quite rapidly as the dimensionless times become small. Fortunately, for a large range of t^+, the required number of terms is quite modest, that is, less than 7 for $r^+ = 0$ and for $t^+ > 1$ to obtain eight-significant-figure accuracy. Also the number of additional terms required to go from three to eight significant figures is not large. The series solution, however, is not appropriate for very small dimensionless times. The limiting appropriate dimensionless times are about $t^+ = 0.01$, 0.05, and 0.1 for $r^+ = 0$, 1 and 2, respectively. For $r^+ \geq 1$, a convenient limiting time expression is

$$\frac{t^+}{(r^+)^2} \geq 0.05 \quad (8.44)$$

Temperatures for $r^+ = 0$, 0.25, 0.5, 0.75, 0.9, and 1.0 are plotted in Fig. 8.9. For the small dimensionless time values at $r^+ = 0$, temperatures were calculated utilizing Eq. (8.36). The $r^+ = 1$ curve for small dimensionless time was found using

$$T(1, 0, t^+) \approx \left(\frac{t^+}{\pi}\right)^{1/2} - \frac{t^+}{2\pi} \left[1 + \frac{t^+}{8} + \frac{9(t^+)^2}{96} \right] \quad (8.45)$$

where $t^+ = \alpha t/a^2$. This expression is accurate to five significant figures for $t^+ < 0.1$. For very small t^+ values (about 10^{-4}) the T given by Eq. (8.45) is one-half the center value given by Eq. (8.36).

8.5.4 Average Temperature

The temperature averaged over position is of interest for determining the contact conductance and for other purposes. For the average temperature between $r^+ = 0$

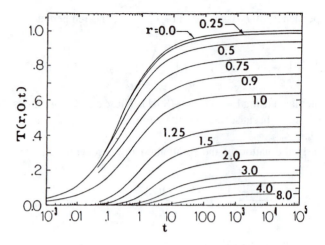

Figure 8.9 Local temperature versus time at $z = 0$ on semi-infinite body.

and $r^+ = c$ (where c is an arbitrary dimensionless radial location), one can multiply $T(r, 0, t)$ by $2\pi r dr$, integrate from $r/a = 0$ to c, and divide by πc^2. The result is

$$\overline{T}(c, 0, t) = \overline{T}(c, 0, \infty) - \overline{I}_2(c, t^+) \tag{8.46}$$

where $\overline{I}_2(c, t^+)$ is exactly the same expression as given by Eq. (8.39) except the inner summation has kj in the denominator instead of simply k and in Eq. (8.41b), r^+ is replaced by c. The term $\overline{T}(c, 0, \infty)$ in Eq. (8.46) for $0 < c \leq 1$ is given by

$$\overline{T}(c, 0, \infty) = \frac{4}{3\pi c^2} [(1 + c^2)E(c) - (1 - c^2)K(c)] \tag{8.47a}$$

and for $1 \leq c \leq \infty$

$$\overline{T}(c, 0, \infty) = \frac{4}{3\pi c^2} [(1 + c^2)E(c^{-1}) - (1 - c^2)K(c^{-1})] \tag{8.47b}$$

where $E(\cdot)$ and $K(\cdot)$ are elliptic integrals defined in Eq. (8.34). At $c = 0$, $\overline{T}(0, 0, \infty) = 1$, and at $c = 1$, $\overline{T}(1, 0, \infty) = 8/3\pi$. For large c values, Eq. (8.47b) can be approximated by

$$\overline{T}(c, 0, \infty) \approx \frac{1}{c} \left(1 - \frac{1}{8c^2} - \frac{1}{64c^2} \right) \tag{8.47c}$$

An expanded form of Eq. (8.46) for a few terms is

$$\overline{T}(c, 0, t) = \overline{T}(c, 0, \infty) - \frac{1}{2\sqrt{\pi t^+}}$$

$$\times \left[1 - \frac{1 + c^2}{24t^+} + \frac{1}{480(t^+)^2} (1 + 3c^2 + c^4) \right.$$

$$\left. - \frac{1}{10752(t^+)^3} (1 + 6c^2 + 6c^4 + 6c^6) + \cdots \right] \tag{8.48}$$

For small t, the average temperature from $r^+ = 0$ to 1 can be approximated by (Beck, 1980)

$$\overline{T}(c, 0, t) \approx 2\left(\frac{t^+}{\pi}\right)^{1/2} - \frac{t^+}{\pi}\left[2 - \frac{t^+}{4} - \frac{(t^+/4)^2}{4} - \frac{15(t^+/4)^3}{4}\right] \quad (8.49)$$

which is accurate to five significant digits for $0 < t^+ < 0.1$.

The average temperatures are plotted in Fig. 8.10 (Beck, 1981). The curve of $\overline{T}$ for small t^+ and for $r^+ > 1$ shown in Fig. 8.10 can be obtained by using

$$\overline{T}(c, 0, t) \approx \frac{2}{c^2}\left(\frac{t^+}{\pi}\right)^{1/2} \quad (8.50)$$

This expression becomes more accurate as $t \to 0$ and as c becomes larger. For $c = 1.5$ and $t^+ = 0.2$, it gives a number that is 5% too large, but for $c = 8$ and $t^+ = 4$, the value given by Eq. (8.50) is only 0.2% large.

A comparison of Figs. 8.9 and 8.10 shows that they have the same general shape, but the average curves start to rise sooner and reach larger steady-state values. This is true for all curves except for $c = r^+ = 0$ for which the curves are identical.

8.6 BODIES WITH ANGULAR DEPENDENCE: $T = T(r, \phi, t)$

When the temperature depends on coordinates r and ϕ, the GFs cannot be found by multiplying one-dimensional GFs. Consequently, these GFs are tabulated separately in Appendix RΦ for boundary conditions of type 0, 1, 2, and 3. In this section, two examples are given of cylinders with angular dependence of the temperature.

In the full cylinder for which $0 < \phi < 2\pi$, the GFs contain Bessel functions $J_n(\cdot)$ and $Y_n(\cdot)$ where n is an integer. The GFs for full cylinders are numbered $RIJ\Phi00$ where $I = 0, 1, 2,$ or 3 and $J = 1, 2,$ or 3. These GFs are derived from separation of variable methods (Ozisik, 1980).

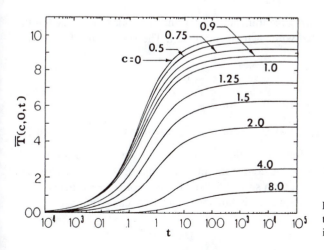

Figure 8.10 Average temperature versus time at $z = 0$ on semi-infinite body.

In the sector of a cylinder for which $0 < \phi < \phi_0$, the GFs contain Bessel functions of fractional order $J_\nu(\cdot)$ and $Y_\nu(\cdot)$ where ν is a rational number. Some other names for the sector of a cylinder are the wedge, the partial cylindrical shell, and the cylinder with a radial slot. The temperature in these bodies is described by the GFs numbered $RIJ\Phi KL$ where J, K, and L are not zero. These functions are listed in table form in Appendix $R\Phi$ in Tables $R\Phi.1$ through $R\Phi.4$.

These Bessel functions are difficult to work with and there are few closed-form solutions that result from the GF method. An attractive alternative to the Bessel functions is the use of Galerkin-based GFs discussed in Chapters 10 and 11. Galerkin-based GFs apply with equal ease to any coordinate system because the GFs are constructed numerically. Since, with the Bessel functions, numerical integration is usually needed to find the temperature distribution, the Bessel function (exact) form of the GF has no advantage of accuracy over the Galerkin-based form.

Example 8.4: Cylinder with initial temperature varying with angle—$R01B0T$-$\Phi00$ case. Find the temperature in the full cylinder with an arbitrary initial condition, $T(r, \phi, t = 0) = F(r, \phi)$ and with specified zero temperature on the boundary $r = a$.

SOLUTION This is the $R01B0\Phi00T$- case. The GF is listed in Appendix $R\Phi$ as

$$G_{R01\Phi00}(r, \phi, t | r', \phi', \tau)$$

$$= \frac{2}{2\pi a^2} \sum_{m=1}^{\infty} e^{-\beta_{m0}^2 \alpha(t-\tau)/a^2} \frac{J_0(\beta_{m0}r/a)J_0(\beta_{m0}r'/a)}{[J_0'(\beta_{m0})]^2}$$

$$+ \frac{2}{\pi a^2} \sum_{n=1}^{\infty} \cos n(\phi - \phi') \sum_{m=1}^{\infty} e^{-\beta_{mn}^2 \alpha(t-\tau)/a^2}$$

$$\times \frac{J_n(\beta_{mn}r/a)J_n(\beta_{mn}r'/a)}{[J_n'(\beta_{mn})]^2} \qquad (8.51)$$

The index on n is for the eigenfunctions $J_n(\cdot)$ and the index on m is for the eigenvalues β_{mn} associated with each eigenfunction. The eigenvalues β_{mn} are the zeroes of

$$J_0(\beta_{m0}) = 0 \qquad \text{for } n = 0$$

$$J_1(\beta_{m1}) = 0 \qquad \text{for } n = 1$$

$$J_2(\beta_{m2}) = 0 \qquad \text{for } n = 2, \text{ and so on} \qquad (8.52)$$

The first 10 eigenvalues for $n = 0$ through $n = 5$ are listed in Ozisik (1980, p. 672). The convergence of the double infinite series in Eq. (8.51) is determined primarily by the exponential term so that for small values of $(t - \tau)$, many terms and many eigenvalues are required; finding the eigenvalues can require significant effort.

The temperature in the cylinder is given by the initial condition term of the GFSE:

$$T(r, \phi, t) = \int_{r'=0}^{a} \int_{\phi'=0}^{2\pi} F(r', \phi')$$

$$\times G_{R01\Phi00}(r, \phi, t | r', \phi, 0) \, r' \, d\phi' \, dr' \qquad (8.53)$$

Note that $dv' = r' d\phi' dr'$ for this integral. Replace the GF into Eq. (8.53) to get the temperature:

$$T(r, \phi, t) = \int_{r'=0}^{a} \int_{\phi'=0}^{2\pi} F(r', \phi') \left\{ \frac{2}{2\pi a^2} \sum_{m=1}^{\infty} \right.$$

$$\times e^{-\beta_{m0}^2 \alpha t/a^2} \frac{J_0(\beta_{m0}r/a)J_0(\beta_{m0}r'/a)}{[J_0'(\beta_{m0})]^2}$$

$$+ \frac{2}{\pi a^2} \sum_{n=1}^{\infty} \cos n(\phi - \phi') \sum_{m=1}^{\infty}$$

$$\left. \times e^{-\beta_{mn}^2 \alpha t/a^2} \frac{J_n(\beta_{mn}r/a)J_n(\beta_{mn}r'/a)}{[J_n'(\beta_{mn})]^2} \right\} r' \, d\phi' \, dr' \qquad (8.54)$$

The integral on r' cannot be evaluated in closed form even if $F(r, \phi) = F(\phi)$, because the integral

$$\int_{r'=0}^{a} J_n\left(\frac{\beta_{mn}r'}{a}\right) r' \, dr'$$

is not available in closed form.

Example 8.5: Cylindrical sector (wedge) heated over the curved surface and insulated elsewhere—R02B-Φ22T0 case. Find the temperature in a sector of a cylinder that is heated over the surface at $r = b$ by a heat flux $q(\phi)$, a steady heat flux that varies with position. The faces of the sector at $\phi = 0$ and $\phi = \phi_0$ are insulated as shown in Fig. 8.11. The initial temperature is zero.

SOLUTION This is the R02B-Φ22T0 geometry. The temperature in the sector due to heat flux $q(\phi)$ at the surface is given by the GFSE

$$T(r, \phi, t) = \frac{\alpha}{k} \int_{\tau=0}^{t} d\tau \int_{\phi'=0}^{2\pi} q(\phi') \, G_{R02\Phi22}(r, \phi, t | b, \phi, \tau) \, bd\phi' \qquad (8.55)$$

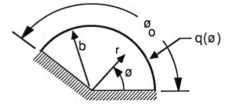

Figure 8.11 Sector of a cylinder heated at $r = b$ and insulated elsewhere, case R02φ22.

The spatial integral extends over the surface at $r = b$. The GF for this case is found from Table RΦ.1 in Appendix RΦ,

$$G(r, \phi, t|r', \phi', \tau) = \frac{R_0(\beta_{00}, r/b)}{N(\beta_{00})\, N(\nu = 0)} + \sum_{m=1}^{\infty} \sum_{\nu}^{\infty} e^{-\beta_{m\nu}^2 \alpha(t-\tau)/a^2}$$

$$\times \frac{R_\nu(\beta_{m\nu}, r/b)\, R_\nu(\beta_{m\nu}, r'/b)}{N(\beta_{m\nu})} \frac{\Phi(\nu, \phi)\, \Phi(\nu, \phi')}{N(\nu)} \quad (8.56)$$

The GF is constructed from the particular R_ν, Φ, N, and eigenconditions listed in Tables RΦ.1 through RΦ.4 in Appendix RΦ. In the $R02\Phi22$ case,

$$R_\nu(\beta_{m\nu}, r/b) = \begin{cases} 1 & m = 0 \\ J_\nu\left(\dfrac{\beta_{m\nu}\, r}{b}\right) & \text{for } m \neq 0 \text{ and } \nu \neq 0 \end{cases} \quad (8.57a)$$

$$\frac{1}{N(\beta_{m\nu})} = \begin{cases} \dfrac{2}{b^2} & \text{for } m = 0; \\[2mm] \dfrac{2\,\beta_{m\nu}^2}{b^2 J_\nu(\beta_{m\nu})\,(\beta_{m\nu}^2 - \nu^2)} & \text{for } m \geq 0 \end{cases} \quad (8.57b)$$

$$\Phi(\nu, \phi) = \begin{cases} 1 & \text{for } \nu = 0 \\ \cos \nu\phi & \text{for } \nu \geq 1 \end{cases} \quad (8.57c)$$

$$\frac{1}{N(\nu)} = \begin{cases} \dfrac{1}{\phi_0} & \text{for } \nu = 0 \\[2mm] \dfrac{2}{\phi_0} & \text{for } \nu \geq 1 \end{cases} \quad (8.57d)$$

and the eigenvalues are given by the roots of

$$J_\nu'(\beta_{m\nu}) = 0 \quad (8.58a)$$

$$\sin(\nu\phi_0) = 0 \quad \left(\text{that is, } \nu = \frac{n\pi}{\phi_0}; n = 0, 1, 2, \ldots\right) \quad (8.58b)$$

The GF may now be assembled by substituting the pieces (8.57) into the general expression (8.56). In the following expression the term for $\nu = 0$ has been separated from the other summation terms:

$$G(r, \phi, t|r', \phi', \tau)$$

$$= \frac{2}{b^2 \phi_0} + \sum_{m=1}^{\infty} 2\, e^{-\beta_{m0}^2 \alpha(t-\tau)/b^2}\, \frac{J_0(\beta_{m0} r/b)\, J_0(\beta_{m0} r'/b)}{b^2 \phi_0}$$

$$+ \sum_{m=1}^{\infty} \sum_{\nu}^{\infty} 4\, e^{-\beta_{m\nu}^2 \alpha(t-\tau)/b^2}$$

$$\times \frac{\beta_{mv}^2 \, J_v(\beta_{mv}r/b) \, J_v(\beta_{mv}r'/b)}{b^2 J_v^2(\beta_{mv})(\beta_{mv}^2 - v^2)} \frac{\cos{(v\phi)} \cos{(v\phi')}}{\phi_0} \qquad (8.59)$$

where now $v = n\pi/\phi_0$ for $n = 1, 2, 3, \ldots$.

The GF may be replaced into Eq. (8.55) to give the temperature expression. After the time integral is evaluated, the temperature is given by

$$
\begin{aligned}
T(r, \phi, t) = & \frac{\alpha t}{b^2} \int_{\phi'=0}^{2\pi} \frac{2 \, q(\phi')}{k \, \phi_0} \, b \, d\phi' \\
& + \frac{\phi b}{k} \int_{\phi'=0}^{2\pi} q(\phi') \left[\sum_{m=1}^{\infty} 2 \, (1 - e^{-\beta_{m0}^2 \alpha t/a^2}) \right. \\
& \left. \times \frac{J_0(\beta_{m0}r/a) \, J_0(\beta_{m0}r'/a)}{\beta_{m0}^2 \, \phi_0} \right] d\phi' \\
& + \frac{\phi b}{k} \int_{\phi'=0}^{2\pi} q(\phi') \left[\sum_{m=1}^{\infty} \sum_v 4 \, (1 - e^{-\beta_{mv}^2 \alpha t/a^2}) \right. \\
& \left. \times \frac{J_v(\beta_{mv}r/b) \, J_v(\beta_{mv}r'/b) \cos{(v\phi)} \cos{(v\phi')}}{J_v^2(\beta_{mv})(\beta_{mv}^2 - v^2) \, \phi_0} \right] d\phi' \qquad (8.60)
\end{aligned}
$$

The units of each term are qb/k which gives temperature as required. The first term in the temperature is the quasi-steady term. As $t \to \infty$, the summation terms drop out and the quasi-steady term causes the temperature to increase linearly with time.

In the special case $\phi_0 = \pi, \pi/2, \pi/3, \ldots$, where the sector is an even fraction of a full cylinder, then the temperature may be found by the method of images on the full cylinder with case $R02B\text{-}\Phi00$. The method of images on the cylinder is discussed in Carslaw and Jaeger (1959), but in brief, the method involves a fictitious full cylinder with a surface heating pattern composed of "images" of the original heating pattern on $0 < \phi < \phi_0$ so that the surfaces $\phi = 0$ and $\phi_0 = 0$ satisfy $\partial T/\partial \phi = 0$ (insulated condition). The temperature in the fictitious full cylinder is found with $R02\Phi00$ analysis and then the desired temperature can be found in the region $0 < \phi < \phi_0$.

8.7 STEADY STATE

Two examples of steady heat transfer in cylindrical coordinates are given in this section. For radial heat flow several steady GFs are given in Appendix R, Table R.1. For other geometries the method of limits may be used to find the steady GF.

Example 8.6: Solid cylinder with internal energy generation—$R03B0G$-case. Find the steady temperature in a solid cylinder with internal heating. The surface of the cylinder is cooled by convection heat transfer and T_∞ is the fluid temperature.

SOLUTION If the temperature is evaluated in the form $(T - T_\infty)$, then the convection boundary condition is homogeneous and this is the *R03B0G-* case. The energy generation term of the GFSE may be used to find the temperature as

$$T(r) - T_\infty = \int_{r'=0}^{b} \frac{g(r')}{k} G_{R03}(r|r') \, 2\pi r' \, dr' \tag{8.61}$$

where $g(r')$ is the volume energy generation. The steady GF is given in Appendix R, Table R.1 as

$$G(r|r') = \begin{cases} \dfrac{\ln (b/r') + 1/B_2}{2\pi} & r < r' \\[3mm] \dfrac{\ln (b/r) + 1/B_2}{2\pi} & r > r' \end{cases} \tag{8.62}$$

where B_2 is hb/k, and h is the heat transfer coefficient. Because the GF is piecewise continuous, the spatial integral in Eq. (8.61) must be carried out in two pieces:

$$T(r) - T_\infty = \int_{r'=0}^{r} \frac{g(r')}{k} \left[\ln \left(\frac{b}{r} \right) + \frac{1}{B_2} \right] r' \, dr'$$

$$+ \int_{r'=r}^{b} \frac{g(r')}{k} \left[\ln \left(\frac{b}{r'} \right) + \frac{1}{B_2} \right] r' \, dr' \tag{8.63}$$

(*a*) *Case R03B0G1.* For uniform heat generation $g(r) = g_0$ the integrals in Eq. (8.63) may be evaluated to give

$$T(r) - T_\infty = \frac{g_0 b^2}{k} \left[\frac{1}{B_2} + \frac{1 - (r/b)^2}{4} \right] \tag{8.64}$$

(*b*) *Case R03B0G5.* For piecewise constant energy generation

$$g(r') = \begin{cases} g_0 & a < r' < b \\ 0 & 0 < r' < a \end{cases} \tag{8.65}$$

the temperature given by Eq. (8.64) must be carried out in two parts depending on the location of the observation point. For $0 < r < a$, the temperature does not depend on location r. Only the first term of Eq. (8.63) is used with limits $a < r' < b$ to give

$$T(r < a) - T_\infty = \frac{g_0 b^2}{2k} \left\{ \frac{1}{B_2} \left[1 - \left(\frac{a}{b} \right)^2 \right] \right.$$

$$+ \frac{1}{2} + \left(\frac{a}{b} \right)^2 \ln \left(\frac{a}{b} \right) - \left(\frac{a}{b} \right)^2 \right\} \tag{8.66a}$$

For $a < r < b$ both terms of Eq. (8.63) are needed and the temperature is

$$T(r > a) - T_\infty = \frac{g_0 b^2}{2k} \left\{ \frac{1}{B_2} \left[1 - \left(\frac{a}{b} \right)^2 \right] \right.$$

$$\left. + \frac{1}{2} + \left(\frac{a}{b} \right)^2 \ln \left(\frac{r}{b} \right) - \left(\frac{r}{b} \right)^2 \right\} \qquad (8.66b)$$

Note that the piecewise continuous temperature distributions are equal at $r = a$.

Example 8.7: Finite cylinder with arbitrary surface temperature on the curved surface—R01B-Z11B00 case. On a finite cylinder of length L find the steady temperature due a specified temperature $f(z)$ over surface $r = b$ and zero temperature the ends $z = 0$ and $z = L$.

SOLUTION This is the $R01B$-$Z11B00$ geometry. The temperature is given by the steady GFSE equation as

$$T(r, z) = - \int_{z'=0}^{L} f(z') \left. \frac{\partial G_{R01Z11}}{\partial n} \right|_{r'=b} 2\pi b \, dz' \qquad (8.67)$$

The steady GF is given by the method of limits combined with the multiplicative property of transient GFs:

$$G(r, z|r', z') = \lim_{t \to \infty} \alpha \int_{t=0}^{t} G_{R01}(r, t|r', \tau) \, G_{Z11}(z, t|z', \tau) \, d\tau \qquad (8.68)$$

The transient GFs are given in Appendixes R and X. The time integral and the limit may be evaluated to give the steady GF:

$$G(r, z|r', z') = \sum_{m=1}^{\infty} \sum_{n=1}^{\infty} \frac{2}{\pi b^2 L} \left[\left(\frac{\beta_m}{b} \right)^2 + \left(\frac{n\pi}{L} \right)^2 \right]^{-1}$$

$$\times \frac{J_0(\beta_m r/b) \, J_0(\beta_m r'/b)}{J_1^2(\beta_m)} \sin \frac{n\pi z}{L} \sin \frac{n\pi z'}{L} \qquad (8.69)$$

The steady GF may be substituted in Eq. (8.67) to find the temperature. The derivative on r' is elementary, and the temperature due to surface temperature distribution $f(z)$ is given by

$$T(r, z) = \sum_{m=1}^{\infty} \sum_{n=1}^{\infty} \frac{4}{b^2 L} \left[\left(\frac{\beta_m}{b} \right)^2 + \left(\frac{n\pi}{L} \right)^2 \right]^{-1} \frac{\beta_m J_0(\beta_m r/b)}{J_1(\beta_m)} \sin \frac{n\pi z}{L}$$

$$\times \int_{z'=0}^{L} f(z') \sin \frac{n\pi z'}{L} \, dz' \qquad (8.70)$$

The integral on z' can be found in closed form for many functions $f(z')$. In the case of uniform surface temperature, $f(z) = T_0$, the integral on z' may be evaluated to give

$$T(r, z) = T_0 \sum_{m=1}^{\infty} \sum_{n=1}^{\infty} \frac{4}{n\pi} \frac{1 - (-1)^n}{\beta_m^2 + (n\pi b/L)^2} \frac{\beta_m J_0(\beta_m r/b)}{J_1(\beta_m)} \sin \frac{n\pi z}{L} \quad (8.71)$$

The double summation in Eq. (8.71) converges somewhat slowly. One technique to improve the numerical convergence of a steady-state problem is to formulate the problem in transient form and then use time partitioning as discussed in Section 5.3. Steady state is just a special case of the large-time solution. See also Sections 5.5 and 5.6. Another technique for steady problems is discussed by Barton (1989) where a single summation form of the two-dimension GF is found from appropriate eigenfunction expansions of the Dirac delta function. Carslaw and Jaeger (1959) also list several two-dimensional, steady-state cylinder solutions based on the single-summation approach.

REFERENCES

Abramowitz, M., and Stegun, I. A., 1964, *Handbook of Mathematical Functions*, National Bureau of Standards, Applied Mathematics Series 55, U.S. Government Printing Office, Washington, D.C.

Barton, G., 1989, *Elements of Green's Functions and Propagation*, Oxford University Press, London.

Beck, J. V., 1980, Average Transient Temperature Within a Body Heated by a Disk Heat Source, *Heat Transfer, Thermal Control and Heat Pipes*, ed. W. B. Olstad, vol. 70, Progress in Astronautics and Aeronautics, pp. 3–24.

Beck, J. V., 1981, Large Time Solutions for Temperatures in a Semi-infinite Body with a Disk Heat Source, *Int. J. Heat Mass Transfer*, vol. 24, pp. 155–164.

Carslaw, H. S., and Jaeger, J. C., 1959, *Conduction of Heat in Solids*, 2d ed., Oxford University Press, New York.

Ozisik, M. N., 1980, *Heat Conduction*, Wiley, New York.

Thomas, P. H., 1957, Some Conduction Problems in the Heating of Small Areas on Large Solids, *Q. J. Mech. Appl. Math.*, vol. 10, pp. 482–493.

PROBLEMS

8.1 Show that $\int_{\phi'=0}^{\phi_0} G_{RIJ\Phi22}(\cdot) \, d\phi' = G_{RIJ}(\cdot)$.

8.2 Show that $\int_{\phi'=0}^{2\pi} G_{RIJ\Phi00}(\cdot) \, d\phi' = G_{RIJ}(\cdot)$.

8.3 Does $G_{RIJ\Phi11}(\cdot)$ for $G = 0$ at $\phi = \phi_0$ equal $G_{RIJ\Phi12}(\cdot)$ for $\partial G/\partial \phi = 0$ at $\phi = \phi_0/2$? Examine both physically and mathematically.

8.4 Does $G_{RIJ\Phi22}(\cdot)|_{\phi_0=2\pi} = G_{RIJ\Phi00}(\cdot)$? Examine both physically and mathematically.

8.5 Refer to Example 8.1 to find the steady-state temperature in a thin-walled tube that is cooled by steady uniform convection inside and heated by incident solar radiation on the outside. Assume that all of the incident radiation is absorbed and that the incident radiation is described by $q = q_0 \cos \phi$ for $-\pi/2 < \phi < \pi/2$, and $q = 0$ otherwise.

8.6 Derive the centerline temperature ($r = 0$) for the disk heat source on a semi-infinite body given by Eq. (8.36). Use the form of G_{R00} that contains the modified Bessel function $I_0(r \, r'/[4\alpha(t - \tau)])$.

8.7 Find an integral expression for the surface temperature caused by a short laser pulse of duration δt on a large flat surface. The surface heating by the laser may be modeled as a uniform disk heat source of radius a. Find the average surface temperature over the laser-heated region as a function of time for $t > \delta t$.

8.8 A more realistic model of laser beam absorption involves a distribution of energy across the beam. Find an integral expression for the transient temperature on the surface of an opaque semi-infinite solid caused by a short laser pulse of duration δt where the incident energy has a Gaussian distribution:

$$q(r, t) = q_0 \, e^{-2(r/a)^2} \qquad \text{for } 0 < t < \delta t$$

where now a is the Gaussian beam radius. The initial temperature is zero. Find the maximum temperature on the surface and the time when it occurs.

8.9 Find an integral expression for the temperature in a half-cylinder (sector $0 < \phi < \pi$) with initial temperature T_0. The surface $r = b$ is insulated and the flat surface ($\phi = 0$ and $\phi = \pi$) is held at a fixed temperature T_0.

8.10 Induction heating is a rapid, highly localized heating method that is used to harden bearing surfaces on crankshafts. If the crankshaft may be modeled as a solid cylinder and the induction heating may be modeled as surface heating, find the transient temperature in the cylinder suddenly heated over a small portion of its length $-a < z < 0$. The remainder of the cylinder surface is insulated. Initially the cylinder has zero temperature.

8.11 A pin fin is a cylinder with a fixed elevated temperature of T_0 at $z = 0$ and convection cooling by a fluid at T_∞ over the other surfaces.

(a) Find an exact expression for the steady heat flow into the pin fin at $z = 0$ (in watts) by analyzing the geometry $R03B0Z11B10$ (assume the temperature at $z = L$ is T_∞).

(b) Find an approximate expression for the steady heat flow into the pin fin at $z = 0$ by analyzing the fin equation $\nabla^2 T - m^2 T = 0$ for geometry $Z11$.

(c) Compare the numerical answer from parts (a) and (b) in the specific cases of a pin fin with length/radius of $L/a = 3$, 10, and comment on the conditions for which the fin approximation is useful.

8.12 Find the transient temperature $T(r, \phi, t)$ in a long circular cylinder initially at zero temperature and with uniform heat flux on a sector of its surface $0 < \phi < \phi_0$. This is a model of a split-film anemometer sensor formed from a platinum heater bonded to a quartz cylinder. Write the temperature as the sum of three terms: the spatial-average (or lumped) term proportional to time, the transient term that dies away as $t \to \infty$, and the quasi-steady term that does not depend on time. If the cylinder properties are $k = 1.4$ W/(m K) and $\alpha = 8.3E\text{-}07$ m^2/s, the cylinder radius is 2.5E-05 m, and $\phi_0 = \pi$, find the time for the transient term to die away to 10% of its initial value; this is a measure of the response time of the split-film anemometer sensor.

CHAPTER NINE

RADIAL HEAT FLOW IN SPHERICAL COORDINATES

9.1 INTRODUCTION

The applications of the Green's function (GF) solution approach to the problems posed in the spherical coordinate system are discussed in this chapter. The general heat conduction equation for linear flow of heat in spherical polar coordinates has the form

$$\frac{1}{r^2}\frac{\partial}{\partial r}\left(r^2\frac{\partial T}{\partial r}\right) + \frac{1}{r^2\sin\theta}\frac{\partial}{\partial\theta}\left(\sin\theta\frac{\partial T}{\partial\theta}\right) + \frac{1}{r^2\sin^2\theta}\frac{\partial^2 T}{\partial\phi^2}$$
$$+ \frac{1}{k}g(r,\theta,\phi,t) = \frac{1}{\alpha}\frac{\partial T}{\partial t} \tag{9.1}$$

where $T = T(r, \theta, \phi, t)$, g represents the generation rate per unit volume (W/m^3) within the spherical region, and k is constant.

As was mentioned in the previous chapter, the applications of the GF solution method to multidimensional problems involve cumbersome analytical work. In addition, for spherical coordinates, unlike the rectangular and cylindrical coordinates, the two- and three-dimensional GFs cannot be obtained from the product of the one-dimensional solutions. Because of these problems and because many heat conduction problems in spherical coordinates involve spherical symmetry (i.e., the temperature does not depend on θ and ϕ) this chapter emphasizes problems with temperature distributions that are functions only of time t and radius r (radial flow of heat). For radial flow of heat, Eq. (9.1) reduces to:

$$\frac{1}{r^2}\frac{\partial}{\partial r}\left(r^2\frac{\partial T}{\partial t}\right) + \frac{g(r,t)}{k} = \frac{1}{r}\frac{\partial^2(rT)}{\partial r^2} + \frac{1}{k}g(r,t) = \frac{1}{\alpha}\frac{\partial T}{\partial t} \tag{9.2}$$

Later in this chapter, we will show how this equation can further be simplified and put into the rectangular form by introducing a new temperature $U(r, t) = rT(r, t)$. Topics covered in the remainder of this chapter include the GFSE for radial flow

253

of heat in spherical coordinates (Section 9.2), the infinite body with radial flow of heat (Section 9.3), methods for obtaining the related GFs (Section 9.4), and how the GF solution method can be used to solve a number of important problems for radial flow of heat in the geometries of solid and hollow spheres and in the region outside a spherical cavity (Sections 9.6–9.8).

9.2 GREEN'S FUNCTION EQUATION FOR RADIAL SPHERICAL HEAT FLOW

From the general GF equation for heat conduction given by Eq. (3.46), one can write down the GF equation for the radial heat flow in spherical coordinates with the exclusion of the term associated with the boundary conditions of the fourth and fifth kinds, as

$$T(r, t) = \int_{r'} G(r, t|r', 0) F(r')4\pi r'^2 \, dr'$$

(for the initial condition)

$$+ \int_{\tau=0}^{t} \int_{r'} \frac{\alpha}{k} G(r, t|r', \tau)g(r', \tau)4\pi r'^2 \, dr' \, d\tau$$

(for volume energy generation)

$$+ \alpha \int_{\tau=0}^{t} \sum_{i=1}^{S} \frac{f_i(r_i, \tau)}{k_i} G(r, t|r_i, \tau) \, 4\pi r_i^2 \, d\tau$$

(for boundary conditions of the second and third kinds)

$$- \alpha \int_{\tau=0}^{t} \sum_{j=1}^{S} f_j(r_j, \tau) \frac{\partial G}{\partial n_j'}\bigg|_{r'=r_j} 4\pi r_j^2 \, d\tau$$

(for boundary condition of the first kind only)   (9.3)

Note that $dv' = 4\pi r'^2 \, dr'$, and that the integrals over boundary surface s_i have been replaced by $4\pi r_i^2$.

The GFs associated with different set of boundary conditions for radial spherical heat flow are denoted by $G_{RSIJ}(\cdot)$, where subscript RS stands for radial spherical according to the heat conduction numbering system. Only boundary conditions of the zeroth through the third kinds are considered here ($I, J = 0, 1, 2, 3$). A listing of the available GFs for radial spherical heat flow is provided in Appendix RS.

9.3 INFINITE BODY

The GF for spherical radial flow of heat in an infinite body is denoted by $G_{RS00}(r, t|r', \tau)$. It is called the fundamental heat conduction solution for spherical radial heat flow and is given by

$$G_{RS00}(r, t|r', \tau) = \frac{1}{8\pi r r'[\pi\alpha(t - \tau)]^{1/2}}$$

$$\times \left\{ \exp\left[-\frac{(r - r')^2}{4\alpha(t - \tau)} \right] - \exp\left[-\frac{(r + r')^2}{4\alpha(t - \tau)} \right] \right\} \quad (9.4)$$

This GF represents the temperature response due to a unit instantaneous spherical surface source of radius r' at time τ in an infinite body with zero initial condition. Do not confuse the spherical-surface source with the point source discussed in Section 4.6. The GF is given by

$$\frac{1}{r}\frac{\partial^2(rG)}{\partial r^2} + \frac{1}{\alpha}\delta(\mathbf{r} - \mathbf{r}')\,\delta(t - \tau) = \frac{1}{\alpha}\frac{\partial G}{\partial t} \quad (9.5a)$$

$$\frac{\partial G}{\partial r}(0, t|r', \tau) = 0 \quad (9.5b)$$

$$G(\infty, t|r', \tau) = 0 \quad (9.5c)$$

$$G(r, 0|r', \tau > 0) = 0 \quad (9.5d)$$

Figure 9.1 shows $r'^3 G_{RS00}(\cdot)$ versus $r^+ = r/r'$ for various values of $t^+ = \alpha(t - \tau)/r'^2$. Note that $G_{RS00}(\cdot)$ is unaffected by the axisymmetric condition of $\partial G/\partial r = 0$ at $r = 0$ for $t^+ < 0.03$ and approaches the Dirac delta function as t^+ goes to zero. For larger values of t^+, the position of the maximum G moves to smaller r^+ values.

It is interesting to note that for the special case where $r' \to 0$, the $G_{RS00}(\cdot)$ becomes:

$$G_{RS00}(r, t|0, \tau) = \frac{1}{[4\pi\alpha(t - \tau)]^{3/2}} \exp\left[-\frac{r^2}{4\alpha(t - \tau)} \right] \quad (9.6)$$

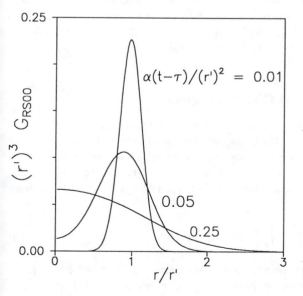

Figure 9.1 The *RS00* GF.

which represents the response due to an instantaneous point source at the origin. It can also be shown that for the case where r' is fixed and $r \to 0$, a similar equation to (9.6) is obtained; that is,

$$G_{RS00}(0, t|r', \tau) = \frac{1}{[4\pi\alpha(t - \tau)]^{3/2}} \exp\left[- \frac{r'^2}{4\alpha(t - \tau)}\right] \tag{9.7}$$

which gives the response at the origin due to an instantaneous spherical surface source at r'. From Eqs. (9.6) and (9.7), it is obvious that the reciprocity relation holds for this GF.

It is also interesting to note that the point source solution, Eq. (9.6), can be represented by the product of three plane heat sources for the x, y, and z directions; that is,

$$G_{RS00}(r, t|0, \tau) = G_{X00}(x, t|0, \tau)\, G_{Y00}(y, t|0, \tau)\, G_{Z00}(z, t|0, \tau) \tag{9.8a}$$

which also shows that the unit of $G_{RS00}(\cdot)$ is $m^{-1}m^{-1}m^{-1} = m^{-3}$. Note that for this case where the source is at the origin $(r' = 0)$, the distance between the impulse and response points is given by

$$r = (x^2 + y^2 + z^2)^{1/2} \tag{9.8b}$$

Similarly, for the case where r' is fixed and $r = 0$, we can write

$$G_{RS00}(0, t|r', \tau) = G_{X00}(0, t|x', \tau)\, G_{Y00}(0, t|y', \tau)\, G_{Z00}(0, t|z', \tau) \tag{9.9a}$$

where

$$r' = (x'^2 + y'^2 + z'^2)^{1/2} \tag{9.9b}$$

represents the distance between the impulse and the response points for this case.

The $RS00$ − GF given by Eq. (9.4) may be employed to obtain GFs for other cases of radial flow of heat in spherical geometry with different types of boundary conditions. For instance, in Section 4.3, we saw how $G_{RS00}(\cdot)$ was used in the Laplace transform approach to obtain G_{RS30} which is the GF for the infinite region outside the spherical cavity $r = a$ with convective boundary condition at $r = a$.

9.3.1 Derivation of the $RS00$ Green's Function

The derivation given here is based on the physical interpretation that $G_{RS00}(\mathbf{r}, t|\mathbf{r}', \tau)$ is equal to the temperature rise due to an instantaneous spherical surface source at time τ and location $\mathbf{r} = \mathbf{r}'$ divided by the strength of the source and multiplied by ρc. Having this in mind, we start with the general form of the GF equation (3.46). From this equation, the temperature due to a distributed energy source is an infinite body with zero initial condition is given by

$$T(\mathbf{r}, t) = \int_{\tau=0}^{t} \int_{R} \frac{\alpha}{k}\, G(\mathbf{r}, t|\mathbf{r}', \tau) g(\mathbf{r}', \tau) dv'\, d\tau \tag{9.10}$$

where $G(\mathbf{r}, t|\mathbf{r}', \tau)$ represents the temperature response at point $\mathbf{r}$ and time t in an infinite body due to an instantaneous impulse at point $\mathbf{r}'$ and time τ. It is given by

$$G(\mathbf{r}, t|\mathbf{r}', t) = \frac{1}{[4\pi\alpha(t - \tau)]^{3/2}} \exp\left[-\frac{R^2}{4\alpha(t - \tau)}\right] \quad (9.11)$$

where R represents the distance between the points $\mathbf{r}$ and $\mathbf{r}'$. In rectangular coordinate system, this distance is given by

$$R^2 = (x - x')^2 + (y - y')^2 + (z - z')^2 \quad (9.12)$$

where (x, y, z) and (x', y', z') are the rectangular coordinates of points $\mathbf{r}$ and $\mathbf{r}'$, respectively. The rectangular coordinates (x, y, z) and (x', y', z') can be transformed into the spherical coordinates (r, θ, ϕ) and (r', θ', ϕ') through the following relations:

$$x = r \sin \theta \cos \phi \quad (9.13a)$$

$$y = r \sin \theta \sin \phi \quad (9.13b)$$

$$z = r \cos \theta \quad (9.13c)$$

Then the distance R in spherical coordinates may be presented by

$$R^2 = r^2 + r'^2 - 2rr'[\sin \theta \sin \theta' \cos (\phi - \phi') + \cos \theta \cos \theta'] \quad (9.14a)$$

For the case of only radial flow of heat in spherical system, there is no temperature variation with θ and ϕ. This implies that the temperature at any point $\mathbf{r}$ over a spherical surface which is at an arbitrary distance from the spherical surface source (at $r = r'$) is the same regardless of the values of θ and ϕ. Accordingly, for simplicity, we choose the spherical coordinates of point $\mathbf{r}$ to be $(0, 0, r)$. Then Eq. (9.14a) simplifies to

$$R^2 = r^2 + r'^2 - 2rr' \cos \theta' \quad (9.14b)$$

The generation term in Eq. (9.10) represents a *continuous distributed volumetric source* and has the unit of W/m³. However, since we are seeking the temperature solution due to an *instantaneous spherical surface source*, $g(\mathbf{r}', \tau)$, in Eq. (9.10) is replaced by

$$g(\mathbf{r}', \tau) = \frac{\delta(\tau - \tau_0)\delta(r' - r_0)g_0}{4\pi r_0^2} \quad (9.15)$$

where r_0 is the radius of the spherical surface source that pulses at time τ_0, and g_0 (Joule) represents the strength of the source. (The strength per unit area is given by $g_0/4\pi r_0^2$ and function $\delta(r - r_0)$ has units of m⁻¹.)

Now by substituting the values of $G(\mathbf{r}, t|\mathbf{r}', \tau)$, R, and $g(\mathbf{r}', \tau)$ from Eqs. (9.11), (9.14b), and (9.15) into Eq. (9.10) and integrating over the appropriate ranges for $r'(0 \rightarrow \infty)$, $\phi'(0 \rightarrow 2\pi)$, and $\theta'(0 \rightarrow \pi)$, one can write

$$T(r, t) = \frac{\alpha}{k} \int_{\tau=0}^{t} \int_{r'=0}^{\infty} \int_{\theta'=0}^{\pi} \int_{\phi'=0}^{2\pi}$$

$$\times \left\{ \frac{1}{[4\pi\alpha(t - \tau)]^{3/2}} \exp\left[-\frac{r^2 + r'^2 - 2rr' \cos \theta'}{4\alpha(t - \tau)}\right] \right.$$

$$\left. \times \frac{\delta(\tau - \tau_0)\delta(r' - r_0)g_0}{4\pi r_0^2} \right\} r'^2 \sin \theta' \, dr' \, d\theta' \, d\phi' \, d\tau \quad (9.16)$$

where dv' in Eq. (9.10) has been replaced by $r'^2 \sin \theta' \, dr' \, d\theta' \, d\phi'$ (see Fig. 9.2). Note that Eq. (9.16) gives the temperature due to an instantaneous spherical surface source at radius r_0 and time τ_0. The integrals over r' and τ can be evaluated easily with the sifting property of the Dirac delta functions. Then Eq. (9.16) reduces to

$$T(r, t) = \frac{\alpha}{k} \frac{g_0/(4\pi)}{[4\pi\alpha(t - \tau_0)]^{3/2}} \exp\left[-\frac{(r^2 + r_0^2)}{4\alpha(t - \tau_0)} \right] \int_{\phi'=0}^{2\pi} d\phi' \int_{\theta'=0}^{\pi}$$

$$\times \exp\left[-\frac{rr_0 \cos \theta'}{2\alpha(t - \tau_0)} \right] \sin \theta' \, d\theta' \tag{9.17}$$

The integral over ϕ' is equal to 2π and the integral over θ' can be evaluated easily by choosing a new variable $\mu = \cos \theta'$ to give

$$T(r, t) = \frac{g_0}{\rho c} \frac{1}{8\pi r r_0 [\alpha\pi(t - \tau_0)]^{1/2}} \left\{ \exp\left[-\frac{(r - r_0)^2}{4\alpha(t - \tau_0)} \right] \right.$$

$$\left. - \exp\left[-\frac{(r + r_0)^2}{4\alpha(t - \tau_0)} \right] \right\} \tag{9.18}$$

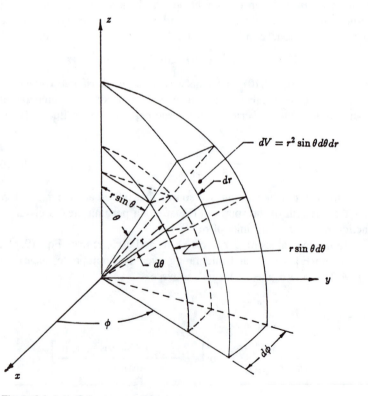

Figure 9.2 Spherical polar coordinate.

Then, the GF is given by the temperature divided by the source strength and multiplied by ρc.

$$G_{RS00}(r, t | r_0, \tau_0) = \frac{T(r, t)}{g_0/\rho c} = \frac{1}{8\pi r r_0 [\alpha\pi(t - \tau_0)]^{1/2}}$$

$$\times \left\{ \exp\left[-\frac{(r - r_0)^2}{4\alpha(t - \tau_0)} \right] - \exp\left[-\frac{(r + r_0)^2}{4\alpha(t - \tau_0)} \right] \right\} \quad (9.19)$$

Finally, by considering the conventional form, that is, the heat source being at (r', τ) instead of at (r_0, τ_0), the same result as in Eq. (9.4) is obtained.

9.4 SEPARATION OF VARIABLES FOR RADIAL FLOW OF HEAT IN SPHERES

In the previous chapters, we saw that the separation of variables method provides an easy and straightforward approach for obtaining the GFs for finite-body problems posed in the Cartesian and cylindrical coordinate systems with arbitrary initial temperature distributions provided that the differential equations and the boundary conditions are homogeneous. This method can also be applied to the radial spherical heat flow problems to obtain the appropriate GFs. However, for radial flow of heat in spheres, there is an alternative approach which is more convenient and involves less analytical work than the separation of variables method; that is, the *RSIJ*-GFs can be obtained from the *XIJ*-GFs through a simple transformation of the variables. Since the method of separation of variables has already been discussed and demonstrated in detail for the problems posed in the Cartesian and cylindrical coordinate systems, in this section we demonstrate only the second approach by considering various examples.

The heat conduction equation for linear radial flow of heat in spherical coordinates is given by Eq. (9.2) as

$$\frac{1}{r} \frac{\partial^2(rT)}{\partial r^2} + \frac{1}{k} g(r, t) = \frac{1}{\alpha} \frac{\partial T}{\partial t} \quad (9.20)$$

This equation can be put into the rectangular form by introducing a new temperature U (the dependent variable) as

$$U(r, t) = rT(r, t) \quad (9.21)$$

Then Eq. (9.20) becomes

$$\frac{\partial^2 U}{\partial r^2} + \frac{1}{k} g^*(r, t) = \frac{1}{\alpha} \frac{\partial U}{\partial t} \quad (9.22)$$

where

$$g^*(r, t) = rg(r, t) \quad (9.23)$$

Note that $U(r, t)$ in Eq. (9.22) is similar to $T(x, t)$ in Eq. (1.40) for the Cartesian coordinate system. The above transformation should also be applied to the boundary conditions and the initial condition of the problem under consideration. Once the problem is completely transformed into the Cartesian coordinate system, the appropriate XIJ-GF's can be found from the separation of variable method or more conveniently from the available tables given in Appendix X. Then, the $RSIJ$ GFs are obtained by transforming the results back into the spherical coordinate system. Note that only the homogeneous part of Eq. (9.22) is considered for derivation of the GFs. The relation between the G_{RSIJ} and G_{XIJ} ($I J = 0, 1, 2, 3$) may vary from case to case depending on the geometry and the type of the boundary conditions. This is best illustrated through the following examples.

Example 9.1: Derivation of G_{RS03}—solid sphere with convective boundary condition. The $RS03$ GF is obtained from the solution to the following homogeneous problem with an arbitrary initial condition described by

$$\frac{1}{r} \frac{\partial^2(rT)}{\partial r^2} = \frac{1}{\alpha} \frac{\partial T}{\partial t} \qquad 0 \leq r \leq b \qquad t > 0 \tag{9.24}$$

$$T \text{ is finite at } r = 0 \qquad t > 0 \tag{9.25a}$$

$$\frac{\partial T(b, t)}{\partial r} + HT(b, t) = 0 \qquad t > 0 \tag{9.25b}$$

where

$$H \equiv \frac{h}{k} \tag{9.25c}$$

$$T(r, 0) = F(r) \qquad 0 \leq r \leq b \tag{9.26}$$

By introducing the new variable U as

$$U(r, t) = r \, T(r, t) \tag{9.27}$$

Eqs. (9.24)–(9.26) become

$$\frac{\partial^2 U(r, t)}{\partial r^2} = \frac{1}{\alpha} \frac{\partial U}{\partial t} \qquad 0 \leq r \leq b \qquad t > 0 \tag{9.28}$$

$$U(0, t) = 0 \qquad t > 0 \tag{9.29a}$$

$$\frac{\partial U(b, t)}{\partial r} + H^*U(b, t) = 0 \qquad t > 0 \tag{9.29b}$$

where

$$H^* = H - \frac{1}{b} \tag{9.29c}$$

$$U(r, 0) = r \, F(r) = F^*(r) \qquad 0 \leq r \leq b \tag{9.30}$$

The transformed problem described by Eqs. (9.28)–(9.30) represents a flat plate problem with an arbitrary initial condition and the homogeneous boundary conditions of the first kind on one side and the third kind on the other side ($X13$). See Figs. 9.3a and b.

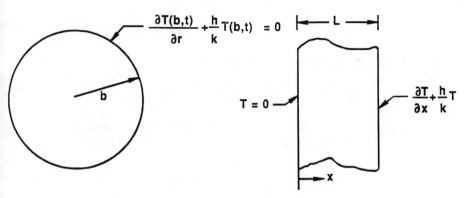

Figure 9.3 (*a*) Solid sphere with convection boundary condition (*RS*03). (*b*) Flat plate with boundary conditions of the first and third kinds (*X*13).

From the separation of variables method or the available tables in Appendix X, the GF for the $X13$ geometry for $\tau = 0$ is given by

$$G_{X13}(x, t|x', 0) = \frac{2}{L} \sum_{m=1}^{\infty} \exp\left(- \beta_m^2 \frac{\alpha t}{L^2}\right)$$

$$\times \frac{(\beta_m^2 + B^2) \sin (\beta_m x/L) \sin (\beta_m x'/L)}{\beta_m^2 + B^2 + B} \qquad (9.31)$$

where

$$\beta_m \cot \beta_m = - B \qquad B = HL = \frac{hL}{k} \qquad (9.32a, b)$$

and L represents the thickness of the plate.

Then with $L \to b$, $x \to r$, $x' \to r'$, and $H \to H^*$, one can write,

$$G_{X13}(r, t|r', 0) = \frac{2}{b} \sum_{m=1}^{\infty} \exp\left(- \beta_m \frac{\alpha t}{b^2}\right)$$

$$\times \frac{(\beta_m^2 + B^2) \sin (\beta_m r/b) \sin (\beta_m r'/b)}{\beta_m^2 + B^2 + B} \qquad (9.33)$$

where

$$\beta_m \cot \beta_m = -B \qquad B = H^*b = Hb - 1 \qquad (9.34a, b)$$

From the first term of Eq. (3.16) which gives the temperature in a flat plate due to a nonuniform initial condition, one can write

$$U(r, t) = \int_{r'=0}^{b} F^*(r')G_{X13}(r, t|r', 0)dr' \qquad (9.35)$$

Replacing for $F^*(r') = r' F(r')$, and transforming $U(r, t)$ back into $T(r, t)$, gives

$$T(r, t) = \frac{1}{r} U(r, t) = \int_{r'=0}^{b} \frac{r'}{r} F(r')G_{X13}(r, t|r', 0)dr' \qquad (9.36)$$

Equation (9.36) can be rearranged to give

$$T(r, t) = \int_{r'=0}^{b} \left[\frac{1}{4\pi rr'} G_{X13}(r, t|r', 0) \right] F(r')4\pi r'^2 \, dr' \qquad (9.37)$$

$T(r, t)$ can also be obtained by solving the initial value problem [Eqs. (9.24)–(9.26)] with the GF equation for radial spherical heat flow [Eq. (9.2)] to give

$$T(r, t) = \int_{r'=0}^{b} G_{RS03}(r, t|r', 0)F(r')4\pi r'^2 \, dr' \qquad (9.38)$$

Now, by comparing Eqs. (9.37) and (9.38), which both represent the same solution, one can conclude that the term in the brackets in Eq. (9.37) must be the $RS03$-GF evaluated at $\tau = 0$; that is,

$$G_{RS03}(r, t|r', 0) = \frac{1}{4\pi rr'} G_{X13}(r, t|r', 0) \qquad (9.39)$$

Finally by substitution of $G_{X13}(r, t|r', 0)$ from Eq. (9.33) and by replacement of $(t - 0)$ by $(t - \tau)$, one can write

$$G_{RS03}(r, t|r', \tau) = \frac{1}{2\pi brr'} \sum_{m=1}^{\infty} \exp\left[-\beta_m^2 \frac{\alpha(t - \tau)}{b^2} \right]$$

$$\times \frac{(\beta_m^2 + B^2) \sin (\beta_m \, r/b) \sin (\beta_m \, r'/b)}{\beta_m^2 + B^2 + B} \qquad (9.40)$$

where

$$\beta_m \cot \beta_m = -B \qquad B = Hb - 1 \qquad (9.41a, b)$$

This GF is also listed in Appendix RS.

Example 9.2: Derivation of G_{RS33}—hollow sphere with convective boundary conditions. Consider the following homogeneous initial-value problem for a hollow sphere as shown in Fig. 9.4a:

$$\frac{1}{r} \frac{\partial^2(rT)}{\partial r^2} = \frac{1}{\alpha} \frac{\partial T}{\partial t} \qquad a \le r \le b \qquad t > 0 \qquad (9.42)$$

$$\frac{\partial T(a, t)}{\partial r} - H_1 T(a, t) = 0 \qquad t > 0 \qquad (9.43a)$$

$$\frac{\partial T(b, t)}{\partial r} + H_2 T(b, t) = 0 \qquad t > 0 \qquad (9.43b)$$

where

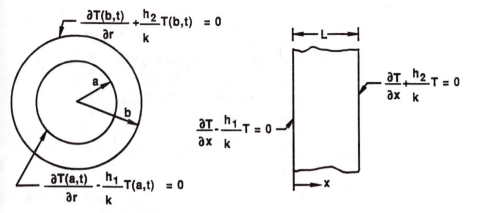

Figure 9.4 (*a*) Hollow sphere with convection boundary conditions (*RS*33). (*b*) Flat plate with convection boundary conditions (*X*33).

$$H_1 = \frac{h_1}{k} \qquad H_2 = \frac{h_2}{k} \tag{9.43c}$$

$$T(r, 0) = F(r) \qquad a \le r \le b \tag{9.44}$$

The solution procedure is the same as that used for Example 9.1. By introducing the new dependent variable $U(r, t) = r\,T(r, t)$, we get a similar differential equation, in terms of U, as that in the previous example. However, the boundary conditions are different from the previous case. For this case the transformation of the variables yields

$$\frac{\partial^2 U}{\partial r^2} = \frac{1}{\alpha}\frac{\partial U}{\partial t} \qquad a \le r \le b \qquad t > 0 \tag{9.45}$$

$$\frac{\partial U(a, t)}{\partial r} - H_1^* U(a, t) = 0 \qquad t > 0 \tag{9.46a}$$

$$\frac{\partial U(b, t)}{\partial r} + H_2^* U(b, t) = 0 \qquad t > 0 \tag{9.46b}$$

where

$$H_1^* = H_1 + \frac{1}{a} \qquad H_2^* = H_2 - \frac{1}{b} \tag{9.46c}$$

$$U(r, 0) = r\,F(r) = F^*(r) \tag{9.47}$$

The transformed equations (9.45)–(9.47) represent a flat plate problem with an arbitrary initial condition and the homogeneous convective boundary conditions on both sides (*X*33) (see Fig. 9.4*b*). From Appendix X, Eq. (*X*33.2), the GF for this case, *X*33 geometry, for $\tau = 0$ is given by

$$G_{X33}(x, t|x', 0) = \frac{2}{L} \sum_{m=1}^{\infty} \exp\left(-\beta_m \frac{\alpha t}{L^2}\right)$$

$$\times \frac{[\beta_m \cos(\beta_m x/L) + B_1 \sin(\beta_m x/L)]}{\times [\beta_m \cos(\beta_m x'/L) + B_1 \sin(\beta_m x'/L)]}{(\beta_m^2 + B_1^2)[1 + B_2/(\beta_m^2 + B_2^2)] + B_1} \quad (9.48)$$

where the β_m values are the positive eigenvalues of

$$\tan \beta_m = \frac{\beta_m(B_1 + B_2)}{\beta_m^2 - B_1 B_2} \qquad B_1 = H_1 L = \frac{h_1 L}{k}$$

$$B_2 = H_2 L = \frac{h_2 L}{k} \qquad (9.49a, b, c)$$

and L represents the plate thickness. Again in a similar manner to that of the previous example, one can show that

$$T(r, t) = \frac{1}{r} U(r, t) = \int_{r'=a}^{b} \left[\frac{1}{4\pi rr'} G_{X33}(r, t|r', 0)\right] F(r')4\pi r'^2 \, dr' \quad (9.50)$$

which yields

$$G_{RS33}(r, t|r', 0) = \frac{1}{4\pi rr'} G_{X33}(r, t|r', 0) \quad (9.51)$$

However, it should be noted that, for this case, the transformation of the Cartesian variables (x, x', L, etc.) to the spherical variables (r, r', b, a, etc.) is not the same as that for the previous example. From Figs. 9.4a and b, one can see that for this case,

$$L \to (b - a) \qquad x \to (r - a) \qquad x' \to (r' - a) \qquad (9.52a, b, c)$$

$$B_1 \to H_1^*(b - a) = \left(H_1 + \frac{1}{a}\right)(b - a) \qquad (9.53a)$$

$$B_2 \to H_2^*(b - a) = \left(H_2 - \frac{1}{b}\right)(b - a) \qquad (9.53b)$$

Finally by substituting for L, x, x', B_1, and B_2 from Eqs. (9.52a, b, c) and (9.53a, b) into Eq. (9.51), and replacing $(t - 0)$ by $(t - \tau)$, one can write

$$G_{RS33}(r, t|r', \tau) = \frac{1}{2\pi rr'(b - a)} \sum_{m=1}^{\infty} \exp\left[-\beta_m^2 \frac{\alpha(t - \tau)}{(b - a)^2}\right]$$
$$\times \frac{\{\beta_m \cos[\beta_m(r - a)/(b - a)] + B_1 \sin[\beta_m(r - a)/(b - a)]\}}{\times \{\beta_m \cos[\beta_m(r' - a)/(b - a)] + B_1 \sin[\beta_m(r' - a)/(b - a)]\}}{(\beta_m^2 + B_1^2)[1 + B_2/(\beta_m^2 + B_2^2)] + B_1} \quad (9.54)$$

where

$$B_1 = \left(\frac{a\,h_1}{k} + 1\right)\left(\frac{b}{a} - 1\right) \tag{9.55a}$$

$$B_2 = \left(\frac{b\,h_2}{k} - 1\right)\left(1 - \frac{a}{b}\right) \tag{9.55b}$$

with

$$\tan \beta_m = \frac{\beta_m(B_1 + B_2)}{\beta_m^2 - B_1 B_2} \tag{9.55c}$$

This GF is also listed in Appendix RS.

The procedure demonstrated in the two previous examples can also be used to obtain G_{RSIJ} from G_{XIJ} for other types of boundary conditions ($I, J = 0, 1, 2, 3$). Table 9.1 gives a summary of how *RSIJ* GFs are obtained from *XIJ* GFs for different values of $I, J = 0, 1, 2, 3$ with the appropriate variable transformations.

9.5 TEMPERATURES IN SOLID SPHERES

In this section, we demonstrate the application of the GF solution method to the solid sphere problems with radial flow of heat numbered by *RS0J* where $J = 1, 2,$ and 3. Three groups of problems are considered: those with a nonzero initial temperature distributions $F(r)$; those with nonhomogeneous boundary conditions; and those containing an energy generation term $g(r, t)$.

The describing partial differential equation for these groups of problems is

$$\frac{1}{r}\frac{\partial^2(rT)}{\partial r^2} + \frac{1}{k}g(r, t) = \frac{1}{\alpha}\frac{\partial T}{\partial t} \qquad 0 \le r \le b \qquad t > 0 \tag{9.56}$$

where $g(r, t)$ represents an energy generation term that makes this equation nonhomogeneous. The boundary conditions at the center of the sphere ($r = 0$) are homogeneous and are given by

$$\frac{\partial T(0, t)}{\partial r} = 0 \qquad \text{or} \qquad T(0, t) \ne \infty \qquad t > 0 \tag{9.57}$$

The condition at the surface of the sphere ($r = b$), can be of the first, second, or third kinds depending on the values of J, that is,

for $J = 1$:
$$T(b, t) = T_b(t) \qquad t > 0 \tag{9.58a}$$

for $J = 2$:
$$k\frac{\partial T(b, t)}{\partial r} = q_b(t) \qquad t > 0 \tag{9.58b}$$

and for $J = 3$:
$$k\frac{\partial T(b, t)}{\partial r} + hT(b, t) = hT_\infty(t) \qquad t > 0 \tag{9.58c}$$

The initial condition is considered to be an arbitrary function of time, given by

$$T(r, 0) = F(r) \qquad 0 \le r \le b \tag{9.59}$$

Table 9.1 Conversion components for derivation of the *RSIJ* Green's function from the *XI'J'* Green's function for $I, J = 0, 1, 2, 3$

$$G_{RSIJ}(r, t|r', \tau) = \frac{1}{4\pi rr'} G_{XI'J'}(x, t|x', \tau)$$

No.	IJ	I'J'	L	x	x'	H	B	B_1	B_2	Characteristic equation
1	00	10	–	r	r'	–	–	–	–	–
2	01	11	b	r	r'	–	–	–	–	–
3†	02	13	b	r	r'	$-\dfrac{1}{b}$	-1	–	–	$\beta_m \cot \beta_m = 1$
4	03	13	b	r	r'	$\dfrac{h}{k} - \dfrac{1}{b}$	$\dfrac{hb}{k} - 1$	–	–	$\beta_m \cot \beta_m = -B$
5	10	10	–	r − a	r' − a	–	–	–	–	–
6	11	11	b − a	r − a	r' − a	–	–	–	–	–
7	12	13	b − a	r − a	r' − a	$-\dfrac{1}{b}$	$\dfrac{a}{b} - 1$	–	–	$\beta_m \cot \beta_m = -B$
8	13	13	b − a	r − a	r' − a	$\dfrac{h}{k} - \dfrac{1}{b}$	$\left(\dfrac{hb}{k} - 1\right)\left(1 - \dfrac{a}{b}\right)$	–	–	$\beta_m \cot \beta_m = -B$
9	20	30	–	r	r' − a	$\dfrac{1}{a}$	–	–	–	–
10	21	31	b − a	r − a	r' − a	$\dfrac{1}{a}$	$\dfrac{b}{a} - 1$	–	–	$\beta_m \cot \beta_m = -B$

(Continued)

Table 9.1 Conversion components for derivation of the *RSIJ* Green's function from the *XI′J′* Green's function for $I, J = 0, 1, 2, 3$ $G_{RSIJ}(r, t|r', \tau) = \dfrac{1}{4\pi r r'} G_{XI'J'}(x, t|x', \tau)$ *(Continued)*

No.	IJ	I′J′	L	x	x′	H	B	B_1	B_2	Characteristic equation
11‡	22	33	$b - a$	$r - a$	$r' - a$	–	–	$\dfrac{b}{a} - 1$	$\dfrac{a}{b} - 1$	$\tan \beta_m = \dfrac{\beta_m(B_1 + B_2)}{\beta_m^2 - B_1 B_2}$
12	23	33	$b - a$	$r - a$	$r' - a$	–	–	$\dfrac{b}{a} - 1$	$\left(\dfrac{bh}{k} - 1\right)\left(1 - \dfrac{a}{b}\right)$	$\tan \beta_m = \dfrac{\beta_m(B_1 + B_2)}{\beta_m^2 - B_1 B_2}$
13	30	30	–	$r - a$	$r' - a$	$\dfrac{h}{k} + \dfrac{1}{a}$	–	–	–	–
14	31	31	$b - a$	$r - a$	$r' - a$	$\dfrac{h}{k} + \dfrac{1}{a}$	$\left(\dfrac{ha}{k} + 1\right)\left(\dfrac{b}{a} - 1\right)$	–	–	$\beta_m \cot \beta_m = -B$
15	32	33	$b - a$	$r - a$	$r' - a$	$\dfrac{h}{k} + \dfrac{1}{a}$	–	$\left(\dfrac{ha}{k} + 1\right)\left(\dfrac{b}{a} - 1\right)$	$\dfrac{a}{b} - 1$	$\tan \beta_m = \dfrac{\beta_m(B_1 + B_2)}{\beta_m^2 - B_1 B_2}$
16	33	33	$b - a$	$r - a$	$r' - a$	–	–	$\left(\dfrac{ah_1}{k} + 1\right)\left(\dfrac{b}{a} - 1\right)$	$\left(\dfrac{bh_2}{k} - 1\right)\left(1 - \dfrac{a}{b}\right)$	$\tan \beta_m = \dfrac{\beta_m(B_1 + B_2)}{\beta_m^2 - B_1 B_2}$

†For this case, the characteristic equation has a zero root; consequently, a term $3/(4b^3)$ has to be added to the value of G_{RS02} obtained from G_{X13}. See Example 9.4.

‡A term $3/[4(b^3 - a^3)]$ has to be added to the value of G_{RS22} obtained from G_{X33}.

From the GFSE for radial flow of heat in spheres, Eq. (9.2), the temperature solution is

$$T(r, t) = \int_{r'=0}^{b} F(r') \, G_{RS0J}(r, t|r', 0) 4\pi r'^2 \, dr'$$

$$+ \frac{\alpha}{k} \int_{\tau=0}^{t} \int_{r'=0}^{b} G_{RS0J}(r, t|r', \tau) \, g(r', t) \, 4\pi r'^2 \, dr' \, d\tau$$

$$- \alpha \int_{\tau=0}^{t} T_b(\tau) \frac{\partial G_{RS01}(r, t|b, \tau)}{\partial r} 4\pi b^2 \, d\tau \quad \text{(for } J = 1 \text{ only)}$$

$$+ \frac{\alpha}{k} \int_{\tau=0}^{t} q_b(\tau) G_{RS02}(r, t|b, \tau) \, 4\pi b^2 \, d\tau \quad \text{(for } J = 2 \text{ only)}$$

$$+ \frac{\alpha}{k} \int_{\tau=0}^{t} hT_\infty(\tau) G_{RS03}(r, t|b, \tau) \, 4\pi b^2 \, d\tau \quad \text{(for } J = 3 \text{ only)} \quad (9.60)$$

Note that since the boundary condition at the center of the sphere is homogeneous, the last two integrals, associated with boundary conditions in Eq. (9.3) are evaluated only at the surface boundary with r_i and r_j equal to b. The $RS0J$- GFs for the cases of $J = 1, 2, 3$ are given in Appendix RS or can be obtained from XIJ GFs through the appropriate transformations provided in Table 9.1. Some example problems are discussed next.

Example 9.3: Solid sphere with arbitrary initial temperature—$RS01B0T$-case. A solid sphere, $0 \le r \le b$, has a known initial temperature distribution $F(r)$. The surface temperature is kept at $T = 0$. See Fig. 9.5. Find the transient temperature distribution in the sphere.

SOLUTION The solution can be obtained from Eq. (9.60) by considering only the first integral on the right-hand side which is due to the nonzero initial temperature distribution. The second through the last integrals vanish since there is no volume energy generation in the above problem, $g(r, t) = 0$, and the boundary conditions are homogeneous. The required GF for this case is $G_{RS01}(r, t|r', 0)$ which is

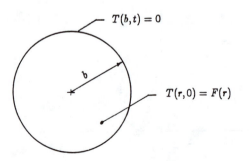

$T(b,t) = 0$

$T(r,0) = F(r)$

Figure 9.5 Solid sphere with temperature boundary condition.

equivalent to G_{X11} in Cartesian coordinates. Following the procedure explained in Section 9.4, from Table 9.1, with $x \to r$, $x' \to r'$, and $L \to b$, one can write

$$G_{RS01}(r, t|r', 0) = \frac{1}{4\pi rr'} G_{X11}(r, t|r', 0)$$

$$= \frac{1}{2\pi brr'} \sum_{m=1}^{\infty} \exp\left[-\frac{m^2\pi^2\alpha t}{b^2} \right]$$

$$\times \sin\left(m\pi \frac{r}{b} \right) \sin\left(m\pi \frac{r'}{b} \right) \qquad (9.61)$$

Note that the expression for $X11$ GF used in the above equation is suitable for large times. Substituting for G_{RS01} from Eq. (9.61) into (9.60) yields

$$T(r, t) = \int_{r'=0}^{b} F(r') \left[\frac{1}{2\pi brr'} \sum_{m=1}^{\infty} \right.$$

$$\times \exp\left(-\frac{m^2\pi^2\alpha t}{b^2} \right) \sin\left(m\pi \frac{r}{b} \right) \left. \sin\left(m\pi \frac{r'}{b} \right) \right] 4\pi r'^2 \, dr' \qquad (9.62)$$

which can be simplified and rearranged to give

$$T(r, t) = \frac{2}{br} \sum_{m=1}^{\infty} \exp\left(-\frac{m^2\pi^2\alpha t}{b^2} \right)$$

$$\times \sin\left(m\pi \frac{r}{b} \right) \int_{r'=0}^{b} F(r') \sin\left(m\pi \frac{r'}{b} \right) r' \, dr' \qquad (9.63)$$

For the special case ($R01B0T1$) where there is a uniform initial temperature distribution, $T(r, 0) = F(r) = T_0$, the integral in solution (9.63) can easily be evaluated as

$$\int_{r'=0}^{b} F(r') \sin\left(m\pi \frac{r'}{b} \right) r' \, dr' = T_0 \int_{r'=0}^{b} \sin\left(m\pi \frac{r'}{b} \right) r' \, dr'$$

$$= -\frac{T_0 b^2}{m\pi} \cos(m\pi) = -\frac{T_0 b^2}{m\pi} (-1)^m \qquad (9.64)$$

since $\int x \sin x \, dx = \sin x - x \cos x$ and $\cos(m\pi) = (-1)^m$. Then the solution becomes

$$T(r, t) = 2T_0 \sum_{m=1}^{\infty} (-1)^{m+1} \exp\left[-m^2\pi^2\alpha t/b^2 \right] \frac{\sin(m\pi r/b)}{(m\pi r/b)} \qquad (9.65)$$

This solution represents a fast convergent series for large times ($\alpha t/b^2$) since the exponential term rapidly decreases as m increases. However, for small times, it takes a large number of terms for convergence. For small values of $\alpha t/b^2$, it is more computationally efficient to use the small-time GF. For small times, the $RS01$ GF can be obtained from the small-time expression for $X11$ GF given by Eq. ($X11.1$) in Appendix X. That is,

$$G_{RS01}^S(r, t|r', 0) = (4\pi r r')^{-1}(4\pi\alpha t)^{-(1/2)} \sum_{n=-\infty}^{+\infty}$$

$$\times \left\{ \exp\left[-\frac{(2nb + r - r')^2}{4\alpha t} \right] - \exp\left[-\frac{(2nb + r + r')^2}{4\alpha t} \right] \right\} \tag{9.66}$$

In a similar manner to that used for the large-time solution, the small-time solution, for $F(r) = T_0$, is obtained by substituting Eq. (9.66) into Eq. (9.60) and integrating from $r' = 0$ to $r' = b$ to give

$$T(r, t) = T_0 - \frac{bT_0}{r} \sum_{n=-\infty}^{\infty} \left\{ \text{erfc}\left[\frac{(2n + 1)b - r}{(4\alpha t)^{1/2}} \right] \right.$$

$$\left. - \text{erfc}\left[\frac{(2n + 1)b + r}{(4\alpha t)^{1/2}} \right] \right\} \tag{9.67}$$

Note that since the complementary error function erfc($\cdot$), decreases rapidly with an increase in its argument, for small times (such as $\alpha t/b^2 < 0.4$), the major contribution to the temperature in the above solution is due to the first two terms of the series ($n = 0, 1$). For smaller times, say $\alpha t/b^2 < 0.1$, even one term in the series is sufficient to give accurate results.

The large- and small-time solutions given by Eqs. (9.65) and (9.67) are applicable for $r > 0$. For the temperature at the center of the sphere, these solutions approach the following expressions as $r \to 0$ at the limit:

$$T(0, t) = 2T_0 \sum_{m=1}^{\infty} (-1)^{m+1} \exp\left[-\frac{m^2\pi^2\alpha t}{b^2} \right] \quad \text{(for large times)} \tag{9.68}$$

$$T(0, t) = T_0 - \frac{bT_0}{(\pi\alpha t)^{1/2}} \sum_{n=-\infty}^{\infty}$$

$$\times \exp\left[-\frac{(2n + 1)^2 b^2}{4\alpha t} \right] \quad \text{(for small times)} \tag{9.69}$$

In the above problem, the boundary condition at the surface $r = b$ was considered to be homogeneous, and consequently, in the derivation of the solution, we did not have to consider the contributions of the last three integrals in the GF Eq. (9.60).

If the boundary condition at $r = b$ is nonhomogeneous but constant at T_b, the problem can be cast as one with homogeneous boundary condition, by defining a new temperature variable $T - T_b$. Therefore, the solutions given by Eqs. (9.65) and (9.67)–(9.69) can still be used by replacing T and T_0 by $T - T_b$ and $T_0 - T_b$, respectively, in these solutions. For the case where the boundary condition at $r = b$ is not constant, the corresponding integral in Eq. (9.60) must be included in the solution. This is best illustrated in the following example.

Example 9.4: Solid sphere heated at surface—RS02B-T- case. A solid sphere, $0 \le r \le b$, has a known initial temperature distribution $F(r)$. The surface of the sphere is heated uniformly by a known heat flux as a function of time, $q_b(t)$. See Fig. 9.6. Find the temperature distribution in the sphere for large times.

SOLUTION The partial differential equation, the initial condition, and the boundary condition at the center of the sphere $r = 0$ for this case are the same as those for Example 9.3. The boundary condition at the surface ($r = b$) is of second kind ($J = 2$) and nonhomogeneous, given by Eq. (9.58b) as

$$k \frac{\partial T(b, t)}{\partial r} = q_b(t) \tag{9.70}$$

From the GF Eq. (9.60), the solution is

$$T(r, t) = \int_{r'=0}^{b} G_{RS02}(r, t|r', 0) F(r') 4\pi r'^2 \, dr'$$

$$+ \frac{\alpha}{k} \int_0^t q_b(t) G_{RS02}(r, t|b, \tau) 4\pi b^2 \, d\tau \tag{9.71}$$

Note that since there is no energy generation in the sphere, the second integral in Eq. (9.60) is not included in the solution. The required GF for this case, $G_{RS02}(r, t|r', \tau)$, can be obtained from the $X13$ GF through the appropriate transformation of the variables given in Table 9.1. It also can be obtained from the $RS03$ GF given by Eq. (9.40) by setting $H = 0$. However, note that when $H = 0$, the corresponding eigenfunction, Eq. (9.41), has a zero root, and consequently, a term $3/(4\pi b^3)$ has to be added to the value of G_{RS03} given by Eq. (9.40) with $H = 0$. Therefore, one can write

$$G_{RS02}(r, t|r', \tau) = \frac{1}{2\pi brr'} \sum_{m=1}^{\infty} \exp\left[-\frac{\beta_m^2 \alpha(t - \tau)}{b^2} \right]$$

$$\times \frac{(\beta_m^2 + 1) \sin\left(\beta_m \frac{r}{b} \right) \sin\left(\beta_m \frac{r'}{b} \right)}{\beta_m^2} + \frac{3}{4\pi b^3} \tag{9.72}$$

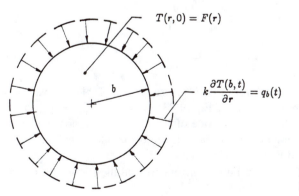

Figure 9.6 Solid sphere with heat flux boundary condition.

where β_m are the roots of the eigenfunction

$$\beta_m \cot \beta_m = 1 \tag{9.73}$$

The first five values of β_m are 4.4934, 7.7253, 10.9041, 14.0662, and 17.2208, respectively.

Note that the $RS02$ GF given here (also listed in Appendix RS) is valid for any time but is best for "large" values of $\alpha(t - \tau)/b^2$. Substituting Eq. (9.72) into Eq. (9.71) yields

$$T(r, t) = \frac{3}{b^3} \int_{r'=0}^{b} r'^2 F(r')dr' + \frac{2}{br} \sum_{m=1}^{\infty}$$

$$\times \exp\left(-\frac{\beta_m^2 \alpha t}{b^2}\right) \frac{\beta_m^2 + 1}{\beta_m^2}$$

$$\times \sin\left(\beta_m \frac{r}{b}\right) \int_{r'=0}^{b} r' F(r')$$

$$\times \sin\left(\beta_m \frac{r'}{b}\right) dr' + \frac{3\alpha}{bk} \int_0^t q_b(\tau)d\tau + \frac{2\alpha}{rk} \sum_{m=1}^{\infty}$$

$$\times \frac{\beta_m^2 + 1}{\beta_m^2} \sin(\beta_m) \sin\left(\beta_m \frac{r}{b}\right) \int_0^t q_b(\tau)$$

$$\times \exp\left[-\frac{\beta_m^2 \alpha(t - \tau)}{b^2}\right] d\tau \tag{9.74}$$

Some special cases are considered next.

1. *Case RS02B1T1.* The initial temperature and the surface heat flux are constant and given by

$$T(r, 0) = F(r) = T_0 \qquad q_b(t) = q_0 \tag{9.75a, b}$$

For this case, the integrals for Eq. (9.74) are given in Tables 5.7 and 5.8 and the solution becomes

$$T(r, t) = T_0 + \frac{3\alpha q_0 t}{bk} + \frac{q_0(5r^2 - 3b^2)}{10kb}$$

$$- \frac{2q_0 b^2}{kr} \sum_{m=1}^{\infty} \frac{\sin(\beta_m r/b)}{\beta_m^2 \sin \beta_m} \exp\left(-\frac{\beta_m^2 \alpha t}{b^2}\right) \tag{9.76}$$

Note that the second integral in Eq. (9.74) (case no. 3 in Table 5.7) vanishes for this case since from the characteristic Eq. (9.73), we have $\sin \beta_m - \beta_m \cos \beta_m = 0$. The third term on the right-hand side of Eq. (9.76) is the nonseries form of the quasi-steady temperature for this case. It can be found by applying the alternative solution method presented in Example 6.5.

2. *Case RS02B2T0.* Zero initial temperature and the surface heat flux is a linear function of time, that is,

$$T(r, 0) = F(r) = 0 \qquad q_b(t) = q_0 t \qquad (9.77)$$

since the initial condition is zero, there is no contribution to the solution due to the initial condition and consequently the first two terms in solution (9.74) vanish. The time integral in the last term can be evaluated using integral number 3 from Table 2.8. Then the solution becomes;

$$T(r, t) = \frac{3\alpha q_0}{2bk} t^2 + \frac{2q_0 b^4}{kr} \sum_{m=1}^{\infty} \frac{\sin (\beta_m r/b)}{\beta_m^4 \sin \beta_m}$$

$$\times \left[\exp \left(- \frac{\beta_m^2 \alpha t}{b^2} \right) + \beta_m \frac{\alpha t}{b^2} - 1 \right] \qquad (9.78)$$

Note that the solutions presented in this example are most efficient for large values of $\alpha t/b^2$. In the next example, time partitioning, introduced in Chapter 5, is used to find a solution that is numerically efficient for both small and large times.

Example 9.5: Solid sphere with convective boundary condition—RS03B1T0 case. A solid sphere, $0 \leq r \leq b$, initially at zero temperature, is suddenly immersed in a fluid at a constant temperature T_∞. The heat transfer coefficient for this process is h, a constant. See Fig. 9.7. Using time partitioning, find expressions for temperature distribution, $T(r, t)$, that are numerically efficient for all values of time.

SOLUTION The temperature due to the convection boundary condition at the surface of the sphere ($r = b$) is given by the last term of Eq. (9.60),

$$T(r, t) = \frac{\alpha}{k} \int_{\tau=0}^{t-\Delta t} hT_\infty G_{RS03}^L (r, t|b, \tau) 4\pi b^2 \, d\tau$$

$$+ \frac{\alpha}{k} \int_{\tau=t-\Delta t}^{t} hT_\infty G_{RS03}^S (r, t|b, \tau) 4\pi b^2 \, d\tau \qquad (9.79)$$

where the integral on τ has been partitioned at $\tau = t - \Delta t$. Note that the first integral in Eq. (9.79) involves the large-time GF, while the second integral contains the small-time GF. The solutions to these integrals are denoted $T^L(r, t, \Delta t)$ and $T^S(r, \Delta t)$, respectively. For small times, the temperature solution $T(r, t)$ can be

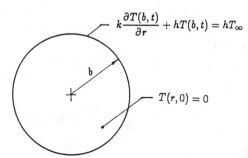

$$-k\frac{\partial T(b, t)}{\partial r} + hT(b, t) = hT_\infty$$

$$T(r, 0) = 0$$

Figure 9.7 Solid sphere with convection boundary condition.

obtained directly from the T^S solution. For large times, however, the expression for the temperature requires both the T^S and the T^L solutions.

First consider T^S, the second integral in Eq. (9.79). The small-time GF, G_{RS03}^S, needed in this integral, can be obtained from G_{X13}^S given by Eq. (X13.1) in Appendix X. Following the procedure explained in Section 9.4 with the appropriate transformation of the variables from Table 9.1, one can get

$$
\begin{aligned}
G_{RS03}^S(r,t|r',\tau) = {}& [4\pi rr']^{-1}[4\pi\alpha(t-\tau)]^{1/2}\left\{\exp\left[-\frac{(r-r')^2}{4\alpha(t-\tau)}\right]\right. \\
& - \exp\left[-\frac{(r+r')^2}{4\alpha(t-\tau)}\right] + \exp\left[-\frac{(2b-r-r')^2}{4\alpha(t-\tau)}\right]\right\} \\
& - (4\pi rr')^{-1}H^*\exp\left[H^*(2b-r-r')\right. \\
& + H^{*2}\alpha(t-\tau)\left]\operatorname{erfc}\left\{\frac{(2b-r-r')}{[4\alpha(t-\tau)]^{1/2}}\right.\right. \\
& + H^*[\alpha(t-\tau)]^{1/2}\Bigg\}
\end{aligned}
$$

(9.80)

where

$$
H^* = H - \frac{1}{b} \qquad H = \frac{h}{k}
$$

(9.81a, b)

Evaluating G_{RS03}^S at $r' = b$ and substituting the result into the second integral of Eq. (9.79) with $t - \tau$ replaced by τ yields

$$
\begin{aligned}
T^S(r,\Delta t) = {}& \frac{\alpha bhT_\infty}{kr}\int_{\tau=0}^{\Delta t}\left((4\pi\alpha\tau)^{-1/2}\right. \\
& \times \left\{2\exp\left[-\frac{(b-r)^2}{4\alpha\tau}\right] - \exp\left[-\frac{(r+b)^2}{4\alpha\tau}\right]\right\} \\
& - H^*\exp\left[H^*(b-r) + H^{*2}\alpha\tau\right] \\
& \times \operatorname{erfc}\left[\frac{(b-r)}{(4\alpha\tau)^{1/2}} + H^*(\alpha\tau)^{1/2}\right]\right)d\tau
\end{aligned}
$$

(9.82)

The above integral can be evaluated using the Laplace transform method. This method has already been demonstrated in Examples 4.1 and 4.2, where similar integrals have been solved for the $X10$ and $X30$ geometries, respectively. Then the small-time solution becomes

$$
\begin{aligned}
T^S(r,\Delta t) = {}& \frac{T_\infty bh}{kr}\left\{\frac{1}{H^*}\operatorname{erfc}\left[\frac{(b-r)}{2(\alpha\Delta t)^{1/2}}\right] - \frac{1}{H^*}\right. \\
& \times \exp\left[(b-r)H^* + \alpha\Delta tH^{*2}\right]\operatorname{erfc}\left[\frac{(b-r)}{2(\alpha\Delta t)^{1/2}} + H^*(\alpha\Delta t)^{1/2}\right] \\
& - (\alpha\Delta t)^{1/2}\operatorname{ierfc}\left[\frac{r+b}{2(\alpha\Delta t)^{1/2}}\right]\right\}
\end{aligned}
$$

(9.83a)

where

$$H* = \frac{h}{k} - \frac{1}{b} \tag{9.83b}$$

Note that this expression is good for r away from the center where the temperature remains unchanged for small values of $\alpha t / b^2$.

Next consider the $T^L(x, t, \Delta t)$ solution, given by the first integral of Eq. (9.79). The required large-time GF is found in Appendix RS, Eq. (RS03.1). Accordingly, T^L is given by

$$T^L(r, t, \Delta t) = \frac{\alpha h T_\infty}{k} \int_{\tau=0}^{t-\Delta t} d\tau \, \frac{2}{r} \sum_{m=1}^{\infty}$$

$$\times e^{-\beta_m^2 \alpha (t-\tau)/b^2} \frac{(\beta_m^2 + B^2) \sin (\beta_m \, r/b) \sin \beta_m}{\beta_m^2 + B^2 + B} \tag{9.84a}$$

where

$$\beta_m \cot \beta_m = -B \quad \text{and} \quad B = \frac{hb}{k} - 1 \tag{9.84b}$$

The integral on τ involves only the exponential term and can easily be evaluated to give (only a few terms are needed for $\alpha \Delta t / b^2 = 0.022$)

$$T^L(r, t, \Delta t) = \frac{2h T_\infty b^2}{kr} \sum_{m=1}^{\infty} \frac{1}{\beta_m^2} (e^{-\beta_m^2 \alpha \Delta t / b^2} - e^{-\beta_m^2 \alpha t / b^2})$$

$$\times \frac{(\beta_m^2 + B^2) \sin (\beta_m \, r/b) \sin \beta_m}{\beta_m^2 + B^2 + B} \tag{9.85}$$

Then, from Eq. (9.79), the temperature for large time is obtained by combining the small-time solution T^S and the large-time solution T^L to give

$$T(r, t) = \frac{T_\infty bh}{kr} \left\{ \frac{1}{H*} \text{erfc} \left[\frac{(b - r)}{2(\alpha \Delta t)^{1/2}} \right] - \frac{1}{H*} \right.$$

$$\times \exp \left[(b - r)H* + \alpha \Delta t H*^2 \right] \text{erfc} \left[\frac{(b - r)}{2(\alpha \Delta t)^{1/2}} + H*(\alpha \Delta t)^{1/2} \right]$$

$$- (\alpha \Delta t)^{1/2} \text{ierfc} \left[\frac{r + b}{2(\alpha \Delta t)^{1/2}} \right] + 2b \sum_{m=1}^{\infty} \frac{1}{\beta_m^2} (e^{-\beta_m^2 \alpha \Delta t / b^2}$$

$$- e^{-\beta_m^2 \alpha t / b^2}) \frac{(\beta_m^2 + B^2) \sin (\beta_m \, r/b) \sin \beta_m}{\beta_m^2 + B^2 + B} \right\} \tag{9.86}$$

for $t > \Delta t$. Note that for small times, $t < \Delta t$, the temperature is obtained simply by replacing Δt by t in the expression for $T^S(r, \Delta t)$, given by Eq. (9.83).

The third group of problems considered in this section are those with an energy generation term. The following example is given to illustrate this case.

Example 9.6: Solid sphere with internal energy generation and insulated surface—*RS02B0T0G*- case. A solid sphere, $0 \leq r \leq b$, is initially at zero temperature. For times $t > 0$, heat is produced within the sphere at the rate $g(r, t)$ per unit time per unit volume while the surface boundary is kept insulated. See Fig. 9.8. Find the temperature distribution within the sphere, $T(r, t)$ for large times.

SOLUTION The temperature due to the heat generation within the sphere is given by the second integral on the right-hand side of Eq. (9.60),

$$T(r, t) = \frac{\alpha}{k} \int_{\tau=0}^{t} \int_{r'=0}^{b} G_{RS02}(r, t|r', \tau) \, g(r', \tau) \, 4\pi r'^2 \, dr' \, d\tau \qquad (9.87)$$

Note that since the initial temperature is zero and the boundary conditions are homogeneous, the other integrals in Eq. (9.60) have no contribution in $T(r, t)$. The large-time form of G_{RS02} is found in Appendix RS and is also given by Eq. (9.72). Substituting G_{RS02} from Eq. (9.72) into Eq. (9.87) yields

$$T(r, t) = \frac{\alpha}{k} \int_{\tau=0}^{t} \int_{r'=0}^{b} \left\{ \frac{1}{2\pi brr'} \sum_{m=1}^{\infty} \exp \left[-\frac{\beta_m^2 \, \alpha(t - \tau)}{b^2} \right] \right.$$

$$\times \frac{(\beta_m^2 + 1) \sin (\beta_m \, r/b) \sin (\beta_m \, r'/b)}{\beta_m^2}$$

$$\left. + \frac{3}{4\pi b^3} \right\} g(r', \tau) \, 4\pi r'^2 \, dr' \, d\tau \qquad (9.88)$$

which can be simplified and rearranged into two different terms as

$$T(r, t) = T_1(r, t) + T_2(r, t) \qquad (9.89)$$

where

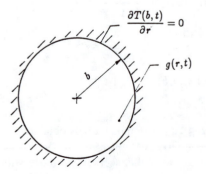

$$\frac{\partial T(b, t)}{\partial r} = 0$$

$$g(r, t)$$

$$b$$

Figure 9.8 Solid sphere with insulated surface and energy generation.

$$T_1(r, t) = \frac{3\alpha}{kb^3} \int_{\tau=0}^{t} d\tau \int_{r'=0}^{b} g(r', \tau) r'^2 \, dr' \tag{9.90a}$$

$$T_2(r, t) = \frac{2\alpha}{kbr} \sum_{m=1}^{\infty} \frac{(\beta_m^2 + 1) \sin (\beta_m r/b)}{\beta_m^2} \int_{\tau=0}^{t}$$

$$\times \exp \left[- \frac{\beta_m^2 \alpha(t - \tau)}{b^2} \right] d\tau \int_{r'=0}^{b} \sin \left(\beta_m \frac{r'}{b} \right)$$

$$\times g(r', \tau) \, r' \, dr' \tag{9.90b}$$

and the eigenvalues β_m are defined by $\beta_m \cot \beta_m - 1 = 0$.

In the above equations, usually, it is more convenient to carry out the integrals over r' first and then over τ. Tables 5.7 and 5.8 give the solutions to many suitable integrals (over space and time, respectively) which appear in the large-time GF solutions. The solution given by Eqs. (9.89) and (9.90) is now examined for some special cases.

1. *Case RS02B0T0G1*. Heat is generated within the sphere at a constant rate, $g(r, t) = g_0 = $ constant. For this case the integrals in Eq. (9.90) are easily evaluated and the solution becomes

$$T(r, t) = T_1(r, t) = \frac{\alpha g_0 t}{k} \tag{9.91}$$

Note that $T_2(r, t)$ is equal to zero in this case since the integration over r' in Eq. (9.90b) results in the term $(\sin \beta_m - \beta_m \cos \beta_m) = 0$.

2. *Case RS02B0T0Gr2*. Heat is generated within the sphere as a linear function of radius given by

$$g(r, t) = \frac{g_0(b - r)}{b} \tag{9.92}$$

The integrals in $T_1(r, t)$ are easily evaluated to give

$$T_1(r, t) = \frac{\alpha t g_0}{4k} \tag{9.93a}$$

The integrals over r' and τ in $T_2(r, t)$ solutions are evaluated using the integrals no. 4 and 2 in Tables 5.7 and 5.8, respectively, to give

$$T_2(r, t) = \frac{2g_0 b^3}{kr} \sum_{m=1}^{\infty} \frac{[2 - (2 + \beta_m^2) \cos \beta_m] (\beta_m^2 + 1) \sin (\beta_m r/b)}{\beta_m^5}$$

$$\times \left[1 - \exp \left(- \frac{\beta_m^2 \alpha t}{b^2} \right) \right] \tag{9.93b}$$

Then the solution becomes

$$T(r, t) = \frac{g_0 b^2}{4k} \left\{ \frac{\alpha t}{b^2} + \frac{8b}{r} \sum_{m=1}^{\infty} \right.$$

$$\times \frac{(\beta_m^2 + 1)[2 - (2 + \beta_m^2) \cos \beta_m] \sin (\beta_m r/b)}{\beta_m^5}$$

$$\times \left. \left[1 - \exp \left(-\frac{\beta_m^2 \alpha t}{b^2} \right) \right] \right\} \tag{9.94}$$

It is interesting to note that $T_1(r, t)$ solution is not a function r and changes linearly with t. This term represents the volume-average temperature in the sphere, defined by

$$T_{av}(t) = \left(\frac{4}{3} \pi b^3 \right)^{-1} \int_{r=0}^{b} T(r, t) 4\pi r^2 \, dr \tag{9.95a}$$

and can be verified by substituting for $T(r, t)$ from Eq. (9.89) and carrying out the integration; that is,

$$T_{av}(t) = \left(\frac{4}{3} \pi b^3 \right)^{-1} \int_{r=0}^{b} [T_1(r, t) + T_2(r, t)] \, 4\pi r^2 \, dr \tag{9.95b}$$

The integration over $T_2(r, t)$ becomes zero, since it involves the eigencondition $\beta_m \cot \beta_m - 1 = 0$. Then, since T_1 is not a function of r, one can write

$$T_{av}(t) = \left(\frac{4}{3} \pi b^3 \right)^{-1} T_1(t) \int_{r=0}^{b} 4\pi r^2 \, dr = T_1(t) \tag{9.96a}$$

or

$$T_{av}(t) = \frac{\alpha t g_0}{4k} \tag{9.96b}$$

It is also interesting to note that the $T_2(r, t)$ solution contains two terms: one is the transient term which decreases exponentially to zero over time, while the other term represents the quasi-steady temperature distribution which does not depend on time. After the transient term becomes zero, the shape of the temperature distribution remains unchanged due to the later term. Note that even though the transient term in T_2 solution dies out with time, since T_1 solution is a function of time, there is no steady-state temperature for this problem. In other words, the average temperature of the sphere increases with time since, due to the insulated surface condition, the heat that is generated has no place to go.

3. *Case—RS02B0T0Gr4.* Heat is generated within the sphere as an exponential function of radius given by

$$g(r, t) = g_0 \, e^{-\gamma r/b} \tag{9.97}$$

The $T_1(r, t)$ solution for this case is given by

$$T_1(r, t) = \frac{3\alpha t g_0}{kb^3} \int_{r'=0}^{b} e^{-\gamma r'/b} r'^2 \, dr' \tag{9.98}$$

The integral over r' in Eq. (9.98) can be evaluated by parts to give

$$T_1(r, t) = \frac{3\alpha t g_0}{k\gamma^3} [2 - e^{-\gamma} (\gamma^2 + 2\gamma + 2)] \tag{9.99}$$

After integrating over time, the $T_2(r, t)$ solution becomes

$$T_2(r, t) = \frac{2g_0 b^3}{kr} \sum_{m=1}^{\infty} \frac{(\beta_m^2 + 1) \sin (\beta_m r/b)}{\beta_m^4}$$

$$\times \left[1 - \exp \left(-\frac{\beta_m^2 \alpha t}{b^2} \right) \right] \int_0^1 \sin \left(\beta_m \frac{r'}{b} \right)$$

$$\times e^{-\gamma r'/b} \left(\frac{r'}{b} \right) d \left(\frac{r'}{b} \right) \tag{9.100}$$

The integral over r' in Eq. (9.100) can be evaluated by using the relation

$$\int x \, e^{Ax} \sin (Bx) \, dx = \frac{x \, e^{Ax}}{A^2 + B^2}$$

$$\times (A \sin Bx - B \cos Bx) - \frac{e^{Ax}}{(A^2 + B^2)^2}$$

$$\times [(A^2 - B^2) \sin Bx - 2AB \cos Bx] \tag{9.101}$$

Then, one can write

$$T_2(r, t) = \frac{2g_0 b^3 \gamma}{kr} \sum_{m=1}^{\infty}$$

$$\times \frac{(1 + \beta_m^2)[2\beta_m - e^{-\gamma} \sin (\beta_m)(2 + 2\gamma + \gamma^2 + \beta_m^2)]}{\beta_m^4 (\gamma^2 + \beta_m^2)^2}$$

$$\times \sin \left(\beta_m \frac{r}{b} \right) \left[1 - \exp \left(-\frac{\beta_m^2 \alpha t}{b^2} \right) \right] \tag{9.102}$$

Finally, the solution for $T(r, t)$ becomes

$$T(r, t) = \frac{3g_0 b^2}{k} \left\{ \frac{\alpha t}{b^2} \left[\frac{2 - e^{-\gamma}(\gamma^2 + 2\gamma + 2)}{\gamma^3} \right] \right.$$

$$+ \frac{2\gamma b}{3r} \sum_{m=1}^{\infty} \frac{(\beta_m^2 + 1)}{\beta_m^4}$$

$$\times \frac{[2\beta_m - e^{-\gamma} \sin (\beta_m)(2 + 2\gamma + \gamma^2 + \beta_m^2)]}{(\gamma^2 + \beta_m^2)^2} \tag{9.103}$$

$$\left. \times \sin \left(\beta_m \frac{r}{b} \right) \left[1 - \exp \left(-\frac{\beta_m^2 \alpha t}{b^2} \right) \right] \right\}$$

Note that for this case, similar to the previous case, the T_1 solution is not a function of position, and the T_2 solution contains a transient decaying term and a quasi-steady term.

4. *Case* RS02B0T0Gt4. Heat is generated within the sphere as an exponential function of time given by

$$g(r, t) = g_0 e^{-\lambda t} \tag{9.104}$$

Similar to case 1, since g is not a function of r, $T_2(r, t)$ becomes zero. Then, the solution is given by

$$T(r, t) = T_1(r, t) = \frac{\alpha g_0}{k\lambda} (1 - e^{-\lambda t}) \tag{9.105}$$

Note that, there is a steady-state temperature for this case given by $\alpha g_0/k\lambda$.

5. *Case* RS02B0T0Gr6. Heat is generated within the sphere with generation rate given by

$$g(r, t) = \frac{g_0}{r} \sin \frac{\pi r}{b} \tag{9.106}$$

The integrals in the T_1 solution, for this case, can easily be evaluated (using Table 5.7) to give,

$$T_1(t) = \frac{3\alpha g_0 t}{\pi k b} \tag{9.107}$$

After integrating over time, the T_2 solution becomes

$$T_2(r, t) = \frac{2g_0 b^2}{kr} \sum_{m=1}^{\infty}$$

$$\times \frac{(\beta_m^2 + 1) \sin (\beta_m r/b)}{\beta_m^4} \left[1 - \exp \left(- \frac{\beta_m^2 \alpha t}{b^2} \right) \right]$$

$$\times \int_{r'=0}^{'} \sin \left(\beta_m \frac{r'}{b} \right) \sin \left(\pi \frac{r'}{b} \right) d \left(\frac{r'}{b} \right) \tag{9.108}$$

The integral over r' in Eq. (9.108) can be evaluated by using the relation

$$\sin A \sin B = \frac{1}{2}[\cos (A - B) - \cos (A + B)] \tag{9.109}$$

Then, the solution for $T(r, t)$ becomes

$$T(r, t) = \frac{3g_0 b}{\pi k} \left\{ \frac{\alpha t}{b^2} + \frac{2b\pi^2}{3r} \sum_{m=1}^{\infty} \frac{\sin (\beta_m r/b)}{\beta_m^2 (\pi^2 - \beta_m^2)} \right.$$

$$\left. \times \left[1 - \exp \left(- \frac{\beta_m^2 \alpha t}{b^2} \right) \right] \right\} \tag{9.110}$$

Again, there is no steady-state temperature for this case.

9.6 TEMPERATURES IN HOLLOW SPHERES

In this section, we demonstrate, with examples, the application of the GF solution method to the hollow sphere problems with radial flow of heat, denoted by $RSIJ$, for $I, J = 1, 2, 3$. The describing equations and the analytical techniques used for solid spheres in the previous section are also applicable to hollow spheres. However, hollow spheres have one more physical boundary ($I = 1, 2,$ or 3) at the inner surface $r = a$ as compared to solid spheres with no physical boundary ($I = 0$) at the center $r = 0$. Accordingly, for hollow spheres, we may have one of the following boundary conditions at the inner surface $r = a$,

for $I = 1,$ $$T(a, t) = T_a(t) \quad t > 0 \tag{9.111a}$$

for $I = 2,$ $$-k \frac{\partial T(a, t)}{\partial r} = q_a(t) \quad t > 0 \tag{9.111b}$$

and for $I = 3,$ $$k \frac{\partial T(a, t)}{\partial r} + h_1 T(a, t) = h_1 T_\infty(t) \quad t > 0 \tag{9.111c}$$

The appropriate GF for the hollow sphere problems with radial flow of heat can be obtained from the XIJ GFs by following the procedure explained in Section 9.4. See Example 9.2. The required transformations are provided in Table 9.1.

Example 9.7: Hollow sphere heated on the inside surface—$RS21B10T0$ case. A hollow sphere, $a \leq r \leq b$, is initially at zero temperature. For time $t > 0$, the inner surface of the hollow sphere is heated by a constant heat flux q_0, while the outer surface is kept at zero temperature. Find the temperature distribution in the hollow sphere for large times.

SOLUTION This is the $RS21$ geometry with no heat generation and zero initial temperature, shown in Fig. 9.9. The temperature solution is only due to the boundary conditions and is given by

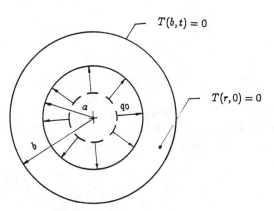

Hollow sphere geometry for case <u>RS21B10T0</u>

Figure 9.9 Hollow sphere with heat flux at inner boundary.

$$T(r, t) = \alpha \int_{\tau=0}^{t} \frac{q_0}{k} G_{RS21}(r, t|a, \tau) 4\pi a^2 \, d\tau \tag{9.112}$$

Note that since the boundary condition at $r = b$ is homogeneous, the $RS21$ GF in Eq. (9.112) is evaluated only at the inner surface boundary, $r = a$, where the heat flux is located. The GF for this case is obtained from the large-time form of G_{X31} with the appropriate transformations given in Table 9.1.

$$G_{RS21}(r, t|r', \tau) = \frac{1}{2\pi(b - a)rr'} \sum_{m=1}^{\infty} \exp\left[-\beta_m^2 \frac{\alpha(t - \tau)}{(b - a)^2}\right]$$

$$\times \frac{[\beta_m^2 + B^2] \sin\{\beta_m[1 - (r - a)/(b - a)]\} \sin\{\beta_m[1 - (r' - a)/(b - a)]\}}{\beta_m^2 + B^2 + B} \tag{9.113}$$

where $B = (b - a)/a$ and the eigenvalues β_m are the positive roots of the characteristic equation

$$\beta_m \cot \beta_m = -B \tag{9.114}$$

Evaluating $G_{RS21}(r, t|r', \tau)$ at $r' = a$ and substituting the result into Eq. (9.112) gives

$$T(r, t) = \frac{2\alpha q_0 a}{k(b - a)r} \sum_{m=1}^{\infty} \frac{(\beta_m^2 + B^2) \sin[\beta_m(b - r)/(b - a)] \sin \beta_m}{\beta_m^2 + B^2 + B}$$

$$\times \int_{\tau=0}^{t} \exp\left[-\beta_m^2 \frac{\alpha(t - \tau)}{(b - a)^2}\right] d\tau \tag{9.115}$$

The time integral in the above equation can be solved as integral no. 2 from Table 5.8 to give

$$T(r, t) = \frac{2q_0 a(b - a)}{kr} \sum_{m=1}^{\infty} \frac{(\beta_m^2 + B^2) \sin[\beta_m(b - r)/(b - a)] \sin \beta_m}{\beta_m^2(\beta_m^2 + B^2 + B)}$$

$$\times \left\{1 - \exp\left[-\frac{\beta_m \alpha t}{(b - a)^2}\right]\right\} \tag{9.116}$$

Note that the steady-state part of the temperature in Eq. (9.116) is given in a series form. The nonseries form of this part can be found by solving the above problem under the steady-state conditions to give

$$T_s(r) = \frac{q_0 a^2(b - a)}{kbr} \tag{9.117}$$

Then the alternative form of the solution (9.116) is given by

$$T(r, t) = \frac{q_0 a^2 (b - r)}{kbr} + \frac{2q_0 a(b - a)}{kr} \sum_{m=1}^{\infty} \frac{(\beta_m^2 + B^2)}{\beta_m (\beta_m^2 + B^2 + B)}$$

$$\times \sin \left[\beta_m (b - r) (b - a)\right] \sin \beta_m \exp \left[- \frac{\beta_m \alpha t}{(b - a)^2} \right] \quad (9.118)$$

Example 9.8: Hollow sphere exposed to convection with large heat transfer coefficient at the inside surface—RS11B10T0 case. A hollow sphere, $a \leq r \leq b$, initially at zero temperature is suddenly exposed to a fluid at a constant temperature T_∞ at its inner surface. The outer surface temperature remains constant at its initial value $T = 0$. The heat transfer coefficient between the fluid and the inner surface is very large. Find the transient temperature distribution within the hollow sphere.

SOLUTION In this problem, the inner surface boundary is exposed to a fluid with a very large heat transfer coefficient, which is equivalent to the case where there is a sudden step change in the surface temperature to the fluid's temperature T_∞. Possible examples might be when the fluid is a liquid metal or is changing phase since these processes usually have very large heat transfer coefficients. Therefore, this is the $RS11$ geometry with a homogeneous boundary condition at the outer surface of the hollow sphere. See Fig. 9.10. Since there is no energy generation within the body, and the initial temperature is zero, the GF solution is only due to the boundary condition at the inner surface and is given by

$$T(r, t) = -\alpha \int_{\tau=0}^{t} T_\infty(\tau) \frac{\partial G_{RS11}(r, t|a, \tau)}{\partial r} 4\pi a^2 \, d\tau \quad (9.119)$$

Note that the above integral is evaluated only at the inner surface with $r' = a$. The derivative of G_{RS11} is listed in Appendix RS, Eq. (RS11.5) as

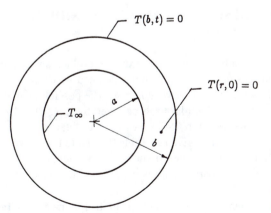

$T(b,t) = 0$

$T(r,0) = 0$

T_∞

a

b

Figure 9.10 Hollow sphere with temperature specified at both boundaries.

$$- \frac{\partial G_{RS11}(r, t \mid a, \tau)}{\partial r} = \frac{1}{2(b - a)^2 ra} \sum_{m=1}^{\infty} m \sin \left(m\pi \frac{r - a}{b - a} \right)$$

$$\times \exp \left[- \frac{m^2 \pi^2 \alpha (t - \tau)}{(b - a)^2} \right] \tag{9.120}$$

Substituting for $\partial G_{RS11}/\partial r$ from Eq. (9.120) into Eq. (9.119) gives

$$T(r, t) = \frac{2\pi \alpha a}{(b - a)^2 r} \sum_{m=1}^{\infty} m \sin \left(m\pi \frac{r - a}{b - a} \right)$$

$$\times \int_0^t \exp \left[- \frac{m^2 \pi^2 \alpha (t - \tau)}{(b - a)^2} \right] d\tau \tag{9.121}$$

The time integral in Eq. (9.121) can easily be evaluated with Table 5.8 to give

$$T(r, t) = \frac{2aT_\infty}{\pi r} \sum_{m=1}^{\infty} \frac{\sin [m\pi (r - a)/(b - a)]}{m}$$

$$\times \left\{ 1 - \exp \left[- \frac{m^2 \pi^2 \alpha t}{(b - a)^2} \right] \right\} \tag{9.122}$$

Note that the steady-state part of the above solution is given in a series form. The nonseries form of this part is obtained by solving the problem under the steady-state conditions. Then the solution becomes

$$T(r, t) = \frac{aT_\infty}{r} \left\{ 1 - \frac{r - a}{b - a} - \frac{2}{\pi} \sum_{m=1}^{\infty} \frac{\sin [m\pi (r - a)/(b - a)]}{m} \right.$$

$$\times \exp \left[- \frac{m^2 \pi^2 \alpha t}{(b - a)^2} \right] \right\} \tag{9.123}$$

9.7 TEMPERATURES IN AN INFINITE REGION OUTSIDE A SPHERICAL CAVITY

In this section, the GF solution method is applied to two example problems with radial flow of heat in an infinite region outside a spherical cavity, denoted by $RSI0$, $I = 1$, 2, 3. There is no physical boundary for this geometry at $r = \infty$. The possible boundary conditions at the inner surface ($r = a$) are similar to those given by Eqs. (9.111a, b, c) for $I = 1$, 2, 3, respectively. The GFs for $RSI0$ cases can be obtained from the $XI0$ GFs with the appropriate transformations given in Table 9.1. Note that these GFs do not involve infinite series; consequently, the solutions are mathematically well behaved, for all values of time. No time partitioning is needed for this geometry.

Example 9.9: Infinite body heated at the surface of a spherical cavity— *RS20B1T0* case. An infinite body bounded internally by the spherical cavity $r =$

a, is initially at zero temperature. For $t > 0$, the surface of the body ($r = a$) is heated uniformly by a known heat flux as a function of time, $q_a(t)$. See Fig. 9.11. Find the transient temperature distribution within the body.

SOLUTION This is the $RS20$ geometry with an arbitrary heat flux boundary condition at $r = a$, given by Eq. (9.111b). The temperature is given by

$$T(r, t) = \frac{\alpha}{k} \int_{\tau=0}^{t} q_a(t) G_{RS20}(r, t | a, \tau) 4\pi a^2 \, d\tau \qquad (9.124)$$

The GF function for this case is obtained from the $X13$ GF with the appropriate transformations given in Table 9.1.

$$G_{RS20}(r, t | r', \tau) = (4\pi r r')^{-1} [4\pi \alpha(t - \tau)]^{-1/2}$$

$$\times \left\{ \exp\left[-\frac{(r - r')^2}{4\alpha(t - \tau)} \right] + \exp\left[-\frac{(r + r' - 2a)^2}{4\alpha(t - \tau)} \right] \right\}$$

$$- (4\pi r r' a)^{-1} \exp\left[\frac{\alpha(t - \tau)}{a^2} + \frac{1}{a}(r + r' - 2a) \right]$$

$$\times \operatorname{erfc}\left\{ \frac{(r + r' - 2a)}{[4\alpha(t - \tau)]^{1/2}} + \frac{1}{a}[\alpha(t - \tau)]^{1/2} \right\} \qquad (9.125)$$

Evaluating $G_{RS20}(r, t | r', \tau)$ at $r' = a$ and substituting the result into Eq. (9.124) gives

$$T(r, t) = \frac{\alpha a}{kr} \int_{0}^{t} q_a(\tau) \left(2[4\pi \alpha(t - \tau)]^{-1/2} \right.$$

$$\times \exp\left[-\frac{(r - r')^2}{4\alpha(t - \tau)} \right] - \frac{1}{a} \exp\left[\frac{\alpha(t - \tau)}{a^2} + \frac{r - a}{a} \right]$$

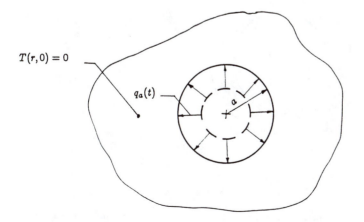

$T(r, 0) = 0$

$q_a(t)$

a

Figure 9.11 Infinite region outside the spherical cavity with heat flux specified at boundary $r = a$.

$$\times \ \text{erfc}\ \left\{ \frac{r - a}{[4\alpha(t - \tau)]^{1/2}} + \frac{1}{a}\ [\alpha(t - \tau)]^{1/2} \right\} \right) d\tau \qquad (9.126)$$

Depending on the functional form of $q_a(t)$, different solutions can be obtained from Eq. (9.126). For special case where $q_a(t) = q_0 = $ constant, the solution becomes

$$T(r,\ t) = \frac{q_0 a^2}{kr} \left\{ \text{erfc}\ \left[\frac{r - a}{(4\alpha t)^{1/2}} \right] - \exp\left(\frac{r - a}{a} + \frac{\alpha t}{a^2} \right) \right.$$

$$\left. \times\ \text{erfc}\ \left[\frac{r - a}{(4\alpha t)^{1/2}} + \frac{(\alpha t)^{1/2}}{a} \right] \right\} \qquad (9.127)$$

Example 9.10: Infinite body with a fixed-temperature spherical cavity with internal energy generation—RS10B00T0Gr5 case. An infinite body bounded internally by the spherical cavity $r = a$ is initially at zero temperature. For time $t > 0$, the body is heated by a volume energy source given by

$$g(r,\ t) = g_0 \qquad \text{for } a \leq r \leq b \qquad (9.128a)$$

$$g(r,\ t) = 0 \qquad \text{for } r > b \qquad (9.128b)$$

and the surface temperature at $r = a$ is kept at its initial value $T = 0$. Find the transient temperature distribution within the body.

SOLUTION This is the $RS10$ geometry with energy generation shown in Fig. 9.12. The GF solution for this case is given by

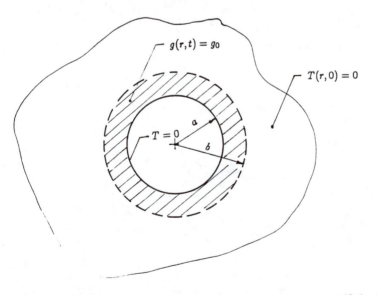

Figure 9.12 Infinite region outside the spherical cavity with temperature specified at boundary $r = a$ and with energy generation.

$$T(r, t) = \frac{\alpha}{k} \int_{\tau=0}^{t} \int_{r'=a}^{b} G_{RS10}(r, t|r', \tau) g_0 4\pi r'^2 \, dr' \, d\tau \quad (9.129)$$

Note that the integral on r' is evaluated from a to b since the generation is zero for $r > b$. The large-time form of G_{RS10} is given in Appendix RS as

$$G_{RS10}(r, t|r', \tau) = (4\pi rr')^{-1} [4\pi\alpha(t - \tau)]^{-1/2}$$

$$\times \left\{ \exp\left[-\frac{(r - r')^2}{4\alpha(t - \tau)} \right] \right.$$

$$\left. - \exp\left[-\frac{(r + r' - 2a)^2}{4\alpha(t - \tau)} \right] \right\} \quad (9.130)$$

Substituting for G_{RS10} into Eq. (9.129) gives

$$T(r, t) = \frac{\alpha g_0}{kr} \int_{\tau=0}^{t} d\tau \int_{r'=a}^{b} [4\pi\alpha(t - \tau)]^{-1/2}$$

$$\times \left\{ \exp\left[-\frac{(r - r')^2}{4\alpha(t - \tau)} \right] \right.$$

$$\left. - \exp\left[-\frac{(r + r' - 2a)^2}{4\alpha(t - \tau)} \right] \right\} r' \, dr' \quad (9.131)$$

The integral over r' is carried out first with the use of Table 5.3 to give

$$T(r, t) = \frac{\alpha g_0}{kr} \int_{\tau=0}^{t} \left(\frac{r}{2} \left\{ \mathrm{erfc} \, \frac{r - b}{[4\alpha(t - \tau)]^{1/2}} - \mathrm{erfc} \, \frac{r + b - 2a}{[4\alpha(t - \tau)]^{1/2}} \right\} \right.$$

$$\left. + a \left\{ \mathrm{erfc} \, \frac{r + b - 2a}{[4\alpha(t - \tau)^{1/2}]} - \mathrm{erfc} \, \frac{r - a}{[4\alpha(t - \tau)]^{1/2}} \right\} \right) d\tau \quad (9.132)$$

Next, the integral over time can be evaluated with Table 5.5 to give

$$T(r, t) = \frac{2\alpha g_0 t}{k} \left[i^2\mathrm{erfc} \, \frac{r - b}{(4\alpha t)^{1/2}} - \left(1 - \frac{2a}{r} \right) \right.$$

$$\left. \times i^2\mathrm{erfc} \, \frac{r + b - 2a}{(4\alpha t)^{1/2}} - \frac{2a}{r} \, i^2\mathrm{erfc} \, \frac{r - a}{(4\alpha t)^{1/2}} \right] \quad (9.133)$$

9.8 STEADY STATE

In this section, three examples of steady heat conduction in radial spherical coordinates are presented. The GFs are listed in Appendix RS, Table RS.1. For two- and three-dimensional heat conduction, the steady GF must be found on a case-by-case basis.

Example 9.11: Hollow sphere heated on the inside surface—*RS21B*10 case. Find the steady temperature in the geometry of Example 9.7, the hollow sphere heated at the inside surface ($r = a$) by heat flux q_0. The outside surface ($r = b$) is maintained at zero temperature.

SOLUTION This is the $RS21B10$ geometry shown in Fig. 9.9. The steady temperature is given by the boundary condition term of the steady GFSE, Eq. (3.94),

$$T(r) = \frac{q_0}{k} G_{RS21}(r|r' = a) \tag{9.134}$$

The steady GF given in Appendix RS Table RS.1 is a piecewise continuous function,

$$G_{RS21}(r|r') = \begin{cases} \dfrac{1/r' - 1/b}{4\pi} & r \leq r' \\ \dfrac{1/r - 1/b}{4\pi} & r \geq r' \end{cases} \tag{9.135}$$

Substitute the above GF evaluated at $r' = a$ into Eq. (9.134) by using the $r \geq r'$ portion of the function to find the temperature,

$$T(r) = \frac{q_0}{k} \left(\frac{1}{r} - \frac{1}{b} \right) a^2 \tag{9.136}$$

Example 9.12: Hollow sphere with temperature fixed on both surfaces—RS11B10 case. Find the steady temperature in the geometry of Example 9.8, the hollow sphere with zero temperature on the outside surface ($r = b$) and with temperature T_∞ maintained at the inside surface ($r = a$).

SOLUTION This is the $RS11$ geometry shown in Fig. 9.10. The steady temperature is driven by the boundary condition at $r = a$ and the steady GFSE, Eq. (3.94), gives

$$T(r) = - T_\infty \left. \frac{\partial G_{RS11}}{\partial n'} \right|_{r'=a} 4\pi a^2 \tag{9.137}$$

The steady GF is given in Appendix RS Table RS.1 as a piecewise continuous function,

$$G_{RS11}(r) = \begin{cases} \dfrac{(b - r')(1 - a/r)}{4\pi r' (b - a)} & r \leq r' \\ \dfrac{(b - r)(1 - a/r')}{4\pi r (b - a)} & r \geq r' \end{cases} \tag{9.138}$$

The derivative in Eq. (9.137) is evaluated at the surface $r' = a$ so that the $r \geq r'$ portion of the GF is used:

$$-\left. \frac{\partial G_{RS11}}{\partial n'} \right|_{r'=a} = \left. \frac{\partial G_{RS11}}{\partial r'} \right|_{r'=a}$$

$$= \left. \frac{(b - r) a/(r')^2}{4\pi r (b - a)} \right|_{r'=a} = \frac{b - r}{4\pi a r (b - a)} \tag{9.139}$$

Then the temperature is given by Eq. (9.137)

$$T(r) = T_\infty \frac{a(b-r)}{r(b-a)} = T_\infty \frac{a}{r}\left(1 - \frac{r-a}{b-a}\right) \qquad (9.140)$$

Example 9.13: Solid sphere with internal energy generation and convective boundary condition. Find the steady temperature in a solid sphere heated by internal energy generation $g(r)$ and cooled by convection from the surface. The heat transfer coefficient is h and the fluid temperature is T_∞.

SOLUTION This geometry is number $RS03B0G-$ if the temperature is normalized in the form $T(r) - T_\infty$. The temperature is given by the energy generation term of the steady GFSE, Eq. (3.94),

$$T(r) - T_\infty = \frac{1}{k}\int_{r'=0}^{b} g(r')\, G_{RS03}(r|r')\, 4\pi(r')^2\, dr' \qquad (9.141)$$

The steady GF is given by

$$G_{RS03}(r|r') = \begin{cases} \dfrac{1/r' + (1/B_2 - 1)/b}{4\pi} & r \le r' \\[2ex] \dfrac{1/r + (1/B_2 - 1)/b}{4\pi} & r \ge r' \end{cases} \qquad (9.142)$$

where $B_2 = hb/k$, the Biot number. Because G_{RS03} is piecewise continuous, the integral in Eq. (9.141) must be carried out in two pieces:

$$T(r) - T_\infty = \frac{1}{k}\int_{r'=0}^{r} g(r')\left[\frac{1}{r} + \frac{1/B_2 - 1}{b}\right](r')^2\, dr'$$

$$+ \frac{1}{k}\int_{r'=r}^{b} g(r')\left[\frac{1}{r'} + \frac{1/B_2 - 1}{b}\right](r')^2\, dr' \qquad (9.143)$$

A symbolic mathematics computer program is very helpful in finding the correct solution to these integrals. When the GF is piecewise continuous, it is particularly important to get all the signs correct because there are usually terms with opposite sign that cancel out.

    a. Case RS03B0G1. In the simple case when the internal energy generation is constant, $g(r) = g_0$, the temperature in Eq. (9.143) is given by a second-order polynomial:

$$T(r) - T_\infty = \frac{g_0 b^2}{3k}\left[\frac{1}{2} + \frac{1}{B_2} - \frac{(r/b)^2}{2}\right] \qquad (9.144)$$

Note that Eq. (9.143) contains six polynomial terms but the solution (9.144) contains only three terms. Two terms canceled and two terms were summed together.

    b. Case RS03B0G4. If the energy generation is a maximum at the surface $r = b$ and is attenuated exponentially inside the body as in microwave heating, then the energy generation term may be written

$$g(r) = g_0 e^{-c(1-r/b)} \tag{9.145}$$

where c is the attenuation parameter (dimensionless) and g_0 is the maximum energy generation (W/m^3). The steady temperature is given by Eq. (9.143) and (9.145) as

$$T(r) - T_\infty = \frac{g_0 b^2}{kc^3} \left[\left(\frac{2b}{r} - c \right) e^{-c(1-b/r)} + \left(1 - \frac{b}{r} - \frac{1}{B_2} \right) \right.$$

$$\left. \times e^{-c} + c - 2 + \frac{c^2 - 2c + 2}{B_2} \right] \tag{9.146}$$

In the limiting case as $B_2 \to \infty$, the temperature at the surface is $T(r = b) = T_\infty$.

PROBLEMS

9.1 A solid sphere, $0 \le r \le b$, is initially at a uniform temperature T_0 when its surface temperature is suddenly changed to T_b and maintained at this value for times $t > 0$. Using the GF method, find the transient temperature distribution in the sphere for small and large times.

9.2 A solid sphere, $0 \le r \le b$, is initially at a uniform temperature of T_0. For times $t > 0$, its surface temperature changes linearly with time as $T(b, t) = ct$. Using the GF method, find the temperature distribution in the sphere for large times.

9.3 Using the method described in Section 9.4, derive the small-time GF for a solid sphere, $0 \le r \le b$, with a heat flux boundary condition (G_{RS02}).

9.4 A solid sphere, $0 \le r \le b$, has an initial temperature distribution given by

$$T(r, 0) = \begin{cases} T_0 & \text{for } 0 \le r \le a \\ 0 & \text{for } a \le r \le b \end{cases}$$

For times $t > 0$, the surface temperature is kept at its initial value. Using the GF method, find the transient temperature distribution in the sphere.

9.5 A solid sphere, $0 \le r \le b$, is initially at zero temperature. For times $t > 0$, heat is generated uniformly within the sphere with a constant rate of g_0 W/m^3, while the surface temperature is kept at its initial value. Using the GF method, find the transient temperature distribution in the sphere.

9.6 Using the method described in Section 9.4, derive the small-time and the large-time GFs for a hollow sphere, $a \le r \le b$, with a prescribed temperature boundary condition at $r = a$ and a convection boundary condition at $r = b$ (G_{RS13}).

9.7 A hollow sphere, $a \le r \le b$, is initially at temperature T_0. For times $t > 0$, the boundary at $r = a$ is kept at zero temperature while the boundary at $r = b$ is dissipating heat by convection into a medium at zero temperature with a constant heat transfer coefficient h. Using the GF method, find the transient temperature distribution in the hollow sphere.

9.8 A spherical capsule contains a fluid with volumetric heat capacity $\rho_f c_f$. The inside and outside radii of the capsule wall are r_i and r_o. The boundary conditions are

$$-\frac{k \partial T}{\partial r} = h(T - T_\infty) \qquad \text{at } r = r_0$$

$$-\left(\frac{3k}{r_i}\right) \frac{\partial T}{\partial r} = \rho_f c_f \frac{\partial T}{\partial t} \qquad \text{at } r = r_i$$

$$T = T_i \qquad \text{at } t = 0$$

Find an expression for the temperature at $r = r_i$. If the encapsulated fluid is an ideal gas at a quasi-uniform temperature and the initial pressure is p_i, find p/p_i as a function of time where p is the instantaneous pressure within the capsule.

9.9 Lead shot is sometimes manufactured in shot towers, where molten lead falls through the air to solidify and then is quenched in a liquid to cool.

(a) Suppose the molten lead droplet of radius a starts falling at the solidification temperature T_s, and the latent heat of fusion is f_0 (J/m^3). If the heat transfer coefficient is h_a and the air temperature is T_a, find an expression for the distance the droplet must fall to solidify. (*Hint*: Use a lumped analysis on the droplet; neglect air friction.)

(b) If $T_s = 327°C$, $T_a = 30°C$, and $h_a = 50$ W/(m^2 K), what is the largest size shot that can be dropped in a 50-m tower? [$f_0 = 23$ kJ/(kg K); $\rho c = 1330$ kJ/(m^3 K).]

(c) Now the shot is quenched in a liquid at T_∞ with heat transfer coefficient h. Find an expression for the transient temperature at the surface of the shot.

9.10 In a pulverized-coal furnace coal particles are blown in with preheated air at a temperature T_i. A new coal particle does not begin to burn until its surface temperature reaches the combustion temperature T_c. If the primarily radiant heat transfer in the furnace may be modeled as a uniform heat flux q_0, on the surface of an approximately spherical coal particle of radius a,

(a) Find a small-time expression for the surface temperature of the particle.

(b) Find an approximate expression for the time it takes the coal particle to begin burning. (*Hint*: Use only one term of the series for the temperature.)

9.11 One technique for handling radioactive waste is encapsulating it in ceramic. Suppose a sphere of the radioactive material of radius a is covered by ceramic to form an encapsulated sphere of radius $b > a$. If the radioactive material produces heat at the rate g_0 W/m^3 and if the outer surface of the sphere (at $r = b$) is cooled by convection (temperature T_∞, heat transfer coefficient h),

(a) Find the steady-state temperature in the ceramic shell, $a < r < b$.

(b) Find an expression for the maximum temperature in the radioactive material.

9.12 In food processing of prepared foods like soup, the heating must continue until the largest chunk in the soup reaches a specified temperature (to kill bacteria).

(a) Find an expression for the center temperature of a (spherical) dumpling initially at T_i and suddenly exposed to heated liquid at T_∞. (Assume a heat transfer coefficient h.)

(b) Find an approximate expression for the time it takes to raise the center temperature of the dumpling to specified temperature T_f. (*Hint*: Use one term of the large-time form of the temperature.)

CHAPTER TEN

GALERKIN-BASED GREEN'S FUNCTIONS AND SOLUTIONS

10.1 INTRODUCTION

The Green's functions (GFs) for regularly shaped bodies, such as plates, cylinders, and spheres, can be obtained by classical methods. These regularly shaped bodies shall be called orthogonal bodies. A normal at any point on the boundary of an orthogonal body is parallel to the direction of a coordinate axis. The solution methods and the derivation of the GFs for various orthogonal bodies are discussed elsewhere in this book. The formulation of the GFs for nonorthogonal bodies, as investigated by Haji-Sheikh and Lakshminarayanan (1987) and Haji-Sheikh and Beck (1988), are in this chapter.

The objective is to provide a methodology for solving the diffusion equation in various orthogonal and nonorthogonal bodies. The orthogonal bodies include plates, solid cylinders, hollow cylinders, solid spheres, and hollow spheres. The examples in this chapter consist of one-dimensional conduction in isotropic media problems that have exact solutions. The procedure is described, convenient and appropriate expressions are provided, and the accuracy of the results is compared with the exact solution. The study of multidimensional conduction in orthogonal and nonorthogonal bodies and related examples are included in Chapter 11. Also, the utility of this method when applied to conduction in heterogeneous problems is demonstrated.

The solution method discussed in this chapter is a Galerkin-based integral method and it is referred to as the GBI method. The range of its usefulness encompasses thermal conduction problems with homogeneous or nonhomogeneous boundary conditions. The diffusion equation, Eq. (3.28), can be written in a generalized form

$$\nabla \bullet [k(\mathbf{r})\nabla T] + g(\mathbf{r}, t) - m(\mathbf{r})^2 T = \rho(\mathbf{r})c_p(\mathbf{r})u(\mathbf{r})\frac{\partial T}{\partial t} \qquad (10.1)$$

where $T = T(\mathbf{r}, t)$ is temperature, $\mathbf{r}$ is position vector, and t is time. The thermophysical properties $\rho(\mathbf{r})$, $c_p(\mathbf{r})$, and $k(\mathbf{r})$ are position-dependent density, specific heat, and thermal conductivity, respectively. The term $m(\mathbf{r})^2 T$ is the fin convection effect. The function $u(\mathbf{r})$ is designated as the velocity function. When dealing with pure conduction, the function $u(\mathbf{r})$ is equal to 1. Section 11.5 deals with flow in ducts where the functional values of $u(\mathbf{r})$ are considered and t is replaced by the axial coordinate. The nonhomogeneous boundary conditions are accommodated by using the GF solution method.

An outline of the remainder of this chapter follows. First, in Section 10.2, the standard derivation of the Green's function solution equation (GFSE) given in Chapter 3 is modified to account for the position-dependent thermophysical properties. Section 10.3 presents an alternative derivation of the GFSE which, when available, provides a rapidly converging temperature solution. This method uses a set of basis functions that need not be orthogonal. In Section 10.4, we demonstrate that, unlike the exact solutions, the functional form of the basis functions for one-dimensional solutions remains unchanged in different coordinates; however, the boundary conditions affect the values of each basis function.

Examples 10.1 and 10.2 are presented mainly to elaborate on the mathematical steps and to introduce the numerical steps in the integral method. Example 10.3 shows the use of the alternative GF solution when the surface temperature is prescribed. Example 10.4 uses a unified solution and compares the accuracy of different GF solutions.

The extension of this method to deal with multidimensional conduction problems in heterogeneous materials and steady-state conduction are in Chapter 11. A study of heat transfer in the thermal entrance region of ducts is also included in Chapter 11.

10.2 GREEN'S FUNCTIONS AND GREEN'S FUNCTION SOLUTION METHOD

The GF method permits the solution of diffusion problems with nonhomogeneous boundary conditions. The GF solution method described in Chapter 3 is modified. The modifications allow the properties ρ, c_p, and k to be position dependent, and the results are useful for the study of conduction of heat in homogeneous as well as heterogeneous bodies. The GF for a body with given boundary conditions describes the temperature effect at point $\mathbf{r}$ at time t if there is an impulsive point energy source of strength unity located at point $\mathbf{r}'$ and released at time τ. The GFs become the solutions of Eq. (10.1) if the term $g(\mathbf{r}, t)$ in Eq. (10.1) is replaced by a point energy source mathematically described by the following delta functions

$$g(\mathbf{r}, t) = \rho(\mathbf{r})c_p(\mathbf{r})\delta(\mathbf{r} - \mathbf{r}')\delta(t - \tau) \tag{10.2}$$

Accordingly, the GF is defined so that it satisfies homogeneous boundary conditions and it is the solution of the following auxiliary equation:

$$\nabla \cdot [k(\mathbf{r})\nabla G(\mathbf{r}, t|\mathbf{r}', \tau)] + C(\mathbf{r})\delta(\mathbf{r} - \mathbf{r}')\delta(t - \tau) \tag{10.3}$$

$$- m(\mathbf{r})^2 G(\mathbf{r}, t|\mathbf{r}', \tau) = C(\mathbf{r})u(\mathbf{r})\frac{\partial G(\mathbf{r}, t|\mathbf{r}', \tau)}{\partial t}$$

where

$$C(\mathbf{r}) = \rho(\mathbf{r})c_p (\mathbf{r}) \qquad (10.4)$$

where $\rho(\mathbf{r})$, $c_p(\mathbf{r})$, $k(\mathbf{r})$, $u(\mathbf{r})$, and $m(\mathbf{r})^2$ are position-dependent density, specific heat, thermal conductivity, velocity function, and fin effect as described for Eq. (10.1). Based on the above-mentioned descriptions of the GF, Eq. (10.1) for temperature and Eq. (10.3) for the GF are the same, except that in Eq. (10.3) the functional value of $g(\mathbf{r}, t)$ is specified. The function $G(\mathbf{r}, t|\mathbf{r}', \tau)$ is called the GF (Ozisik, 1980). A formal solution of Eq. (10.3) based on the Galerkin method yields the GF and is presented in this section. Following the discussion of the properties of the GFs, the GF solution method is presented.

10.2.1 Galerkin-Based Integral Method

The solution of the diffusion equation in a relatively general form, Eq. (10.1), or the auxiliary equation for the GFs, Eq. (10.3), is derived using a Galerkin-based integral (GBI) method. It is assumed that the thermal conductivity, density, specific heat, velocity function, and fin effect are independent of temperature; however, no other restriction as to spatial variation of these thermophysical properties is needed. As described earlier, when the boundary conditions are homogeneous, $T(\mathbf{r}, 0) = 0$, and $g(\mathbf{r}, t) = C(\mathbf{r})\delta(\mathbf{r} - \mathbf{r}')\delta(t - \tau)$, the function $T(\mathbf{r}, t)$ is equal to the GF $G(\mathbf{r}, t|\mathbf{r}', \tau)$. Therefore, the value of the GF is readily available after a generalized solution of Eq. (10.1) is accomplished. The GBI solution described here was used by LeCroy and Eraslan (1969) in the study of temperature development in the entrance region of an MHD parallel plate channel.

To solve a differential equation with a nonhomogeneous term, the solution is frequently broken into two parts, complementary and particular. The complementary form of Eq. (10.1), that is, in essence, the diffusion equation in the absence of energy generation, is (Haji-Sheikh and Mashena, 1987)

$$\nabla \cdot (k\nabla\Theta) - m(\mathbf{r})^2\Theta = \rho(\mathbf{r})c_p (\mathbf{r})u(\mathbf{r})\frac{\partial\Theta}{\partial t} \qquad (10.5)$$

The boundary conditions for Eq. (10.5) are the same as those for Eq. (10.1) or (10.3) and must be homogeneous. They are of the first kind (prescribed temperature), the second kind (prescribed heat flux), and the third kind (convective). It is also permissible for different parts of the boundary to have different kinds of boundary conditions. A solution to Eq. (10.5) can be written as

$$\Theta = \sum_{n=1}^{N} c_n\psi_n(\mathbf{r}) \exp(-\gamma_n t) \qquad (10.6)$$

where γ_n is the nth eigenvalue and is independent of $\mathbf{r}$, and c_n is a constant to be evaluated. For convenience, assume that the body has finite dimensions. Because Θ is the complementary solution, it is not necessary to specify the initial condition at this time. The function $\psi_n(\mathbf{r})$ is selected so that (1) the homogeneous boundary conditions are satisfied, and (2) Eq. (10.6) is a solution of Eq. (10.5). The former condition is exactly satisfied if $\psi_n(\mathbf{r})$ satisfies the boundary conditions. The latter is accommodated if Eq. (10.6) is substituted in Eq. (10.5), resulting in

$$\nabla \cdot [k\nabla\psi_n(\mathbf{r})] - m(\mathbf{r})^2 \psi_n + \rho(\mathbf{r})c_p (\mathbf{r})u(\mathbf{r})\gamma_n\psi_n(\mathbf{r}) = 0 \qquad (10.7)$$

for every n value. The diffusion equation now becomes an eigenvalue problem and the function $\psi_n(\mathbf{r})$ is the eigenfunction.

When an exact solution does not exist or a simpler approximate solution is preferred, Eq. (10.7) will be approximately satisfied. A function $\psi_n(\mathbf{r})$ is to be constructed as a linear combination of a properly selected set of basis functions. A properly selected set of basis functions is a complete set, its members are linearly independent, each member satisfies exactly the same homogeneous boundary conditions as those given for Θ, and not all members become zero at any interior point. The function $\psi_n(\mathbf{r})$, for $n = 1, 2, \ldots, N$, is chosen to be a linear combination of N basis functions,

$$\psi_n(\mathbf{r}) = \sum_{j=1}^{N} d_{nj}f_j(\mathbf{r}) \qquad (10.8)$$

where $f_j(\mathbf{r})$ is an element of a set of basis functions and the d_{nj}'s are constants to be evaluated.

The Galerkin procedure (Kantorovich and Krylov, 1960) is now used; that is, both sides of Eq. (10.7) are multiplied by $f_i dV$ and integrated over the volume V to get

$$\int_V f_i \nabla \cdot (k\nabla\psi_n)dV - \int_V m(\mathbf{r})^2 f_i\psi_n dV + \gamma_n \int_V \rho(\mathbf{r})c_p (\mathbf{r})u(\mathbf{r})f_i\psi_n dV = 0 \qquad (10.9)$$

Substituting ψ_n from Eq. (10.8) into Eq. (10.9) yields

$$\sum_{j=1}^{N} d_{nj} \left[\int_V f_i \nabla \cdot (k\nabla f_j)dV - \int_V m(\mathbf{r})^2 f_i f_j dV \right.$$

$$\left. + \gamma_n \int_V \rho(\mathbf{r})c_p (\mathbf{r})u(\mathbf{r})f_i f_j dV \right] = 0 \qquad (10.10)$$

in which $i = 1, 2, \ldots, N$. The matrix form of Eq. (10.10) is

$$(\mathbf{A} + \gamma_n \mathbf{B})\mathbf{d}_n = \mathbf{0} \qquad (10.11)$$

where $\mathbf{A}$ and $\mathbf{B}$ are square matrices of size N with the elements

$$a_{ij} = \int_V f_i \nabla \cdot (k\nabla f_j) \, dV - \int_V m(\mathbf{r})^2 f_i f_j \, dV \qquad (10.12)$$

and

$$b_{ij} = \int_V \rho(\mathbf{r})c_p (\mathbf{r})u(\mathbf{r})f_i f_j \, dV \qquad (10.13)$$

The coefficients $d_{n1}, d_{n2}, \ldots, d_{nN}$ in Eq. (10.8) are the member elements of the vector $\mathbf{d}_n$ in Eq. (10.11). The second integral in Eq. (10.12) vanishes in the absence of the fin effect. The fin effect, $m(\mathbf{r})^2$, influences only the elements of matrix $\mathbf{A}$.

An examination of Eq. (10.13) reveals that matrix $\mathbf{B}$ is symmetric; that is, $b_{ij} = b_{ji}$. When i and j are switched, the second integral on the right side of Eq. (10.12) will not be affected. Matrix **A** is also symmetric if the first term on the right side of Eq. (10.12) is symmetric. This is accomplished by using the identities 1, 2, and 3 in Note 1 at the end of this chapter to show that

$$\int_V f_i \nabla \cdot (k\nabla f_j) dV = \int_V \nabla \cdot (k f_i \nabla f_j) dV - \int_V k\nabla f_i \cdot \nabla f_j dV$$

$$= \int_S k f_i \nabla f_j \cdot \mathbf{n} dS - \int_V k\nabla f_i \cdot \nabla f_j dV$$

$$= \int_S k f_i \left(\frac{\partial f_j}{\partial n} \right) dS - \int_V k\nabla f_i \cdot \nabla f_j dV \qquad (10.14)$$

When dealing with homogeneous boundary conditions of the first kind (prescribed temperature $f_j = 0$) or the second kind (prescribed heat flux $\partial f_j/\partial n = 0$), the first term on the right side of Eq. (10.14) is zero while the second term is always symmetric. For homogeneous boundary conditions of the third kind (convective, $-k\partial f_j/\partial n = hf_j$), the first term on the right side of Eq. (10.14) becomes

$$\int_S k f_i (\partial f_j/\partial n) dS = -\int_S h f_i f_j dS \qquad (10.15)$$

which is also symmetric when i and j are switched. Inasmuch as the boundary conditions for f_i and f_j are always homogeneous, matrix **A** is always symmetric.

The calculation procedure for temperature distribution is summarized below:

a. It is important to select a complete set of basis functions that are linearly independent. A complete set requires that all contributing members of the set be included. The members of a set are linearly independent if no member of the set is a linear combination of the other members.
b. The computations of the values of a_{ij} and b_{ij} in Eqs. (10.12) and (10.13) are the major analytical or numerical computational tasks. For $N = 1$ and 2 and simple geometries, the computations are not difficult. For some complex geometries, it is convenient to utilize a symbolic software to carry out the analytical integrations which result in more accurate values and often require less computation time. When exact integration is not possible, numerical integrations can be used.
c. The next step is to calculate the eigenvalues and eigenvectors of Eq. (10.11) to be used in Eq. (10.6). When $N = 1$ or $N = 2$, the procedure is discussed in Example 10.2. The details, when N is large, are given after Example 10.2.
d. Following the calculation of eigenvalues and eigenvectors, the eigenfunctions, Eq. (10.8), are known. The solution for temperature is complete after calculation of c_n in Eq. (10.6). The initial temperature distribution is used to calculate the c_n values.

Examples 10.1 and 10.2 demonstrate the steps itemized above. Notice that the boundary conditions are homogeneous. The nonhomogeneous boundary conditions will be included using the Green's function solution method in Section 10.2.3.

Example 10.1 Consider an infinite homogeneous plate with thickness L and having boundary conditions $T(0, t) = T(L, t) = 0$ when $t > 0$. The thermal properties have constant values, $u(\mathbf{r}) = 1$, and $m(\mathbf{r}) = 0$. Furthermore, the initial temperature distribution is $F(x) = T_0$. Find the temperature distribution using orthogonal basis functions.

SOLUTION The number for this case is $X11B00T1$. A mathematical statement of this problem is

$$\alpha \frac{\partial^2 T}{\partial x^2} = \frac{\partial T}{\partial t} \qquad \text{for } 0 < x < L \text{ and } t > 0 \qquad (10.16)$$

$$T(0, t) = 0 \qquad T(L, t) = 0 \qquad \text{and} \qquad T(x, 0) = T_0$$

If the set of orthogonal basis functions has the two members, $f_1 = \sin(\pi x/L)$, and $f_2 = \sin(2\pi x/L)$, the function ψ_n, using Eq. (10.8), is

$$\psi_n = d_{n1} \sin\left(\frac{\pi x}{L}\right) + d_{n2} \sin\frac{2\pi x}{L} \qquad (10.17)$$

Both $f_1 = \sin(\pi x/L)$ and $f_2 = \sin(2\pi x/L)$ functions satisfy the homogeneous boundary conditions ($f_1 = f_2 = 0$ at $x = 0$ and L). Here, the energy equation in its integral form, Eq. (10.10), must be satisfied instead of Eq. (10.7).

It is convenient to designate $f_i = \sin(i\pi x/L)$, for $i = 1, 2$, and $f_j = \sin(j\pi x/L)$, for $j = 1, 2$. Then, when k is a constant, $\nabla \cdot (k\nabla f_j) = k\nabla^2 f_j = -k(j\pi/L)^2 \sin(j\pi x/L)$. The elements of matrix $\mathbf{A}$, using Eq. (10.12), become

$$a_{ij} = -k\left(\frac{j\pi}{L}\right)^2 \int_0^L \sin\left(\frac{j\pi x}{L}\right) \sin\left(\frac{i\pi x}{L}\right) dx$$

$$\text{for } i = 1, 2, \text{ and } j = 1, 2 \qquad (10.18)$$

For off-diagonal elements where i and j are not the same, this equation yields $a_{12} = a_{21} = 0$. When $i = j$, $a_{jj} = -k(j\pi)^2/2L$, resulting in $a_{11} = -k\pi^2/2L$ and $a_{22} = -2k\pi^2/L$. Similarly, the elements of matrix $\mathbf{B}$ using Eq. (10.13) are

$$b_{ij} = \rho c_p \int_0^L \sin\left(\frac{j\pi x}{L}\right) \sin\left(\frac{i\pi x}{L}\right) dx$$

$$\text{for } i = 1, 2, \text{ and } j = 1, 2 \qquad (10.19)$$

Because the basis functions are orthogonal, only the diagonal terms have nonzero values; for example, $b_{12} = b_{21} = 0$. However, the diagonal terms are $b_{11} = b_{22} = \rho c_p L/2$. Then, Eq. (10.11) takes the following dimensionless form for $n = 1$ and $n = 2$:

$$\begin{bmatrix} -\dfrac{\pi^2}{2} + \dfrac{L^2\gamma}{\alpha}\dfrac{1}{2} & 0 \\ 0 & -(2\pi^2) + \dfrac{L^2\gamma}{\alpha}\dfrac{1}{2} \end{bmatrix} \begin{bmatrix} d_{n1} \\ d_{n2} \end{bmatrix} = 0 \qquad (10.20)$$

where $\alpha = k/\rho c_p$ is the thermal diffusivity.

The eigenvalues γ_1 and γ_2 are chosen to make the determinant of this matrix equal to zero. Because all the off-diagonal terms are zero, the determinant is the product of the diagonal terms; the eigenvalues are obtained by setting each diagonal element equal to zero, $\gamma_1 = \pi^2\alpha/L^2$ and $\gamma_2 = 4\pi^2\alpha/L^2$. Since the simultaneous equations resulting from Eq. (10.20) are homogeneous, one of the coefficients, d_{n1} or d_{n2}, can be arbitrarily selected. By choosing $d_{11} = d_{22} = 1$, for both $n = 1$ and 2, the other unknowns become $d_{12} = d_{21} = 0$. Inasmuch as the differential equation for T is the same as that for Θ, then $T = \Theta$ and the solution using Eq. (10.6) is

$$T = c_1 \sin\left(\frac{\pi x}{L}\right) \exp\left(-\frac{\pi^2\alpha t}{L^2}\right)$$

$$+ c_2 \sin\left(\frac{2\pi x}{L}\right) \exp\left(-\frac{4\partial^2\alpha t}{L^2}\right) \tag{10.21}$$

Substitute $t = 0$ and the initial temperature $T(x, 0) = t_0$ into Eq. (10.21), to obtain

$$T_0 = c_1 \sin\left(\frac{\pi x}{L}\right) + c_2 \sin\frac{2\pi x}{L} \tag{10.22}$$

Analogous to the exact solution and the Fourier series expansion, both sides of this equation are multiplied by $\sin(\pi x/L)$, and then integrated over x from 0 to L to yield $c_1 = 4T_0/\pi$. Repeating the calculation but using $\sin(2\pi x/L)$ produces $c_2 = 0$. The final temperature solution is

$$\frac{T}{T_0} = \frac{4}{\pi} \sin\left(\frac{\pi x}{L}\right) \exp\left(-\frac{\pi^2\alpha t}{L^2}\right) \tag{10.23}$$

The generalization of this procedure is discussed in Section 10.2.3 and later verified in Section 10.4.

Equation (10.23) is identical to the first two terms of the exact solution. The procedure used to approximately satisfy the initial condition is not required when calculating the GF. However, it is used in Haji-Sheikh and Mashena (1987) in the integral solution as a standard procedure of dealing with the initial condition. It is used here to show the equivalence of the GF solution method and the Galerkin-based integral solution as they deal with the initial temperature distribution.

Example 10.2 Repeat the procedure used in Example 10.1 and use nonorthogonal basis functions.

SOLUTION Because the boundaries of the slab are at $x = 0$ and $L - x = 0$ surfaces, the function $(L - x)x$ will vanish on both surfaces. Also, the product of $(L - x)x$ and a member of a polynomial series (e.g., 1, x, x^2, . . .) will vanish on $x = 0$ and $x = L$ surfaces. For this two-term solution, both $(L - x)x$ and $(L - x)x^2$ functions satisfy the boundary conditions. These functions will be designated as the basis functions. More details concerning the method of selecting these basis functions are given in Section 10.4. Then, in the dimensionless form, one may write $f_1 = (1 - x/L)(x/L)$ and $f_2 = (1 - x/L)(x/L)^2$. The eigenfunction ψ_n becomes

$$\psi_n = d_{n1}\left(1 - \frac{x}{L}\right)\frac{x}{L} + d_{n2}\left(1 - \frac{x}{L}\right)\left(\frac{x}{L}\right)^2 \qquad \text{for } n = 1 \text{ and } 2 \quad (10.24)$$

Equation (10.12) is used to compute a_{ij} using $f_i = (1 - x/L)(x/L)^i$, for $i = 1$ and 2, and $f_j = (1 - x/L)(x/L)^j$, for $j = 1$ and 2. When $m(\mathbf{r}) = 0$ and $k = $ constant, Eq. (10.12) for a one-dimensional Cartesian system is

$$a_{ij} = k\int_0^L f_i\left(\frac{d^2 f_j}{dx^2}\right)dx \qquad (10.25a)$$

in which

$$\frac{d^2 f_j}{dx^2} = \frac{j(j - 1)(x/L)^{j-2} - (j + 1)j(x/L)^{j-1}}{L^2} \qquad (10.25b)$$

resulting in

$$a_{ij} = -k\left(\frac{1}{L}\right)^2 \int_0^L \left(1 - \frac{x}{L}\right)\left(\frac{x}{L}\right)^i$$

$$\times \left[j(j - 1)\left(\frac{x}{L}\right)^{j-2} - (j + 1)j\left(\frac{x}{L}\right)^{j-1}\right]dx$$

$$= \frac{k}{L}\left[\frac{j(j - 1)}{i + j - 1} - \frac{j(j - 1)}{i + j} - \frac{(j + 1)j}{i + j}\right.$$

$$\left. + \frac{(j + 1)j}{i + j + 1}\right] \qquad \text{for } i = 1, 2, \text{ and } j = 1, 2 \quad (10.25c)$$

Substituting for i and j results in $a_{11} = -k/3L$, $a_{12} = a_{21} = -k/6L$, $a_{22} = -2k/15L$. Similarly, the substitution of f_i and f_j in Eq. (10.13) produces (set $u = 1$)

$$b_{ij} = \int_0^L \rho c_p\left[\left(1 - \frac{x}{L}\right)\left(\frac{x}{L}\right)^i\right]$$

$$\times \left[\left(1 - \frac{x}{L}\right)\left(\frac{x}{L}\right)^j\right]dx$$

$$= \rho c_p L\left[\frac{1}{i + j + 1} - \frac{2}{i + j + 2} + \frac{1}{i + j + 3}\right]$$

$$\text{for } i = 1, 2, \text{ and } j = 1, 2 \qquad (10.26)$$

which results in $b_{11} = \rho c_p L/30$, $b_{12} = b_{21} = \rho c_p L/60$, $b_{22} = \rho c_p L/105$. After substituting matrices **A** and **B** in Eq. (10.11) and putting parameters in the dimensionless form, one obtains

$$
\begin{bmatrix}
-\dfrac{1}{3} + \dfrac{\gamma L^2}{\alpha}\dfrac{1}{30} & -\dfrac{1}{6} + \dfrac{\gamma L^2}{\alpha}\dfrac{1}{60} \\[2ex]
-\dfrac{1}{6} + \dfrac{\gamma L^2}{\alpha}\dfrac{1}{60} & -\dfrac{2}{15} + \dfrac{\gamma L^2}{\alpha}\dfrac{1}{105}
\end{bmatrix}
\begin{bmatrix} d_{n1} \\[2ex] d_{n2} \end{bmatrix} = 0
\qquad (10.27)
$$

Notice that this square matrix is symmetric, as discussed in the derivation of Eqs. (10.14) and (10.16). Unlike Example 10.1, the off-diagonal elements are not equal to zero; therefore, the basis functions f_1 and f_2 are not orthogonal. However, it is easy to show that ψ_n's are orthogonal; see Problem 10.7.

Since the two equations described by Eq. (10.27) are homogeneous, the values of d_{n1} and d_{n2} exist if the determinant of their coefficients is zero, that is,

$$
\left(-\frac{1}{3} + \frac{\gamma L^2/\alpha}{30} \right)\left(-\frac{2}{15} + \frac{\gamma L^2/\alpha}{105} \right) - \left(-\frac{1}{6} + \frac{\gamma L^2/\alpha}{60} \right)^2
$$

$$
= \left(\frac{1}{3150} - \frac{1}{3600} \right)\left(\frac{\gamma L^2}{\alpha} \right)^2 + \left(\frac{1}{180} - \frac{1}{315} - \frac{1}{225} \right)
$$

$$
\times \left(\frac{\gamma L^2}{\alpha} \right) + \left(\frac{2}{45} - \frac{2}{36} \right) = 0
\qquad (10.28)
$$

The solution of this quadratic equation yields $\gamma_1 L^2/\alpha = 10$, $\gamma_2 L^2/\alpha = 42$. When $n = 1$, the value of γ_1 is substituted in Eq. (10.27). Note that Eq. (10.27) is homogeneous and one of the d's can be selected arbitrarily. After selecting $d_{11} =$

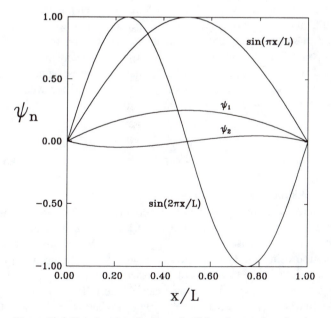

Figure 10.1 Eigenfunctions for Example 10.2 and exact eigen functions.

1, either one of two equations yields $d_{12} = 0$. Repeating this process, but using $n = 2$ and $d_{22} = 1$, gives $d_{21} = -\frac{1}{2}$. The eigenfunctions ψ_1 and ψ_2, using Eq. (10.24) are plotted on Fig. 10.1. For comparison, the corresponding eigenfunctions of the exact solution are plotted on the same figure. Except for a scale factor, the shape of the ψ_1 and $\sin(\pi x/L)$ are similar. Notice that ψ_2 and $\sin(2\pi x/L)$ have opposite signs which will be accounted for when calculating the coefficient c_2. The function $T = \Theta$ using Eq. (10.24) in Eq. (10.6) becomes

$$T = \sum_{n=1}^{N} c_n \psi \exp(-\gamma_n t)$$

$$= c_1(d_{11}f_1 + d_{12}f_2) \exp(-\gamma_1 t)$$

$$+ c_2(d_{21}f_1 + d_{22}f_2) \exp(-\gamma_2 t)$$

$$= c_1(1)\left(1 - \frac{x}{L}\right)\frac{x}{L} \exp\left(-\frac{10\alpha t}{L^2}\right)$$

$$+ c_2\left[-\frac{1}{2}\left(1 - \frac{x}{L}\right)\frac{x}{L}\right.$$

$$\left. + (1)\left(1 - \frac{x}{L}\right)\left(\frac{x}{L}\right)^2\right]\exp\left(-\frac{42\alpha t}{L^2}\right) \qquad (10.29)$$

The solution is complete, except for the evaluation of c_1 and c_2 which are found from the initial condition.

Applying the initial condition, $T(0, x) = T_0$ at $t = 0$, to Eq. (10.29) and multiplying the resulting relation by $f_1 = (1 - x/L)(x/L)$ and then integrating over x from 0 to L results in one equation. Repeating this process but using $f_2 = (1 - x/L)(x/L)^2$ produces a second equation. The simultaneous solution of these two equations yields $c_1 = 5T_0$ and $c_2 = 0$. Because ψ_1 is smaller than $\sin(\pi x/L)$ in Example 10.1, the calculated value of c_1 is larger than the corresponding value of $4t_0/\pi$ obtained in Example 10.1. The solution in this example, as well as in Example 10.1, is for $T(0, x) = T_0$, which gives $c_2 = 0$. Therefore, the resulting solution

$$\frac{T}{T_0} = 5\left(1 - \frac{x}{L}\right)\left(\frac{x}{L}\right)\exp\left(-\frac{10\alpha t}{L^2}\right) \qquad (10.30)$$

is also a one-term solution.

Table 10.1 provides a comparison of the temperatures at $x = 0.5L$ obtained from the one-term solutions given in Examples 10.1 and 10.2 with the exact solution. Also, the polynomial-based solutions for $N = 3$ and $N = 5$ are recorded. The one-term solution using the GBI method is usually less accurate when $\alpha t/L^2 > 0.06$ than the results of the first term of the exact solution. However, when $N = 3$ (actually a two-term solution), the accuracy of the GBI solution is substantially improved (within 0.1% when $\alpha t/L^2 \geq 0.06$). When $N = 5$, the GBI solution exhibits extremely good accuracy (within 0.013% at $\alpha t/L^2 \geq 0.06$).

Table 10.1 Comparison of $T(0.5L, t)$ using GBI solution and exact solution for a slab in Examples 10.1 and 10.2

$\dfrac{\alpha t}{L^2}$	GBI Solution			Exact solution (one term)	Exact solution
	$N = 1$	$N = 3$	$N = 5$		
0.02	1.0234	0.9927	0.97605	1.0452	0.97516
0.04	0.8379	0.8505	0.84629	0.8579	0.84580
0.06	0.6860	0.7028	0.70229	0.7043	0.70220
0.08	0.5617	0.5775	0.57777	0.5781	0.57775
0.10	0.4598	0.4741	0.47449	0.4745	0.47449
0.15	0.2789	0.2894	0.28971	0.2897	0.28971
0.20	0.1692	0.1767	0.17687	0.1769	0.17687
0.25	0.1026	0.1079	0.10798	0.1080	0.10798
0.30	0.0622	0.0659	0.06592	0.0659	0.06592

10.2.2 Numerical Calculation of Eigenvalues

As a generalized and formal procedure, the computation of the temperature T and subsequent determination of the GF can be accomplished by algebraic manipulation of $N \times N$ square matrices **A** and **B**. The next step is the evaluation of the needed eigenvalues. The eigenvalues $\gamma_1, \gamma_2, \ldots, \gamma_N$ can be obtained analytically if N is small; otherwise, numerical steps may become necessary. In Example 10.2, the method of calculating eigenvalues and eigenvectors when $N = 2$ was discussed. However, when N is larger than four, the eigenvalues must be computed numerically. There are many numerical methods available in the literature (Carnahan et al., 1969) with various degrees of efficiency. In order to utilize these eigenvalue-solving routines, Eq. (10.11) should be reduced to the following form:

$$(\overline{\mathbf{A}} + \gamma_n \mathbf{I}) \overline{\mathbf{d}}_n = 0 \tag{10.31}$$

where **I** is the identity matrix and $\overline{\mathbf{d}}_n$ is a column vector with N elements.

The symmetric nature of matrices **A** and **B** permits accurate and fast numerical computation of eigenvalues and eigenvectors, e.g., by the Jacobi method (Carnahan et al., 1969). An accurate method is to use the Cholesky decomposition (Forsyth and Moler, 1967) to decompose matrix **B** into $\mathbf{L} \cdot \mathbf{L}^T$, where **L** is a lower triangular matrix and $\mathbf{L}^T$ is its transposed matrix. A small FORTRAN subroutine that will perform algebra for matrix decomposition is given in Note 2. This decomposition of matrix **B** is instrumental in reducing Eq. (10.11), following some elementary matrix algebra, to

$$(\mathbf{L}^{-1} \cdot \mathbf{A} \cdot \mathbf{L}^{-T} + \gamma_n \mathbf{I}) \cdot \overline{\mathbf{d}}_n = 0 \tag{10.32}$$

where $\mathbf{L}^{-1}$ and $\mathbf{L}^{-T}$ are the inverses of **L** and $\mathbf{L}^T$, respectively. Now, Eq. (10.32) has an acceptable form of Eq. (10.31) for all available eigenvalue solvers. Since matrix $\mathbf{L}^{-1} \cdot \mathbf{A} \cdot \mathbf{L}^{-T}$ is symmetric, the computationally efficient Jacobi method (Carnahan et al., 1969), can be used to find the eigenvalues and eigenvectors of Eq. (10.32). The eigenvectors $\mathbf{d}_n$, computed using Eq. (10.32), are different from but related to the eigenvectors of Eq. (10.11) through the relation

$$\mathbf{d}_n = \mathbf{L}^{-T} \cdot \bar{\mathbf{d}}_n \qquad (10.33)$$

A FORTRAN subroutine that uses the Jacobi method to compute the eigenvalues and eigenvectors is given in Carnahan et al. (1969).

Once the eigenvalues are found, the values of the coefficient d_{nn}, for $n = 1, 2, \ldots, N$, may be selected equal to unity without any loss of generality. For convenience of analysis, matrix $\mathbf{D}$ is defined so that its nth row has the components of the eigenvector $\mathbf{d}_n$; the components are $d_{n1}, d_{n2}, \ldots, d_{nN}$

$$\mathbf{D} = \begin{bmatrix} \mathbf{d}_1^T \\ \mathbf{d}_2^T \\ \cdot \\ \cdot \\ \mathbf{d}_N^T \end{bmatrix} = \begin{bmatrix} d_{11} & d_{12} & \cdots & d_{1N} \\ d_{21} & d_{22} & \cdots & d_{2N} \\ \cdots\cdots\cdots\cdots\cdots\cdots \\ \cdots\cdots\cdots\cdots\cdots\cdots \\ d_{N1} & d_{N2} & \cdots & d_{NN} \end{bmatrix} \qquad (10.34)$$

Often, standard subroutine packages place eigenvectors in the columns of a matrix which must be transposed to obtain matrix $\mathbf{D}$.

10.2.3 Nonhomogeneous Solution

The objective of the following derivation is to solve Eq. (10.3) which yields an expression for the GF. Equation (10.3) is essentially the same as Eq. (10.1), except the volume energy source term in Eq. (10.1) is specified in Eq. (10.3). A solution for the nonhomogeneous equation, Eq. (10.1), is now proposed by considering c_n in Eq. (10.6) to be time dependent. The variation of parameters method is used to solve the nonhomogeneous, first-order, ordinary differential equations. Now, a general solution is considered as

$$T = \sum_{n=1}^{N} c_n(t)\psi_n(\mathbf{r})e^{-\gamma_n t} \qquad (10.35)$$

Equation (10.35) is an acceptable solution if it can satisfy the basic differential equation, Eq. (10.1). The substitution of Eq. (10.35) into Eq. (10.1), followed by multiplying both sides of the resulting equation by f_i, for $i = 1, 2, \ldots, N$, and then integrating over the volume yields

$$\sum_{n=1}^{N} c_n \left[\int_V \nabla \cdot (k\nabla\psi_n) f_i dV - \int_V m(\mathbf{r})^2 f_i \psi_n dV \right.$$

$$\left. + \gamma_n \int_V \rho(\mathbf{r})c_p(\mathbf{r})u(\mathbf{r})\psi_n f_i dV \right] e^{-\gamma_n t}$$

$$+ \int_V f_i g(\mathbf{r}, t) dV - \sum_{n=1}^{N} \left[\frac{dc_n(t)}{dt} \right] e^{-\gamma_n t}$$

$$\times \int_V \rho(\mathbf{r})c_p(\mathbf{r})u(\mathbf{r})\psi_n f_i dV = 0 \qquad (10.36)$$

The above procedure is called the Galerkin method (Kantorovich and Krylov, 1960). The first summation term is zero for any value of n because of Eq. (10.9). The remaining two terms constitute a system of N ordinary differential equations

$$\sum_{n=1}^{N} \left[\frac{dc_n(t)}{dt} \right] e^{-\gamma_n t} \int_V \rho(\mathbf{r}) c_p(\mathbf{r}) u(\mathbf{r}) \psi_n f_i dV = g_i^* \qquad (10.37)$$

where

$$g_i^* = \int_V [g(\mathbf{r}, t)] f_i(\mathbf{r}) dV \qquad (10.38)$$

for $i = 1, 2, \ldots, N$. Note that the homogeneous partial differential equation, Eq. (10.5) and the nonhomogeneous partial differential equation, Eq. (10.1), are approximated by the Galerkin integral procedure. Also, the solution for T satisfies homogeneous boundary conditions.

Once the expression for $\psi_n(\mathbf{r})$ from Eq. (10.8) is substituted into Eq. (10.37), the result will be (see Problem 10.8)

$$\sum_{n=1}^{N} e_{in} \left[\frac{dc_n(t)}{dt} \right] e^{-\gamma_n t} = g_i^* (t) \qquad (10.39)$$

where

$$e_{in} = \sum_{j=1}^{N} d_{nj} b_{ji} \qquad (10.40)$$

and $i = 1, 2, \ldots, N$. Therefore, e_{in} in Eq. (10.40) is an element of the square of matrix **E**. Matrix **E** is also obtained if **D**, Eq. (10.34), is multiplied by **B**, whose elements are defined by Eq. (10.13), and the resulting matrix transposed

$$\mathbf{E} = (\mathbf{DB})^T \qquad (10.41)$$

Let

$$\chi_n = \left[\frac{dc_n(t)}{dt} \right] \exp(-\gamma_n t) \qquad (10.42a)$$

in Eq. (10.39); then the following set of N simultaneous equations

$$\sum_{n=1}^{N} e_{in} \chi_n = g_i^* \qquad \text{for } i = 1, 2, \ldots, N \qquad (10.42b)$$

are obtained. They can be presented in matrix form as

$$\mathbf{E} \cdot \{X\} = \{\mathbf{g}^*\} \qquad (10.43)$$

The notation $\{\cdot\}$ indicates that the arrays X and $\mathbf{g}^*$ in Eq. (10.43) are column vectors.

Since the elements of the vector $\mathbf{g}^* = \{g_1^*, g_2^*, \ldots, g_N^*\}$ are known, the elements of the array X can be calculated if both sides of Eq. (10.43) are premultiplied by $\mathbf{E}^{-1}$, the inverse of matrix **E**, to obtain

$$\{X\} = \mathbf{E}^{-1} \cdot \{\mathbf{g}^*\} \tag{10.44}$$

The matrix $\mathbf{E}^{-1}$ may be designated as $\mathbf{P}$

$$\mathbf{P} = \mathbf{E}^{-1} \tag{10.45}$$

with elements p_{ni}. Then the elements of array X given by Eq. (10.42a) are determined by Eq. (10.44) as

$$\chi_n = \left[\frac{dc_n(t)}{dt}\right] e^{-\gamma_n t} = \sum_{i=1}^{N} p_{ni}\, g_i^*\,(t) \qquad n = 1, 2, \ldots, N \tag{10.46}$$

which can be solved for $dc_n(t)/dt$ to obtain

$$\frac{dc_n(t)}{dt} = \sum_{i=1}^{N} p_{ni}\, g_i^*\,(t) e^{\gamma_n t} \qquad n = 1, 2, \ldots, N \tag{10.47}$$

in which p_{ni}'s are constants given by Eq. (10.45).

The integration of Eq. (10.47) yields the function $c_n(t)$ as

$$c_n(t) = A_n + \sum_{i=1}^{N} p_{ni} \int_0^t g_i^*\,(t')e^{\gamma_n t'}\,dt' \qquad n = 1, 2, \ldots, N \tag{10.48}$$

where A_n represents the constant of integration. The expression for $c_n(t)$ is now known. Substitution of Eq. (10.48) into Eq. (10.35) yields the final form of the solution

$$T = \sum_{n=1}^{N} \psi_n(\mathbf{r})e^{-\gamma_n t}\left[A_n + \sum_{i=1}^{N} p_{ni} \int_0^t g_i^*\,(t')\exp{(\gamma_n t')}dt'\right] \tag{10.49}$$

The second term in the square bracket represents the contribution of the internal energy source. The solution presented by Eq. (10.49) is completed after A_n is evaluated. The initial condition (the $t = 0$ temperature distribution), $F(\mathbf{r})$, can be utilized to compute the constant A_n. When $t = 0$, the integral in Eq. (10.49) vanishes and the resulting equation is

$$F(\mathbf{r}) = \sum_{n=1}^{N} \psi_n(\mathbf{r})A_n \tag{10.50}$$

When calculating the GF, the initial temperature $F(\mathbf{r}) = 0$; hence, $A_n = 0$. However, to show the equivalence between the GF solution and the GBI solution when boundary conditions are homogeneous, the calculation of A_n is necessary.

The calculation of A_n when $F(\mathbf{r})$ is nonzero can be carried out by the GBI method (Section 10.2.6 shows that the following procedure agrees with the GFSE). The procedure to determine A_n is to multiply both sides of Eq. (10.50) by $\rho c_p u(\mathbf{r}) f_i(\mathbf{r})dV$ and to integrate over the volume. Then, using Eq. (10.8) for $\psi_n(\mathbf{r})$ results in a set of N linear algebraic equations for evaluating $A_1, A_2, \ldots, A_N$,

$$\sum_{n=1}^{N} A_n e_{in} = \lambda_i \qquad \text{for } i = 1, 2, \ldots, N \tag{10.51}$$

where

$$\lambda_i = \int_V \rho(\mathbf{r})c_p(\mathbf{r})u(\mathbf{r})F(\mathbf{r})f_i(\mathbf{r})dV \qquad (10.52)$$

in which e_{in} is the element of matrix $\mathbf{E}$ and defined in Eq. (10.40). The inverse of matrix $\mathbf{E}$ is given as $\mathbf{P}$ by Eq. (10.45). The coefficients A_1, A_2, A_N, are obtained when matrix $\mathbf{P}$ is multiplied by a column vector whose elements are $\lambda_1, \lambda_2, \ldots, \lambda_N$ as

$$A_n = \sum_{i=1}^{N} p_{ni}\lambda_i$$

$$= \sum_{i=1}^{N} p_{ni} \int_V \qquad (10.53)$$

$$\times \rho(\mathbf{r})c_p(\mathbf{r})u(\mathbf{r})F(\mathbf{r})f_i(\mathbf{r})dV \qquad \text{for } n = 1, 2, \ldots, N$$

The coefficients $A_1, A_2, \ldots, A_N$ are analogous to the Fourier coefficients in the exact solutions.

Equation (10.49), following the substitution of g_i^* from Eq. (10.38) and A_n from Eq. (10.53), becomes

$$T = \sum_{n=1}^{N} \sum_{i=1}^{N} p_{ni}\psi_n(\mathbf{r}) \int_V e^{-\gamma_n t} \rho(\mathbf{r}^*)c_p(\mathbf{r}^*)u(\mathbf{r}^*)F(\mathbf{r}^*)f_i(\mathbf{r}^*) \, dV^*$$

$$+ \sum_{n=1}^{N} \sum_{i=1}^{N} p_{ni}\psi_n(\mathbf{r}) \int_0^t \int_V e^{-\gamma_n(t-t')} g(\mathbf{r}^*, t')f_i(\mathbf{r}^*) \, dV^* \, dt' \qquad (10.54)$$

where $\mathbf{r}^*$ and t' are dummy variables of integration, dV^* is the volume element in $\mathbf{r}^*$ space, and $\psi_n(\mathbf{r})$ is obtained from Eq. (10.8).

10.2.4 Green's Functions Expression

It is now possible to obtain an expression for the GF. Equation (10.54) is the solution of Eq. (10.1) when the boundary conditions are homogeneous and the initial temperature distribution is $F(\mathbf{r})$. The temperature T in Eq. (10.54) is identical to $G(\mathbf{r}, t|\mathbf{r}', \tau)$ if $F(\mathbf{r}) = 0$ and $g(\mathbf{r}, t) = \rho(\mathbf{r})c_p(\mathbf{r})\delta(\mathbf{r} - \mathbf{r}')\delta(t - \tau)$; see Eq. (10.2). Because $F(\mathbf{r}) = 0$, the first term on the right side of Eq. (10.54) is zero. The next step is to replace the variable $\mathbf{r}$ and t in $g(\mathbf{r}, t)$ by $\mathbf{r}^*$ and t', and insert $g(\mathbf{r}^*, t') = \rho(\mathbf{r}^*)c_p(\mathbf{r}^*)\delta(\mathbf{r}^* - \mathbf{r}')\delta(t' - \tau)$ in Eq. (10.54). After performing the integration over $\mathbf{r}^*$ and t' and using the Identity 6 in Note 1, the GF becomes

$$G(\mathbf{r}, t|\mathbf{r}', \tau) = C(\mathbf{r}') \sum_{n=1}^{N} \sum_{j=1}^{N} \sum_{i=1}^{N} d_{nj}p_{ni} \exp\left[-\gamma_n(t - \tau)\right]f_j(\mathbf{r})f_i(\mathbf{r}') \qquad (10.55)$$

where $C(\mathbf{r}') = \rho(\mathbf{r}')c_p(\mathbf{r}')$, and d_{nj} and p_{ni} are numbers.

10.2.5 Properties of Green's Functions

The GF defined by Eq. (10.55) has the following three properties:

1. If t is replaced by $-\tau$ and τ by $-t$, the following GF property applies:

$$G(\mathbf{r}, t|\mathbf{r}', \tau) = G(\mathbf{r}, -\tau|\mathbf{r}', -t) \tag{10.56}$$

   This can readily be proved by replacing t by $-\tau$ and τ by $-t$ in Eq. (10.55).

2. The GF remains the same if $\mathbf{r}$ is changed to $\mathbf{r}'$ and $\mathbf{r}'$ to $\mathbf{r}$, provided $C(\mathbf{r})$ or $\rho(\mathbf{r})c_p(\mathbf{r})$ is constant,

$$G(\mathbf{r}, t|\mathbf{r}', \tau) = G(\mathbf{r}', t|\mathbf{r}, \tau) \tag{10.57}$$

3. It is also possible to derive the following GF relation when $C(\mathbf{r})$ is variable.

$$\frac{G(\mathbf{r}, t|\mathbf{r}', \tau)}{C(\mathbf{r}')} = \frac{G(\mathbf{r}', t|\mathbf{r}, \tau)}{C(\mathbf{r})} \tag{10.58}$$

The above GF properties are useful in the derivation of the GF solution discussed in Section 10.2.6.

The derivation of the second and third properties of the GF is accomplished by considering that the temperature at point $\mathbf{r}'$ is caused by an energy source located at the point $\mathbf{r}$. The temperature distribution is the solution of the equation

$$\nabla_0 \bullet [k(\mathbf{r}')\nabla_0 G(\mathbf{r}', t|\mathbf{r}, \tau)] + C(\mathbf{r}')\delta(\mathbf{r}' - \mathbf{r})\delta(t - \tau)$$

$$- m(\mathbf{r}')^2 G(\mathbf{r}', t|\mathbf{r}, \tau) = C(\mathbf{r}')u(\mathbf{r}')\frac{\partial G(\mathbf{r}', t|\mathbf{r}, \tau)}{\partial t} \tag{10.59}$$

which is Eq. (10.3) with $\mathbf{r}$ and $\mathbf{r}'$ interchanged; the del operator ∇_0 uses the components of the $\mathbf{r}'$ position vector. The solution of Eq. (10.59) is identical to that of Eq. (10.3), except $\mathbf{r}$ and $\mathbf{r}'$ have switched places. Repeating the same algebraic steps that led to the derivation of Eq. (10.55) yields

$$G(\mathbf{r}', t|\mathbf{r}, \tau) = C(\mathbf{r}) \sum_{n=1}^{N} \sum_{j=1}^{N} \sum_{i=1}^{N} d_{nj}p_{ni} \exp\left[-\gamma_n(t - \tau)\right]f_j(\mathbf{r}')f_i(\mathbf{r}) \tag{10.60}$$

Because the volume integrals and other algebraic operations used to compute p_{ni} and d_{nj} are not affected by switching $\mathbf{r}$ and $\mathbf{r}'$, one can conclude from a comparison of Eqs. (10.55) and (10.60) that Eq. (10.58) is valid. Equation (10.58) implies that the GF remains the same when $\mathbf{r}$ and $\mathbf{r}'$ are switched if $C(\mathbf{r})$ is a constant and Eq. (10.57) is valid.

In the above analysis, it is possible to modify the source term, Eq. (10.2), [by omitting $C(\mathbf{r})$] so that the GF remains symmetric in $\mathbf{r}$ and $\mathbf{r}'$ even when the properties are variable. However, the GF is not modified here in order to adhere to the existing technical literature. The three properties of the GF described above are essential when deriving the GFSE.

10.2.6 Green's Function Solution Equation

The purpose of this section is to derive an equation for the temperature distribution in terms of the GF. The solution will consider the effects of nonzero initial conditions, distributed volumetric energy source, and nonhomogeneous boundary conditions of the first, second, and third kinds. The body may be nonhomogeneous (i.e., composed of several different materials with different k and ρc_p values) and have irregular shapes. As usual, the solution is restricted to linear problems which means that k and ρc_p cannot be functions of temperature.

The del operator ∇ in Eq. (10.3) uses the components of $\mathbf{r}$ (not $\mathbf{r}'$). The del operator ∇_0 is defined earlier that uses the components of $\mathbf{r}'$. If $\mathbf{r}$ is now replaced by $\mathbf{r}'$, $\mathbf{r}'$ by $\mathbf{r}$, t by $-\tau$, and τ by $-t$, Eq. (10.3) becomes

$$\nabla_0 \cdot [k(\mathbf{r}')\nabla_0 G(\mathbf{r}', -\tau | \mathbf{r}, -t)]$$

$$+ C(\mathbf{r}')\delta(\mathbf{r}' - \mathbf{r})\delta(\tau - t) - m(\mathbf{r}')^2 G(\mathbf{r}', -\tau | \mathbf{r}, -t)$$

$$= -C(\mathbf{r}')u(\mathbf{r}') \frac{\partial G(\mathbf{r}', -\tau | \mathbf{r}, -t)}{\partial \tau} \tag{10.61}$$

The diffusion equation, Eq. (10.1), using $\mathbf{r}'$ and τ as the independent variables can be written as

$$\nabla_0 \cdot [k(\mathbf{r}')\nabla_0 T(\mathbf{r}', \tau)] + g(\mathbf{r}', \tau) - m(\mathbf{r}')^2 T(\mathbf{r}', \tau)$$

$$= C(\mathbf{r}')u(\mathbf{r}') \frac{\partial T(\mathbf{r}', \tau)}{\partial \tau} \tag{10.62}$$

where $g(\mathbf{r}', \tau)$ is the contribution of a distributed volumetric energy source. To shorten the equations, the function $G(\mathbf{r}', -\tau | \mathbf{r}, -t)$ will be designated as G.

Equation (10.62) is now multiplied by G, and Eq. (10.61) multiplied by T. The resulting equations are then subtracted from each other to produce

$$T\nabla_0 \cdot [k(\mathbf{r}')\nabla_0 G] - G\nabla_0 \cdot [k(\mathbf{r}')\nabla_0 T]$$

$$+ C(\mathbf{r}')T\delta(\mathbf{r}' - \mathbf{r})\delta(\tau - t) - Gg(\mathbf{r}', \tau) = -C(\mathbf{r}')u(\mathbf{r}') \frac{\partial(TG)}{\partial \tau} \tag{10.63}$$

The following two relations, derived using Identity 1 in Note 1,

$$\nabla_0 \cdot [Tk(\mathbf{r}')\nabla_0 G] = T\nabla_0 \cdot [k(\mathbf{r}')\nabla_0 G] + k(\mathbf{r}')\nabla_0 T \cdot \nabla_0 G \tag{10.64}$$

$$\nabla_0 \cdot [Gk(\mathbf{r}')\nabla_0 T] = G\nabla_0 \cdot [k(\mathbf{r}')\nabla_0 T] + k(\mathbf{r}')\nabla_0 G \cdot \nabla_0 T \tag{10.65}$$

provide the expressions for the first two terms on the left side of Eq. (10.63) when Eq. (10.64) is subtracted from Eq. (10.65). Upon substituting the results in Eq. (10.63), integrating in $\mathbf{r}'$ space over the volume V, and over τ from 0 to $t^* = t + \epsilon$ where ϵ has a small but positive value, one obtains

$$\int_{\tau=0}^{t^*} \int_V \{\nabla_0 \cdot [k(\mathbf{r}')T\nabla_0 G] - \nabla_0 \cdot [k(\mathbf{r}')G\nabla_0 T]\}dV' \, d\tau$$

$$+ \int_{\tau=0}^{t^*} \int_V C(\mathbf{r}')T\delta(\mathbf{r}' - \mathbf{r})\delta(\tau - t)dV' \, d\tau$$

$$- \int_{\tau=0}^{t^*} \int_V Gg(\mathbf{r}', \tau)dV' \, d\tau = -\int_{\tau=0}^{t^*} \int_V$$

$$\times \left[C(\mathbf{r}')u(\mathbf{r}') \frac{\partial(TG)}{\partial\tau} \right] d\tau \, dV' \qquad (10.66)$$

Various terms in Eq. (10.66) are now considered. Green's theorem, Identity 3 in Note 1, can be used to reduce the first volume integral on the left side of Eq. (10.66) to a surface integral. In addition, Identity 6 in Note 1 reduces the second term on the left side of Eq. (10.66) to become $C(\mathbf{r})T$. Furthermore, the term on the right side of Eq. (10.66) can be readily integrated over τ. Note that

$$\int_{\tau=0}^{t^*} \left[\frac{\partial(GT)}{\partial\tau} \right] d\tau = G(\mathbf{r}', -t^*|\mathbf{r}, -t)T(\mathbf{r}', t^*) - G|_{\tau=0} T(\mathbf{r}', 0) \quad (10.67a)$$

and the value of the GF, G, at the upper limit when $\tau = t^*$ is [see Eqs. (10.56) and (10.58)]

$$G(\mathbf{r}', -t^*|\mathbf{r}, -t) = \frac{G(\mathbf{r}, t|\mathbf{r}', t^*)C(\mathbf{r})}{C(\mathbf{r}')} = 0 \qquad (10.67b)$$

which is the value of temperature at time t when a pulse appears at a later time, $t^* = t + \epsilon$; hence, the first term on the right side of Eq. (10.67a) is zero. Equation (10.66) then becomes

$$C(\mathbf{r})T(\mathbf{r}, t) = \int_V C(\mathbf{r}')u(\mathbf{r}')G|_{\tau=0} F(\mathbf{r}')dV'$$

$$+ \int_{\tau=0}^{t} d\tau \int_V g(\mathbf{r}', \tau)G \, dV' + \int_{\tau=0}^{t} d\tau \int_S$$

$$\times k(S') \left(G\frac{\partial t}{\partial n} - T\frac{\partial G}{\partial n} \right)_{S'} dS' \qquad (10.68)$$

where $C(\mathbf{r}) = \rho(\mathbf{r})c_p(\mathbf{r})$ and $G = G(\mathbf{r}', -\tau|\mathbf{r}, -t)$. Equation (10.68) is the basic GFSE for heterogeneous and homogeneous materials.

The operator $\partial/\partial n$ designates differentiation along the outer normal to the external surface and $F(\mathbf{r}')$ is the initial temperature distribution. The first term on the right side of Eq. (10.68) is the contribution of the initial temperature distribution. The influence of the volumetric energy source is included in the second term. The boundary conditions for G in Eq. (10.68) are homogeneous. The third term on the right side describes the boundary condition effects. When the surface temperature $T|_{S'}$ is prescribed (boundary condition of the first kind), then $G|_{S'} = 0$. When the heat flux is given (boundary condition of the second kind), $\partial T/\partial n = -(q/k)|_{S'}$, then $\partial G/\partial n|_{S'} = 0$. For convective boundary conditions (boundary conditions of the third kind), the boundary conditions are

$$-k\frac{\partial T}{\partial n} = h(T - T_\infty) \quad \text{on } S' \tag{10.69}$$

and

$$-k\frac{\partial G}{\partial n} = hG \quad \text{on } S' \tag{10.70}$$

where k and h may vary with position. When Eq. (10.69) is multiplied by $G|_S$, and Eq. (10.70) by $T|_S$, and the resulting equations are subtracted from each other, the following relation is obtained

$$\left[G\left(\frac{\partial T}{\partial n}\right) - T\left(\frac{\partial G}{\partial n}\right)\right]\Bigg|_{S'} = \left(\frac{h}{k}\right)GT_\infty|_{S'} \tag{10.71}$$

The right side of Eq. (10.71) then replaces the term in square brackets in Eq. (10.68).

In the derivation of Eq. (10.54) the boundary conditions were considered to be homogeneous. It is of interest to compare Eqs. (10.68) and (10.54). For homogeneous boundary conditions, the third term on the right side of the GFSE, Eq. (10.68), is equal to 0. Then, taking the Green's function, $G = G(\mathbf{r}', -\tau|\mathbf{r}, -t) = G(\mathbf{r}', t|\mathbf{r}, \tau)$, from Eq. (10.60) and substituting it into Eq. (10.68) yields Eq. (10.54). This indicates the procedure used to include the initial conditions in Eq. (10.54) and in Examples 10.1 and 10.2 are consistent with the derivation of the GFSE.

The last term in the GFSE, Eq. (10.68), contains the contribution of nonhomogeneity of the boundary conditions. Boundary conditions of the first and (or) second kinds are nonhomogeneous if the surface temperature and (or) the surface heat flux are nonzero. Boundary conditions of the third kind (convective) are nonhomogeneous if the ambient temperature is nonzero. The convergence of Eq. (10.68), in some cases, is slow. For instance, when the surface temperature is prescribed, the term $f_j(\mathbf{r})$ in the GF, Eq. (10.60), takes the value of zero after $\partial G/\partial n$ is computed over S'. The temperature solution at the surface becomes singular (cannot be computed) and is inaccurate in the vicinity of the surface. Similar situations also exist for other boundary conditions. A GF expression that behaves more favorably and converges more rapidly for nonhomogeneous boundary conditions is derived in the next section.

10.3 ALTERNATIVE FORM OF THE GREEN'S FUNCTION SOLUTION

As discussed in the previous section, when the temperature is prescribed on the external surface, there is a singularity associated with using Eq. (10.68). Equation (10.68) yields a value of zero for the surface temperature because $f_j(\mathbf{r})$ has a zero value at the surface. The implication is that Eq. (10.68) may provide inaccurate temperature values in the vicinity of the wall and erroneous heat flux at the wall. When the wall heat flux is prescribed, the convergence at the boundaries can be very slow.

The following procedure removes this singularity and improves the convergence of the GF solution for the temperature distribution (Haji-Sheikh, 1988; Haji-Sheikh and Beck, 1988). It begins by defining a differentiable temperature function that satisfies the boundary conditions used in Eq. (10.68). This new function is designated as T^*. It is usually possible to find a function T^* such as

$$T^* = c_1 u_p + c_2 \qquad (10.72a)$$

For example, consider a body bounded by two surfaces as shown in Fig. 10.2a. When the surface temperature is prescribed, the steady-state temperature is approximated by

$$T^* = (T_2 - T_1) \frac{\ln (r/r_1)}{\ln (r_2/r_1)} + T_1 \qquad (10.72b)$$

where $r_1 = r_1(\theta, z)$ and $r_2 = r_2(\theta, z)$ are coordinates of two arbitrarily selected surfaces whose respective temperatures are $T_1 = T_1(\theta, z, t)$ and $T_2 = T_2(\theta, z, t)$. For a different geometry, shown in Fig. 10.2b, the following form is sometimes preferred:

$$T^* = (T_2 - T_1) \frac{x - x_1(y, z)}{x_2(y, z) - x_1(y, z)} + T_1 \qquad (10.72c)$$

The function T^* is called the quasi-steady solution if it satisfies the Laplace equation and the prescribed boundary conditions. In this part of the analyses, any internal source can be ignored. However, T^* given by Eqs. (10.72b) or (10.72c) does not always satisfy the Laplace equation, but it will satisfy the Laplace equation in cylindrical coordinates if r_1 and r_2 are constants. Also, T^* given by Eq. (10.72c) will satisfy the Laplace equation in Cartesian coordinates if x_1 and x_2 are constants. In one-dimensional coordinates, except when dealing with prescribed heat flux at both surfaces, it is possible to use Eq. (10.72a) to derive an equation for T^* that satisfies the boundary conditions. The function u_p takes the value of x in Cartesian coordinates and $\ln r$ or $-1/r$ in the radial cylindrical or spherical coordinates, respectively.

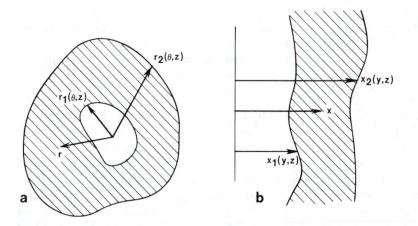

Figure 10.2 (a) Geometry for Eq. (10.72b), and (b) geometry for Eq. (10.72c).

Except when the heat flux is prescribed on both surfaces, the constants c_1 and c_2 can be determined by applying the appropriate boundary conditions. The calculation of c_1 and c_2, for nonhomogeneous boundary conditions of the second and third kinds, is included in Examples 10.5 and 10.6. However, when the heat flux on both surfaces is given, the constant c_2 in Eq. (10.72a) should be replaced by $c_2 r^2$ before calculating c_1 and c_2; see Example 10.7. Although there are numerous conduction problems for which a T^* can be computed, it is sometimes impossible or cumbersome to find this function for many problems, e.g., locally varying heat flux and heat transfer coefficients in multidimensional bodies. In the absence of a suitable T^*, the partitioning of the GF discussed in Chapter 5 is a logical approach.

Whenever an auxiliary function T^* is available, a function $f^*(\mathbf{r}', \tau)$ is defined (Haji-Sheikh and Beck, 1988) so that

$$\nabla_0 \cdot [k\nabla_0 T^*(\mathbf{r}', \tau)] = f^*(\mathbf{r}', \tau) \tag{10.73}$$

The function $f^*(\mathbf{r}', \tau)$ defined by Eq. (10.73) is unrelated to the basis functions $f_j(\mathbf{r})$. When Eq. (10.73) is multiplied by G and Eq. (10.3) is multiplied by $T^* = T^*(\mathbf{r}', \tau)$, then, after subtracting the former from the latter, the following equation is obtained.

$$T^*\nabla_0 \cdot (k\nabla_0 G) - G\nabla_0 \cdot (k\nabla_0 T^*) + T^*C(\mathbf{r}')\delta(\mathbf{r}' - \mathbf{r})\delta(\tau - t)$$

$$= -Gf^* - T^*C(\mathbf{r}')u(\mathbf{r}') \frac{\partial G}{\partial \tau} \tag{10.74}$$

Integration of Eq. (10.74) over τ is carried out between the limits of 0 and $t^* = t + \epsilon$, where ϵ is a small positive number. Then integration with respect to $\mathbf{r}'$ over the entire volume, application of the Green's theorem (see Note 1), and reduction of some algebraic terms, results in

$$\int_{\tau=0}^{t^*} d\tau \int_V k\left(G \frac{\partial T^*}{\partial n} - T^* \frac{\partial G}{\partial n}\right)\Bigg|_{S'} dS'$$

$$= C(\mathbf{r})T^* + \int_V C(\mathbf{r}')u(\mathbf{r}') \left[\int_{\tau=0}^{t^*} T^*\left(\frac{\partial G}{\partial \tau}\right)d\tau\right]dV'$$

$$+ \int_{\tau=0}^{t^*} d\tau \int_V Gf^* \, dV' \tag{10.75}$$

Integrating by parts and then letting ϵ go to zero, the term in the square bracket in Eq. (10.75) becomes

$$\lim_{\epsilon\to 0} \int_{\tau=0}^{t^*} T^*\left(\frac{\partial G}{\partial \tau}\right)d\tau = \lim_{\epsilon\to 0}\left[GT^*|_{\tau=0}^{t^*} - \int_{\tau=0}^{t^*} G\left(\frac{\partial T^*}{\partial \tau}\right)d\tau\right]$$

$$= -G|_{\tau=0}\, T^*(\mathbf{r}, 0) - \int_{\tau=0}^{t} G\left(\frac{\partial T^*}{\partial \tau}\right)d\tau \tag{10.76}$$

The substitution of Eq. (10.76) into Eq. (10.75) followed by the substitution of the resulting equation into Eq. (10.68), the basic GFSE for heterogeneous materials, and some minor algebraic simplifications produces the alternative form of the GF solution (Haji-Sheikh and Beck, 1988) for $T(\mathbf{r}, t)$,

$$
\begin{aligned}
C(\mathbf{r})T(\mathbf{r}, t) = C(\mathbf{r})T^*(\mathbf{r}, t) &+ \int_{\tau=0}^{t} d\tau \int_{V} G \Bigg[g(\mathbf{r}', \tau) \\
&- C(\mathbf{r}')u(\mathbf{r}')\frac{\partial T^*(\mathbf{r}', \tau)}{\partial \tau} \Bigg] dV' \\
&+ \int_{V} C(\mathbf{r}')u(\mathbf{r}')G \big|_{\tau=0} [F(\mathbf{r}') - T^*(\mathbf{r}', 0)] dV' \\
&+ \int_{\tau=0}^{t} d\tau \int_{V} Gf^* \, dV' \qquad (10.77)
\end{aligned}
$$

An expression for the function $G = G(\mathbf{r}', -\tau | \mathbf{r}, -t) = G(\mathbf{r}', t | \mathbf{r}, \tau)$ is given by Eq. (10.60). The function T^* contains the contribution of nonhomogeneous boundary conditions. If $f^*(\mathbf{r}', \tau)$ is zero, then $T^*(\mathbf{r}', \tau)$ satisfies the Laplace equation and it is the quasi-steady solution; accordingly, the term that contains f^* in Eq. (10.77) is equal to zero.

As a special case, when the temperature of the entire surface T_s has a constant value and is different from $F(\mathbf{r}) = T_0 = $ constant and $g = 0$, then $T^* = T_s, f^* = 0$, and Eq. (10.77) reduces to

$$
C(\mathbf{r}) \left[\frac{T(\mathbf{r}, \tau) - T_s}{T_0 - T_s} \right] = \int_{V} C(\mathbf{r}')u(\mathbf{r})G \big|_{\tau=0} \, dV' \qquad (10.78)
$$

Example 10.3 Consider a slab of isotropic and homogeneous material with thickness L and which is initially at zero temperature, $F(x) = 0$. Assume thermophysical properties are constants and there is no volumetric energy source, $g = 0$. The boundary conditions are $T(0, t) = 0$ and $T(L, t) = T_L \sin(\omega t)$. Use Eq. (10.77) to calculate temperature at $x = L/2$ and compare the results with the exact solution.

SOLUTION A function T^*, using Eq. (10.72a), that satisfies both boundary conditions is $T^* = T_L(x/L) \sin(\omega t)$. In this example, the value of $\nabla^2 T^* = d^2T^*/dx^2 = f^*/k = 0$; see Eq. (10.73). The last term in Eq. (10.77) is zero and the function T^* is called the quasi-steady solution. The solution using Eq. (10.77) is

$$
T = T^* - \int_0^t d\tau \int_0^L G\left(\frac{\partial T^*}{\partial \tau}\right) dx' \qquad (10.79)
$$

since $g = 0$ and $F(x) - T^*(x, 0) = 0$. The basis functions, $X11$ case, are presented in Example 10.2 as $f_j = (1 - x/L)(x/L)^j$. The elements of matrices $\mathbf{A}$ and $\mathbf{B}$ are given by Eqs. (10.25) and (10.26). The eigenvalues are calculated as discussed in Example 10.2. For $N = 3$, the eigenvalues are $\gamma_1 = 9.86975\alpha/L^2$, $\gamma_2 = 42\alpha/L^2$, and $\gamma_3 = 102.13\alpha/L^2$. Using Eqs. (10.11) matrix $\mathbf{D}$ is

$$\mathbf{D} = \begin{bmatrix} 1 & 1.1331 & -1.1331 \\ -0.5 & 1 & 0 \\ 0.21584 & -1 & 1 \end{bmatrix} \tag{10.80}$$

Equations (10.41) and (10.45) are used to calculate the elements of matrix $\mathbf{P}$:

$$\mathbf{P} = \begin{bmatrix} 19.395 & 21.977 & -21.977 \\ -420 & 840 & 0 \\ 3802 & -17615 & 17615 \end{bmatrix} \tag{10.81}$$

The GF is obtained from Eq. (10.60) as

$$G(x', -\tau|x, -t) = C(\mathbf{r}) \sum_{n=1}^{N} \sum_{j=1}^{N} \sum_{i=1}^{N}$$

$$\times d_{nj}p_{ni} \exp\left[-\gamma_n(t - \tau)\right]f_j(x')f_i(x) \tag{10.82}$$

where $f_j = (1 - x/L)(x/L)^j$. The final solution is obtained by substituting the GF from Eq. (10.82) into Eq. (10.79) as

$$\frac{T}{T_L} = \frac{x}{L}\sin(\omega t) - \omega \sum_{n=1}^{N}$$

$$\times \left[\sum_{j=1}^{N} d_{nj}\left(1 - \frac{x}{L}\right)\left(\frac{x}{L}\right)^j\right]\left[\sum_{i=1}^{N} p_{ni}\left(\frac{1}{i+2} - \frac{1}{i+3}\right)\right]$$

$$\times \frac{\gamma_n \cos(\omega t) + \omega \sin(\omega t) - \gamma_n \exp(-\gamma_n t)}{\gamma_n^2 + \omega^2} \tag{10.83}$$

For $N = 2$ and $N = 3$, the results obtained by this equation are given in Table 10.2. The exact solution is [Ozisik, 1980, Eq. (5–50), p. 203]

$$\frac{T}{T_L} = \frac{x}{L}\sin(\omega t) + \frac{2\omega}{\pi}\sum_{n=1}^{\infty}\frac{(-1)^n}{n}\sin\left(\frac{n\pi x}{L}\right)$$

$$\times \frac{\alpha(n\pi/L)^2\{\cos(\omega t) - \exp[-(n\pi/L)^2 \alpha t]\} + \omega \sin(\omega t)}{\alpha^2(n\pi/L)^4 + \omega^2} \tag{10.84}$$

which is used to check the accuracy of the alternative GF solution. The entries in Table 10.2 are for $x = 0.5L$ and $L^2\omega/\alpha = \pi/10$. The results for $N = 2$ and $N = 3$ are quite accurate. Table 10.2 shows that a three-term solution yields results comparable to 10 terms of the exact solution. Even for $N = 2$, the solution closely agrees with the exact solution when the dimensionless time is larger than 0.2. A significantly better agreement with the exact solution is attributed to the lack of step change in the surface temperature. In addition to removing the singularity at the surface associated with Eq. (10.68), the alternative form of the

Table 10.2 Results for Example 10.3 for $L^2\omega/\alpha = \pi/10$. Comparison of $T(0.5L, t)$ using the alternative Green's function solution and exact solution for a slab

$\dfrac{\alpha t}{L^2}$	AGFS[†], Eq. (10.83)		Exact solution, Eq. (10.84)	
	$N = 2$	$N = 3$	$N = 10$	$N = 30$
0.1	0.00330	0.00362	0.00362	0.00365
0.2	0.01443	0.01459	0.01458	0.01459
0.5	0.05888	0.05889	0.05888	0.05889
1	0.13566	0.13566	0.13565	0.13566
2	0.27766	0.27765	0.27764	0.27765
3	0.39248	0.39246	0.39245	0.39246
4	0.46888	0.46886	0.46885	0.46886
5	0.49938	0.49936	0.49936	0.49936
6	0.48098	0.48098	0.48098	0.48098

[†]Alternative GF solution, Eq. (10.77).

GF solution, Eq. (10.77), has another advantage; it provides a faster converging solution than Eq. (10.68) when t becomes large. For further discussion, see Example 10.4.

10.4 BASIS FUNCTIONS AND SIMPLE MATRIX OPERATIONS

A major step in obtaining an integral solution is to construct a set of basis functions. The set must contain linearly independent elements, and each element must satisfy all homogeneous boundary conditions. If the boundary conditions are nonhomogeneous, the basis functions must be homogeneous and of the same type. Consideration is given to two types of problems. First, the basis functions for one-dimensional and regular geometries are presented. It is shown that a unified solution procedure is possible for regular-shaped bodies. Then, the method of finding basis functions for some irregular-shaped bodies is presented; an irregular-shaped body refers to a nonorthogonal body. Although obtaining the basis functions for many irregular-shaped bodies is a simple task, for many others it can become cumbersome. The reason is that each irregular-shaped body must be treated differently. After the basis functions are determined, Eqs. (10.12) and (10.13) yield matrices **A** and **B**. Next, Eq. (10.11) yields the eigenvalues and eigenvectors. The computation of matrix **P** completes the variables needed for calculation of the GF using Eq. (10.60).

10.4.1 One-Dimensional Bodies

The method for establishing the basis functions for one-dimensional problems is an interesting feature of the GBI method. When these basis functions are established, the remaining steps for finding temperature solutions follow the same procedures as discussed in Examples 10.1–10.3. Moreover, the basis functions for multidimensional regular geometries can be constructed as a product of one-dimensional basis functions. The product method of finding the basis functions is valid even when the GF cannot be obtained using a product of the corresponding one-dimensional GF; see Example 10.8. This subsection describes a method of obtaining the basis functions for one-dimensional bodies subject to boundary conditions of the first, second, and third kinds. The basis functions must satisfy homogeneous boundary conditions whether the actual boundary conditions are homogeneous or nonhomogeneous. For example, if a boundary condition for temperature is nonhomogeneous, the basis functions must satisfy the homogeneous boundary condition of the same kind. The variable z used in this derivation stands for axial, radial, or angular coordinates.

A generalized set of basis functions that satisfies the homogeneous boundary conditions $k_1 df_j/dz = h_1 f_j$ at $z = a$ and $-k_2 df_j/dz = h_2 f_j$ at $z = b$ (where $b > a$) is

$$f_j = (\delta_j z^2 + \beta_j z + \eta_j) z^{j-1} \quad \text{for } j = 1, 2, \ldots, N \quad (10.85)$$

The variable z stands for the specific coordinate system; for example, z is x in Cartesian coordinates, XIJ, or r in cylindrical, RIJ, and spherical coordinates, $RSIJ$.

Two equations for determining the three coefficients, δ_j, β_j, and η_j, are obtained by evaluating Eq. (10.85) at the two boundaries. Since one of the coefficients δ_j, β_j, or η_j can be selected arbitrarily, the coefficient δ_j is set equal to the determinant of the coefficients in the two equations. The resulting expressions for δ_j, β_j, and η_j are

$$\delta_j = a(j - aB_1)(j - 1 + bB_2) - b(j + bB_2)(j - 1 - aB_1) \quad (10.86a)$$

$$\beta_j = a^2(aB_1 - j - 1)(j - 1 + bB_2)$$
$$+ b^2(bB_2 + j + 1)(j - 1 - aB_1) \quad (10.86b)$$

$$\eta_j = -ab^2(j - aB_1)(bB_2 + j + 1) - ba^2(j + bB_2)(aB_1 - j - 1) \quad (10.86c)$$

for $j = 1, 2, 3, \ldots, N$

The parameters B_1 and B_2 appearing in Eqs. (10.86a–c) are h_1/k_1 and h_2/k_2, respectively. The parameters B_1 and B_2 are finite for $X33$, $R33$, and $RS33$ cases. If the surface $z = a$ is insulated, then $B_1 = 0$; $X23$, $R23$, or $RS23$. Similarly, $B_2 = 0$ if the $z = b$ surface is insulated; $X32$, $R32$, or $RS32$. For $X22$, $R22$, and $RS22$ problems, B_1 and B_2 are set equal to 0 in Eqs. (10.86a–c). Equation (10.85) holds for any one-dimensional conduction problem in a finite domain. Modifications are necessary when B_1, B_2, or both, are infinite; boundary conditions of the first kind, see Table 10.3a. In special cases when $a = 0$ or both a and B_1 are equal to zero, the coefficients δ_j, β_j, and η_j also must be modified. The values of δ_j, β_j, and η_j for special cases are found in Table 10.3b.

10.4.2 Matrices A and B for One-Dimensional Problems

After expressions for δ_j, β_j, and η_j are available, Eqs. (10.12), and (10.13) yield the values of a_{ij} and b_{ij}. Next, two indefinite integrals useful for calculating the values of a_{ij} and b_{ij} are presented.

Matrix A. When calculating the elements of matrix **A**, and the thermal conductivity is constant, the following integral can be used as a computational aid:

$$I_a(z) = \int f_i(\nabla^2 f_j)z^p dz = \sum_{k=1}^{5} P_k \frac{z^{i+j+p+2-k}}{i+j+p+2-k} \qquad (10.94)$$

where

$$P_1 = \delta_i\delta_j(j+1)(j+p) \qquad (10.95a)$$

$$P_2 = \beta_i\delta_j(j+1)(j+p) + \beta_j\delta_i(j+p-1)j \qquad (10.95b)$$

$$P_3 = \eta_i\delta_j(j+1)(j+p)$$
$$+ \beta_i\beta_j(j+p-1)j + \eta_j\delta_i(j+p-2)(j-1) \qquad (10.95c)$$

$$P_4 = \eta_i\beta_j(j+p-1)j + \eta_j\beta_i(j-1)(j+p-2) \qquad (10.95d)$$

$$P_5 = \eta_i\eta_j(j-1)(j+p-2) \qquad (10.95e)$$

Note that when $p = 0$ and $j = 1$, the term containing P_4 is zero. Also, when $j = 1$ or $j + p = 2$, the term containing P_5 is zero. When there is a fin effect, additional terms are necessary [see Eq. (10.12)].

Matrix B. The following integral can be used as a computational aid to calculate the elements of matrix **B** when $\rho c_p u(\mathbf{r})$ is constant.

$$I_b(z) = \int f_i f_j z^p \, dz = \sum_{k=1}^{5} Q_k \frac{z^{i+j+p+4-k}}{i+j+p+4-k} \qquad (10.96)$$

where

Table 10.3a Coefficients δ_j, β_j, and η_j in Eq. (10.85) when B_1 or B_2 are infinite

$B_1 = \infty$; Prescribed T at $z = a$, i.e., boundary conditions of the first kind at $z = a$; $X13$[†]

$$\delta_j = -a(j-1+bB_2) + b(j+bB_2) \qquad (10.87a)$$
$$\beta_j = a^2(j-1+bB_2) - b^2(bB_2+j+1) \qquad (10.87b)$$
$$\eta_j = ab^2(bB_2+j+1) - ba^2(j+bB_2) \qquad (10.87c)$$

for $j = 1, 2, 3, \ldots, N$

$B_2 = \infty$; Prescribed T at $z = b$, i.e., boundary conditions of the first kind at $z = b$; $X31$[†]

$$\delta_j = a(j-aB_1) - b(j-1-aB_1) \qquad (10.88a)$$
$$\beta_j = a^2(aB_1 - j - 1) + b^2(j-1-aB_1) \qquad (10.88b)$$
$$\eta_j = -ab^2(j-aB_1) - ba^2(aB_1 - j - 1) \qquad (10.88c)$$

for $j = 1, 2, 3, \ldots, N$

$B_1 = \infty$ and $B_2 = \infty$; boundary conditions of the first kind at $z = a$ and $z = b$; $X11$[†]

$$\delta_j = 1 \qquad (10.89a)$$
$$\beta_j = -(a+b) \qquad (10.89b)$$
$$\eta_j = ab \qquad (10.89c)$$

for $j = 1, 2, 3, \ldots, N$

[†]Also for *RIJ* and *RSIJ* cases.

Table 10.3b Coefficients δ_j, β_j, and η_j in Eq. (10.85) for special cases when $a = 0$

$a = 0$, B_1 and B_2 finite, and $B_1 > 0$; $X33^\dagger$

$$\delta_1 = B_1 + B_2 + bB_1B_2 \tag{10.90a}$$
$$\eta_1 = -2b - b^2B_2 \tag{10.90b}$$
$$\beta_1 = -2bB_1 - b^2B_1B_2 \tag{10.90c}$$

and for $j > 1$

$$\delta_j = j + bB_2 \tag{10.90d}$$
$$\eta_j = 0 \tag{10.90e}$$
$$\beta_j = -b(j + 1) - b^2B_2 \tag{10.90f}$$

for $j = 2, 3, \ldots, N$

$a = 0$, $B_1 = 0$, and B_2 finite; $X23^\dagger$

$$\delta_j = j - 1 + bB_2 \tag{10.91a}$$
$$\beta_j = 0 \tag{10.91b}$$
$$\eta_j = -b^2(j + 1) - b^3B_2 \tag{10.91c}$$

for $j = 1, 3, 5, \ldots, N$

$a = 0$, $B_1 = 0$, and $B_2 = \infty$; $X21^\dagger$

$$\delta_j = 1 \tag{10.92a}$$
$$\beta_j = 0 \tag{10.92b}$$
$$\eta_j = -b^2 \tag{10.92c}$$

for $j = 1, 3, 5, \ldots, N$

$a = 0$ and $B_1 = \infty$; $X13^\dagger$

$$\delta_j = j + bB_2 \tag{10.93a}$$
$$\beta_j = -b^2B_2 - (j + 1)b \tag{10.93b}$$
$$\eta_j = 0 \tag{10.93c}$$

for $j = 1, 2, 3, \ldots, N$

$\dagger$Also for *RIJ* and *RSIJ* cases.

$$Q_1 = \delta_i\delta_j \tag{10.97a}$$

$$Q_2 = \beta_i\delta_j + \beta_j\delta_i \tag{10.97b}$$

$$Q_3 = \eta_i\delta_j + \beta_i\beta_j + \eta_j\delta_i \tag{10.97c}$$

$$Q_4 = \beta_i\eta_j + \beta_j\eta_i \tag{10.97d}$$

$$Q_5 = \eta_i\eta_j \tag{10.97e}$$

Also this integral provides the contribution of the fin effect for the elements of matrix **A** when $m(\mathbf{r}) = $ constant.

10.4.3 Matrix Operations When $N = 1$ and $N = 2$

Following the computation of the values of the components of matrices **A** and **B**, the eigenvalues and the eigenvectors are computed using Eq. (10.11). After the computation of p_{ni}'s, using Eqs. (10.41) and (10.45), the GF is obtained using Eq. (10.60).

The mathematical procedure is to solve Eq. (10.11) when $N = 1$ to obtain $\gamma_1 = -a_{11}/b_{11}$, $d_{11} = 1$, and then Eqs. (10.41) and (10.45) yield $p_{11} = 1/b_{11}$. The one-term GF is

$$G(z', -\tau|z, -t) = \rho c_p d_{11} p_{11} \exp[-\gamma_1(t - \tau)]f_1(z')f_1(z) \qquad (10.98)$$

Expressions for finding γ_n, d_{nj} and p_{ni}, when $N = 2$, follow the procedure presented in Example 10.2. The eigenvalues are the roots of a quadratic equation

$$\text{Det}|A + \gamma B| = D_1\gamma^2 + D_2\gamma + D_3 = 0 \qquad (10.99)$$

where

$$D_1 = \text{Det}(B) = b_{11}b_{22} - b_{12}b_{21} \qquad (10.100a)$$

$$D_2 = (a_{11}b_{22} - a_{12}b_{21}) + (b_{11}a_{22} - b_{12}a_{21}) \qquad (10.100b)$$

$$D_2 = \text{Det}(A) = a_{11}a_{22} - a_{12}a_{21} \qquad (10.100c)$$

The elements of matrix **D** are computed after arbitrarily selecting $d_{11} = d_{22} = 1$. The reason for the arbitrary choice of d_{11} and d_{22} is that c_n in Eq. (10.6) is yet to be determined and, at this stage, c_n can be multiplied or divided by a constant. The other elements are

$$d_{12} = -\frac{a_{12} + \gamma_1 b_{12}}{a_{22} + \gamma_1 b_{22}} \qquad (10.101a)$$

$$d_{21} = -\frac{a_{12} + \gamma_2 b_{12}}{a_{11} + \gamma_2 b_{11}} \qquad (10.101b)$$

Since the transpose of matrix **BD** is matrix **E**, Eq. (10.41), and matrix **P** is the inverse of matrix **E**, Eq. (10.45), the elements of matrix **P** are for $N = 2$

$$p_{11} = \frac{e_{22}}{\det(E)} \qquad (10.102a)$$

$$p_{12} = -\frac{e_{12}}{\det(E)} \qquad (10.102b)$$

$$p_{21} = -\frac{e_{21}}{\det(E)} \qquad (10.102c)$$

$$p_{22} = \frac{e_{11}}{\det(E)} \qquad (10.102d)$$

The GF is obtained when **r** and **r**' are replaced by z and z', and N is set equal to 2 in Eq. (10.60).

Example 10.4 A homogeneous hollow cylinder (Fig. 10.3), with inner radius a and outer radius $b = 2a$, is considered with the boundary conditions $kdT/dr = h_1T$ at $r = a$ and $-kdT/dr = q(t)$. Furthermore, it is assumed that the initial temperature distribution is zero, $T(r, 0) = 0$. The heat flux $q(t)$ at $r = b$ is given

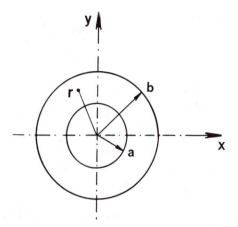

Figure 10.3 Hollow cylinder in Example 10.4; boundary conditions denoted $R32$.

by the relation $q(t) = q_0 t$ so that q varies linearly with t. The $h_1 a/k$ ratio is selected to be 1. Compare the alternative GF solution with the exact solution.

SOLUTION The notation for this case is $R32B02T0$. The temperature at $r = b$ is to be found as a function of time. The step-by-step procedure for obtaining the GF is presented in this example for $N = 2$. The basic procedure, except for the method of obtaining the basis functions, is applicable to all transient, one-dimensional conduction problems.

1. It is necessary to introduce a set of basis functions, f_j. A general set of basis functions that satisfies the homogeneous convective conditions $kdf_j/dr = h_1 f_j$ at $r = a$ and $kdf_j/dr = 0$ at $r = b$ are obtained from Eq. (10.85) when z is replaced by r. Using $j = 1$ and 2, the coefficients δ_j, β_j, and η_j in Eq. (10.85) are available from Eqs. (10.86a–c) for $b = 2a$ as $\delta_1 = 1$, $\beta_1 = -4$, $\eta_1 = 1$, $\delta_2 = 1$, $\beta_2 = -2$, and $\eta_2 = -4$. Because the basis functions are not unique, the function f_j as given by Eq. (10.85) is divided by δ_j and then made dimensionless. The basis functions f_1 and f_2, with $B_1 = 1$ and $B_2 = 0$ are

$$f_1 = \left(\frac{r}{a}\right)^2 - 4\frac{r}{a} + 1 \tag{10.103a}$$

$$f_2 = \left(\frac{r}{a}\right)^3 - 2\left(\frac{r}{a}\right)^2 - 4\frac{r}{a} \tag{10.103b}$$

2. Once the basis functions are available, Eqs. (10.103a and b), the elements of matrices **A** and **B** are calculated using Eqs. (10.12) and (10.13). Because the conduction is one-dimensional and there is no fin effect, the indefinite integrals can be evaluated using Eqs. (10.94)–(10.96) to yield

$$a_{11} = -\frac{34\pi}{3a} \tag{10.104a}$$

$$a_{12} = a_{21} = -\frac{446\pi}{15a} \tag{10.104b}$$

$$a_{22} = - \frac{1181\pi}{15a} \tag{10.104c}$$

and

$$b_{11} = \frac{337\pi a}{15} \tag{10.105a}$$

$$b_{12} = b_{21} = \frac{1231\pi a}{21} \tag{10.105b}$$

$$b_{22} = \frac{21421\pi a}{140} \tag{10.105c}$$

3. The eigenvalues and the eigenvectors are computed using Eq. (10.11). Equation (10.99) is used (for $N = 2$) to obtain

$$\gamma_1 = \frac{0.50276\alpha}{a^2} \tag{10.106a}$$

$$\gamma_2 = \frac{11.9830\alpha}{a^2} \tag{10.106b}$$

4. The elements of matrix **D** are computed after arbitrarily selecting

$$d_{11} = d_{22} = 1 \tag{10.107a}$$

The other elements according to Eqs. (10.101a and b) are

$$d_{12} = - \frac{a_{12} + \gamma_1 b_{12}}{a_{22} + \gamma_1 b_{22}} = -0.14495 \tag{10.107b}$$

$$d_{21} = - \frac{a_{12} + \gamma_2 b_{12}}{a_{11} + \gamma_2 b_{11}} = -2.6085 \tag{10.107c}$$

5. Following computation of matrix **E** from Eq. (10.40) or Eq. (10.41), matrix **P** is the inverse of matrix **E**. When $N = 2$, Eqs. (10.102a–d) provide the elements of matrix **P** as

$$p_{11} = \frac{e_{22}}{\det(\mathbf{E})} = 3.6638 \tag{10.108a}$$

$$p_{12} = - \frac{e_{12}}{\det(\mathbf{E})} = -0.0053107 \tag{10.108b}$$

$$p_{21} = - \frac{e_{21}}{\det(\mathbf{E})} = -13.637 \tag{10.108c}$$

$$p_{22} = \frac{e_{11}}{\det(\mathbf{E})} = 5.2279 \tag{10.108d}$$

6. All the parameters needed to calculate the GF relation are now available and Eq. (10.60) becomes

$$G(r', -\tau|r, -t) = C(r) \sum_{n=1}^{2} \sum_{j=1}^{2} \sum_{i=1}^{2} d_{nj}p_{ni} \exp[-\gamma_n(t-\tau)]f_j(r')f_i(r) \quad (10.109)$$

The temperature distribution is given by Eq. (10.68) or Eq. (10.77). When using Eq. (10.77), T^* is not a unique function. A function such as $T^* = -bq_0t$ $[k/(ha) + \ln(r/a)]/k$ would satisfy the nonhomogeneous boundary conditions. For this case, the value of f^* is zero; see Eq. (10.73). Therefore, the analysis is simpler. The temperature solution when $ha/k = 1$ is obtained using Eq. (10.77):

$$\frac{kT}{q_0} = -bt\left(1 + \ln\frac{r}{a}\right) + \int_0^t d\tau \int_a^b 4\pi G\left(1 + \ln\frac{r'}{a}\right)dr' \quad (10.110)$$

Since $b = 2a$ and G is given by Eq. (10.109), Eq. (10.77) becomes

$$\frac{kT}{a^3q_0/\alpha} = \frac{-2\alpha t}{a^2}\left(1 + \ln\frac{r}{a}\right) + 4\pi \sum_{n=1}^{N}$$

$$\times \left\{\sum_{j=1}^{N} d_{nj}\left[\delta_j\left(\frac{r}{a}\right)^2 + \beta_j\frac{r}{a} + \eta_j\right]\left(\frac{r}{a}\right)^{j-1}\right\}$$

$$\times \left\{\sum_{i=1}^{N} p_{ni}\left[\delta_i\frac{2^{i+3}(1 + \ln 2) - 1}{i + 3}\right.\right.$$

$$+ \beta_i\frac{2^{i+2}(1 + \ln 2) - 1}{i + 2} + \eta_i\frac{2^{i+1}(1 + \ln 2) - 1}{i + 1}$$

$$- \delta_i\frac{2^{i+3} - 1}{(i + 3)^2} - \beta_i\frac{2^{i+2} - 1}{(i + 2)^2}$$

$$\left.\left.- \eta_i\frac{2^{i+1} - 1}{(i + 1)^2}\right]\right\}\frac{1 - \exp(-\gamma_n t)}{a^2\gamma_n/\alpha} \quad (10.111)$$

Notice that $\partial T^*/\partial t = -bq_0[k/(ha) + \ln(r/a)]/k$ is multiplied by $G = G(r', -\tau|r, -t)$ and $dA = 2\pi r\, dr$ and then integrated between a and $b = 2a$.

Table 10.4 shows the numerical values of the GFs. The values of the GFs using the integral method agree well with the exact values when $N \geq 5$ and the $\alpha(t - \tau)/(b - a)^2 > 0.04$. Table 10.4 suggests using the small-time solution when $\alpha(t - \tau)/(b - a)^2 < 0.12$. This latter number is used for the time partitioning of the integral solution results appearing in Table 10.5 when $N = 2$.

The dimensionless temperature solution, $kT/(a^3q_0/\alpha)$, at $r = 2a$ and for $N = 2$ and 7 is presented in Table 10.5. The results for the alternative GF solution, which uses Eq. (10.77), agree closely with the exact solution. However, the standard GF solution using Eq. (10.68) is much less accurate for $N = 7$. When $N < 7$ (not shown in this table), the accuracy further decreases. Notice the remarkable accuracy of the solution using the time-partitioned GF when $N = 2$. The eigenvalues in this example were computed using Cholesky's decomposition of matrix **B** and applying the Jacobi method (Carnahan et al., 1969, p. 250; see p. 255 for FORTRAN subroutine) to Eq. (10.32).

As discussed earlier, the alternative GF solution requires an auxiliary equa-

Table 10.4 Values of the Green's functions $G(b, t|b, \tau)$ for the integral method, exact, and small-time asymptotic solutions ($R32$ case with $B_1 = 1$)

$\dfrac{\alpha(t - \tau)}{(b - a)^2}$	Galerkin-based integral method				Small-time solution	Exact solution
	$N = 2$	$N = 3$	$N = 5$	$N = 7$		
0.01	0.26457	0.38672	0.49460	0.47371	0.46974	0.46975
0.02	0.24802	0.32655	0.34709	0.33841	0.33862	0.33863
0.05	0.20838	0.22820	0.22280	0.22275	0.22275	0.22277
0.1	0.16600	0.16691	0.16495	0.16498	0.16491	0.16499
0.2	0.12670	0.12632	0.12588	0.12588	0.12480	0.12588
0.5	0.096912	0.097070	0.097042	0.097042	0.091206	0.097042
1	0.075063	0.075199	0.075185	0.075184	0.076939	0.075184
2	0.045402	0.045485	0.045477	0.045477	–	0.045477
5	0.010047	0.010066	0.010064	0.010064	–	0.010064

tion which is usually available for one-dimensional geometries. However, in multidimensional and complex geometries, this auxiliary equation either is unavailable or difficult to obtain. The partitioning of the Green's function, as given in Chapter 5, Eq. (5.17), is an attractive option. Equation (R02.5) in Appendix R provides the small-time GF while Eq. (10.60) is used to obtain the large-time GF. As shown in Table 10.5, partitioning of the GF and alternative GF solutions exhibit good accuracy. Even when $N = 2$, the time-partitioned solution produces accurate results.

Example 10.5 Derive Eq. (10.72a) and calculate c_1 and c_2 when the boundary conditions are

Table 10.5 Partitioned and alternative Green's function solution for dimensionless surface temperature $-kT/(a^3q_0/\alpha)$ at $r/a = 2$

Time $\alpha t/a^2$	$N = 2$		$N = 7$			Exact series (first 14,000 terms)
	AGFS[†]	PGFS[‡]	AGFS[†]	GFS[§]	PGFS[‡]	
0.01	0.001372	0.000765	0.000765	0.000608	0.000765	0.000765
0.02	0.003072	0.002179	0.002179	0.001870	0.002179	0.002179
0.05	0.009945	0.008740	0.008740	0.007973	0.008740	0.008739
0.1	0.026438	0.025135	0.025136	0.023607	0.025136	0.025134
0.2	0.074303	0.072923	0.072872	0.069817	0.072870	0.072868
0.5	0.30721	0.30574	0.30518	0.29755	0.30518	0.30517
1	0.93291	0.93117	0.93001	0.91475	0.93000	0.92999
2	2.8434	2.8406	2.8393	2.8088	2.8393	2.8392
5	11.243	11.235	11.237	11.161	11.237	11.237
10	27.715	27.696	27.709	27.557	27.709	27.709

[†]Alternative Green's function solution, Eq. (10.111).
[‡]Partitioned Green's function solution, see Chapter 5.
[§]Green's function solution, Eq. (10.68).

$$\frac{\partial T}{\partial z} = \frac{h_1}{k_1} (T - T_{\infty 1}) \qquad \text{at } z = a \qquad (10.112a)$$

$$\frac{\partial T}{\partial z} = -\frac{h_2}{k_2} (T - T_{\infty 2}) \qquad \text{at } z = b \qquad (10.112b)$$

SOLUTION The generalized form of the Laplace equation in one-dimensional bodies is

$$\frac{1}{z^p} \frac{d}{dz} \left(z^p \frac{dT^*}{dz} \right) = 0 \qquad (10.113)$$

where $p = 0$ in Cartesian coordinates
$p = 1$ in cylindrical coordinates
$p = 2$ in spherical coordinates

Integrate twice to obtain

$$T^* = c_1 \int \frac{dz}{z^p} + c_2 = c_1 u_p(z) + c_2 \qquad (10.114)$$

This equation assumes different forms in different coordinate systems:

In Cartesian coordinates, $z = x$, $p = 0$, and $u_p(z) = x$
In cylindrical coordinates, $z = r$, $p = 1$, and $u_p(z) = \ln r$
In spherical coordinates, $z = r$, $p = 2$, and $u_p(z) = -1/r$

The first term on the right side of Eq. (10.77) contains T^*, which must satisfy the nonhomogeneous boundary conditions because the remaining terms in Eq. (10.77) only satisfy the homogeneous boundary conditions. Then, the boundary conditions for T^* are

$$\frac{\partial T^*}{\partial z} = \frac{h_1}{k_1} (T^* - T_{\infty 1}) \qquad \text{at } z = a \qquad (10.115a)$$

$$\frac{\partial T^*}{\partial z} = -\frac{h_2}{k_2} (T^* - T_{\infty 2}) \qquad \text{at } z = b \qquad (10.115b)$$

Introducing T^* from Eq. (10.114) in Eqs. (10.115a and b) results in the following two simultaneous equations:

$$\left[u_p(a) - \frac{k_1}{h_1 a^p} \right] c_1 + c_2 = t_{\infty 1} \qquad (10.116a)$$

$$\left[u_p(b) + \frac{k_2}{h_2 b^p} \right] c_1 + c_2 = T_{\infty 2} \qquad (10.116b)$$

The solutions for c_1 and c_2 are

$$c_1 = \frac{T_{\infty 2} - T_{\infty 1}}{u_p(b) - u_p(a) + k_1/(h_1 a^p) + k_2/(h_2 b^p)} \qquad (10.117a)$$

$$c_2 = \frac{T_{\infty 2}[u_p(b) + k_2/(h_2 b^p)] - T_{\infty 1}[u_p(a) - k_1/(h_1 a^p)]}{u_p(b) - u_p(a) + k_1/(h_1 a^p) + k_2/(h_2 b^p)} \quad (10.117b)$$

The above choice of T^* forces f^* to become equal to zero. The basis functions, as usual, must satisfy homogeneous boundary conditions, and they are given by Eqs. (10.85) and (10.86) when h_1 or h_2 are nonzero. The case when $h_1 = 0$ or $h_2 = 0$ is trivial. When $h_1 = 0$, $T^* = T_{\infty 2}$, and when $h_2 = 0$, $T^* = T_{\infty 1}$. Equations (10.117a and b) can be used to calculate c_1 and c_2 when either h_1 or h_2, or both are infinite.

Example 10.6 In this example, T^*, using Eq. (10.72a), is to be calculated when heat flux is prescribed on one surface and the other surface is exposed to a convective boundary condition.

SOLUTION First consider the following boundary conditions:

$$\frac{\partial T^*}{\partial z} = \frac{q_1}{k_1} \qquad \text{at } z = a \qquad (10.118a)$$

$$\frac{\partial T^*}{\partial z} = -\frac{h_2}{k_2}(T^* - T_{\infty 2}) \qquad \text{at } z = b \qquad (10.118b)$$

Using these boundary conditions, the following two simultaneous equations are obtained

$$\frac{c_1}{a^p} = +\frac{q_1}{k} \qquad (10.119a)$$

$$\left[u_p(b) + \frac{k_2}{h_2 b^p}\right] c_1 + c_2 = T_{\infty 2} \qquad (10.119b)$$

The solutions for c_1 and c_2 are

$$c_1 = +\frac{q_1 a^p}{k_1} \qquad (10.120a)$$

and

$$c_2 = -\frac{[k_2/(h_2 b^p) + u_p(b)]q_1 a^p}{k_1} + T_{\infty 2} \qquad (10.120b)$$

when the boundary conditions at $z = a$ and $z = b$ are switched, that is,

$$\frac{\partial T^*}{\partial z} = \frac{h_1}{k_1}(T^* - T_{\infty 1}) \qquad \text{at } z = a \qquad (10.121a)$$

$$\frac{\partial T^*}{\partial z} = -\frac{q_2}{k_2} \qquad \text{at } z = b \qquad (10.121b)$$

The two simultaneous equations are

$$c_1 = -\frac{q_2 b^p}{k_2} \tag{10.122a}$$

$$c_2 = -\frac{[k_1/(h, a^p) - u_p(a)]q_2 b^p}{k_2} + T_{\infty 1} \tag{10.122b}$$

In this example, similar to Example 10.5, $f^* = 0$ because the Laplace equation is satisfied.

Example 10.7 Consider a one-dimensional conduction problem when heat flux is prescribed on both surfaces. The goal is to calculate the values of T^* and f^*.

SOLUTION As before, the nonhomogeneous boundary conditions are assigned to T^*. The boundary conditions are

$$\frac{\partial T^*}{\partial z} = \frac{q_1}{k_1} \qquad \text{at } z = a \tag{10.123a}$$

$$\frac{\partial T^*}{\partial z} = -\frac{q_2}{k_2} \qquad \text{at } z = b \tag{10.123b}$$

The proposed auxiliary solution T^* is

$$T^* = c_1 u_p(z) + c_2 z^2 \tag{10.124}$$

This equation must satisfy the boundary conditions given by Eqs. (10.123a and b). After substituting T^* from Eq. (10.124) in Eqs. (10.123a and b), the following two simultaneous relations are obtained:

$$\frac{k_1 u_p(a)}{a^p} c_1 + 2ac_2 = q_1 \tag{10.125a}$$

$$\frac{k_2 u_p(b)}{b^p} c_1 + 2bc_2 = -q_2 \tag{10.125b}$$

The constants c_1 and c_2 are calculated as

$$c_1 = \frac{q_2 a + q_1 b}{k_1 ba^{-p} u_p(a) - k_2 ab^{-p} u_p(b)} \tag{10.126a}$$

and

$$c_2 = -\frac{q_2 + k_2 b^{-p} u_p(b) c_1}{2b} \tag{10.126b}$$

The value of f^* is obtained from the relation

$$f^* = \frac{1}{z^p} \frac{d}{dz}\left(z^p \frac{dT}{dz}\right) = \frac{1}{z^p} \frac{d}{dz}\left[z^p\left(\frac{c_1}{z^p} + 2c_2 z\right)\right]$$

$$= 2c_2(p + 1) \tag{10.127}$$

Using Eq. (10.77), the term that contains f^* behaves as a uniform energy source that liberates $2c_2(p + 1)$ units of energy per unit time and per unit volume. Indeed, as a general rule, the function $f^*(\mathbf{r'}, \tau)$ can be lumped together with $g(\mathbf{r'}, \tau)$.

The basis functions needed for calculating the GF are given by Eq. (10.127) for which the values of δ_j, β_j, and η_j are given by Eqs. (10.86a–c) as

$$\delta_j = j(j - 1)(a - b) \tag{10.128a}$$

$$\beta_j = (j^2 - 1)(b^2 - a^2) \tag{10.128b}$$

$$\eta_j = abj(j + 1)(a^2 - b^2) \tag{10.128c}$$

For a special case when $a = 0$, Eqs. (10.91a–c) yield

$$\delta_j = j - 1 \qquad \beta_j = 0 \qquad \text{and} \qquad \eta_j = -b^2(j + 1) \tag{10.129}$$

Example 10.8 Consider a finite cylinder with boundary conditions

$$\frac{\partial T}{\partial x} = 0 \qquad \text{at } x = 0 \tag{10.130a}$$

$$\frac{\partial T}{\partial x} = -\frac{h_2}{k}(T - T_\infty) \qquad \text{at } x = L \tag{10.130b}$$

$$\frac{\partial T}{\partial r} = 0 \qquad \text{at } r = 0 \tag{10.130c}$$

$$T = T_\infty \qquad \text{at } r = r_0 \tag{10.130d}$$

and find a set of basis functions.

SOLUTION The basis functions must satisfy homogeneous boundary conditions of the same types as the boundary conditions on temperature

$$\frac{\partial f_j}{\partial x} = 0 \qquad \text{at } x = 0 \tag{10.131a}$$

$$\frac{\partial f_j}{\partial x} = -\frac{h_2}{k}f_j \qquad \text{at } x = L \tag{10.131b}$$

$$\frac{\partial f_j}{\partial r} = 0 \qquad \text{at } r = 0 \tag{10.131c}$$

$$f_j = 0 \qquad \text{at } r = r_0 \tag{10.131d}$$

The contribution of the x direction to the basis functions is obtained using Eq. (10.85) for which δ_j, β_j, and η_j coefficients are given by Eqs. (10.91a–c). The contribution of the r direction to the basis functions is computed, in a similar manner, using Eq. (10.85). Setting $\beta_2 = h_2/k_2$, the basis functions become

$$f_j = [(m_j - 1 + LB_2)x^2 - L^2(m_j + 1)$$

$$- L^3 B_2](r^2 - r_0^2)x^{m_j - 1} r^{n_j - 1} \tag{10.132}$$

The variables m_j and n_j replaced j in Eq. (10.85) to account for all combinations of $m_j = 1, 2, 3, \ldots$, and $n_j = 1, 2, 3, \ldots$, for example

$$j = 1 \qquad m_j = 1 \qquad \text{and} \qquad n_j = 1$$

$$j = 2 \qquad m_j = 2 \qquad \text{and} \qquad n_j = 1$$

$$j = 3 \qquad m_j = 1 \qquad \text{and} \qquad n_j = 2$$

$$j = 4 \qquad m_j = 3 \qquad \text{and} \qquad n_j = 1$$

$$j = 5 \qquad m_j = 2 \qquad \text{and} \qquad n_j = 2$$

$$j = 6 \qquad m_j = 1 \qquad \text{and} \qquad n_j = 3$$

$$\cdots \qquad \cdots\cdots\cdots\cdots\cdots\cdots\cdots$$

To include all relevant powers of x and r, use $m_j = 1$ and $n_j = 1$ for a one-term solution. For more accuracy, $N = 3$ should be used for which the polynomial coefficients are $x^0 r^0$, $x^1 r^0$, $x^0 r^1$. The accuracy can be further improved by including higher order polynomial coefficients $x^2 r^0$, $x^1 r^1$, and $x^0 r^2$ with N taking the value of 6. The next higher level of accuracy is achieved when $N = 10, 15$, and so on.

Instead of finding the set of two-dimensional basis functions and then calculating the GF, the two-dimensional GF can be computed as a product of two one-dimensional GFs. However, the two-dimensional basis functions, Eq. (10.132), can be used to calculate the temperature field in a finite cylinder if it also contains inclusions with different thermophysical properties; see Section 11.3. In this latter case, the GF is not a product of two one-dimensional GFs.

The product method can be used for all multidimensional regular bodies with a few exceptions. The noted exception is the case when the regular geometry is cylindrical or spherical and there is a surface that convects heat in angular directions. However, boundary conditions of the first and second kinds in an angular direction can be accommodated by the product method.

10.5 FINS AND FIN EFFECT

The GF solution, Eq. (10.77), is modified to solve temperature and heat flux in bodies with fin effect (Haji-Sheikh et al., 1991). The bodies can be single layer or multilayers, although bodies with multilayers are treated in Chapter 11. For multidimensional conduction with fin effect, Eq. (10.77) is valid. For quasi-one-dimensional conduction Eq. (10.77) is written as

$$T(r, t) = T^*(r, t) + \frac{1}{\rho(r)c_p(r)} \left\{ \left[\int_{\tau=0}^{t} d\tau \int_{V} \right. \right.$$

$$\times G \left[f^* + g(r', \tau) - \rho(r')c_p(r') \frac{\partial T^*(r', \tau)}{\partial \tau} \right] dV'$$

$$+ \int_{V} \rho(r')c_p(r')G|_{\tau=0} [F(r') - T^*(r', 0)] dV' \right\} \qquad (10.133)$$

where r is the axial coordinate.

The auxiliary function T^* must satisfy the nonhomogeneous boundary conditions but it is not necessarily the steady-state or quasi-steady-state solution. The function T^* contains only the contribution of nonhomogeneous boundary conditions. The function f^*, appearing as a source term in Eq. (10.133), compensates for the arbitrary nature of T^* and is given by the modified form of Eq. (10.73) as

$$f^*(r', \tau) = \nabla_0 \cdot [k\nabla_0 T^*(r', \tau)] - m(r)^2 T^* \qquad (10.134)$$

where ∇_0 implies the derivatives are in r' space. If $f^*(r', \tau) = 0$ and $m(r)^2 = 0$, then $T^*(r', \tau)$ satisfies the Laplace equation and it is the quasi-steady solution. The function T^* is chosen in the same manner as discussed in the examples with no fin effect.

Example 10.9 Calculate the fin efficiency in a straight cylindrical fin. The boundary conditions are: $T = T_b = 1$ at $r = r_1$ and $q = 0$ at $r = r_2$. Finally, compare the results with the exact values.

SOLUTION The selection of the basis functions for this problem is exactly the same as earlier examples with no fin effect. Also, the function T^* is selected in a similar manner using Eq. (10.72a) mainly to satisfy the boundary conditions; here, $T^* = 1$ and $f^* = -m^2$ where m^2 is $2h/\delta^*$, and δ^* is the fin thickness. Equation (10.1) in cylindrical coordinates, when $T = T(r, t)$, $g(\mathbf{r}) = 0$, $u(\mathbf{r}) = 1$ and thermophysical properties are constant, is

$$\frac{1}{r} \frac{\partial}{\partial r} \left(r \frac{\partial T}{\partial r} \right) - \frac{2h}{k\delta^*} T = \frac{1}{\alpha} \frac{\partial T}{\partial t} \qquad (10.135)$$

The basis functions are given by Eq. (10.85) and the coefficients δ_j, β_j, and η_j by Eqs. (10.89a–c). The fin efficiency is defined as the ratio of heat transfer from an actual fin to the heat transfer from an isothermal fin at $T = T_b$. Figure 4 shows the efficiency, when $r_2/r_1 = 2$, as a function of dimensionless time, $\alpha t/r_1^2$, for different values of $m^* = r_1(2h/k\delta^*)^{0.5}$.

To show the accuracy obtainable with the single-equation solution, the steady-state efficiency for different r_2/r_1 ratios, for a range of values of m are shown in Table 10.6. The data compare well with the exact solution; usually up to five significant figures. All entries in Table 10.6 are for $N = 9$. Table 10.7 contains the efficiency calculated for different values of N. Only one value of r_2/r_1 is used

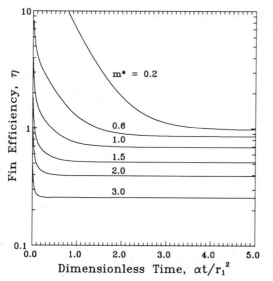

Fin Efficiency, η

$m^* = 0.2$

0.6
1.0

1.5

2.0

3.0

Dimensionless Time, $\alpha t/r_1{}^2$

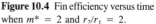

Figure 10.4 Fin efficiency versus time when $m^* = 2$ and $r_2/r_1 = 2$.

in this presentation. Table 10.7 shows that, when $N = 5$, sufficient accuracy is achieved for nearly all practical applications.

10.6 CONCLUSIONS

The Galerkin-based GF solution method discussed in this chapter has many advantages. Generally, for one-dimensional problems, a two-term solution provides results accurate enough for most applications. The methodology, especially for one-dimensional bodies, is universal, and a single computer program can be used for different-shaped bodies with different boundary conditions. This is a unique feature that is not shared

Table 10.6 Fin efficiency using GBI method and comparison with the exact solution for cylindrical fins

	$r_2/r_1 = 1.2$		$r_2/r_1 = 2.0$		$r_2/r_1 = 3.0$	
m^*	GBI	Exact	GBI	Exact	GBI	Exact
0.0	1.00000	1.00000	1.00000	1.00000	1.00000	1.00000
0.4	0.99766	0.99767	0.93025	0.93024	0.73696	0.73700
0.8	0.99074	0.99075	0.77434	0.77434	0.43493	0.43494
1.2	0.97946	0.97946	0.61464	0.61464	0.27932	0.27934
1.6	0.96416	0.96417	0.48705	0.48706	0.19933	0.19935
2.0	0.94531	0.94531	0.39335	0.39332	0.15338	0.15340
2.4	0.92341	0.92342	0.32543	0.32542	0.12415	0.12419
5.0	0.74443	0.74448	0.14607	0.14609	0.05460	0.05479
8.0	0.55316	0.55334	0.08821	0.08840	0.03249	0.03315

Table 10.7 Fin efficiency using GBI method with different N values for straight cylindrical fins, $r_2/r_1 = 2$

m^*	GBI Solution			Exact solution
	$N = 2$	$N = 5$	$N = 7$	
0.0	1.00000	1.00000	1.00000	1.00000
0.4	0.90400	0.93008	0.93021	0.93024
0.8	0.74667	0.77418	0.77432	0.77434
1.2	0.58523	0.61449	0.61463	0.61464
1.6	0.45599	0.48688	0.48706	0.48706
2.0	0.36067	0.39312	0.39332	0.39332
2.4	0.29133	0.32516	0.32541	0.32542
5.0	0.10673	0.14515	0.14603	0.14609
8.0	0.04999	0.08590	0.08807	0.08840

by the exact solution. A computer program that uses the Galerkin-based GF is available. It utilizes a personal computer with screen graphics to display the results (Haji-Sheikh, 1989).

Two solution methods were covered in this chapter: a GF solution and an alternative GF solution. The general solution method requires time partitioning of the GF to achieve a high degree of accuracy. The alternative GF solution, when available, is simple and accurate if time is not extremely small. If time is extremely small, the small-time solution yields the temperature and time partitioning of the GF is not a prerequisite. The time-partitioned Galerkin-based GF solution has flexibility, accuracy, and computational speed, which are the features of an efficient computational method. The unique feature is that a single solution is used for nearly all one-dimensional conduction problems. This feature was successfully incorporated in a computer program.

Many GFs for two- or three-dimensional solutions of regular bodies are the products of appropriate one-dimensional GFs; others must be computed in multidimensional space, see Chapter 11.

The derivation of the GF solution presented in this chapter applies to heterogeneous as well as homogeneous bodies. However, all examples given in this chapter are for homogeneous and regular bodies. Further discussion of this solution method and additional examples are presented in Chapter 11.

REFERENCES

Carnahan, B., Luther, H. A., and Wilkes, J. O., 1969, *Applied Numerical Methods*, p. 250, Wiley, New York.

Forsythe, G., and Moler, C. B., 1967, *Computer Solution of Linear Algebraic Systems*, Prentice-Hall, Englewood Cliffs, N.J.

Haji-Sheikh, A., 1988, Heat Diffusion in Heterogeneous Media Using Heat-Flux-Conserving Basis Functions, *ASME J. Heat Transfer*, vol. 110, pp. 276–282.

Haji-Sheikh, A., 1989, Conduction Solver, Lab. Report, Mech. Engineering, University of Texas at Arlington.

Haji-Sheikh, A., and Beck, J.V., 1988, Green's Function Partitioning in Galerkin-Based Solution of the Diffusion Equation, *Proceedings of the 1988 National Heat Transfer Conference*, ed. H. R. Jacobs, ASME HTD-vol. 96, pp. 257–266 in vol. 3.

Haji-Sheikh, A., and Lakshminarayanan, R., 1987, Integral Solution of Diffusion Equation: Part 2-Boundary Conditions of Second and Third Kinds, *ASME J. Heat Transfer*, vol. 109, pp. 557–562.

Haji-Sheikh, A., and Mashena, M., 1987, Integral Solution of Diffusion Equation: Part 1-General Solution, *ASME J. Heat Transfer*, vol. 109, pp. 551–556.

Haji-Sheikh, A., Yan, L., and Kinsey, S. P., 1991, A Single-Equation Solution for Conduction in Fins, *Int. J Heat Mass Transfer*, vol. 34, no. 1, pp. 159–165.

Hay, G. E., 1953, *Vector and Tensor Analysis*, Dover, New York.

Kantorovich, L. V., and Krylov, V. I., 1960, *Approximate Methods of Higher Analysis*, Wiley, New York.

Kaplan, W., 1956, *Advanced Calculus*, Addison-Wesley, Reading, Mass.

LeCroy, R. C., and Eraslan, A. H., 1969, The Solution of Temperature Development in the Entrance Region of an MHD Channel by the G. B. Galerkin Method, *ASME J. Heat Transfer*, vol. 91, pp. 212–220.

Ozisik, M. N., 1980, *Heat Conduction*, Wiley, New York.

PROBLEMS

10.1 Verify Eq. (10.33).

10.2 Repeat Example 10.2, except assume the surface at $x = 0$ is insulated.

10.3 Find the eigenvalues and write a two-term expression for temperature distribution in a solid cylinder with radius r_0. The initial temperature is T_0 and the surface temperature is suddenly reduced to T_s. (Answer: $\gamma_1 = 5.784\alpha/r_0^2$, $\gamma_2 = 36.88\alpha/r_0^2$.)

10.4 Consider the GF in Problem 3 to be the large-time GF. Find a small-time GF to solve for temperature distribution using the time partitioning of the GF.

10.5 A solid cylinder with radius r_0 is initially at temperature T_0. If there is a prescribed heat flux at the rate $q(t)$, find the basis functions. Retaining only two eigenvalues, $N = 2$, calculate the eigenvalues, and derive an expression for the temperature distribution. (Answer: $\gamma_1 = 0$, $\gamma_2 = 15\alpha/r_0^2$.)

10.6 Use the GF partitioning to derive an equation for the temperature at the surface of the cylinder described in Problem 5.

10.7 Show that the functions $\psi_1, \psi_2, \psi_3, \ldots$, are orthogonal, that is,

$$\int_V \rho c_p \psi_m \psi_n dV = 0$$

when m and n are not equal. [*Hint*: Substitute ψ_m and ψ_n in Eq. (10.5), then use Identities 1–3 in Note 1.]

10.8 Verify Eqs. (10.39) and (10.40).

10.9 Verify Eq. (10.77).

10.10 Find the eigenvalues and write a two-term expression for temperature distribution in a solid sphere with radius r_0. The initial temperature is T_0 and the surface temperature is suddenly reduced to T_s. (Answer: $\gamma_1 = 9.875\alpha/r_0^2$, $\gamma_2 = 50.12\alpha/r_0^2$.)

10.11 Find the eigenvalues and write a two-term expression for temperature distribution in a hollow cylinder whose inner radius is r_1 and outer radius is r_2. The initial temperature is T_0, the surface temperature is suddenly reduced to T_s at $r = r_1$, and the surface at $r = r_2$ is insulated. (Answer: $\gamma_1 = 7.407\alpha/r_2^2$, $\gamma_2 = 88.61\alpha/r_2^2$.)

10.12 Find the eigenvalues and write a two-term expression for temperature distribution in a hollow sphere whose inner radius is r_1 and outer radius is r_2. The initial temperature is T_0 and the surface at $r = r_2$ is insulated. There is convection to a zero temperature fluid at $r = r_1$ surface and $hr_2/k = 1$. (Answer: $\gamma_1 = 7.480\alpha/r_2^2$, $\gamma_2 = 47.02\alpha/r_2^2$.)

10.13 A straight fin with constant cross-sectional area A and length L is insulated at the tip while the base temperature is T_b. Find T^* and use the alternative GFSE to derive a solution. Compare the fin efficiency for $N = 2$ with its exact value when $L(hP/kA)^{0.5} = 1$; P is the perimeter. (Answer: Fin efficiency when $N = 2$ is 0.75312.)

10.14 For a circular pin fin write the temperature solution. The radius r varies as x^2, the tip at $x = 0$ may be considered insulated, and the base is at a constant temperature.

NOTE 1: MATHEMATICAL IDENTITIES

Consider v to be a scalar and $\mathbf{W}$ to be a vector.

Identity 1   (Hay, 1953, p. 117)

$$\nabla \cdot (v\mathbf{W}) = v(\nabla \cdot \mathbf{W}) + (\nabla v) \cdot \mathbf{W}$$

Identity 2

$$\nabla v \cdot \mathbf{n} = \frac{\partial v}{\partial n}$$

Identity 3

The generalization of the Green's theorem for line integrals is called the Green's theorem in space (Hay, 1953, p. 143), the Green's theorem, the divergence theorem, or Gauss's theorem (Kaplan, 1956, p. 269).

$$\int_V \nabla \cdot \mathbf{W} \, dV = \int_S \mathbf{W} \cdot \mathbf{n} \, dS$$

Identity 4

$$\delta(z - b) = 0 \quad \text{when } z \neq b \text{ and } \int_{-\infty}^{+\infty} \delta(z)dz = 1$$

Identity 5

$$v(z)\delta(z - b) = v(b)\delta(z - b)$$

Identity 6

$$\int_{-\infty}^{+\infty} v(z)\delta(z - b)dz = v(b)$$

NOTE 2: DECOMPOSITION OF MATRIX B

A FORTRAN subroutine that will decompose a symmetric matrix $\mathbf{B}$ into $\mathbf{LL}^T$ is discussed here. Matrix $\mathbf{L}$ is a lower triangular matrix and $\mathbf{L}^T$ is its transpose. The elements of matrix $\mathbf{AL}$ in this subroutine are elements of matrix $\mathbf{L}$ except for the diagonal elements which are set equal to 1. However, the array D in this subroutine contains the square of the diagonal elements of matrix $\mathbf{L}$. If any element of array D is negative, the subsequent computation of eigenvalues will be unsuccessful. In order to obtain matrix $\mathbf{L}$, the diagonal elements of matrix **AL** must be replaced by the square root of the corresponding elements of array D. Similarly, matrix $\mathbf{L}^{-1}$ is obtained if the diagonal elements of matrix *ALINV* are replaced by the inverse of square root of the corresponding elements of array D.

```
      SUBROUTINE LDLTS(B,AL,ALINV,D,N,NX)
C     B IS THE INPUT MATRIX, N IS THE SIZE OF B, NX IS THE DIMENSION OF
C     B IN THE MAIN PROGRAM, AND AL, ALINV, AND D ARE THE OUTPUTS. SEE
C     INSTRUCTIONS FOR OUTPUT PARAMETERS. Programmed by A. Haji-Sheikh
      IMPLICIT REAL*8 (A-H,O-Z)
      DIMENSION B(NX,*),AL(NX,*),ALINV(NX,*),D(*)
      DO 1 I=1,N
      DO 1 J=1,N
      AL(I,J) = 0.0D+00
    1 ALINV(I,J) = 0.0D+00
      D(1)=B(1,1)
      DO 200 I=1, N
      AL(I,1) = B(I,1)/D(1)
  200 AL(I,I) = 1.0D+00
      DO 350 J=2, N
      J1 = J-1
      D(J) = B(J,J)
      DO 250 L1=1,J1
  250 D(J) = D(J)-AL(J,L1)*AL(J,L1)*D(L1)
      I1 = J+1
      D0 340 I=I1,N
      AL(I,J) = B(I,J)
      DO 310 L1=1,J1
  310 AL(I,J) = AL(I,J)-AL(I,L1)*AL(J,L1)*D(L1)
  340 AL(I,J) = AL(I,J)/D(J)
  350 CONTINUE
      DO 900 I=1,N
      ALINV(I,I) = 1.0D+00
      I1 = I+1
      IF (I1.GT.N) GO TO 900
      DO 860 J= I1, N
      X = 0.0D+00
      DO 850 IR = I, J
  850 X = X-AL(J,IR)*ALINV(IR,I)
  860 ALINV(J,I) = X
  900 CONTINUE
      RETURN
      END
```

APPLICATIONS OF THE GALERKIN-BASED GREEN'S FUNCTIONS

11.1 INTRODUCTION

The Galerkin-based Green's function (GF) solution of the diffusion equation is presented in Chapter 10, which also contains simple one-dimensional examples that demonstrate the method of solution and that discuss the accuracy of the results. In this chapter, the Galerkin-based GF solution method is extended to more advanced problems. Thermal conduction in multidimensional bodies is presented in Section 11.2. In Section 11.3 the basis functions are modified so that conduction in heterogeneous materials can be accommodated. Then, in Section 11.4, the GF solution is developed and applied to steady-state conduction problems. Finally, a study of heat transfer in the thermal entrance region of ducts is included in Section 11.5.

Sections 11.2–11.6 each include one or more examples that demonstrate the procedure. Except for selection of basis functions, the same mathematical procedure applies to simple and complex problems. The major difficulty in dealing with complex problems is the selection of a complete and linearly independent set of basis functions that satisfy the boundary conditions. Unlike the one-dimensional problems studied in Chapter 10, there is no generalized form for the basis functions; hence, each multidimensional body must be treated differently.

The GF for a few orthogonal multidimensional bodies are products of one-dimensional GFs. However, it is not difficult to define basis functions for most orthogonal multidimensional bodies. The basis functions are usually the products of one-dimensional basis functions. It is also possible to find basis functions for irregular bodies when boundary conditions are of the first kind. However, the basis functions for nonorthogonal bodies with boundary conditions of second and third kinds are sometimes difficult to obtain. Once the basis functions are defined, the computation of matrices **A** and **B** may require numerical integration. After the matrices **A** and **B** are determined, the calculation of parameters in the GF is exactly the same as that for one-dimensional bodies. Indeed, the matrix algebra, the GFs, and the GF solution method are the same for one-dimensional or multidimensional, and orthogonal or nonorthogonal bodies. In this chapter, emphasis is placed on finding the basis functions.

11.2 BASIS FUNCTIONS IN SOME COMPLEX GEOMETRIES

As discussed earlier, the derivation of the GF solution method is in Chapter 10. The algebraic steps leading to the computation of the GF and the GF solution method are the same for bodies of different shapes. The method of finding the basis functions and the analytical (or numerical) efforts needed to compute matrices **A** and **B** elucidate the complexity of the problem. The procedure for finding the basis functions is not unique and any properly defined basis functions, as discussed in Section 10.2, are acceptable; a few methods of selecting basis functions for nonorthogonal bodies are discussed. The procedure includes the basis functions that satisfy boundary conditions of the first kind (prescribed temperature, $f_j = 0$), the second kind (prescribed heat flux, $\partial f_j/\partial n = 0$), or the third kind (convective, $-k\partial f_j/\partial n = hf_j$).

11.2.1 Boundary Conditions of the First Kind

A universal relation to give a_{ij} and b_{ij} for many regular geometries is derived in Chapter 10. For more complex geometries, the necessary integrations using Eqs. (10.12) and (10.13) may require a symbolic software or numerical quadrature. By using time partitioning, it may be possible to reduce the computations in these cases. The remaining matrix operation is independent of the dimensions of the body and the boundary conditions. When a multidimensional body has a regular shape, the GF is a product of one-dimensional GFs. To obtain a reasonably accurate solution for irregular multidimensional bodies, the number of basis functions is usually larger than 2. Numerical matrix operation becomes necessary when dealing with complex multidimensional problems.

The method of selecting the basis functions for boundary conditions of the first kind is available in the literature (Kantorovich and Krylov, 1960; Ozisik, 1980; Haji-Sheikh and Mashena, 1987). If a region is bounded by M surfaces $\phi_1, \phi_2, \ldots, \phi_M$ (Fig. 11.1), the first member of the set of basis functions is

$$f_1(\mathbf{r}) = \phi_1\phi_2\phi_3 \cdots \phi_M \tag{11.1}$$

Each subsequent member of the set of basis functions is obtained by multiplying $f_1(\mathbf{r})$ by an element of a complete set, for example, in a Cartesian coordinate system

$$f_2(\mathbf{r}) = f_1(\mathbf{r})x \tag{11.2a}$$

$$f_3(\mathbf{r}) = f_1(\mathbf{r})y \tag{11.2b}$$

$$f_4(\mathbf{r}) = f_1(\mathbf{r})z \tag{11.2c}$$

$$f_5(\mathbf{r}) = f_1(\mathbf{r})x^2 \tag{11.2d}$$

$$f_6(\mathbf{r}) = f_1(\mathbf{r})xy \tag{11.2e}$$

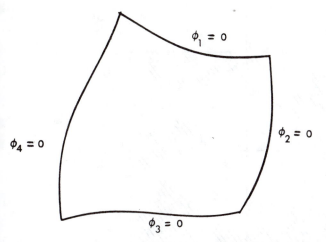

Figure 11.1 Generalized configuration for use with Eq. (11.1).

$$f_7(\mathbf{r}) = f_1(\mathbf{r})xz \tag{11.2f}$$

$$f_8(\mathbf{r}) = f_1(\mathbf{r})y^2 \tag{11.2g}$$

$$f_9(\mathbf{r}) = f_1(\mathbf{r})yz \tag{11.2h}$$

$$f_{10}(\mathbf{r}) = f_1(\mathbf{r})z^2 \tag{11.2i}$$

Each basis function is required to vanish only over the exterior boundaries. Some, but not all, basis functions may become zero at any interior point. This can be ensured if $f_1(\mathbf{r})$ is not zero within the region. Whenever all basis functions vanish at an interior point, the region can be subdivided into different subregions. The basis functions are constructed for each subregion and then are matched at the common boundary of the subregions (Kantorovich and Krylov, 1960).

Example 11.1 Consider a two-dimensional solid bounded by the surfaces $a - x = 0$, $a + x = 0$, $b - y = 0$, $b + y = 0$, and a circular surface $c^2 - x^2 - y^2 = 0$ (Fig. 11.2) and find the basis functions.

SOLUTION The first basis function, for this example, is a product of all the functions representing the surfaces of this body,

$$f_1 = (a^2 - x^2)(b^2 - y^2)(c^2 - x^2 - y^2) \tag{11.3a}$$

The other basis functions are obtained when f_1 is multiplied by polynomial terms in the ascending order (1 has already been used),

$$f_2 = (a^2 - x^2)(b^2 - y^2)(c^2 - x^2 - y^2)x \tag{11.3b}$$

$$f_3 = (a^2 - x^2)(b^2 - y^2)(c^2 - x^2 - y^2)y \tag{11.3c}$$

$$f_4 = (a^2 - x^2)(b^2 - y^2)(c^2 - x^2 - y^2)x^2 \tag{11.3d}$$

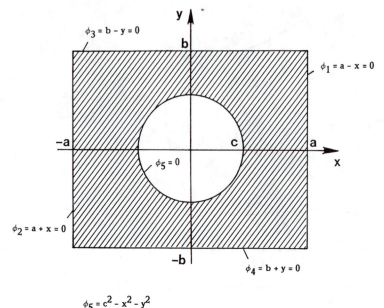

Figure 11.2 Two-dimensional body for Example 11.1 with boundary conditions of the first kind.

$$f_5 = (a^2 - x^2)(b^2 - y^2)(c^2 - x^2 - y^2)xy \qquad (11.3e)$$

$$f_6 = (a^2 - x^2)(b^2 - y^2)(c^2 - x^2 - y^2)y^2 \qquad (11.3f)$$

In the absence of the circular hole within this rectangular body, clearly the basis functions are the products of one-dimensional basis functions; see Problem 11.1 in this chapter.

Example 11.2 Calculate temperature distribution in a spheroidal body (see the inset of Fig. 11.3) whose contour is defined by

$$1 - r^2 - \frac{z^2}{b^2} = 0 \qquad (11.4a)$$

The initial temperature is zero and, at $t \geq 0$, the surface temperature is equal to one.

SOLUTION First, the function f_1 is chosen as

$$f_1 = 1 - r^2 - \frac{z^2}{b^2}$$

and then the remaining basis functions are

$$f_2 = \left(1 - r^2 - \frac{z^2}{b^2}\right) r^2 \qquad (11.4b)$$

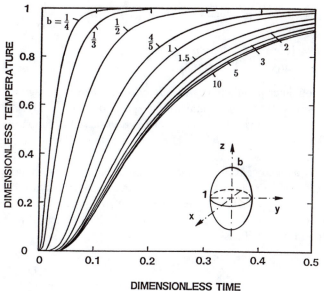

Figure 11.3 Example 11.2 results for temperature at $r = 0$ versus $\alpha t/a^2$.

$$f_3 = \left(1 - r^2 - \frac{z^2}{b^2}\right) z^2 \tag{11.4c}$$

$$f_4 = \left(1 - r^2 - \frac{z^2}{b^2}\right) r^4 \tag{11.4d}$$

$$f_5 = \left(1 - r^2 - \frac{z^2}{b^2}\right) r^2 z^2 \tag{11.4e}$$

$$f_6 = \left(1 - r^2 - \frac{z^2}{b^2}\right) z^4 \tag{11.4f}$$

The spheroidal solid is homogeneous with constant thermophysical properties. The coordinates and time are viewed as dimensionless. Using Eqs. (10.12), and (10.13), the elements of the matrices **A** and **B** for a three-term solution are (Haji-Sheikh, 1986)

$$a_{11} = \frac{4(2b^2 + 1)}{5b^2} \qquad a_{12} = \frac{8(2b^2 + 1)}{35b^2} \qquad a_{13} = \frac{4(2b^2 + 1)}{35}$$

$$a_{22} = \frac{32(4b^2 + 1)}{(315b^2)} \qquad a_{23} = \frac{8(2b^2 + 1)}{315} \qquad a_{33} = \frac{4b^2(4b^2 + 11)}{315}$$

and

$$b_{11} = \frac{8}{35} \qquad b_{12} = \frac{16}{315} \qquad b_{13} = \frac{8b^2}{315} \qquad b_{22} = \frac{64}{3465}$$

$$b_{23} = \frac{16b^2}{3465} \qquad b_{33} = \frac{8b^4}{1155}$$

Symbolic computer programming was used to calculate the integrals. Each of these values is divided by the volume of the spheroid. Additionally, the integration of the function f_i over the volume is needed in the GF solution method, Eq. (10.77). The corresponding values, after they are divided by the volume of the spheroid, are 2/5, 4/35, and $2b^2/35$.

The steps for a one-eigenvalue solution are discussed mainly to show that the procedure is independent of the shape of the domain and complexity of the problem. Equation (10.11) yields the eigenvalue $\gamma_1 = a_{11}/b_{11} = 7(2b^2 + 1)/(2b^2)$ and the eigenvector $d_{11} = 1$. Equations (10.41) and (10.45) yield $p_{11} = 35/8$ and the GF using Eq. (10.60) is

$$G(r', z', t|r, z, \tau) = \frac{35}{8} f_1(r, z) f_1(r', z') \exp\left[-\gamma_1(t - \tau)\right]$$

Finally, the solution using Eq. (10.77) is

$$T(r, z, \tau) = 1 - \frac{7}{4}\left(1 - r^2 - \frac{z^2}{b^2}\right) \exp\left[-\frac{7(2b^2 + 1)t}{2b^2}\right]$$

A one-eigenvalue solution is a crude approximation to the exact solution. A 10-term solution yields four accurate digits except at small time. For example when $b = 2$ and $t = 0.3$ the exact solution is 0.7758, while a one-term solution gives 0.835 and a three-term solution is 0.780.

A desktop computer is adequate to perform similar calculations using more eigenvalues. For instance, the calculated values of temperature at the point $(0, 0)$ are computed with speed and efficiency using a small personal computer, and the results are plotted in Fig. 11.3. As many as 21 eigenvalues are used to compute the data. The large-time data agree with the exact solution within five significant digits; this accuracy diminishes as t becomes small (Haji-Sheikh, 1986).

11.2.2 Boundary Conditions of the Second Kind

We now focus on the insulated boundaries. The selection of the basis functions becomes simple if a flat section of boundary is insulated. As an illustration, Fig. 11.4b shows a flat section of the boundary described by $\phi_1 = 0$, which is insulated. For this planar surface, a condition of symmetry about that surface is implied. Then, the original region can be replaced by a new region that includes itself and its mirror image (Fig. 11.4a). If, for instance, x is selected perpendicular to the $\phi_1 = 0$ surface, then $\phi_3 = 0$ is $\phi_2 = 0$ except the variable x is replaced by $-x$. Therefore, the basis function f_1 is obtained using the boundary conditions of the first kind, by utilizing Eq. (11.1), as $f_1 = \phi_2 \phi_3$. Then, the remaining basis functions are defined by using Eq. (11.2) and

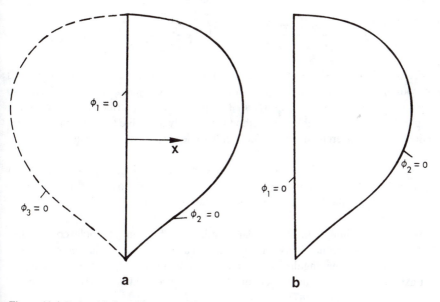

Figure 11.4 Body with flat surface insulated at $\phi_1 = 0$.

retaining the terms with x to the power of even numbers. This will automatically result in $\partial f_j / \partial n = 0$ along the x axis on the $\phi_1 = 0$ line for all the basis functions.

There are other methods of finding the basis functions that satisfy the boundary conditions of the second kind (Lee and Haji-Sheikh, 1991). One way of finding the basis functions for boundary conditions of the second kind can be illustrated through a simple example. Consider that the geometry depicted in Fig. 11.4b has the following boundary conditions: $f_j = 0$ on the $\phi_1 = 0$ line and $\partial f_j / \partial n = 0$ on the $\phi_2 = 0$ line. The basis function $f_j^{(2)}$ that satisfies the boundary conditions of the second kind is considered to be of the form

$$f_j^{(2)} = f_j^{(1)} (\phi_2 H - 1) \tag{11.5}$$

The term -1 in the parentheses is for the convenience of analysis and has no effect on the final solution because $f_j^{(2)}$ can be multiplied by a constant without loss of generality, and H is yet to be determined. The function $f_j^{(1)}$ satisfies the boundary condition of the first kind everywhere except on the $\phi_2 = 0$ surface, which is insulated

$$f_j^{(1)} = \phi_1 x^{m_j} y^{n_j} z^{l_j} \tag{11.6}$$

Since the surface $\phi_2 = 0$ is insulated, then the relation $\partial f_j^{(2)} / \partial n = 0$ on the $\phi_2 = 0$ surface requires that

$$\frac{-\partial f_j^{(1)}}{\partial n} + f_j^{(1)} \left(\frac{\partial \phi_2}{\partial n} \right) H = 0 \qquad \text{on } \phi_2 = 0 \text{ surface} \tag{11.7}$$

which yields a relation for function H to be used in Eq. (11.5) as

$$H = \left(\frac{\partial f_j^{(1)}/\partial n}{f_j^{(1)} \partial \phi_2 / \partial n} \right)\Bigg|_{\phi_2 = 0} = \left(\frac{\nabla f_j^{(1)} \cdot \nabla \phi_2}{f_j^{(1)} \nabla \phi_2 \cdot \nabla \phi_2} \right)\Bigg|_{\phi_2 = 0} \tag{11.8}$$

For some geometric configurations, it is possible to define a set of basis functions in the polynomial form with free constants. The free constants can be evaluated so that $\nabla f_j \cdot \nabla \phi = 0$ on the $\phi = 0$ surface. This procedure is described in a forthcoming example. However, no established method is presently available to determine the basis functions for all different-shaped bodies with some (or all) walls insulated.

Example 11.3 Consider a homogeneous spheroidal solid, Fig. 11.5, whose boundary in the cylindrical coordinate system is given by the equation $\phi_1 = 1 - r^2/a^2 - z^2/b^2$ and find the basis functions when the external surface is insulated.

SOLUTION Although this is a regular or orthogonal body in spheroidal coordinates, it is a nonorthogonal body in the cylindrical coordinates. The temperature is independent of the angular coordinate and is symmetric about the $z = 0$ plane. Basis functions for boundary conditions of the second kind are (Haji-Sheikh and Lakshminarayanan, 1987)

$$f_j^{(2)} = r^{m_j} z^{n_j} (B_1 r^2 + B_2 z^2 + B_3) \qquad j = 1, 2, \ldots, N \tag{11.9}$$

where $m_j = 0, 2, 4, \ldots$, and $n_j = 0, 2, 4, \ldots$. However, if the temperature is not symmetric about the $z = 0$ plane, the odd n_j's must be included. The basis functions, Eq. (11.9), must satisfy the boundary condition

$$\frac{\partial f_j^{(2)}}{\partial n} = 0 \qquad \text{when } \phi_1 = 0 \tag{11.10}$$

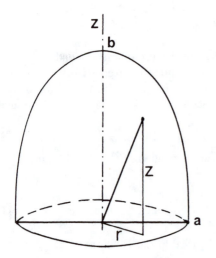

Figure 11.5 Spheroidal body for Examples 11.3 and 11.4.

which can be written as

$$\nabla f_j^{(2)} \cdot \nabla \phi_1 = 0 \qquad \text{when } \phi_1 = 0 \qquad (11.11)$$

Introducing Eq. (11.9) into Eq. (11.11) and deleting z using the relation $\phi_1 = 0$, the following second-degree polynomial equation is obtained:

$$\left[B_1 \left(\frac{m_j + 2}{a^2} + \frac{n_j}{b^2} \right) - B_2 \left(\frac{m_j}{a^2} + \frac{n_j + 2}{b^2} \right) \frac{b^2}{a^2} \right] r^2$$

$$+ \left[B_2 \left(\frac{m_j}{a^2} + \frac{n_j + 2}{b^2} \right) b^2 + B_3 \left(\frac{m_j}{a^2} + \frac{n_j}{b^2} \right) \right] r^0 = 0 \quad (11.12)$$

Since Eq. (11.12) is satisfied at all r's, then the coefficients that multiply by r to any power must be zero, or

$$B_2 = - \frac{B_3(m_j/a^2 + n_j/b^2)}{b^2(m_j/a^2 + (n_j + 2)/b^2)} = B_3 \left[\frac{2/b^2}{m_j b^2/a^2 + n_j + 2} - \frac{1}{b^2} \right] \quad (11.13)$$

and

$$B_1 = \frac{B_2 \, b^2/a^2 [m_j/a^2 + (n_j + 2)/b^2]}{(m_j + 2)/a^2 + n_j/b^2}$$

$$= B_3 \frac{1}{a^2} \left[\frac{2b^2/a^2}{(m_j + 2)b^2/a^2 + n_j} - 1 \right] \quad (11.14)$$

One of the three coefficients can be selected arbitrarily (e.g., $B_3 = 1$).

11.2.3 Boundary Conditions of the Third Kind

We will demonstrate that the basis functions satisfying the boundary conditions of the third kind can be constructed from the basis functions that satisfy the boundary conditions of the second kind. For this presentation, $f_j^{(2)}$ will designate the basis functions that satisfy the boundary conditions of the second kind on the $\phi_1 = 0$ surface. The basis functions satisfying the boundary conditions of the third kind are obtained from the simple relation,

$$f_j^{(3)} = f_j^{(2)} \left(\phi_1 H - \frac{k}{h} \right) \qquad j = 1, 2, \dots, N \qquad (11.15)$$

The method of calculating H in Eq. (11.15) is similar to that for Eq. (11.5). The function $f_j^{(3)}$ must satisfy the relation $-k \partial f_j^{(3)}/\partial n = h f_j^{(3)}$ on the surface $\phi_1 = 0$. This leads to

$$-k f_j^{(2)} \frac{H \partial \phi_1}{\partial n} = h f_j^{(2)} \left(\frac{-k}{h} \right) \qquad \text{when } \phi_1 = 0 \qquad (11.16)$$

The function H to be used in Eq. (11.15) then becomes

$$H = \left(\frac{1}{\partial \phi_1 / \partial n} \right) \Bigg|_{\phi_1 = 0} \tag{11.17}$$

in which $\phi_1 = 0$ designates the convective surface. It is also possible to obtain, for some geometries, the basis functions using series expansion as discussed for boundary conditions of the second kind.

Example 11.4 Spheroidal bodies with a convective surface have many interesting applications in aerospace, food, and agricultural industries. The spheroidal body defined in Example 11.3 is subject to convective boundary conditions. The initial temperature is T_0 and the ambient temperature is T_∞ when $t \geq 0$. Calculate the temperature at the point $r = 0$ and $z = 0$.

SOLUTION Although, in theory, a spheroid with a convective surface submits to an exact solution, such an exact solution has not been found. This is due to the complexity of the exact mathematical and subsequent numerical procedures. Using Eq. (11.15), it is possible to find a set of basis functions that satisfy the convective boundary conditions. After $f_j^{(2)}$ from Eq. (11.9) is inserted into Eq. (11.15), the basis functions for convective spheroids are

$$f_j^{(3)} = f_j^{(2)} \left\{ \frac{\text{Bi}[(b^2/a^2)(1 - r^2/a^2) - z^2/a^2]}{2(b/a)[(b^2/a^2)(r^2/a^2) - r^2/a^2 + 1]^{1/2}} - 1 \right\} \tag{11.18}$$

Next, the function $f_j^{(3)}$ must replace f_j in Eqs. (10.12) and (10.13) to compute matrices **A** and **B**. The analytical integrations of the resulting equations, if possible, are complicated and are not cost effective. Numerical quadrature was used by Haji-Sheikh and Lakshminarayanan (1987) to compute a_{ij}'s and b_{ij}'s. The remaining steps are identical to those described in Example 11.2. In fact, the same computer program is used to solve for temperature here and for Example 11.2, except matrices **A** and **B** in this example are computed numerically, while symbolic computer algebra was used to calculate **A** and **B** in Example 10.2. The computed temperature results for a range of Biot numbers, ha/k, and aspect ratios are shown in Fig. 11.6. The solid lines are generally in good agreement with the Monte Carlo data (Haji-Sheikh and Sparrow, 1967). Previous comparisons, in Example 10.2, with the exact solution imply that any small discrepancy can be attributed to the sampling error in the Monte Carlo solution.

11.3 HETEROGENEOUS SOLIDS

The derivation of the GF solution, Eq. (10.68) or the alternative GF solution, Eq. (10.77) permits the computation of thermal conduction in heterogeneous bodies. The only difference between solutions for homogeneous and heterogeneous solids is the selection of a set of basis functions. First, a set of basis functions must be defined that satisfies the boundary conditions on the external surfaces and perfect or imperfect

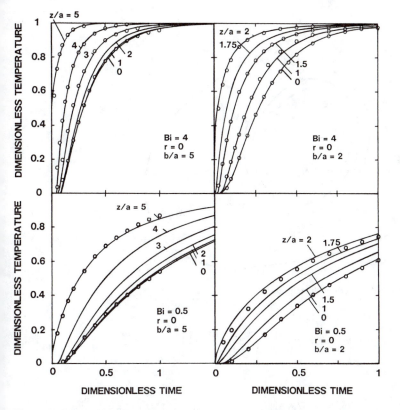

Figure 11.6 Example 11.4 results for temperature, $(T - T_0)/(T_x - T_0)$, at $r = 0$ versus $\alpha t/a^2$.

contact relations between adjacent materials. Then, the basis functions are used to solve a numerical example.

Let the subscript e identify an inclusion of different material enclosed in the main body (Fig. 11.7), and let m denote the main domain. The basis function $f_{j,m}$, which satisfies the boundary conditions of the main body, is selected ignoring the inclusion; therefore, f_j is $f_{j,m}$ in the main body. However, the basis function should be modified as it crosses the boundary of the inclusion. The formulation of the basis functions in the absence of contact conductance is given by Haji-Sheikh (1988). The formulation is then modified to include the effect of finite contact conductance as (Haji-Sheikh and Beck, 1990)

$$f_j = f_{j,m} \quad \text{(in the main domain)} \tag{11.19a}$$

and

$$f_j = f_{j,m} + U + \phi_e H \quad \text{(in the } i\text{th inclusion)} \tag{11.19b}$$

for $j = 1, 2, \ldots, N$. The continuity condition that $k_m(\partial f_j/\partial n)_m = k_e(\partial f_j/\partial n)_e$ and the jump condition $f_{j,e} = f_{j,m} - (k_m/C)(\partial f_{j,m}/\partial n)$ at the boundary of the inclusion ($\phi_e = 0$ surface is different for different inclusions) permit the calculation of U and H as

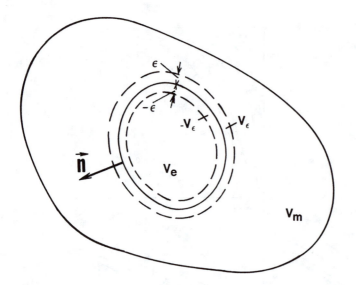

Figure 11.7 Composite body with inclusion.

$$U = - \left(\frac{k_m}{C}\right) \left(\frac{\partial f_{j,m}}{\partial n}\right)\bigg|_{\phi_e = 0} \tag{11.20}$$

and

$$H = \frac{[(\nabla f_{j,m} \cdot \nabla \phi_e)|_{\phi_e = 0}(k_m/k_e - 1) - \nabla U \cdot \nabla \phi_e|_{\phi_e = 0}]}{(\nabla \phi_e \cdot \nabla \phi_e)|_{\phi_e = 0}} \tag{11.21}$$

where C is the contact conductance. A linear combination of the basis function f_j satisfies the continuity of heat flux, $k_m(\partial T/\partial n)_m = k_e(\partial T/\partial n)_e$, and temperature jump, $T_e = T_m - (k_m/C)(\partial T_{j,m}/\partial n)$, on the boundary of inclusion e. When the inclusion has other boundaries in addition to the $\phi_e = 0$ surface, other modifications to the values of U and H become necessary (e.g., see Example 11.5).

Example 11.5 To illustrate the method for accommodating the contribution of contact conductance, consider two plates: one has a thickness of a and the other $L - a$ (see Fig. 11.8). It is convenient to let subscripts e and m stand for the regions designated using these letters in Fig. 11.8. The composite slab is initially at temperature T_0 and has the following boundary conditions

$$-k_m \frac{\partial T}{\partial x} = hT \qquad \text{at } x = L \text{ and when } t > 0 \tag{11.22a}$$

and

$$\frac{\partial T}{\partial x} = 0 \qquad \text{at } x = 0 \text{ and when } t > 0 \tag{11.22b}$$

Write an equation for temperature distribution.

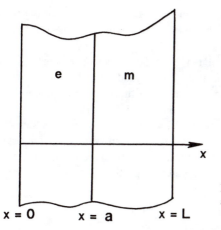

Figure 11.8 Two-layer wall insulated at $x = 0$ and convective surface at $x = L$ for Example 11.5.

SOLUTION The basis function, f_j, that satisfies the conditions $-k_m \partial f_{j,m}/\partial x = h f_{j,m}$ at $x = L$ and $\partial f_{j,m}/\partial x = 0$ at $x = 0$, for $j = 1, 2, 3, \ldots$, is

$$f_j = f_{j,m} = \left(\delta_j - \frac{x^2}{a^2} \right) \left(\frac{x}{a} \right)^{2(j-1)} \qquad a < x < L \qquad (11.23a)$$

where $\delta_j = (L/a)^2 (L/a + 2j/\text{Bi})/[L/a + 2(j - 1)/\text{Bi}]$ and $\text{Bi} = ha/k_m$. A function $\phi(x) = 1 - x^2/a^2$ is selected to satisfy the boundary condition given by Eq. (11.22b) and to vanish at $x = a$; then

$$f_j = f_{j,e} = \left(\delta_j - \frac{x^2}{a^2} \right) \left(\frac{x}{a} \right)^{2(j-1)}$$

$$+ U + \left(1 - \frac{x^2}{a^2} \right) H \qquad 0 < x < a \qquad (11.23b)$$

The value of $U = 2[j - (j - 1)\delta_j]/(Ca/k_m)$ is computed so that the equation $f_{j,e} = f_{j,m} - (k_m/C)(\partial f_{j,m}/\partial x)$ at $x = a$ is satisfied, where C is the contact conductance. The continuity of heat flux at $x = a$ yields $H = (k_m/k_e - 1)[j - (j - 1)\delta_j]$. The solution when $N = 2$, $L/a = 2$, $k_e/k_m = 2$, $ha/k_m = 1$, $Ca/k_m = 1$, and $\rho_e c_{pe}/\rho_m c_{pm} = 1$ is

$$\frac{T}{T_0} = 0.15048 \psi_1(x) \exp\left(-\frac{1.5407\alpha t}{a^2} \right)$$

$$+ 0.006762 \psi_2(x) \exp\left(-\frac{11.449\alpha t}{a^2} \right) \qquad (11.24)$$

where

$$\psi_1(x) = f_1 - 0.040305 f_2 \qquad (11.25a)$$

$$\psi_2(x) = f_1 + 20.999 f_2 \qquad (11.25b)$$

When $x \leq a$ (that is, in region e), the functions $f_{1,e}$ and $f_{2,e}$ replace the f_1 and f_2 functions. However, when $x \geq a$, the functions $f_{1,m}$ and $f_{2,m}$ replace the f_1 and f_2 functions.

Whenever the boundary condition at $x = 0$ is convective, U must also satisfy the convective boundary condition at $x = 0$. In addition, the coefficient $(1 - x^2/a^2)$ that multiplies H should be replaced by a function that becomes 0 when $x = a$ and satisfies the convective condition at $x = 0$.

When calculating a_{ij} from Eq. (10.12), the function f_j suffers a step change at and along the contact surface. The derivatives of f_j across the inclusion boundary are singular, and it can be shown that the value of the volume integral over the singularity zone is zero. The integral given in Eq. (10.12) for this example is

$$\int_V f_i \, \nabla \cdot (k\nabla f_j) \, dV = \int_{V_e - V_\epsilon} f_{i,e} \, \nabla \cdot (k\nabla f_{j,e}) \, dV$$

$$+ \int_{V_e - V_\epsilon}^{V_e + V_\epsilon} f_i \, \nabla \cdot (k\nabla f_j) \, dV$$

$$+ \int_{V - V_e - V_\epsilon} f_{i,m} \, \nabla \cdot (k\nabla f_{j,m}) \, dV \qquad (11.26)$$

The contact zone (Fig. 11.7) is divided into $+V_\epsilon$ and $-V_\epsilon$. The first integration on the right side is over the inclusion up to the contact zone. The second integration on the right side is over the contact zone, and the last integration is over the entire domain less the inclusion and the contact zone. It can be shown that, as $\epsilon \rightarrow 0$, the first and third integrals yield the value of a_{ij} if the second integral (over the contact zone) on the right side of Eq. (11.26) vanishes. Assuming the thickness of the contact zone is extremely small, it is possible to ignore the derivatives of f_j in the directions perpendicular to the normal to the contact surface $\mathbf{n}$ (see Fig. 11.7). Then, the second term on the right side is integrated by parts

$$\int_{-\epsilon}^{+\epsilon} f_i \frac{d}{dn} \left(k \frac{df_j}{dn} \right) dn = f_i \left(\frac{kdf_j}{dn} \right) \Bigg|_{-\epsilon}^{+\epsilon}$$

$$- \int_{-\epsilon}^{+\epsilon} k \left(\frac{df_i}{dn} \right) \left(\frac{df_j}{dn} \right) dn \qquad (11.27)$$

At the limit as $\epsilon \rightarrow 0$, Eq. (11.27) reduces to

$$\lim_{\epsilon \rightarrow 0} q_j[(f_{i,m}(\epsilon) - f_{i,e}(-\epsilon)] - q_j[(f_{i,m}(\epsilon) - f_{i,e}(-\epsilon)] = 0 \qquad (11.28)$$

where $q_j = k_m \partial f_{j,m}/\partial n = k_e \partial f_{j,e}/\partial n$ is a constant in V_ϵ.

When f_j or its normal derivative on the exterior surfaces is zero, it is possible to substitute Eq. (10.14) in Eq. (10.12) to obtain

$$a_{ij} = -\int_V k\nabla f_i \cdot \nabla f_j \, dV \qquad (11.29)$$

If this equation is utilized to compute a_{ij} instead of Eq. (10.12), the value of the integral over the contact zone, where the derivatives of f_j are singular, is not zero and should be evaluated.

At this stage, it is appropriate to solve a three-dimensional nonorthogonal problem to illustrate the strength of this Galerkin-based integral (GBI) solution. The following example does not have an exact solution and numerical computation of the temperature is a formidable task. The procedure discussed in Example 11.5 is applied to a more complex problem to demonstrate the possibility of accommodating difficult thermal conduction problems.

Example 11.6 Consider a spherical inclusion whose radius is equal to a, centrally located in a cubical body with dimensions $2b \times 2b \times 2b$ (Fig. 11.9). The initial temperature is 0 and the external surface temperature is maintained at 1 when $t \geq 0$. Find the temperature distribution.

SOLUTION Although the shape of the body is simple, it contains all the complexities one expects in a conduction problem. The procedure described in this example and the previous example can be applied to other geometries. The basis functions are

$$f_{j,m} = (b^2 - x^2)(b^2 - y^2)(b^2 - z^2)x^{m_j}y^{n_j}z^{l_j} \tag{11.30}$$

in which m_j, n_j, and l_j take values of 0, 1, 2, ... The function ϕ_e is given by

$$\phi_e = a^2 - x^2 - y^2 - z^2 \tag{11.31}$$

which unconditionally vanishes on the surface of the inclusion. The function f_j in the inclusion is $f_{j,e}$ obtained from Eq. (11.19b) assuming a perfect contact between materials ($U = 0$). Due to symmetry, only even values of m_j, n_j, and l_j need be

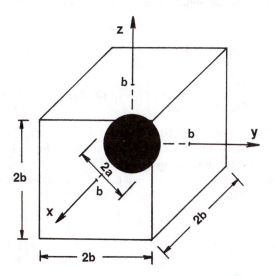

Figure 11.9 Cubical body with centrally located spherical inclusion.

considered. However, for a one-term solution, $m_j = n_j = l_j = 0$, the functions $f_{1,m}, f_{1,e}$ are

$$f_{1,m} = (b^2 - x^2)(b^2 - y^2)(b^2 - z^2) \tag{11.32a}$$

$$f_{1,e} = f_{1,m} + \left(\frac{k_m}{k_e} - 1\right)(a^2 - x^2 - y^2 - z^2)$$

$$\times [x^2(b^2 - y^2)(b^2 - z^2) + y^2(b^2 - x^2)(b^2 - z^2)$$

$$+ z^2(b^2 - x^2)(b^2 - y^2)] \tag{11.32b}$$

A solution with a higher degree polynomial is a four-term solution, and the next higher degree polynomial yields a 10-term solution. Problems with high degree polynomials are ideally suited to symbolic algebra software because the exact integrations leading to the computation of matrices $\mathbf{A}$ and $\mathbf{B}$ are repetitive and lengthy. To show the mathematical steps, the elements of matrices $\mathbf{A}$ and $\mathbf{B}$, when $N = 1$, are evaluated using Eqs. (10.12) and (10.13):

$$a_{11} = I_{a1} + \left(\frac{k_e}{k_m} - 1\right)(I_{a2} - I_{a3}) + \frac{k_e}{k_m}\left(\frac{k_m}{k_e} - 1\right)^2 I_{a4} \tag{11.33}$$

and

$$b_{11} = I_{b1} + \left(\frac{C_e}{C_m} - 1\right)I_{b2} + \frac{C_e}{C_m}\left[\left(\frac{k_m}{k_e} - 1\right)I_{b3}\right.$$

$$\left. + \left(\frac{k_m}{k_e} - 1\right)^2 I_{b4}\right] \tag{11.34}$$

where C_m and C_e stand for ρc_p of the main region and inclusion, respectively, and the values of $I_{a2}, I_{a3}, I_{a4}, I_{b1}, I_{b2}, I_{b3}$, and I_{b4} in Eqs. (11.33) and (11.34) are in Table 11.1. The alternative GF solution, Eq. (10.77) is used to compute the temperature distribution

$$\frac{T - T_0}{T_s - T_0} = 1 - p_{11}\left[I_{c1} + \left(\frac{C_e}{C_m} - 1\right)I_{c2}\right.$$

$$\left. + \frac{C_e}{C_m}\left(\frac{k_m}{k_e} - 1\right)I_{c3}\right]f_1 \exp(-\gamma_1 t) \tag{11.35}$$

The integrals I_{c1}, I_{c2}, and I_{c3} are also presented in Table 11.1. The function f_1 is f_j when $j = 1$ given by Eq. (11.30) outside of the inclusion and by Eq. (11.32) inside of the inclusion. When the initial temperature $F(\mathbf{r}) = T_0$, the surface temperature T_s is a constant, $b/a = 3$, $k_e/k_m = 10$, and $C_e/C_m = 1$, the following dimensionless parameters are obtained:

$$a_{11} = 1.848 \times 10^6 \tag{11.36a}$$

$$b_{11} = 2.168 \times 10^6 \tag{11.36b}$$

Table 11.1 Values of the integrals in Eqs. (11.33)–(11.35)

Integrals[†]	$b/a = 1.5$	$b/a = 2.0$	$b/a = 2.5$	$b/a = 3$	$b/a = 5$
I_{a1}	-2.214×10^2	-9.321×10^3	-1.695×10^5	-1.814×10^6	-1.389×10^9
I_{a2}	-1.138×10^2	-2.488×10^3	-2.545×10^4	-1.657×10^5	-2.945×10^7
I_{a3}	-1.098×10^2	-2.444×10^3	-2.517×10^4	-1.644×10^5	-2.936×10^7
I_{a4}	-9.063×10^0	-1.014×10^2	-6.374×10^3	-2.820×10^3	-1.751×10^5
I_{b1}	6.643×10^1	4.971×10^3	1.413×10^5	2.177×10^6	4.630×10^9
I_{b2}	3.895×10^1	1.576×10^3	2.566×10^4	2.430×10^5	1.217×10^8
I_{b3}	7.199×10^0	1.505×10^2	1.508×10^3	9.713×10^3	1.699×10^6
I_{b4}	3.828×10^{-1}	4.165×10^0	2.582×10^1	1.134×10^2	6.962×10^3
I_{c1}	1.139×10^1	1.517×10^2	1.130×10^3	5.832×10^3	5.787×10^5
I_{c2}	4.471×10^0	2.864×10^1	1.158×10^2	3.564×10^2	7.981×10^3
I_{c3}	4.104×10^{-1}	1.357×10^0	3.383×10^1	7.092×10^0	5.560×10^1

[†]$I_{a1} = -256a^{13}/225$, $I_{b1} = (8a^5/15)^3$, and $I_{c1} = (2a^3/3)^3$.

$$\frac{a^2\gamma_1}{\alpha_m} = -\frac{a_{11}}{b_{11}} = 0.8525 \tag{11.36c}$$

$$d_{11} = 1 \tag{11.36d}$$

$$p_{11} = \frac{1}{b_{11}} \tag{11.36e}$$

and the temperature solution using Eqs. (11.36c–e) is

$$\frac{T - T_0}{T_s - T_0} = 1 - 2.687 \times 10^{-3} f_1 \exp\left(-\frac{0.8525\alpha t}{a^2}\right) \tag{11.37}$$

Similar calculations, using many basis functions, were carried out by Nomura and Haji-Sheikh (1988). The computed temperature when $k_e/k_m = 10$ and for $a = 3$ is shown in Fig. 11.10. It is evident that, because of the high thermal conductivity of the inclusion, the temperature change within the inclusion is extremely small. Figure 11.10 shows the temperature at the center of the inclusion, point $(0, 0, 0)$, is nearly the same as the temperature at the contact point $(1, 0, 0)$. The differentiations and integrations required for calculating a_{ij} can be done manually; however, manual integrations are too time consuming. Nomura and Haji-Sheikh (1988) performed the integrations with the aid of the symbolic software, REDUCE-3 (Hearn, 1983). Note that it is mathematically and numerically feasible to add inclusions of various shapes to the main domain.

11.4 STEADY-STATE CONDUCTION

The GFs and GF solutions for steady-state conduction can be deduced by modifying the GF and GF solutions for transient conduction. The steady state is defined as being

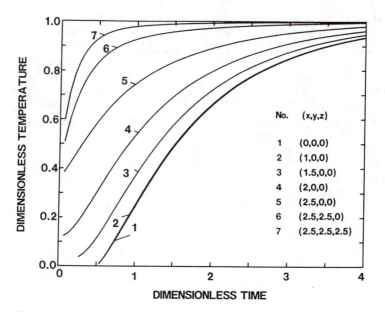

Figure 11.10 Dimensionless temperature, $(T - T_0)/(T_s - T_0)$, versus dimensionless time $\alpha t/a^2$ for a sphere in cubical body.

independent of time. The modification is equally applicable to the GF solution, Eq. (10.68), and the alternative GF solution, Eq. (10.77). The transient solution approaches the steady-state solution as $t \to \infty$. Accordingly, the contribution of the initial temperature distribution in the GF solution will not influence the steady-state solution. If G_{ss} is defined as the steady-state GF, then, using Eq. (10.60),

$$G_{ss} = G(\mathbf{r}'|\mathbf{r})$$

$$= \lim_{t \to \infty} \int_{\tau=0}^{t} G(\mathbf{r}', -\tau|\mathbf{r}, t) \, d\tau$$

$$= \sum_{n=1}^{N} \sum_{j=1}^{N} \sum_{i=1}^{N} \frac{d_{nj} \, p_{ni} \rho(\mathbf{r}) c_p(\mathbf{r}) f_j(\mathbf{r}') f_i(\mathbf{r})}{\gamma_n} \tag{11.38}$$

The GF solution then becomes

$$\rho(\mathbf{r}) c_p(\mathbf{r}) T(\mathbf{r}) = \int_V g(\mathbf{r}') G_{ss} \, dV' + \int_S k(S') \left(\frac{G_{ss} \partial T}{\partial n} - T \frac{\partial G_{ss}}{\partial n} \right)_{S'} dS' \tag{11.39}$$

Similarly, the alternative GF solution reduces to

$$\rho(\mathbf{r}) c_p(\mathbf{r}) T(\mathbf{r}) = \rho(\mathbf{r}) c_p(\mathbf{r}) T^*(\mathbf{r}) + \int_V G_{ss} [g(\mathbf{r}') + f^*] \, dV' \tag{11.40}$$

One can show analytically that Eq. (11.40) reduces to the standard Galerkin solution (Kantorovich and Krylov, 1960)

$$T = T^* - [\{\mathbf{A}^{-1} \cdot \{g^*\}\}^T]\{\mathbf{f}\} \tag{11.41}$$

where $\{\mathbf{f}\}$ is a column vector with elements $f_1, f_2, \ldots, f_N$ and $\{\mathbf{g}^*\}$ is another column vector whose members are

$$g_i^* = \int_V [g(\mathbf{r}') + f^*(\mathbf{r}')] f_i(\mathbf{r}') \, dV' \tag{11.42}$$

When the boundary conditions are nonhomogeneous, the standard Galerkin solution of Poisson's equation is possible if an auxiliary function, T^*, exists.

Example 11.7 Consider a cylindrical pipe with radius $r = a$ centrally placed in a long square box $2b \times 2b$ (Fig. 11.11). The boundary conditions are $T = T_0$ at $r = a$ and $T = 0$ at $x = b$ and at $y = b$. Calculate the temperature field and plot the isotherms.

SOLUTION This example shows the method of calculating the steady-state temperature using the quasi-steady temperature T^*. The method is applicable to numerous conduction problems for which an exact solution does not exist. The computation begins by utilizing Eq. (10.72b) which satisfies the boundary conditions $T = T^* = T_0$ at $r = a$ and $T = T^* = 0$ on the surface of the square box

$$\frac{T^*}{T_0} = 1 - \frac{\ln[(x^2 + y^2)/a^2]}{\ln[(b^2 + x^2 y^2/b^2)/a^2]} \tag{11.43}$$

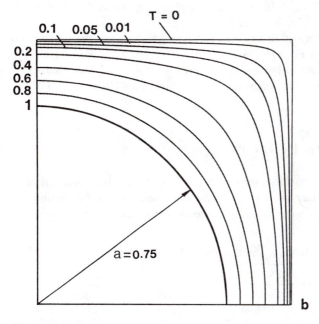

T = 0

0.1   0.05   0.01

0.2
0.4
0.6
0.8
1

a = 0.75

b

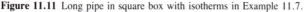

Figure 11.11 Long pipe in square box with isotherms in Example 11.7.

The basis functions are

$$f_j = \left(1 - \frac{x^2}{b^2}\right)\left(1 - \frac{y^2}{b^2}\right)\left(\frac{x^2}{b^2} + \frac{y^2}{b^2} - \frac{a^2}{b^2}\right)\left(\frac{x}{b}\right)^{m_j}\left(\frac{y}{b}\right)^{n_j} \quad (11.44)$$

When $N = 3$, $a/b = 0.75$, $g = 0$, and f^* is defined by substituting T^* from Eq. (11.43) in Eq. (10.73), then Eq. (11.42) yields

$$\mathbf{g}^* = \{0.10750, 0.050747, 0.050747\} \quad (11.45)$$

Note that when $j = 1$, $m_j = n_j = 0$; when $j = 2$, $m_j = 2$ and $n_j = 0$; when $j = 3$, $m_j = 0$ and $n_j = 2$. After the basis functions defined by Eq. (11.44) are inserted in Eq. (10.12), matrix $\mathbf{A}$ becomes (fin effect is neglected)

$$\mathbf{A} = \begin{bmatrix} 0.13268 & 0.065463 & 0.065463 \\ 0.065463 & 0.046750 & 0.024989 \\ 0.065463 & 0.024989 & 0.046750 \end{bmatrix} \quad (11.46)$$

Here, matrices $\mathbf{A}$ and $\mathbf{B}$ can be evaluated analytically, but numerical quadrature is needed to evaluate the elements of $\mathbf{g}^*$. Then Eq. (11.41) provides the temperature distribution

$$\frac{T}{T_0} = T^* + (b^2 - x^2)(b^2 - y^2)(x^2 + y^2 - a^2)(d_1 + d_2 x^2 + d_3 y^2) \quad (11.47)$$

where d_2 and d_3 are identical. The isotherms are computed by this method and plotted in Fig. 11.11. Table 11.2 also supplies temperature distribution for other values of $b/a = 0.25$, 0.5, and 1.

The availability of an auxiliary function T^* eliminates the need to compute matrix $\mathbf{B}$ because Eq. (11.41) yields the same results as Eq. (11.40), yet the number of algebraic and matrix operations are substantially less. However, when T^* is not available, the steady-state formulation of the GF solution, Eq. (11.39), should be used.

11.5 FLUID FLOW IN DUCTS

A knowledge of heat transfer in the entrance region of ducts is essential in the design of compact heat exchangers. The analytical steps described in this section apply to entrance flow in ducts with various cross-sectional shapes; hence, the geometric re-

Table 11.2 Coefficients d_1, d_2, and $d_3 = d_2$ in Eq. (11.47)

d's	$a/b = 0.25$	$a/b = 0.5$	$a/b = 0.75$	$a/b = 0.9$	$a/b = 1$
d_1	0.15837	0.33238	1.1268	4.7624	19.092
d_2	-0.055156	-0.087483	-0.32086	-2.0544	-10.001

striction to obtain a solution is essentially eliminated. The restrictions are that the flow must be hydrodynamically fully developed. The velocity profile is the solution of the momentum equation written for hydrodynamically fully developed, laminar, and Newtonian flow as

$$\frac{\partial^2 W}{\partial X^2} + \frac{\partial^2 W}{\partial Y^2} + 1 = 0 \qquad (11.48\text{a})$$

where

$$W = \frac{-w}{(a^2/\mu)(\partial P/\partial z)} \qquad (11.48\text{b})$$

P is pressure, w is local velocity in the z direction, a is the characteristic length, $X = x/a$, $Y = y/a$, and μ is the viscosity coefficient. After defining the basis functions so that $f_i = 0$ at the wall (note $w = 0$ at the wall), Eq. (11.48a) yields the value of W using Eq. (11.41). The parameters g_i^* are obtained from Eq. (11.42) after substituting $g = 1$ and $f^* = 0$ as

$$g_i^* = \frac{1}{A_c} \int_{A_c} f_i \, dA \qquad (11.49)$$

where A_c is the cross-sectional area of the duct. Equation (11.41), which is the standard Galerkin method, is used to solve for the velocity distribution. The auxiliary function T^* is zero since the boundary conditions are homogeneous. The $\mathbf{A}^{-1} \cdot \{\mathbf{g}^*\}$ in Eq. (11.41) results in coefficients $d_1, d_2, \ldots, d_N$, and the solution for W is

$$W = \sum_{i=1}^{N} d_i f_i \qquad (11.50)$$

The standard definition for average velocity is used to calculate the value of W_{av} as

$$W_{av} = \frac{1}{A_c} \int_{A_c} W \, dA = \sum_{i=1}^{N} d_i g_i^* \qquad (11.51)$$

The friction factor C_f is defined as $-D_e(\partial P/\partial z)/(\rho w_{av}^2/2)$ and it can be written as

$$C_f \, \text{Re} = \frac{2D_e^2}{aW_{av}} = \frac{2\bar{D}_e^2}{W_{av}} \qquad (11.52)$$

where $\text{Re} = \rho D_e w_{av}/\mu$, and D_e/a is designated as the dimensionless hydraulic diameter. Then, the dimensionless velocity is

$$\frac{W}{W_{av}} = \frac{w}{w_{av}} = \frac{C_f \, \text{Re}}{2D_e^2/a^2} \sum_{j=1}^{N} d_j f_j \qquad (11.53)$$

After calculating w (or W), attention must be focused on the computation of temperature. The value of $w(\mathbf{r})$ replaces $u(\mathbf{r})$ and z replaces t in Eq. (10.11) to yield the energy equation for incompressible fluid flowing at a constant rate in a duct [$g(\mathbf{r}, t) = 0$ and $m(\mathbf{r}) = 0$] as

$$\rho(\mathbf{r})c_p(\mathbf{r})w(\mathbf{r})\frac{\partial T}{\partial z} = \frac{\partial}{\partial x}\left[k(\mathbf{r})\frac{\partial T}{\partial x}\right] + \frac{\partial}{\partial y}\left[k(\mathbf{r})\frac{\partial T}{\partial y}\right] \tag{11.54}$$

Here, the effect of axial conduction is neglected. A solution method that includes the effect of axial conduction is reported in Lakshminarayanan (1988) and Lakshminarayanan and Haji-Sheikh (1988).

The volume integrals in Eqs. (10.12) and (10.13) become surface integrals once V is replaced by the cross-sectional area A_c, and the variable t is replaced by the axial coordinate z. Equations (10.12) and (10.13) yield the elements of matrices $\mathbf{A}$ and $\mathbf{B}$

$$a_{ij} = \frac{1}{A_c}\int_{A_c} f_i \, \nabla \cdot (\nabla f_j) \, dA \tag{11.55a}$$

which, for boundary conditions of the first or second kind, reduces to

$$a_{ij} = -\frac{a^2}{A_c}\int_{A_c} \nabla f_i \cdot \nabla f_j \, dA \tag{11.55b}$$

and

$$b_{ij} = \frac{1}{A_c}\int_{A_c} \frac{w}{w_{av}} f_i f_j \, dA \tag{11.56}$$

The thermophysical properties in the definition of a_{ij} and b_{ij} are omitted to make a_{ij} and b_{ij} dimensionless. The quantity

$$\overline{\gamma}_n = \frac{a^2 w_{av}\gamma_n}{\alpha} \tag{11.57}$$

is now the dimensionless eigenvalue, since dimensionless variables in the mathematical formulations of a_{ij} and b_{ij} are being used. The conservation of energy, $dQ_s = hdA_s(T_s - T_b) = \rho w_{av}A_c c_p dT_b$, at any z, dictates that

$$\frac{4h}{\rho c_p w_{av}D_e} = -\frac{d(T_b - T_s)/dz}{T_b - T_s} \tag{11.58}$$

where T_b is the bulk temperature defined by

$$T_b = \frac{1}{A_c}\int_{A_c} (w/w_{av}) T \, dA \tag{11.59}$$

Also, T_s, Q_s, and A_s are the surface temperature, surface heat flux, and surface area, respectively. Equation (10.77) when $g = 0$ and $T^* = T_s$ provides the temperature distribution which can be substituted in Eq. (11.59) to obtain the bulk temperature T_b. The substitution of the bulk temperature in Eq. (11.58) results in the value of the circumferentially averaged heat transfer coefficient, $h = h(z)$. As z approaches infinity, the contribution of all eigenvalues will diminish except the first eigenvalue and the left side of Eq. (11.58) becomes γ_n. Then, using Eq. (11.57) to evaluate γ_n, Eq. (11.58) reduces to

$$\text{Nu} = \frac{hD_e}{k} = \frac{\overline{D}_e^2 \overline{\gamma}_1}{4} \tag{11.60}$$

Therefore, the first eigenvalue is proportional to the thermally fully developed Nusselt number. Table 11.3 gives the analytical expressions of the components of matrices **A** and **B** and vector **g*** needed to solve the velocity and temperature fields in selected ducts. The entries in Table 11.3 are for the prescribed surface temperature. Equation (11.55b) is used to calculate the elements of matrix **A**.

Example 11.8 For a laminar and fully developed flow of an incompressible and Newtonian fluid in a circular pipe, calculate temperature distribution and the heat transfer coefficient. Fluid at temperature of 0 enters a heated pipe and the surface temperature of the pipe is maintained at temperature of 1.

SOLUTION The well-known Graetz problem is selected to illustrate how to use the integral method; it leads to the solution of the heat transfer coefficient for flow in circular pipes. The basis functions are

$$f_j = \left(1 - \frac{r^2}{r_0^2}\right) \left(\frac{r}{r_0}\right)^{2(j-1)} \qquad \text{for } j = 1, 2, \ldots, N \tag{11.61}$$

The velocity profile for fully developed flow is $u/u_{av} = 2(1 - r^2/r_0^2)$. This velocity profile can be obtained from the exact solution or the Galerkin method. The elements of matrix **B** that use this parabolic velocity profile and the elements of matrix **A** are in Table 11.3. Again, Eq. (10.11) yields the eigenvalues. As discussed earlier, the first eigenvalue in the duct flow problems is of special importance. It provides the fully developed heat transfer coefficient. When $N = 1$, Nu $= 3$, whereas a two-term solution (i.e., $N = 2$) yields Nu $= 20[1 - (2/3)^{1/2}] = 3.6701$; this is very close to the value of 3.6568 obtained from the exact solution. When N is increased to 3, a very accurate value of the Nu $= 3.6570$ is obtained.

After computation of the eigenvalues and matrices **D** and **P**, the alternative GF solution, Eq. (10.77), yields the temperature distribution. The values of the Nusselt number within the entrance region of the pipe are computed by Lakshminarayanan (1988) and Lakshminarayanan and Haji-Sheikh (1986), and compared with the exact solution in Table 11.4. The agreement between the two solutions is generally excellent. The GF solution method permits the inclusion of position dependent wall temperature and locally variable volumetric heat generation in the solution. The boundary condition of second and third kinds can be accommodated using the one-dimensional basis functions already defined in Chapter 10. The calculation can be extended to ducts with more complex cross sections. For instance, the heat transfer coefficients for various isosceles and right triangular ducts are calculated and reported by Lakshminarayanan (1988) and Lakshminarayanan and Haji-Sheikh (1986). The matrices **A** and **B** for the above-mentioned ducts are in Table 11.3.

Table 11.3 Matrices A and B and vector g* for selected ducts

Circular duct

$$w/w_{av} = 2(1 - r^2), \quad \phi = 1 - r^2$$

$$a_{ij} = -(8i - 4)\left(\frac{1}{i + j - 1} - \frac{1}{i + j}\right) + 4(i - 1)^2\left(\frac{1}{i + j - 2} - \frac{2}{i + j - 1} + \frac{1}{i + j}\right)$$

and

$$b_{ij} = 2\left(\frac{1}{i + j - 1} - \frac{3}{i + j} + \frac{3}{i + j + 1} - \frac{1}{i + j + 2}\right)$$

Right triangular ducts

$$f_j = (x/a)(y/a - b/a)[y/a$$
$$\quad - (b/a)(x/a)](x/a)^{m_j}(y/b)^{n_j}$$

$j = 1;$ $m_1 = 0$ and $n_1 = 0$

$j = 2;$ $m_2 = 1$ and $n_2 = 0$

$j = 3;$ $m_3 = 0$ and $n_3 = 1$

$j = 4;$ $m_4 = 2$ and $n_4 = 0$

$j = 5;$ $m_4 = 1$ and $n_4 = 1$

$$a_{ij} = -2\left(\frac{b}{a}\right)^{n_i + n_j + 4}\left[\frac{(m_i + 1)(m_j + 1)}{m_i + m_j + 1} - \frac{2m_im_j + 3(m_i + m_j) + 4}{m_i + m_j + 2}\right.$$
$$\left. + \frac{(m_i + 2)(m_j + 2)}{m_i + m_j + 3}\right]\left(\frac{1}{l + 6} - \frac{2}{l + 5} + \frac{1}{l + 4}\right)$$
$$- 2\left(\frac{b}{a}\right)^{n_i + n_j + 2}\left(\frac{G_1}{m_i + m_j + 3} - \frac{G_2}{m_i + m_j + 4} + \frac{G_3}{m_i + m_j + 5}\right)$$

where

$$G_1 = \frac{n_in_j + 2(n_i + n_j) + 4}{l + 6} - \frac{2n_in_j + 3(n_i + n_j) + 4}{l + 5} + \frac{n_in_j + (n_i + n_j) + 1}{l + 4}$$

$$G_2 = \frac{2n_in_j + 3(n_i + n_j) + 4}{l + 6} - \frac{4n_in_j + 4(n_i + n_j) + 2}{l + 5} + \frac{2n_in_j + n_i + n_j}{l + 4}$$

$$G_3 = \frac{n_in_j + n_i + n_j + 1}{l + 6} - \frac{2n_in_j + n_i + n_j}{l + 5} + \frac{n_in_j}{l + 4}$$

$$b_{ij} = 2\frac{C_f Re}{2D_e^2/a^2}\sum_{k=1}^{M} d_k\left(\frac{b}{a}\right)^{v_1 + 6}\left(\frac{1}{\mu_1 + 4} - \frac{3}{\mu_1 + 5} + \frac{3}{\mu_1 + 6} - \frac{1}{\mu_1 + 7}\right)$$
$$\times \left(\frac{1}{v + 11} - \frac{3}{v + 10} + \frac{3}{v + 9} - \frac{1}{v + 8}\right)$$

$$\psi_j = \frac{2(b/a)^{n_j + 2}}{(m_j + 2)(m_j + 3)(m_j + n_j + 4)(m_j + n_j + 5)}$$

Isosceles triangular ducts

$$f_j = (y/a - b/a)[(y/a)^2$$
$$\quad - (b/a)^2(x/a)^2](x/a)^{m_j}(y/b)^{n_j}$$

$j = 1;$ $m_1 = 0$ and $n_1 = 0$

$j = 2;$ $m_2 = 0$ and $n_2 = 1$

$j = 3;$ $m_3 = 2$ and $n_3 = 0$

$j = 4;$ $m_4 = 0$ and $n_4 = 2$

$$a_{ij} = -2\left(\frac{b}{a}\right)^{n_i+n_j+6}\left[\frac{m_i m_j}{m_i + m_j - 1} - \frac{2(m_i m_j + m_i + m_j)}{m_i + m_j + 1}\right.$$

$$\left. + \frac{(m_i + 2)(m_j + 2)}{m_i + m_j + 3}\right]\left(\frac{1}{l+6} - \frac{2}{l+5} + \frac{1}{l+4}\right) - 2\left(\frac{b}{a}\right)^{n_i+n_j+4}$$

$$\times \left(\frac{G_1}{m_i + m_j + 1} - \frac{G_2}{m_i + m_j + 3} + \frac{G_3}{m_i + m_j + 5}\right)$$

where

$$G_1 = \frac{(n_i + 3)(n_j + 3)}{l+6} - \frac{(n_i + 2)(n_j + 3) + (n_i + 3)(n_j + 2)}{l+5} + \frac{(n_i + 2)(n_j + 2)}{l+4}$$

$$G_2 = 2\left[\frac{n_i n_j + 2(n_i + n_j) + 3}{l+6} - \frac{2n_i n_j + 3(n_i + n_j) + 2}{l+5} + \frac{n_i n_j + n_i + n_j}{l+4}\right]$$

$$G_3 = \frac{n_i n_j + n_i + n_j + 1}{l+6} - \frac{2n_i n_j + n_i + n_j}{l+5} + \frac{n_i n_j}{l+4}$$

$$b_{ij} = 2\frac{C_f Re}{2D_c^2/a^2}\sum_{k=1}^{M} d_k\left(\frac{b}{a}\right)^{v_1+9}\left(\frac{1}{\mu_1 + 1} - \frac{3}{\mu_1 + 3} + \frac{3}{\mu_1 + 5}\right.$$

$$\left. - \frac{1}{\mu_1 + 7}\right)\left(\frac{1}{v + 11} - \frac{3}{v + 10} + \frac{3}{v + 9} - \frac{1}{v + 8}\right)$$

$$g_j^* = \frac{4(b/a)^{n_j+3}}{(m_j + 1)(m_j + 3)(m_j + n_j + 4)(m_j + n_j + 5)}$$

Nomenclature of indices

i, j, k, m, n	indices
l	$m_i + n_i + m_j + n_j$
μ_1	$m_i + m_j + m_k$
v_1	$n_i + n_j + n_k$
v	$\mu_1 + v_1$

11.6 CONCLUSION

The multidimensional applications discussed in this chapter show that many complex geometries can be accommodated using the Galerkin-based GF. The success of this method depends on the availability of the basis functions for a given application or one's ability to find a set of basis functions. Because the number of basis functions needed to provide an accurate solution is usually small, numerical computation can be used to compute the elements of matrices **A** and **B**. Various symbolic software programs are widely available and are valuable tools to assist in the mathematical differentiation of the basis functions. Also, the symbolic integration, whenever possible, results in high-speed computer operation by providing virtually error-free mathematical equations.

Table 11.4 Local Nusselt number in circular ducts

$\dfrac{z/D_e}{Pe}$	Integral method, $N = 12$	Results Kays and Perkins (1973)	Shah and London (1978)
0.00001	59.621	—	61.877
0.0001	28.148	—	28.254
0.001	12.824	12.86	12.824
0.004	8.036	7.91	8.036
0.01	6.002	5.99	6.002
0.04	4.172	4.18	4.172
0.08	3.769	3.79	3.769
0.1	3.710	3.71	3.710
0.2	3.658	3.66	3.658
0.5	3.657	3.66	3.657

We showed that the application of the Galerkin-based GF solution to heterogeneous bodies is possible and the generalized formulation of the GF can be used once the basis functions are available. In addition, we showed that the generalized GF solution can be modified for steady-state conduction problems. However, the steady-state solution, using the alternative formulation of the GF, reduces to the standard Galerkin method.

The Galerkin-based solution can also be used to solve for the heat transfer coefficient in the entrance region of ducts. The usefulness of the GF solution method given in Chapter 10 and utilized in this chapter is limited to the case when the thermal conduction in the flow direction is negligible (large Péclet number). However, it is possible to modify the Galerkin-based integral method so that the effect of axial conduction can be included in the analysis.

REFERENCES

Haji-Sheikh, A., 1986, "On Solution of Parabolic Partial Differential Equations Using Galerkin Functions," in *Integral Methods in Science and Engineering*, Eds. F. R. Payne et al., Hemisphere, Washington.

Haji-Sheikh, A., 1988, "Heat Diffusion in Heterogeneous Media Using Heat-Flux-Conserving Basis Functions," ASME J. Heat Transfer, vol. 110, pp. 276–282.

Haji-Sheikh, A., and Beck, J. V., 1990, "Green's Function Partitioning in Galerkin-Based Solution of the Diffusion Equation," ASME J. Heat Transfer, vol. 112, pp. 28–34.

Haji-Sheikh, A., and Lakshminarayanan, R., 1987, "Integral Solution of Diffusion Equation: Part 2-Boundary Conditions of Second and Third Kinds," ASME J. Heat Transfer, vol. 109, pp. 557–552.

Haji-Sheikh, A., and Sparrow, E. M., 1967, "The Solution of Heat Conduction Problems by Probability Methods," ASME J. Heat Transfer, vol. 89, pp. 121–131.

Haji-Sheikh, A., Mashena, M., and Haji-Sheikh, M. J., 1983, "Heat Transfer Coefficient in Ducts with Constant Wall Temperature," ASME J. Heat Transfer, vol. 105, pp. 878–883.

Hearn, A., 1983, REDUCE-3 User's Manual, Rand Corporation.

Kantorovich, L. V., and Krylov, V. I., 1960, *Approximate Methods of Higher Analysis*, Wiley, New York.

Kays, W. M., and Perkins, H. C., 1973, Forced Convection, Internal Flow in Ducts, in *Handbook of Heat Transfer*, p. 22, ed. W. M. Rohsenow and J. P. Hartnett, McGraw-Hill, New York.

Lakshminarayanan, R., 1988, Integral Solutions to Thermal and Hydrodynamic Entrance Problems in Ducts, Ph.D. thesis, University of Texas at Arlington, Arlington, Tex.

Lakshminarayanan, R., and Haji-Sheikh, A., 1986, A Generalized Closed-Form Solution to Laminar Thermal Entrance Problems, in *Heat Transfer 86*, pp. 861–876, ed. C. L. Tien, et al., Hemisphere, Washington, D.C.

Lakshminarayanan, R., and Haji-Sheikh, A., 1988, Extended Graetz Problems in Irregular Ducts, in *Proceedings of the 1988 National Heat Transfer Conference*, ed. H. R. Jacobs, ASME HTD-vol. 96, pp. 475–482 in vol. 1.

Lee, Y.-M., and Haji-Sheikh, 1991, Temperature Field in Heterogeneous Bodies: A Non-Orthogonal Solution, *Proceedings of the National Heat Transfer Conference*, eds. M. Imber and M.M. Yovanovich, ASME HTD-vol. 173, pp. 1–9.

Nomura, S., and Haji-Sheikh, A., 1988, Analysis of Transient Heat Conduction in Complex Shaped Composite Materials, *ASME J. Eng. Materials Technol.*, vol. 110, pp. 110–112.

Ozisik, M. N., 1980, *Heat Conduction*, Wiley, New York.

Shah, R. K., and London, A. L., 1978, in *Advances in Heat Transfer*, ed. T. F. Irvine and J. P. Hartnett, Academic Press, New York.

PROBLEMS

11.1 A square bar has dimensions 1×1. When the boundary conditions are of the first kind, use the product method to compute the basis functions. Repeat the steps using the method used in Example 11.1.

11.2 A finite cylinder with radius r_0 is subject to convective heat transfer at $r = r_0$ while the temperature is prescribed on other surfaces. Find the GF using the product method.

11.3 A hemisphere of radius r_0 has prescribed convection on $r = r_0$ surface while the temperature is prescribed at the other surface. Use the product method to define the basis functions. Comment on the case when convection is prescribed for all surfaces.

11.4 Consider a spheroidal solid whose surface is given by equation $r^2/a^2 + z^2/b^2 = 1$. For boundary conditions of the first kind, show that $a_{11} = -96(19b^2 + 13)V/945b^2$ and $b_{11} = 384V/2079$, where V is the volume of the spheroid. Find matrices $\mathbf{D}$ and $\mathbf{P}$ and the GF. Propose a small-time GF for the purpose of partitioning.

11.5 Equation (11.5) and (11.9) give the basis functions for a spheroidal solid with insulated external surface. A spheroid, $a = 1$ and $b = 6$, receives heat from a heat source at the rate of $q(t)$. Is it possible to have a one-term solution using Eq. (10.68)? What is the smallest number of terms for a reasonable solution? Show that the first eigenvalue is $\gamma_1 = 0$.

11.6 Repeat Example 11.5, except let $T = 0$ at $x = 0$. Redefine $f_{j,m}$, U, and H so that the boundary conditions are satisfied.

11.7 An isosceles right triangular solid bar is externally insulated. The central portion of this long bar, in a circular zone, has thermophysical properties different from the rest of the bar. Find the parametric relations for the basis functions.

11.8 Use a one-term solution to show that the alternative GF solution becomes identical to the Galerkin solutions as $t \to \infty$. [*Hint:* When $j = 1$, Eqs. (11.40) and (11.41) are identical.]

11.9 Show that Eqs. (11.40) and (11.41) produce the same results for any number of terms.

11.10 Calculate the Nusselt number for a fully developed laminar flow in an elliptical duct. The duct's wall temperature is constant. Find a solution that uses the GF for an arbitrarily selected surface temperature.

11.11 Reproduce the data near the entry point of a circular duct using time partitioning of the GF and compare with the entries in Table 11.4.

11.12 The GFs in solid right-triangular rods are needed. Show that the elements of matrix $\mathbf{A}$ for boundary conditions of the first kind are the same as those given in Table 11.3. Calculate a similar relation for matrix $\mathbf{B}$. [*Caution:* The entries in Table 11.3 are from Eqs. (11.51b) and (11.52).]

11.13 Fluid passes through an annulus whose external surface is elliptical and the internal surface is circular. Consider that the flow is laminar and the boundary conditions are of the first kind. Find the GFs assuming: (a) slug flow, and (b) viscous flow.

11.14 Use Example 11.8 and find a two-term temperature solution when heat generates at the rate of g W/m^2. The inlet and wall temperatures are maintained at zero.

11.15 Repeat Example 11.8, except, now the surface heat flux is prescribed instead of the wall temperature.

11.16 Repeat Problem 11.14, except now the surface heat flux instead of surface temperature is prescribed.

11.17 A 10-cm diameter steel pipe 3 mm thick, $k = 60$ W/mK, $\rho = 7850$ kg/m^3, and $c_p = 434$ J/kgK, is carrying a gas. It has a 5-cm-thick insulation with thermophysical properties $k = 0.04$ W/mK, $\rho = 100$ kg/m^3, and $c_p = 1200$ J/kgK. Inside and outside fluid temperatures are 600 K and 300 K, the corresponding heat transfer coefficients are 100 W/m^{2}K and 50 W/m^{2}K, and the contact conductance between two layers is 10 W/m^{2}K. If the initial temperature is 300 K, using the COND program, display the surface temperatures and calculate the variation of external and internal heat flux with time.

11.18 A straight fin has dimensionless quantities $L = 0$, $T = 1$ at $x = 0$, $q = 0$ at $x = 1$. When $m = \sqrt{hP/kA} = 0.5$ and fin is initially at zero temperature, use the COND program to calculate the dimensionless heat flux per unit area of the base at $\alpha t/L^2 = 0.2, 0.4, 0.8, 1, 2, \infty$.

11.19 The radius r of a pin fin varies as x^2 when $0.5 < x/x_2 < 1$. Also, the perimeter varies as x^2. The initial and ambient temperatures are 0. The boundary condition at $x = x1 = x_2/2$ is convective so that $h_1 x_2/k = 0.02$. The heat transfer coefficient, h, on the fin surface varies as $r^{-0.25}$ so that $x_2^2(hP/kA) = 2x^{-2.5}$. Find temperature distribution as a function of $\alpha t/L^2$ at $x = x_1$ using the COND program.

11.20 When initial temperature is 1, $a = 0$, $b = 1$, $q_1/k = -1$, $q_2/k = -2$, and $\alpha = 1$, for Example 10.7, use the COND program to display temperature distribution at $x = 0, 0.2, 0.4, 0.6, 0.8$, and 1. Explain the nature of steady-state solution. Repeat the calculations but consider that the plate has a uniform volumetric heat source, $g = 3$.

Note: Materials in Sections 10.3, 10.5, 11.3, and Tables 10.3a–b were used to develop the computer program COND. It solves and graphically displays temperature data for one- and two-layer, one-dimensional bodies. The geometries are plates, solid, and hollow cylinders; solid and hollow spheres; and various fins. Also, COND permits heat flux calculations at any point within the domain. The values of the basis functions, and matrices **D** and **P** needed in the definition of the Green's function are automatically provided by this software. For more difficult applications, a user can employ the Green's functions data to calculate the temperature field in one-dimensional or some multidimensional bodies.

TWELVE

UNSTEADY SURFACE ELEMENT METHOD

12.1 INTRODUCTION

The unsteady surface element (USE) method is a boundary discretization method for solution of linear transient two- and three-dimensional heat transfer problems. Its development originated with the need to calculate interface temperatures and heat fluxes for similar and dissimilar geometries connected over a relatively small portion of their surface boundaries. Examples of bodies connected over a small area occur in contact conductance problems such as the case of two semi-infinite cylinders in contact over only a central circular region, as shown in Fig. 12.1a. An example involving dissimilar geometries is the intrinsic thermocouple problem which involves a semi-infinite cylinder attached to a semi-infinite body (Fig. 12.1b) or to an infinite plate (Fig. 12.1c). Other related examples are those associated with the electrical contacts, cooling of electronic systems, fins, and conjugated problems.

The above-mentioned problems may involve transient heat transfer and differing thermophysical properties. The solution is difficult because the separate regions are coupled by simultaneous interfacial boundary conditions that may vary with time in some unspecified manner. Numerical methods are the primary means to solve such problems, even though for certain problems it is sometimes possible to obtain approximate solutions by relaxing the conditions that the coupled regions must satisfy.

Closely related to the USE method is the boundary element (BE) method, which has been used in a variety of engineering problems such as solid mechanics, fluid flow, soil mechanics, water waves, heat conduction, electrical problems and a broad range of other applications (Banerjee and Butterfield, 1979; Banerjee and Shaw, 1982; Banerjee and Mukherjee, 1984; and Banerjee and Watson, 1985). The BE method involves Green's theorem to formulate the problem described by a partial differential equation in a given region with some specific boundary conditions as an integral equation which applies only to the boundary of the region. Basic building blocks used in the BE method are source solutions (Green's functions, GFs) for infinite homogeneous bodies.

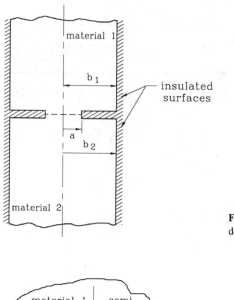

Figure 12.1 (*a*) Two connected semi-infinite cylinders simulating contact conductance problem.

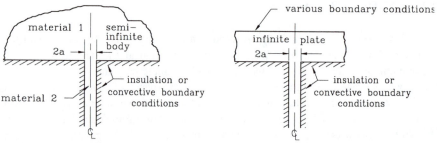

Figure 12.1 (*b, c*) Some geometries for intrinsic thermocouple problem.

The BE method is well suited for solving steady-state problems with infinite domain and irregular-shaped boundaries. A number of papers have been written for steady-state heat conduction problems (Schneider, 1979; Schneider and Ledain, 1979; Khader, 1980; Khader and Hanna, 1981).

Application of the BE method to transient problems has received less attention compared to the steady-state problems. This is due to the complexity of having the independent variable of time. There are two basic ways of handling the effects of time. One is to temporarily eliminate time as an independent variable by utilizing the Laplace transform and then solving the problem in the transform space by using the BE method. The time solution is then obtained by numerical transform inversion. This is the approach taken by Rizzo and Shippy (1970) to solve the problem of heat conduction in an infinite cylinder of an isotropic medium. The other approach is to treat the time directly in the same manner as the spatial coordinates are treated, integrating numerically over the time as well as over the boundary of the body. Shaw (1974) utilized the direct approach to investigate heat conduction in a circular sector of an isotropic medium. A similar approach was taken by Chang et al. (1973) to treat anisotropic heat conduction in the transient case with heat generation. Wrobel and Brebbia (1981) employed this approach to solve three-dimensional axisymmetric transient heat conduction problems of a solid cylinder, a prolate spheroid, and a solid sphere, all with time-dependent boundary conditions.

In the USE method, only the interface between the contacted bodies (or the active part of the boundary) requires discretization as compared to the discretization of the whole domain required in the finite-difference and finite-element methods or discretization of the whole boundary in the BE method. This, in turn, reduces the size of numerical computations, especially for three-dimensional problems. Another aspect of the USE method is that, unlike the above-mentioned alternative methods, it does not require any modifications or special handling of points near the domain boundaries. The USE method uses Duhamel's theorem and involves the inversion of a set of Volterra integral equations, one for each surface element. Though the method is limited to linear regions it can be used for nonlinear boundary conditions.

Two types of kernels ("building blocks" or influence functions) can be employed in the USE method: temperature based and heat flux based. The method requires that these kernels or influence functions be known for the basic geometries under consideration. For many geometries, the influence functions are known or can be obtained by analytical methods or through the use of GFs.

Yovanovich and Martin (1980) suggested the name "surface element method" and did early work on a steady-state form of this method. Keltner and Beck (1981) were the first to employ the surface element method for transient problems. They considered only one element along the interface and utilized the Laplace transform technique to obtain "early" and "late" time approximate analytical solutions for two arbitrary bodies suddenly brought into thermal contact over a small area. The multinode form of USE (numerical approach) was originally developed by Litkouhi and Beck (1985, 1986) and applied to contact between large bodies over a small circular area and the intrinsic thermocouple problem. Cole and Beck (1987, 1988) have extended the USE method to a conjugated heat transfer problem.

The objective of this chapter is to introduce the basic mathematical concepts and formulations of the USE method and to demonstrate its applications by presenting some example problems. Duhamel's theorem and its relation to the GF function method is presented in Section 12.2. Section 12.3 is devoted to formulation and development of the USE equations and the related numerical solutions. The approximate analytical solutions of the USE equations are discussed in Section 12.4, and finally, to illustrate the application of the USE method, some example problems are given and discussed in Section 12.5.

12.2 DUHAMEL'S THEOREM AND GREEN'S FUNCTION METHOD

When the boundary condition is a function of time, solution of a linear heat conduction problem may be deduced from the well-known Duhamel's theorem. Duhamel's theorem employs a fundamental (or a "building block") solution which is used with the superposition principle to obtain temperature at any point **r**, and time t. Briefly, it states that if $\psi(\mathbf{r}, t)$ is the solution to a linear system initially at zero temperature, due to a unit stepwise input, then the solution to the same system initially at zero temperature due to a time-varying input $F(t)$ (instead of a unit step) is given by

$$T(\mathbf{r}, t) = \frac{\partial}{\partial t} \int_0^t F(\tau)\psi(\mathbf{r}, t - \tau) \, d\tau \tag{12.1a}$$

where **r** is the position vector, t is time, and τ is a dummy variable for integration. The input function $F(t)$ can be any type of time-dependent boundary condition (such as prescribed surface temperature, ambient temperature, or prescribed surface heat flux) or heat generation. An alternative form of Eq. (12.1a) can be obtained with the Leibnitz's rule for differentiation of an integral,

$$T(\mathbf{r}, t) = \int_0^t F(\tau) \frac{\partial\psi(\mathbf{r}, t - \tau)}{\partial t} \, d\tau \tag{12.1b}$$

Equations (12.1a and b) represent a form of Duhamel's theorem where the input function $F(t)$ varies only with time. The derivation of this form of Duhamel's theorem is given by several authors using different approaches. The approach presented by Ozisik (1980, p. 197) and Luikov (1968, p. 344) uses Laplace transformations. Myers (1987, p. 153) uses the concept of superposition to derive Duhamel's theorem for prescribed surface temperature boundary condition; while Beck et al. (1985a, p. 81) employ the same principle to derive Duhamel's theorem for heat flux boundary conditions. It is also conventional to treat problems with spatially varying boundary conditions by using Duhamel's theorem with integration over space (Eckert and Drake, 1972, p. 322) and (Kays and Crawford, 1980, p. 118). The following derivation of Duhamel's theorem involves simultaneous variation of both time and space conditions for an arbitrary two-dimensional geometry.

12.2.1 Derivation of Duhamel's Theorem for Time- and Space-Variable Boundary Conditions

Consider the boundary value problem of heat conduction for an arbitrary two-dimensional region R initially at zero temperature, with a time- and space-variable heat flux over boundary S as shown in Fig. 12.2. For simplicity, it is assumed that $q(s, t)$ is nonzero only over the portion of boundary S from $s = 0$ to $s = L$, and the other portion of the boundary is insulated [$q(s, t) = 0$, for $s > L$]. The objective is to find an expression for the solution of above problem using Duhamel's theorem.

In the first step the solution to the fundamental problem is found. The fundamental problem is identical to the above problem with the exception that the variable flux boundary condition, $q(s, t)$, is replaced by a special unit step function. It is described by the following equations:

$$\nabla^2\psi_q = \frac{1}{\alpha}\frac{\partial\psi_q}{\partial t} \tag{12.2}$$

$$\psi_q(x, y, 0) = 0 \tag{12.3}$$

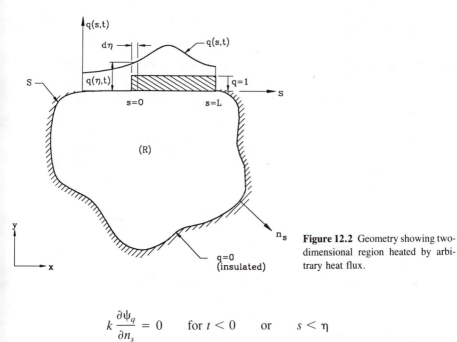

Figure 12.2 Geometry showing two-dimensional region heated by arbitrary heat flux.

$$k \frac{\partial \psi_q}{\partial n_s} = 0 \qquad \text{for } t < 0 \qquad \text{or} \qquad s < \eta$$

$$= 1 \qquad \text{for } t > 0 \qquad \text{and} \qquad \eta < s < L \qquad (12.4)$$

where η is a dummy length variable along the boundary S between $s = 0$ to $s = L$, and $\psi_q(x, y, \eta, t)$ is the temperature rise at position (x, y) and time t caused by a unit step change of heat flux at time $t = 0$, from $s = \eta$ to $s = L$ as shown in Fig. 12.2 by the cross-hatched portion. It is called the flux-based fundamental solution (FBFS). Notice that, for fixed (x, y) and t, $\psi_q(x, y, \eta, t)$ decreases as η increases, that is,

$$\psi_q(x, y, \eta, t) > \psi_q(x, y, \eta + d\eta, t) \qquad (12.5)$$

Temporarily, let η be fixed and consider the variation of the heat flux with time only. From the fundamental solution, the temperature rise at position (x, y) and time t due to a unit step change of heat flux at time τ is

$$\psi_q(x, y, \eta, t - \tau) \qquad (12.6a)$$

where $t - \tau$ is the time that has elapsed since the step at τ. Also the temperature rise at time t due to a unit step change of heat flux at time $\tau + d\tau$ is

$$\psi_q[x, y, \eta, t - (\tau + d\tau)] \qquad (12.6b)$$

Then from Eqs. (12.6a and b), the temperature rise at position (x, y) and time t due to a unit step change in q for $\tau < t < \tau + d\tau$ is

$$-d_\tau \psi_q(x, y, \eta, t - \tau) = \psi_q(x, y, \eta, t - \tau) - \psi_q[x, y, \eta, t - (\tau + d\tau)] \quad (12.7)$$

where d_τ is a differentiation operator for τ. Notice that $\psi_q(x, y, \eta, t - \tau)$ is greater than $\psi_q[x, y, \eta, t - (\tau + d\tau)]$. Using Eq. (12.7), the temperature rise at position (x, y) and time t due to the value $q(\eta, t)$ for $\tau < t < \tau + d\tau$ and η being fixed is

$$-q(\eta, \tau)d_\tau\psi_q(x, y, \eta, t - \tau) = -q(\eta, \tau)\frac{\partial\psi_q(x, y, \eta, t - \tau)}{\partial\tau}\,d\tau \quad (12.8)$$

for small $d\tau$. Since the problem is linear, superposition can be employed and the total effect of all step changes of heat flux over small $d\tau$'s from time zero to time t is simply found by integrating Eq. (12.8) from 0 to t. Denoting the result $\psi_q'(x, y, \eta, t)$, one can write

$$\psi_q'(x, y, \eta, t) = -\int_0^t q(\eta, \tau)\frac{\partial\psi_q(x, y, \eta, t - \tau)}{\partial\tau}\,d\tau \quad (12.9a)$$

From the relation

$$\frac{\partial\psi_q(x, y, \eta, t - \tau)}{\partial\tau} = -\frac{\partial\psi_q(x, y, \eta, t - \tau)}{\partial t}$$

Eq. (12.9a) can be written as

$$\psi_q'(x, y, \eta, t) = \int_0^t q(\eta, \tau)\frac{\partial\psi_q(x, y, \eta, t - \tau)}{\partial t}\,d\tau \quad (12.9b)$$

Note that ψ_q' is the temperature rise for the case that the time-variable q is zero for $s < \eta$, and is uniformly distributed over space for $\eta < s < L$.

In a similar way, one can show that the temperature rise for the case that the flux q is zero for $s < \eta + d\eta$, and is uniformly distributed for $\eta + d\eta < s < L$, is

$$\psi_q'(x, y, \eta + d\eta, t) = \int_0^t q(\eta, \tau)\frac{\partial\psi_q(x, y, \eta + d\eta, t - \tau)}{\partial t}\,d\tau \quad (12.10)$$

Using Eqs. (12.9b) and (12.10), the temperature rise due to a uniform heat flux q, between $s = \eta$ and $s = \eta + d\eta$ and for $t > 0$ is

$$-d_\eta\psi_q'(x, y, \eta, t) = \psi_q'(x, y, \eta, t) - \psi_q'(x, y, \eta + d\eta, t)$$

$$= -\frac{\partial\psi_q'(x, y, \eta, t)}{\partial\eta}\,d\eta \quad (12.11)$$

where d_η is a differentiation operator for η. Notice that $\psi_q'(x, y, \eta, t)$ is greater than $\psi_q'(x, y, \eta + d\eta, t)$. Introducing Eqs. (12.9b) and (12.10) into Eq. (12.11) yields

$$-d_\eta\psi_q'(x, y, \eta, t) = -\int_0^t q(\eta, \tau)\frac{\partial^2\psi_q(x, y, \eta, t - \tau)}{\partial t\,\partial\eta}\,d\tau\,d\eta \quad (12.12)$$

Again superposition can be employed and the total effect of the variation of heat flux from $s = 0$ to $s = L$ can be found by integrating Eq. (12.12) over space from 0 to L, to give

$$T(x, y, t) = -\int_0^L\int_0^t q(\eta, \tau)\frac{\partial^2\psi_q(x, y, \eta, t - \tau)}{\partial t\,\partial\eta}\,d\tau\,d\eta \quad (12.13)$$

In this problem, it was assumed that only a portion of the surface boundary is exposed to heat flux with the remainder being insulated. However, if none of the boundary S is insulated, the first integral in Eq. (12.13) extends over the entire boundary S. Furthermore, if the initial temperature of the system is T_0 instead of being zero, the solution becomes

$$T(x, y, t) - T_0 = - \int_S \int_0^t q(\eta, \tau) \frac{\partial^2 \psi_q(x, y, \eta, t - \tau)}{\partial t\, \partial \eta} \, d\tau \, d\eta \quad (12.14)$$

In Eq. (12.14), the input function $q(\eta, \tau)$ is the heat flux along the boundary (surface heat flux) which varies with both space and time, and the solution is in terms of the FBFS, ψ_q. If, however, the surface temperature is known along the boundary as the input function (instead of heat flux), then in a similar manner to that described above, the solution in terms of the temperature-based fundamental solution (TBFS), ψ_T, can be obtained as

$$T(x, y, t) - T_0 = - \int_S \int_0^t [T_s(\eta, \tau) - T_0] \frac{\partial^2 \psi_T(x, y, \eta, t - \tau)}{\partial \eta\, \partial t} \, d\tau \, d\eta \quad (12.15)$$

Equations (12.14) and (12.15) are rather general expressions for the case that the input function varies with both space and time in a two-dimensional region. Both equations can be employed to obtain the temperature history at any position (x, y) of the region. However, depending on the type of boundary condition, one might be more appropriate than the other. To compare the two approaches and discuss their utility for each particular type of boundary condition, both forms of solutions are examined below.

For problems with boundary conditions of the first kind, where the temperature is specified everywhere along the boundary, the right-hand side of Eq. (12.15) is known, and one can solve for the temperature history of any interior point of R, by direct integration. If, however, Eq. (12.14) is employed instead of Eq. (12.15), the direct evaluation of $T(x, y, t)$ is not possible because of the unknown heat flux q in the right-hand side of this equation. In this case, an inverse integration must first be performed to solve for the unknown surface heat flux which is the information needed by Eq. (12.14) to find $T(x, y, t)$ at any interior point. Therefore, in problems with the first kind boundary conditions, Eq. (12.15) is more appropriate than Eq. (12.14).

On the other hand, if the boundary condition is of the second kind where q is specified along the boundary S, then the right hand side of Eq. (12.14) is known which leads to evaluation of a direct integral. In this case Eq. (12.14) is more appropriate than Eq. (12.15).

For boundary conditions of the third kind where neither the surface temperature nor its normal derivative are completely known over the entire boundary S (mixed boundary conditions), none of the above equations can be used directly to obtain temperature history for any interior point. An example is given to illustrate this case better.

Consider the homogeneous convective boundary condition given by

$$k \frac{\partial T(s, t)}{\partial \eta_s} + h_s T_s(s, t) = 0 \qquad \text{on } S \quad (12.16)$$

where $\partial/\partial \eta_s$ denotes differentiation with respect to the outward pointing normal to the surface boundary S as shown in Fig. 12.2. Substituting for q in Eq. (12.14) from Eq. (12.16), one can write

$$T(x, y, t) - T_0 = - \int_S \int_0^t h_s T_s(\eta, \tau) \frac{\partial^2 \psi_q(x, y, \eta, t - \tau)}{\partial t \, \partial \eta} \, d\tau \, d\eta \quad (12.17)$$

Equation (12.17) cannot directly be integrated for $T(x, y, t)$, since $T_s(s, t)$ inside the integral is unknown. In other words, the number of unknown functions in Eq. (12.17) is more than one, $T_s(s, t)$ and $T(x, y, t)$. However, for a point along the boundary S, Eq. (12.17) reduces to

$$T_s(s, t) - T_0 = - \int_S \int_0^t h_s T_s(\eta, \tau) \frac{\partial^2 \psi_q(\eta, t - \tau)}{\partial t \, \partial \eta} \, d\tau \, d\eta \quad (12.18)$$

which is a Volterra integral equation of the second kind with the only unknown function, $T_s(s, t)$, both inside and outside the integral. In an inverse manner, Eq. (12.18) can be solved numerically for $T_s(s, t)$. Once the surface temperature, $T_s(s, t)$, has been determined, the solution to the interior temperature history, $T(x, y, t)$, can be obtained by substituting $T_s(s, t)$ into Eq. (12.17).

Hence, for the problems with mixed or convective boundary conditions, the temperature history at any interior point can be determined in two steps:

1. Find the boundary information by solving an inverse integral equation.
2. Using the boundary data obtained in step 1, find the interior temperature history by using a direct integration.

Equations (12.14) and (12.15) are the flux-based and temperature-based forms of Duhamel's theorem. They are used as the basic building blocks in the development of the USE formulation in the following sections.

12.2.2 Relation to the Green's Function Method

The Duhamel's theorem (sometimes called Duhamel's integral) approach given herein is related to the GF method. One advantage of the Duhamel's theorem approach is that it follows from the well-known concepts of superposition in a more direct manner than the GF method. Another advantage is that there are no singularities in the fundamental solutions; that is, the $\psi(\cdot)$ functions are finite for $t - \tau \to 0$, while the GFs go to infinity. The GF method, however, has the advantage that the GFs are more accessible and easier to obtain than the $\psi(\cdot)$ functions.

To show the relationship between the Duhamel's theorem approach and the GF method, two examples are given below. One example demonstrates the application of Eq. (12.1b), where the boundary condition is only a function of time (no spatial variation), while the other example shows the applications of Eqs. (12.14) and (12.15), where the boundary conditions vary with both time and space.

Example 12.1: X11B-0T0 case. Consider a one-dimensional flat plate geometry initially at zero temperature with an arbitrary time-dependent prescribed surface temperature at $x = 0$ as shown in Fig. 12.3. For $t > 0$, the surface temperature at $x = L$ is kept at its initial value $T_0 = 0$. The describing equations are

$$\frac{\partial^2 T(x, t)}{\partial x^2} = \frac{1}{\alpha} \frac{\partial T(x, t)}{\partial t} \tag{12.19}$$

$$T(0, t) = f(t) \qquad \text{for } t > 0 \tag{12.20a}$$

$$T(L, t) = 0 \qquad \text{for } t > 0 \tag{12.20b}$$

$$T(x, 0) = 0 \tag{12.20c}$$

The fundamental solution for this problem is $\psi_T(x, t)$, which represents the temperature at point x and at time t in the flat plate geometry, with zero initial temperature and with a unit-step temperature at the boundary $x = 0$. It is a solution of the problem

$$\frac{\partial^2 \psi_T(x, t)}{\partial x^2} = \frac{1}{\alpha} \frac{\partial \psi_T(x, t)}{\partial t} \tag{12.21}$$

$$\psi_T(0, t) = 0 \qquad \text{for } t < 0$$

$$\qquad\qquad = 1 \qquad \text{for } t > 0 \tag{12.22a}$$

$$\psi_T(L, t) = 0 \qquad \text{for } t > 0 \tag{12.22b}$$

$$\psi_T(x, 0) = 0 \tag{12.22c}$$

Note that the subscript T in $\psi_T(\cdot)$ function indicates that it is a temperature-based fundamental solution.

From the Duhamel's integral Eq. (12.1b), the transient temperature distribution in the plate is given by

$$T(x, t) = \int_{\tau=0}^{t} f(\tau) \frac{\partial \psi_T(x, t - \tau)}{\partial t} d\tau \tag{12.23}$$

The GF solution to this problem is given by Eq. (3.46) with zero initial condition and no energy generation within the body,

$$T(x, t) = \alpha \int_{\tau=0}^{t} f(\tau) \frac{\partial G(x, t|x', \tau)}{\partial x'} \bigg|_{x'=0} d\tau \tag{12.24}$$

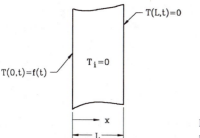

Figure 12.3 A flat plate with prescribed surface temperatures.

Note that since n_i' in Eq. (3.46) represents *outward normal* from the body, the $(\partial G/\partial n_i')|_{x'=x_i}$ term in this equation is replaced by $-(\partial G/\partial x')|_{x'=0}$ in Eq. (12.24). Also, since $T(L, t)$ is equal to zero for $t > 0$, there is no contribution due to the boundary condition at $x = L$.

A comparison of the GF solution, Eq. (12.24), and the Duhamel's theorem solution, Eq. (12.23), reveals that

$$\frac{\partial \psi_T(x, t - \tau)}{\partial t} = -\frac{\partial \psi_T(x, t - \tau)}{\partial \tau} = \alpha \left. \frac{\partial G(x, t|x', \tau)}{\partial x'} \right|_{x'=0} \tag{12.25}$$

and, then, integrating Eq. (12.25) over time gives

$$\psi_T(x, t - \tau) = -\int_{t'=\tau}^{t} \alpha \left. \frac{\partial G(x, t|x', t')}{\partial x'} \right|_{x'=0} dt' \tag{12.26}$$

Thus, Duhamel's theorem for this case is the same as the GF equation for a specified boundary temperature, where the fundamental solution is related to the GF function by Eqs. (12.25) and (12.26).

Example 12.2: $X20B(x$-t-$)T0$ case. Consider a semi-infinite body initially at zero temperature exposed to a time- and space-variable heat flux boundary condition over a portion of its surface boundary from $x = 0$ to $x = L$, with the rest of the surface boundary being insulated (see Fig. 12.4). From the flux-based Duhamel's integral Eq. (12.14) the temperature at any point (x, z) of the semi-infinite body and at any time t is given by

$$T(x, z, t) = -\int_0^L \int_0^t q(\eta, \tau) \frac{\partial^2 \psi_q(x, z, \eta, t - \tau)}{\partial t \, \partial \eta} \, d\tau \, d\eta \tag{12.27}$$

Here η is a dummy length variable along the surface boundary between $x = 0$ to $x = L$, and ψ_q is the flux-based fundamental solution for this problem, described by the following equations:

$$\nabla^2 \psi_q = \frac{1}{\alpha} \frac{\partial \psi_q}{\partial t} \tag{12.28}$$

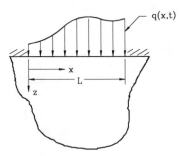

Figure 12.4 Semi-infinite body exposed to time- and space-variable boundary condition.

$$k \frac{\partial \psi_q}{\partial z} = 0 \qquad \text{for } t < 0 \qquad \text{or} \qquad x < \eta \qquad z = 0$$

$$= 1 \qquad \text{for } t > 0 \qquad \text{and} \qquad \eta < x < L \qquad z = 0 \qquad (12.29a)$$

$$\psi_q(x, \infty, \eta, t) = 0 \qquad (12.29b)$$

$$\psi_q(x, z, \eta, 0) = 0 \qquad (12.29c)$$

Similar to that of the previous example, the solution to this example problem can also be obtained from the GF Eq. (3.46) for the boundary condition of the second kind, with zero initial temperature and no energy generation as

$$T(\mathbf{r}, t) = \frac{\alpha}{k} \int_{\tau=0}^{t} \int_{S_i} q(\mathbf{r}_i', \tau) \, G(\mathbf{r}, t | \mathbf{r}_i', \tau) \, ds_i \, d\tau \qquad (12.30)$$

where S_i is the heated surface. For the coordinates $\mathbf{r} = (x, z)$, the heat flux at $z = 0$, and heating from $x' = \eta = 0$ to L, Eq. (12.30) can be written as

$$T(x, z, t) = \int_{\eta=0}^{L} \int_{\tau=0}^{t} \frac{\alpha}{k} q(\eta, \tau) \, G(x, z, t | \eta, 0, \tau) \, d\tau \, d\eta \qquad (12.31)$$

Comparing Eqs. (12.27) and (12.31) yields

$$-\frac{\partial^2 \psi_q(x, z, \eta, t - \tau)}{\partial t \, \partial \eta} = \frac{\alpha}{k} G(x, z, t | \eta, 0, \tau) \qquad (12.32)$$

and then, integrating Eq. (12.32) twice (over time and space), gives

$$\psi_q(x, z, \eta, t - \tau) = \int_{t'=\tau}^{t} \int_{x'=\eta}^{L} \frac{\alpha}{k} G(x, z, t | x', 0, t') \, dx' \, dt' \qquad (12.33)$$

which demonstrates the relationship between the flux-based fundamental solution and the GF for the semi-infinite body problem given above.

Both Duhamel's theorem and the GF equation are convolution integrals because they involve a product of two functions, one a function of τ and the other a function of $t - \tau$. Duhamel's theorem can be thought of as a boundary condition term of the GF equation, a special case of the general method of GF.

12.3 UNSTEADY SURFACE ELEMENT FORMULATIONS

There are two different formulations of the USE method. One is the single-node formulation, which uses the Laplace transform technique to obtain an approximate analytical solution. The other one is the multinode formulation (numerical solution) which is more general and can be applied to a variety of problems. The multinode formulation allows for spatial variation of surface heat flux and temperature by dividing the surface boundary into several surface elements, while in the single-node formulation, the surface heat flux and temperature are considered spatially constant.

Both formulations may be used with either heat flux-based or temperature-based fundamental solutions. The multinode formulation is given in this section. In Section 12.4, the single-node analytical solution is given as a special case where there is only one element along the surface boundary.

12.3.1 Surface Element Discretization

To numerically solve the Duhamel's integral Eqs. (12.14) and (12.15), the surface boundary is divided into N finite surface elements, Δs_j, as shown in Fig. 12.5a and b. Notice that only the parts of the boundary with nonzero values of heat flux [for Eq. (12.14)] and with a temperature different from the initial temperature [for Eq. (12.15)] need to be discretized. Then, Eqs. (12.14) and (12.15) can be written as

$$T(x, y, t) - T_0 = -\int_0^t \left[\sum_{j=1}^N \int_{\Delta s_j} q(\eta, \tau) \frac{\partial^2 \psi_q(x, y, \eta, t - \tau)}{\partial t\, \partial \eta} d\eta \right] d\tau \quad (12.34a)$$

and

$$T(x, y, t) - T_0 = -\int_0^t \left\{ \sum_{j=1}^N \int_{\Delta s_j} [T_s(\eta, \tau) - T_0] \right.$$

$$\left. \times \frac{\partial^2 \psi_T(x, y, \eta, t - \tau)}{\partial t\, \partial \eta} d\eta \right\} d\tau \quad (12.34b)$$

By assuming uniform heat flux and temperature over each surface element in Eqs. (12.34a) and (12.34b), respectively, one can write, for flux-based equations,

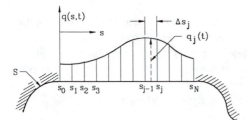

Figure 12.5 (a) Geometry showing discretization over heated portion of surface boundary.

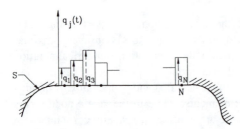

Figure 12.5 (b) Uniform heat flux assumption over each surface element.

$$T(x, y, t) - T_0 = - \int_0^t \left\{ \sum_{j=1}^N q_j(\tau) \frac{\partial}{\partial t} \left[\psi_q(x, y, \eta, t - \tau) \big|_{s_{j-1}}^{s_j} \right] \right\} d\tau$$

$$= - \int_0^t \left\{ \sum_{j=1}^N q_j(\tau) \frac{\partial}{\partial t} \left[\Delta \psi_{qj}(x, y, t - \tau) \right] \right\} d\tau \qquad (12.35a)$$

and, for temperature-based equations,

$$T(x, y, t) - T_0 = - \int_0^t \left\{ \sum_{j=1}^N [T_{sj}(\tau) - T_0] \frac{\partial}{\partial t} \left[\psi_{Tj}(x, y, t - \tau) \big|_{s_{j-1}}^{s_j} \right] \right\} d\tau$$

$$= - \int_0^t \left\{ \sum_{j=1}^N [T_{sj}(\tau) - T_0] \frac{\partial}{\partial t} \left[\Delta \psi_{Tj}(x, y, t - \tau) \right] \right\} d\tau \qquad (12.35b)$$

where

$$\Delta \psi_{qj} = \psi_q(x, y, s_j, t) - \psi_q(x, y, s_{j-1}, t) \qquad (12.36a)$$

$$\Delta \psi_{Tj} = \psi_T(x, y, s_j, t) - \psi_T(x, y, s_{j-1}, t) \qquad (12.36b)$$

Further, if the temperature rise at position (x, y) due to a unit step increase in heat flux and temperature at the element j are denoted $\phi_j(x, y, t)$ and $\theta_j(x, y, t)$, respectively, then it can be shown that

$$- \Delta \psi_{qj}(x, y, t) \equiv \phi_j(x, y, t) \qquad (12.37a)$$

$$- \Delta \psi_{Tj}(x, y, t) \equiv \theta_j(x, y, t) \qquad (12.37b)$$

Using Eq. (12.37a) into Eq. (12.36a) and Eq. (12.37b) into Eq. (12.36b) gives,

$$T(x, y, t) - T_0 = \sum_{j=1}^N \int_0^t q_j(\tau) \frac{\partial \phi_j(x, y, t - \tau)}{\partial t} d\tau \qquad (12.38a)$$

and

$$T(x, y, t) - T_0 = \sum_{j=1}^N \int_0^t [T_{sj}(\tau) - T_0] \frac{\partial \theta_j(x, y, t - \tau)}{\partial t} d\tau \qquad (12.38b)$$

Equations (12.38a and b) are the Duhamel's integral forms of the flux-based and the temperature-based USE equations for a single two-dimensional body.

Equation (12.38a) gives the temperature rise at location (x, y) and time t due to the effect of N surface heat flux histories $q_1(t), q_2(t), \ldots, q_N(t)$; while Eq. (12.38b) gives the temperature rise at the same location and time due to the effect of N time-varying surface temperatures $T_{s1}(t), T_{s2}(t), \ldots, T_{sN}(t)$. The functions ϕ_j and θ_j are the basic building-block solutions needed in the above expressions and are termed as the flux-based and the temperature-based *influence functions*. They are called influence functions because they give the influence of the jth surface element on the body. The USE method requires that the influence functions be known for the geometries under consideration. For instance, for the geometry of Fig. 12.4, the flux-based influence function ϕ is the solution to the problem of a semi-infinite body heated by a constant heat flux over an infinite strip as shown in Fig. 12.6.

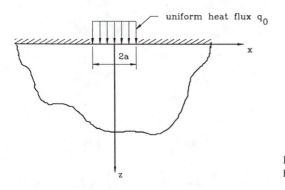

Figure 12.6 Geometry of semi-infinite body heated by a uniform over an infinite strip.

The USE Eqs. (12.38a and b) can be written in their general forms by replacing (x, y) in these equations with a position vector $\mathbf{r}$; that is,

Flux-based USE:
$$T(\mathbf{r}, t) - T_0 = \sum_{j=1}^{N} \int_0^t q_j(\tau) \frac{\partial \phi_j(\mathbf{r}, t - \tau)}{\partial t} d\tau \qquad (12.39a)$$

Temperature-based USE:
$$T(\mathbf{r}, t) - T_0 = \sum_{j=1}^{N} \int_0^t [T_{sj}(\tau) - T_0]$$
$$\times \frac{\partial \theta_j(\mathbf{r}, t - \tau)}{\partial t} d\tau \qquad (12.39b)$$

12.3.2 Green's Function Form of the USE Equations

The flux-based and temperature-based Eqs. (12.39a and b) represent the Duhamel's integral forms of the USE equations. The GF forms of the USE equations can be obtained by discretizing the GF Eq. (3.46) over N surface elements. For convenience, it is assumed that there is only one nonhomogeneous boundary (with a nonzero value of heat flux or a temperature different from the initial temperature). Accordingly, the summation terms in Eq. (3.46) which are for more than one nonhomogeneous boundary condition are dropped and one can write, for boundary condition of the second kind (flux-based equations),

$$T(\mathbf{r}, t) - T_0 = \int_0^t \left[\sum_{j=1}^{N} \frac{\alpha}{k} \int_{\Delta s_j} q(\mathbf{r}_j', \tau) G(\mathbf{r}, t|\mathbf{r}_j', \tau) ds_j \right] d\tau \qquad (12.40a)$$

and for boundary condition of the first kind (temperature-based equations):

$$T(\mathbf{r}, t) - T_0 = \int_0^t \left\{ \sum_{j=1}^{N} \alpha \int_{\Delta s_j} [T_s(\mathbf{r}', \tau) - T_0] \frac{\partial G(\mathbf{r}, t|\mathbf{r}_j', \tau)}{\partial n_j'} ds_j \right\} d\tau \qquad (12.40b)$$

By assuming uniform heat flux and temperature over each surface element in Eqs. (12.40a and b), respectively, one can get for the flux-based equation:

$$T(\mathbf{r}, t) - T_0 = \sum_{j=1}^{N} \int_0^t q_j(\tau) \overline{G}_j(\mathbf{r}, t|\mathbf{r}_j', \tau) d\tau \qquad (12.41a)$$

for the temperature-based equation:

$$T(\mathbf{r}, t) - T_0 = \sum_{j=1}^{N} \int_0^t [T_{sj}(\tau) - T_0] \, \overline{G}_j'(\mathbf{r}, t|\mathbf{r}_j', \tau) \, d\tau \qquad (12.41\text{b})$$

The notations $\overline{G}_j$ and $\overline{G}_j'$ are defined as

$$\overline{G}_j(\mathbf{r}, t|\mathbf{r}_j', \tau) = \frac{\alpha}{k} \int_{\Delta s_j} G(\mathbf{r}, t|\mathbf{r}_j', \tau) \, ds_j \qquad (12.42\text{a})$$

$$\overline{G}_j'(\mathbf{r}, t|\mathbf{r}_j', \tau) = -\alpha \int_{\Delta s_j} \frac{\partial G(\mathbf{r}, t|\mathbf{r}', \tau)}{\partial n_j'} \, ds_j \qquad (12.42\text{b})$$

Equations (12.41a and b) are the GF forms of the flux-based and temperature-based USE equations. Note that $\overline{G}_j$ and $\overline{G}_j'$ appearing in these equations correspond to the time derivatives of the flux-based and temperature-based influence functions, ϕ_j and θ_j, given in Eqs. (12.39a and b), respectively.

In the derivation of Eqs. (12.41a and b), it was assumed that there is only one nonhomogeneous boundary in the problem under consideration. For the problems with more than one nonhomogeneous boundary condition, a summation term should be added to each of these equations. Furthermore, it should be noted that Eqs. (12.41a and b) are, respectively, for the problems with the second kind (prescribed surface heat flux) and first kind (prescribed surface temperature) boundary conditions. For the problems with mixed boundary conditions, these two equations can be superimposed.

The USE Eqs. (12.39a, b) and (12.41a, b) can be applied to two- or three-dimensional geometries in the rectangular, cylindrical, or spherical coordinate systems. In the two-dimensional USE equations, the surface elements are infinite strips that may be treated as one-dimensional elements (or line elements). In the three-dimensional USE equations, the surface elements are two-dimensional elements and can be chosen in different shapes such as triangular, rectangular, or circular, depending on the nature of the problem under consideration.

12.3.3 Time Integration of the USE Equations

The time integration of the USE Eqs. (12.39a, b) and (12.41a, b), can be performed directly by dividing the entire time domain into M equal small time intervals, Δt, that is, for Duhamel's integral fluxed-based equation:

$$T(\mathbf{r}, t) - T_0 = \sum_{j=1}^{N} \sum_{i=1}^{M} \int_{t_{i-1}}^{t_i} q_j(\tau) \frac{\partial \phi_j(\mathbf{r}, t - \tau)}{\partial t} \, d\tau \qquad (12.43\text{a})$$

for Duhamel's integral temperature-based equation:

$$T(\mathbf{r}, t) - T_0 = \sum_{j=1}^{N} \sum_{i=1}^{M} \int_{t_{i-1}}^{t_i} [T_{sj}(\tau) - T_0] \frac{\partial \theta_j(\mathbf{r}, t - \tau)}{\partial t} \, d\tau \qquad (12.43\text{b})$$

for the GF flux-based equation:

$$T(\mathbf{r}, t) - T_0 = \sum_{j=1}^{N} \sum_{i=1}^{M} \int_{t_{i-1}}^{t_i} q_j(\tau) \, \overline{G}_j(\mathbf{r}, t | \mathbf{r}'_j, \tau) \, d\tau \qquad (12.44a)$$

for the GF temperature-based equation:

$$T(\mathbf{r}, t) - T_0 = \sum_{j=1}^{N} \sum_{i=1}^{M} \int_{t_{i-1}}^{t_i} [T_{sj}(\tau) - T_0] \overline{G}'_j(\mathbf{r}, t | \mathbf{r}'_j, \tau) \, d\tau \qquad (12.44b)$$

By assuming the elemental heat flux and temperature histories being uniform over each time interval, for temperature at time $t_M = M\Delta t$, one can write, for the Duhamel's integral flux-based USE equation:

$$T(\mathbf{r}, t_M) - T_0 = \sum_{j=1}^{N} \sum_{i=1}^{M} q_{ji} \, \Delta\phi_{j,M-i}(\mathbf{r}) \qquad (12.45a)$$

for the Duhamel's integral temperature-based equation:

$$T(\mathbf{r}, t_M) - T_0 = \sum_{j=1}^{N} \sum_{i=1}^{M} [T_{sji} - T_0] \Delta\theta_{j,M-i}(\mathbf{r}) \qquad (12.45b)$$

for the GF flux-based equation:

$$T(\mathbf{r}, t_M) - T_0 = \sum_{j=1}^{N} \sum_{i=1}^{M} q_{ji} \Delta\overline{G}_{j,M-i}(\mathbf{r}) \qquad (12.46a)$$

for the GF temperature-based equation:

$$T(\mathbf{r}, t_M) - T_0 = \sum_{j=1}^{N} \sum_{i=1}^{M} (T_{sji} - T_0) \Delta\overline{G}'_{j,M-i}(\mathbf{r}) \qquad (12.46b)$$

where q_{ji} and T_{sji} represent the jth surface heat flux and temperature evaluated at time t_i, and

$$\Delta\phi_{j,M-i}(\mathbf{r}) = \phi_j(\mathbf{r}, t_{M+1-i}) - \phi_j(\mathbf{r}, t_{M-i}) \qquad (12.47a)$$

$$\Delta\theta_{j,M-i}(\mathbf{r}) = \theta_j(\mathbf{r}, t_{M+1-i}) - \theta_j(\mathbf{r}, t_{M-i}) \qquad (12.47b)$$

$$\Delta\overline{G}_{j,M-i}(\mathbf{r}) = \int_{t_{i-1}}^{t_i} \overline{G}(\mathbf{r}, t_M | \mathbf{r}'_j, \tau) \, d\tau \qquad (12.47c)$$

$$\Delta\overline{G}'_{j,M-i}(\mathbf{r}) = \int_{t_{i-1}}^{t_i} \overline{G}'(\mathbf{r}, t_M | \mathbf{r}'_j, \tau) \, d\tau \qquad (12.47d)$$

Notice that $\Delta\overline{G}$ and $\Delta\overline{G}'$ each involve two integrations, one over the element and the other over the time step. For certain geometries, these integrations can be performed analytically (see Problem 12.3). With problems for which the analytical evaluations of these integrals are not possible, the GF USE equations could still be used by replacing the integrals with suitable quadrature formulas.

12.3.4 Flux-Based USE Equations for Bodies in Contact

For convenience, in the further development of the USE formulation, given in the rest of this section, only the flux-based Eqs. (12.38a) and (12.39a) are considered. The temperature-based USE formulation can be developed in a similar manner as the flux-based case.

Consider two arbitrary bodies, initially at uniform but different temperatures (T_{01} and T_{02}), brought into perfect contact over a portion of their boundaries, as shown in Fig. 12.7. The bodies may have different conductivities, k, and different density-specific heats ρc. For simplicity, a two-dimensional geometry is assumed; there is no variation of temperature or heat flux in the z direction. To apply the USE method, the interface is divided into N surface elements, each being an infinite strip, Δs_j. It is assumed that the flux and temperature are uniform over each surface element. The temperature associated with element j will be taken as the temperature at the center of the element, located at point $s'_j = s_j - \Delta s_j/2$. The average temperature over the element may also be used as the temperature associated with the element but it complicates the analysis slightly and will not be discussed here. The heat flux $q_j(t)$, associated with element j, which leaves body 2 in Fig. 12.7, is the same heat flux that enters body 1 over the region $s = s_{j-1}$ to $s = s_j$, that is,

$$- k_1 \frac{\partial T_1}{\partial n_j} = - k_2 \frac{\partial T_2}{\partial n_j} \quad \text{for } t > 0, \, s_{j-1} \leq s \leq s_j \quad \text{on } S \quad (12.48)$$

where n_j is the outward normal to element j. Using Eq. (12.38a), the temperature at element k in body 1 and time t can be given by

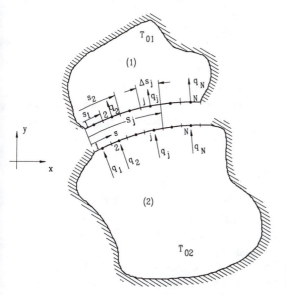

Figure 12.7 Possible distribution of surface elements for connected geometries.

$$T_{k1}(t) = T_{01} + \sum_{j=1}^{N} \int_0^t q_j(\tau) \frac{\partial \phi_{kj}^{(1)}(t - \tau)}{\partial t} d\tau \tag{12.49}$$

where $\phi_{kj}^{(1)}(t)$ is the temperature rise at element k and time t due to unit step heat flux over element j of surface 1. Similar to Eq. (12.49), an integral equation can be given for the kth surface element of body 2.

$$T_{k2}(t) = T_{02} - \sum_{j=1}^{N} \int_0^t q_j(\tau) \frac{\partial \phi_{kj}^{(2)}(t - \tau)}{\partial t} d\tau \tag{12.50}$$

where $\phi_{kj}^{(2)}(t)$ is the influence function for body 2. Note that the minus sign before the summation in Eq. (12.50) is used because the heat flux is pointing outward from body 2. For the case where the bodies are in perfect contact, one can write

$$T_{k1}(t) = T_{k2}(t) \qquad \text{for} \quad k = 1, 2, \ldots, N \tag{12.51}$$

Then, by introducing Eqs. (12.49) and (12.50) into Eq. (12.51), a set of integral equations for $k = 1, 2, \ldots, N$ is obtained as

$$T_{02} - T_{01} = \sum_{j=1}^{N} \int_0^t q_j(\tau) \frac{\partial \phi_{kj}(t - \tau)}{\partial t} d\tau \qquad \text{for } k = 1, 2, 3, \ldots, N \tag{12.52}$$

where

$$\phi_{kj} = \phi_{kj}^{(1)}(t) + \phi_{kj}^{(2)}(t) \tag{12.53}$$

Equation (12.52) is the flux-based USE equation for two bodies in perfect contact. It represents a set of Volterra equations of the first kind with the unknown heat fluxes, $q_k(t)$'s appearing inside the integrals.

Even though the perfect contact is a common interface assumption, it will only be valid for very intimate contact, such as a soldered joint. For a more general case of imperfect contact, Eq. (12.51) is replaced by

$$q_k(t) = h_k(t)[T_{k2}(t) - T_{k1}(t)] \qquad \text{for } k = 1, 2, \ldots, N \tag{12.54}$$

where $h_k(t)$ is the time-variable contact conductance for surface element k. The above relation tends to the case of perfect contact as $h_k \to \infty$. It also includes the cases of convection, prescribed heat flux, and prescribed temperature boundary conditions. By introducing Eqs. (12.49) and (12.50) into Eq. (12.54), a set of integral equations for $k = 1, 2, \ldots, N$, is obtained:

$$T_{02} - T_{01} = \frac{q_k(t)}{h_k(t)} + \sum_{j=1}^{N} \int_0^t q_j(\tau) \frac{\partial \phi_{kj}(t - \tau)}{\partial t} d\tau \qquad \text{for } k = 1, 2, \ldots, N \tag{12.55}$$

Equation (12.55) is the flux-based USE equation for two bodies with imperfect contacts. It represents a set of Volterra equations of the second kind with the unknown heat fluxes, $q_k(t)$'s, appearing both inside and outside the integrals.

The sets of integral equations presented by the USE equations (12.52) and (12.55) can be solved simultaneously for N unknown heat flux histories $q_1(t), q_2(t), \ldots,$

$q_N(t)$. The method of solution is described for the case of imperfect contact which includes the other cases as well.

12.3.5 Numerical Solution of the USE Equations for Bodies in Contact

In a similar manner to that discussed in Section 12.3.3, the flux-based USE equation (12.55) can be approximated by a system of linear algebraic equations by replacing the integrals with summations. As the first step, the time region 0 to t is divided into M equal small time intervals, Δt, so that t_M represents the value of t at the end point of the Mth interval ($t_M = M\Delta t$). Then, Eq. (12.55) can be written as

$$T_{02} - T_{01} = \frac{q_k(t_M)}{h_k(t_M)} + \sum_{j=1}^{N} \sum_{i=1}^{M} \int_{t_{i-1}}^{t_i} q_j(\tau) \frac{\partial \phi_{kj}(t_M - \tau)}{\partial t} d\tau$$

$$\text{for } k = 1, 2, \ldots, N \quad (12.56a)$$

where

$$t_0 \equiv 0 \quad (12.56b)$$

In the simplest form of approximation the heat flux histories $q_j(t)$ are assumed to have constant values in each time interval so that

$$T_0 = \frac{q_{kM}}{h_{kM}} + \sum_{j=1}^{N} \sum_{i=1}^{M} q_{ji} \Delta\phi_{kj.M-i} \quad \text{for } k = 1, 2, \ldots, N \quad (12.57a)$$

where

$$T_0 = T_{02} - T_{01} \qquad \Delta\phi_{kj.M-i} = \phi_{kj.M+1-i} - \phi_{kj.M-i} \quad (12.57b, c)$$

and

$$q_{ji} \equiv q_j(t_i) \qquad \phi_{kj.i} \equiv \phi_{kj}(t_i) \quad (12.57d, e)$$

In the form given by Eq. (12.57a), the heat fluxes q_{jM}'s (for $j = 1, 2, \ldots, N$) can be determined at different time intervals one after another, by marching forward in time for $M = 1, 2, 3, \ldots$. While calculating each new time component, the fluxes at previous times, $q_{j1}, q_{j2}, q_{j3}, \ldots, q_{j.M-2}, q_{j.M-1}$ are known for $j = 1, 2, \ldots, N$. Thus for each time step, Eq. (12.57a) represents a system of N equations with N unknowns $q_{1M}, q_{2M}, q_{3M}, \ldots, q_{NM}$. The objective is to solve this system for the unknowns q_{jM}, for $j = 1, 2, \ldots, N$. Rearranging Eq. (12.57a) in standard form with unknowns, q_{jM}'s, on the left, and knowns on the right, and noting that $\phi_{kj0} = 0$, one can write

$$\frac{q_{kM}}{h_{kM}} + \sum_{j=1}^{N} q_{jM}\phi_{kj1} = T_0 - \sum_{j=1}^{N} \sum_{i=1}^{M-1} q_{ji} \Delta\phi_{kj.M-i} \quad (12.58)$$

Expressing Eq. (12.58) in matrix form gives

$$(\bar{\bar{H}}_M + \bar{\bar{\Phi}}_1)\bar{q}_M = \bar{T}_0 - \sum_{i=1}^{M-1} \Delta\bar{\bar{\Phi}}_{M-i}\bar{q}_i \quad (12.59)$$

where $\overline{T}_0$ is the initial temperature vector, $\overline{\overline{H}}_M$ is the conductance matrix, $\overline{\overline{\Phi}}_i$ and $\overline{q}_i$ are the influence matrix and the heat flux vector at time t_i, respectively.

$$\overline{\overline{\Phi}}_i \equiv \begin{bmatrix} \phi_{11i} & \phi_{12i} & \cdots & \phi_{1Ni} \\ \phi_{21i} & \phi_{22i} & & \phi_{2Ni} \\ \vdots & & & \\ \phi_{N1i} & \phi_{N2i} & & \phi_{NNi} \end{bmatrix} \tag{12.60a}$$

$$\overline{\overline{H}}_M \equiv \text{diag} \left[\frac{1}{h_{1M}} \frac{1}{h_{2M}} \cdots \frac{1}{h_{NM}} \right] \tag{12.60b}$$

$$\overline{q}_i \equiv \begin{bmatrix} q_{1i} \\ q_{2i} \\ \vdots \\ q_{Ni} \end{bmatrix} \qquad \overline{T}_0 \equiv \begin{bmatrix} T_0 \\ T_0 \\ \vdots \\ T_0 \end{bmatrix} \tag{12.60c, d}$$

If further $\overline{\overline{C}}_M$ and $\overline{D}_M$ are defined to be the matrices

$$\overline{\overline{C}}_M = \overline{\overline{H}}_M + \overline{\overline{\Phi}}_1 \qquad \overline{D}_M = \overline{T}_0 + \overline{E}_M - \overline{F}_M \tag{12.61a, b}$$

where

$$\overline{E}_M = \sum_{i=1}^{M-1} \overline{\overline{\Phi}}_{M-i}\, \overline{q}_i \tag{12.62a}$$

and

$$\overline{F}_M = \sum_{i=1}^{M-1} \overline{\overline{\Phi}}_{M+1-i}\, \overline{q}_i \tag{12.62b}$$

Then Eq. (12.59) can be written as

$$\overline{\overline{C}}_M\, \overline{q}_M = \overline{D}_M \tag{12.63}$$

solving Eq. (12.63) for $\overline{q}_M$, gives

$$\overline{q}_M = \overline{\overline{C}}_M^{-1}\, \overline{D}_M \tag{12.64}$$

The $\overline{\overline{C}}_M$ matrix, multiplier of $\overline{q}_M$, has to be calculated at each time step if the diagonal matrix $\overline{\overline{H}}_M$ is a function of time. However, if contact conductances do not change with time, the $\overline{\overline{C}}_M$ matrix needs to be calculated only once during the entire solution and an alternative form of solution can be given as (see Note 1 at end of chapter).

$$q_1 = \overline{\overline{C}}^{-1}\overline{T}_0 \tag{12.65a}$$

$$\overline{q}_M = M\overline{q}_1 + \overline{\overline{B}} \left[\sum_{i=1}^{M-1} \overline{q}_i \right] - \overline{\overline{C}}^{-1}\overline{F}_M \qquad \text{for } M = 2, 3, \ldots \tag{12.65b}$$

where

$$\bar{\bar{B}} = \bar{\bar{H}}^{-1}\bar{\bar{\Phi}}_1 \tag{12.66}$$

Notice that, since $\bar{\bar{C}}$ and $\bar{\bar{H}}$ are not functions of time in Eqs. (12.65) and (12.66), the subscript M is dropped. For the case of perfect contact where $h_{kM} \to \infty$, the diagonal conductance matrix, $\bar{\bar{H}}_M$, becomes zero, which implies that

$$\bar{\bar{C}} = \bar{\bar{\Phi}}_1 \tag{12.67}$$

Introducing Eqs. (12.61b), (12.62a, b) and (12.67) into Eq. (12.64) results in a simpler form of solution as

$$\bar{q}_1 = \bar{\bar{C}}^{-1}\bar{T}_0 \tag{12.68a}$$

$$\bar{q}_M = M\bar{q}_1 - \bar{\bar{C}}^{-1}\bar{F}_M \qquad \text{for } M = 2, 3, \ldots \tag{12.68b}$$

The elements of the $[N \times N]$ influence matrix $\bar{\bar{\Phi}}_i$ are

$$\phi_{kji} = \phi_{kji}^{(1)} + \phi_{kji}^{(2)} \tag{12.69}$$

If the two bodies in contact have the same geometry and thermal properties, then

$$\phi_{kji} = 2\phi_{kji}^{(1)} = 2\phi_{kji}^{(2)} \tag{12.70}$$

It is helpful to display the expression for $\bar{q}_M$ more explicitly. To illustrate, the case of perfect contact at the interface with only two elements is considered ($N = 2$). In other words there are two heat flux histories, $q_1(t)$ and $q_2(t)$, to be determined. For simplicity, only three time steps are considered ($M = 3$). At the first time step, Eq. (12.68a) becomes

$$\begin{bmatrix} q_{11} \\ q_{12} \end{bmatrix} = \begin{bmatrix} C_{11} & C_{12} \\ C_{21} & C_{22} \end{bmatrix}^{-1} \begin{bmatrix} T_0 \\ T_0 \end{bmatrix} \tag{12.71}$$

where

$$C_{kj} = \phi_{kj1} = \phi_{kj1}^{(1)} + \phi_{kj1}^{(2)} \tag{12.72}$$

Solving the above system, Eq. (12.71) for q_{11} and q_{21} yields

$$q_{11} = \frac{T_0(C_{22} - C_{12})}{\Delta} \tag{12.73a}$$

$$q_{21} = \frac{T_0(C_{11} - C_{21})}{\Delta} \tag{12.73b}$$

where

$$\Delta = C_{11}C_{22} - C_{12}C_{21} \tag{12.73c}$$

For the second time step, $M = 2$, Eq. (12.68b) becomes

$$
\begin{bmatrix} q_{12} \\ q_{22} \end{bmatrix} = 2 \begin{bmatrix} q_{11} \\ q_{21} \end{bmatrix} - \begin{bmatrix} C_{11} & C_{12} \\ C_{21} & C_{22} \end{bmatrix}^{-1} \begin{bmatrix} F_{12} \\ F_{22} \end{bmatrix}
\tag{12.74}
$$

Solving Eq. (12.74) for q_{12} and q_{22} yields

$$
q_{12} = \frac{(2T_0 - F_{12})C_{22} - (2T_0 - F_{22})C_{12}}{\Delta}
$$

$$
= 2q_{11} - \frac{F_{12}C_{22} - F_{22}C_{12}}{\Delta}
\tag{12.75a}
$$

$$
q_{22} = \frac{(2T_0 - F_{22})C_{11} - (2T_0 - F_{12})C_{21}}{\Delta}
$$

$$
= 2q_{21} - \frac{F_{22}C_{11} - F_{12}C_{21}}{\Delta}
\tag{12.75b}
$$

where

$$
F_{12} = \phi_{112}q_{11} + \phi_{122}q_{21}
\tag{12.76a}
$$

$$
F_{22} = \phi_{212}q_{11} + \phi_{222}q_{21}
\tag{12.76b}
$$

In a similar manner, for the third time step, $M = 3$, one can write

$$
q_{13} = 3q_{11} - \frac{F_{13}C_{22} - F_{23}C_{12}}{\Delta}
\tag{12.77a}
$$

$$
q_{23} = 3q_{21} - \frac{F_{23}C_{11} - F_{13}C_{21}}{\Delta}
\tag{12.77b}
$$

where

$$
F_{13} = \sum_{i=1}^{2} (\phi_{11,4-i}q_{1i} + \phi_{12,4-i}q_{2i})
\tag{12.78a}
$$

$$
F_{23} = \sum_{i=1}^{2} (\phi_{21,4-i}q_{1i} + \phi_{22,4-i}q_{2i})
\tag{12.78b}
$$

Notice that F_{1M} and F_{2M} are the only terms that should be evaluated at each time step.

Because of convolution behavior of the summations given in Eqs. (12.62a and b), the influence matrices, $\bar{\bar{\Phi}}_i$'s, need to be calculated at each time step. Consequently, most of the computation effort is in the evaluation of column matrix $\bar{D}$, particularly as the value of M becomes larger.

12.3.6 Influence Functions

An influence function, $\phi_{kj}(t)$, is the temperature rise at time t and element k due to a unit step heat flux at $t = 0$ at element j. When providing the influence functions, there are two cases to consider: when $k = j$ (temperature rise at location of heating) and $k \neq j$ (temperature rise at other than the heating location). The more important and more difficult to obtain is for $k = j$, particularly for small times. A number of influence functions for $\phi_{kk}(t)$ are described and referenced in this section. For the case of $k \neq j$, the $\Delta\phi_{kj,M}$ values, given by Eq. (12.57c), are efficiently obtained through the use of GFs.

The simplest influence functions are for one-dimensional cases, such as shown in Fig. 12.8a–c. The first is for constant heat flux q_0 equal to unity over the surface of a semi-infinite body. The $\phi_{kk}(t)$ expression at the heated surface of a semi-infinite body shown in Fig. 12.8a is simply

$$\phi_{kk}(t) = 2\left(\frac{t}{\pi k \rho c}\right)^{1/2} \tag{12.79}$$

Figure 12.8b is for a solid cylinder or sphere. Another basic case is for the region outside a radius of a and with the heat flux of $q_0 = 1$ at $r = a$; this can be for both cylindrical and spherical geometries and is illustrated by Fig. 12.8c. For early times, the geometries shown in Figs. 12.8b and c have $\phi_{kj}(t)$ values that contain additive curvature corrections (Beck et al., 1985b) to Eq. (12.79).

Two cases having two-dimensional heat transfer for the semi-infinite geometries are shown in Figs. 12.8d and e. Figure 12.8d is for a heated strip 2a wide (Litkouhi and Beck, 1982) and Fig. 12.8e has a circular source of radius a (Beck, 1980). These two cases have "edge" corrections for small times that are additive to Eq. (12.79). Large-time behavior of these two cases are also known; Fig. 12.8d approaches an $\ln(t)$ variation that is typical of a line source. The circular heat source case of Fig. 12.8e goes to steady state for large times. There is also a principle of additivity for large times, which is discussed by Beck et al. (1985b) in detail. Another case of circular source is when it is centered in the surface of a semi-infinite cylinder (Beck, 1981b); see Fig. 12.8f. A case of rectangular heat source on the surface of a semi-infinite body (Keltner et al., 1988) is depicted by Fig. 12.8g. A finite geometry is shown by Fig. 12.8h; a circular source is applied at the end of finite cylinder (Beck and Keltner, 1987). Figures 12.8i and j show two solutions that can be constructed by the principles of additivity that are discussed by Beck et al. (1984). Many other solutions can be constructed in a similar manner.

12.4 APPROXIMATE ANALYTICAL SOLUTION (SINGLE ELEMENT)

It is sometimes possible to obtain approximate analytical solutions by considering only one surface element along the interface between the connected bodies. This is known as the single-node USE approach. In this approach the coupling interfacial boundary conditions is relaxed so that neither temperature nor heat flux need simultaneously match for all points along the interface and at all times. Instead, a less stringent requirement equates average heat fluxes between the coupled regions while still requiring simultaneous matching of area-average interfacial temperatures.

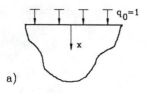

a)

b)

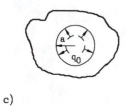

c)

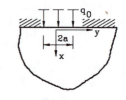

d)

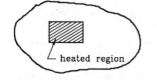

e)

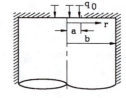

f)

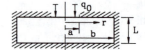

g)

h)

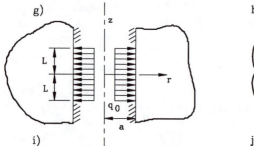

i)

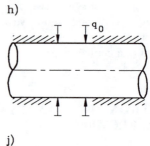

j)

Figure 12.8 Geometries and boundary conditions for various influence functions.

For one surface element, the sets of integral equations represented by Eqs. (12.39a and b) reduce to two single integral equations given, for flux-based equations, by

$$T(\mathbf{r}, t) = \int_0^t q(\tau) \frac{\partial \phi(\mathbf{r}, t - \tau)}{\partial t} \, d\tau \qquad (12.80a)$$

and, for temperature-based equations:

$$T(\mathbf{r}, t) - T_0 = \int_0^t [T_s(\tau) - T_0] \frac{\partial \theta(\mathbf{r}, t - \tau)}{\partial t} \, d\tau \qquad (12.80b)$$

These Duhamel's integral equations can be written in their alternative forms [see Eqs. (12.1a, b)] as

flux-based equation: $\qquad T(\mathbf{r}, t) = \dfrac{\partial}{\partial t} \displaystyle\int_0^t q(\tau) \phi(\mathbf{r}, t - \tau) \, d\tau \qquad (12.81a)$

temperature-based equation: $\quad T(\mathbf{r}, t) - T_0 = \dfrac{\partial}{\partial t} \displaystyle\int_0^t [T_S(\tau) - T_0]$

$$\times \, \theta(\mathbf{r}, t - \tau) \, d\tau \qquad (12.81b)$$

Taking the Laplace transform of Eqs. (12.81a and b), analytical solutions may be obtained that yield relatively accurate results for a certain class of problems. The ease or difficulty of obtaining such solutions depends entirely on the particular expressions for the influence functions ϕ or θ. This procedure is best illustrated with the following example.

Example 12.3 Consider the specific classic case of two homogeneous semi-infinite bodies initially at different temperatures T_{01} and T_{02} brought together as shown in Fig. 12.9. The objective is to find approximate analytical solutions for the interface temperature and/or heat flux by utilizing the Laplace transformations. Both temperature-based and heat flux-based solutions are considered here.

TEMPERATURE-BASED SOLUTION From the temperature-based Eq. (12.81b), the temperature at position x in body 1 and at time t is given by

$$T_1(x, t) = T_{01} + \frac{\partial}{\partial t} \int_0^t [T_1(0, \tau) - T_0] \theta^{(1)}(x, t - \tau) \, d\tau \qquad (12.82)$$

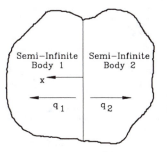

Figure 12.9 Two homogeneous semi-infinite bodies at different initial temperatures brought into thermal contact.

where $\theta^{(1)}(x, t)$ is the temperature-based influence function for body 1. It represents the temperature rise at position x in body 1 due to a unit step increase in temperature at the surface of the body. The heat flow through the surface region of body 1 is given by

$$q_1(t)A = \frac{\partial}{\partial t} \left\{ A \int_0^t [T_1(0, \tau) - T_{01}] \theta_q^{(1)}(0, t - \tau) \, d\tau \right\} \qquad (12.83)$$

where A is the surface area and $q_1(t)$ is the surface (or interface) heat flux. The function $\theta_q^{(1)}$ is the area average heat flux for a unit increase in surface temperature. It is given by

$$\theta_q^{(1)}(0, t) = \frac{1}{A} \int_A -k \left. \frac{\partial \theta^{(1)}(x, t)}{\partial x} \right|_{x=0} dA \qquad (12.84)$$

If there are no interface heat sources, the same average heat flux that enters body 1 then leaves body 2 so that

$$Aq_1(t) = -Aq_2(t) \qquad (12.85)$$

For perfect or imperfect contact the influence function $\theta^{(i)}(x, t)$ is (Carslaw and Jaeger, 1959)

$$\theta^{(i)}(x, t) = \mathrm{erfc} \left(\frac{|x|}{2\sqrt{\alpha_i t}} \right) \qquad (12.86)$$

The related heat flux is

$$\theta_q^{(i)}(0, t) = \pm \frac{k_i}{(\pi \alpha_i t)^{1/2}} \exp \left(\frac{-x^2}{4\alpha_i t} \right) \bigg|_{x=0} \qquad (12.87)$$

where the plus sign is for $i = 1$ (body 1) and the minus sign is for $i = 2$ (body 2).

For perfect contact, the interface temperature (at $x = 0$) is identical for both bodies so that $T_1(0, t) = T_2(0, t) = T(0, t)$. Substituting Eq. (12.83) in Eq. (12.85) gives

$$\frac{\partial}{\partial t} \int_0^t [T(0, \tau) - T_{10}] \theta_q^{(1)}(0, t - \tau) \, d\tau$$

$$= \frac{\partial}{\partial t} \int_0^t [T(0, \tau) - T_{20}] \theta_q^{(2)}(0, t - \tau) \, d\tau \qquad (12.88)$$

and then using Eq. (12.87) gives

$$\frac{\partial}{\partial t} \int_0^t [T(0, \tau) - T_{10}] \frac{k_1}{\sqrt{\pi \alpha_1 (t - \tau)}} \, d\tau$$

$$= -\frac{\partial}{\partial t} \int_0^t [T(0, \tau) - T_{20}] \frac{k_2}{\sqrt{\pi \alpha_2 (t - \tau)}} \, d\tau \qquad (12.89)$$

This is a Volterra equation of the first kind where the unknown function is $T(0, t)$.

Taking the Laplace transform of Eq. (12.89) gives

$$s\mathcal{L}[T(0, t) - T_{10}] \cdot \frac{k_1}{\sqrt{\alpha_1 s}} = -s\mathcal{L}[T(0, t) - T_{20}] \cdot \frac{k_2}{\sqrt{\alpha_2 s}} \quad (12.90)$$

where $\mathcal{L}[T(0, t)]$ is the Laplace transform of $T(0, t)$ and s is the Laplace transform parameter. For convenience, let $\hat{T}(0, s) = \mathcal{L}[T(0, t)]$. Without loss of generality, let $T_{10} = T_0$ and $T_{20} = 0$, then

$$\left[\hat{T}(0, s) - \frac{T_0}{s}\right] \cdot \frac{k_1}{\sqrt{\alpha_1 s}} = -\hat{T}(0, s) \cdot \frac{k_2}{\sqrt{\alpha_2 s}} \quad (12.91)$$

Solving for $\hat{T}(0, s)$ gives

$$\hat{T}(0, s) = \frac{T_0}{s} \frac{k_1}{(\alpha_1 s)^{1/2}} \left[\frac{k_1}{(\alpha_1 s)^{1/2}} + \frac{k_2}{(\alpha_2 s)^{1/2}}\right]^{-1} \quad (12.92)$$

which has the inverse Laplace transform of

$$T(0, t) = T_0(1 + \beta)^{-1} \quad (12.93a)$$

where

$$\beta = \left(\frac{k_2 \rho_2 c_2}{k_1 \rho_1 c_1}\right)^{1/2} \quad (12.93b)$$

This is the desired exact solution for the surface temperature for both bodies for the case of perfect contact (Carslaw and Jaeger, 1959).

Next, consider the more complex case of imperfect contact of two semi-infinite bodies. Let there be a contact conductance h between the bodies. The heat fluxes are related by

$$q_1(t) = -q_2(t) = h[T_1(0, t) - T_2(0, t)] \quad (12.94)$$

where now both $T_1(0, t)$ and $T_2(0, t)$ are unknown functions.

Solving Eq. (12.94) for $T_1(0, t)$, in terms of $q_2(t)$, substituting in Eq. (12.83) and again equating the heat flows gives

$$\frac{\partial}{\partial t}\left(\int_0^t [T_2(0, \tau) - T_{10}]\theta_q^{(1)}(0, t - \tau) \, d\tau + \int_0^t \left\{\frac{1}{h}\frac{\partial}{\partial t}\int_0^\tau [T_2(0, \gamma)\right.\right.$$

$$\left.\left. - T_{20}]\theta_q^{(2)}(0, \tau - \gamma) \, d\gamma\right\}\theta_q^{(1)}(0, t - \tau) \, d\tau\right)$$

$$= -\frac{\partial}{\partial t}\int_0^t [T_2(0, \tau) - T_{20}]\theta_q^{(2)}(0, t - \tau) \, d\tau \quad (12.95)$$

This is an integral equation of the Volterra type, except now a double integration is present; the unknown function is $T_2(0, t)$.

The integral equation given by Eq. (12.95) would in most cases be solved numerically, but fortunately in this case the exact solution can be found using the Laplace transform.

$$s\left[\left(\hat{T}_2 - \frac{T_{10}}{s}\right)\hat{\theta}_q^{(1)} + \frac{s}{h}\left(\hat{T}_2 - \frac{T_{20}}{s}\right)\hat{\theta}_q^{(2)}\hat{\theta}_q^{(1)}\right] = -s\left(\hat{T}_2 - \frac{T_{20}}{s}\right)\hat{\theta}_q^{(2)} \quad (12.96)$$

where $\hat{\theta}_q^{(i)} = k_i/(\alpha_i s)^{1/2}$.

Factoring out common terms and letting $T_{10} = T_0$ and $T_{20} = 0$ gives

$$\hat{T}_2 = \frac{T_0}{s}\left(\frac{k_1/\sqrt{\alpha_1 s}}{k_1/\sqrt{\alpha_1 s} + k_1 k_2/h\sqrt{\alpha_1 \alpha_2} + k_2/\sqrt{\alpha_2 s}}\right)$$

$$= \frac{T_0 h}{\sqrt{k_2 \rho_2 c_2}} \cdot \frac{1}{s[\sqrt{s} + h(1/\sqrt{k_1 \rho_1 c_1} + 1/\sqrt{k_2 \rho_2 c_2})]} \quad (12.97)$$

Taking the inverse Laplace transform yields the desired exact interface temperature of

$$T_2(0, t) = \frac{T_0[1 - e^{h^2 b^2 t}\,\text{erfc}\,(hb\sqrt{t})]}{1 + \beta} \quad (12.98a)$$

where β is defined in Eq. (12.93b) and

$$b = (k_1 \rho_1 c_1)^{-1/2} + (k_2 \rho_2 c_2)^{-1/2} \quad (12.98b)$$

Interior temperatures can now be found by introducing the expression given by Eq. (12.98a) into a Duhamel's integral similar to Eq. (12.82) with $\theta(x, t)$ given by Eq. (12.86). The heat flow across the interface can be found by using Eq. (12.98a) in Duhamel's integral similar to Eq. (12.83).

HEAT FLUX-BASED SOLUTION The heat flux-based solution can be obtained in a similar manner as the temperature-based solution. From the heat flux-based integral Eq. (12.81a), the temperature at position x in body 1 for a time variable surface heat flux $q(t)$ is given by

$$T_1(x, t) = T_{10} + \frac{\partial}{\partial t}\int_0^t q(\tau)\phi^{(1)}(x, t - \tau)\,d\tau \quad (12.99)$$

where $\phi^{(1)}(x, t)$ is the flux-based influence function for body 1. It represents the temperature rise at position x in body 1 and at time t due to unit step increase in the surface heat flux at time zero. Similarly, the temperature at any position x in body 2 and at time t is given by

$$T_2(x, t) = T_{20} - \frac{\partial}{\partial t}\int_0^t q(\tau)\phi^{(2)}(x, t - \tau)\,d\tau \quad (12.100)$$

where $\phi^{(2)}(x, t)$ is the influence function for body 2. The assumption of a spatially uniform heat flux is not always compatible with a spatially uniform temperature. The statement in terms of an average heat flux given by Eq. (12.85) is always true.

For the special geometry of two semi-infinite bodies coming into uniform contact over the complete interface, the interface temperatures are not functions of position. The influence function $\phi^{(i)}(x, t)$ is a function of a single space dimension and time,

$$\phi^{(i)}(x, t) = \frac{2t^{1/2}}{(k_i \rho_i c_i)^{1/2}} \text{ ierfc} \left[\frac{x}{2(\alpha_i t)^{1/2}} \right] \tag{12.101}$$

where x is directed inward in each body and the i is 1 or 2. At the surface of the body, $x = 0$ and ierfc $(0) = \pi^{-1/2}$.

Consider the first case of perfect contact for which the interface temperature must be the same for both bodies. Equating Eqs. (12.99) and (12.100) with $T_{10} = T_0$ and $T_{20} = 0$ at $x = 0$ yields

$$T_0 + \frac{\partial}{\partial t} \int_0^t q(\tau) \frac{2(t - \tau)^{1/2} \, d\tau}{(\pi k_1 \rho_1 c_1)^{1/2}} = -\frac{\partial}{\partial t} \int_0^t q(\tau) \frac{2(t - \tau)^{1/2} \, d\tau}{(\pi k_2 \rho_2 c_2)^{1/2}} \tag{12.102}$$

which can be rearranged to the form

$$\frac{\partial}{\partial t} \int_0^t q(\tau) 2b\pi^{-1/2}(t - \tau)^{1/2} \, d\tau = -T_0 \tag{12.103}$$

This is again a Volterra integral equation of the first kind and can be solved for $q(t)$ using the numerical methods in (Beck, 1968). For this simple case, the solution can be obtained as above by utilizing the Laplace transform to get

$$q(t) = -T_0 b^{-1}(\pi t)^{-1/2} \tag{12.104}$$

Utilizing Eq. (12.101) (with $x = 0$) and Eq. (12.104) in Eq. (12.99) yields

$$T_1(0, t) = T_0(1 + \beta)^{-1} \tag{12.105}$$

which is the same as Eq. (12.93) which was derived using the temperature form of Duhamel's theorem.

A comparison of the above procedures for the T- and q-based solutions for the perfect contact example considered shows both approaches yield a Volterra integral equation of the first kind. In the temperature case, the solution is for the interface temperature while in the heat flux case the solution is for $q(t)$. The solutions are similar, although different quantities are found. For the T case, the $q(t)$ function is found by solving Eq. (12.83) given $T_1(0, t)$ and, for the heat flux case, the $T_i(0, t)$ function is found by solving Eqs. (12.99) or (12.100) given $q(t)$. If only the interface temperature is desired, then the T-based method is more direct.

Next, for the q-based approach, consider the imperfect contact case. Equation (12.94) still applies but utilizing Eqs. (12.99)–(12.101) yields

$$-T_0 + \frac{q(t)}{h} = \frac{2b}{\sqrt{\pi}} \frac{\partial}{\partial t} \int_0^t q(\tau)(t - \tau)^{1/2} \, d\tau \tag{12.106}$$

which is again a Volterra integral equation of the second kind since $q(t)$ appears both inside and outside the integral. Equation (12.106) is simpler than the comparable T-based Eq. (12.95) which has a double integral. A solution of Eq. (12.106) for $q(t)$ utilizing the Laplace transform is

$$q(t) = -hT_0 e^{h^2 b^2 t} \text{ erfc} (hbt^{1/2}) \tag{12.107}$$

where b is defined by Eq. (12.98b). If h goes to infinity, Eq. (12.107) reduces to Eq. (12.104). The next step is to use Eqs. (12.99) and (12.100) with Eq. (12.107) to determine the surface temperature histories. Though the integrals are not easy to evaluate, the same results are found as by the temperature-based approach.

From a comparison of the T- and q-based USE integral equations (12.95) and (12.106), the q-based equation has a simpler form and poses less difficulty in numerical solution (which might be required for more complex geometries). Furthermore, the q-based Eq. (12.106), is derived in a much more straightforward manner. Hence, based on the above example, the q-based approach is to be recommended over the T-based approach.

12.5 EXAMPLES

To demonstrate the utility of the USE method, two well-known problems are solved in this section. The first problem is a semi-infinite body with the mixed boundary conditions of a step change of the surface temperature over a disk of radius **a** and insulated elsewhere, as shown in Fig. 12.10a. This problem is similar to the problem of two semi-infinite bodies initially at different uniform temperatures suddenly brought together over a circular area, as shown in Fig. 12.10b (a contact conductance problem). The second problem involves a semi-infinite cylinder attached perpendicularly to a semi-infinite body (an intrinsic thermocouple problem); see Fig. 12.10c. Both single-node and multinode USE solutions are given and the results are compared with other existing analytical and numerical solutions.

Due to the axisymmetric nature of these problems, in each case the interface area is divided into 10 annular variable-spaced surface elements with smaller elements being closer to the edge of the contact area, as shown in Figs. 12.11 and 12.12. The inner and outer radii of each element are denoted by a_{j-1} and a_j, respectively ($j = 1, 10$ and $a_0 = 0$). The heat flux and temperature are approximated to be constant over each surface element and are specified at the points

$$r_1 = 0 \tag{12.108a}$$

$$r_j = \frac{a_j - a_{j-1}}{2} \quad \text{for } j = 2, 10 \tag{12.108b}$$

Since, in each problem, the connected bodies are assumed to be in perfect contact, the simplified form of solution given by Eqs. (12.68a, b) are used. Substituting Eqs. (12.62b) and (12.67) into Eqs. (12.68a, b) yields

$$\bar{q}_1 = \bar{\bar{\Phi}}_1^{-1} \bar{T}_0 \tag{12.109a}$$

$$\bar{q}_M = M\bar{q}_1 - \bar{\bar{\Phi}}_1^{-1} \sum_{i=1}^{M-1} \bar{q}_i \quad \text{for } M = 2, 3, \ldots \tag{12.109b}$$

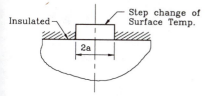

Figure 12.10 (*a*) A semi-infinite body with step change of surface temperature over a circular area and insulated elsewhere.

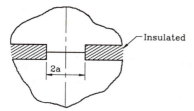

Figure 12.10 (*b*) Two homogeneous semi-infinite bodies at different initial temperatures brought into thermal contact over a circular area (a contact conductance problem).

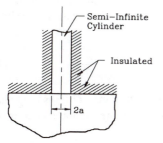

Figure 12.10 (*c*) Semi-infinite cylinder attached to semi-infinite body simulating intrinsic thermocouple problem.

At each time step, Eqs. (12.109a, b) are solved for unknown elemental heat fluxes $q_{1M}, q_{2M}, \ldots, q_{10M}$.

The required influence functions for the above problems are shown in Figs. 12.13*a*, *b*. They are evaluated from the available exact closed-form solutions of a semi-infinite body heated by a constant disk heat source (Beck, 1981a, Fig. 12.14*a*) and a semi-infinite insulated cylinder heated by a constant heat flux over a disk area centered at the end (Beck, 1981b, Fig. 12.14*b*), by simple superposition. That is,

$$\phi_{kji} = \gamma_{kji} - \sum_{n=1}^{j-1} \phi_{kni} \qquad \text{for } j, k = 1, 2, \ldots, 10 \qquad (12.110)$$

where γ_{kji} represents the temperature rise at element $k(r = r_k)$ due to a unit heat flux at the disk with radius a_j and at time t_i. See Fig. 12.15.

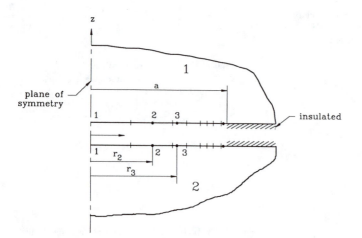

Figure 12.11 Distribution of surface elements for connected semi-infinite bodies.

For the first problem, the USE solutions are compared with other available solutions (Schneider et al., 1977; Sadhel, 1980; Keltner, 1973) on the basis of the dimensionless thermal constriction resistance across the interface area. The transient thermal constriction resistance is defined as "the difference between the average temperature of the contact area and the temperature far from the contact area divided by the total instantaneous heat flow through the contact area" (Schneider, 1979), and is given by

$$R_{c1}(t) = \frac{T_c(t) - T_{10}}{Q_c(t)} \qquad R_{c2}(t) = \frac{T_{20} - T_c(t)}{Q_c(t)} \qquad (12.111a,b)$$

Where $T_c(t)$ is the average temperature of the contact area, $Q_c(t)$ is the total heat flow through the contact area, and the $R_{c1}(t)$ and $R_{c2}(t)$ are the thermal constriction

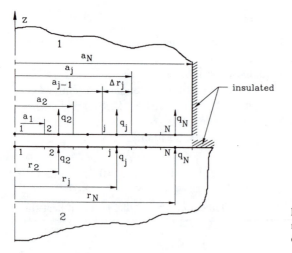

Figure 12.12 Distribution of surface elements for connected semi-infinite cylinder and semi-infinite body.

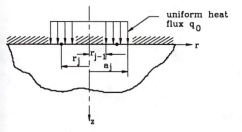

Figure 12.13 (*a*) Semi-infinite body heated at surface over annular-shaped region.

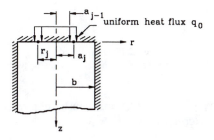

Figure 12.13 (*b*) Semi-infinite insulated cylinder heated with constant annular-shaped heat source.

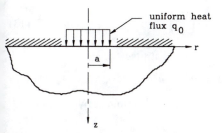

Figure 12.14 (*a*) Semi-infinite body heated over circular area.

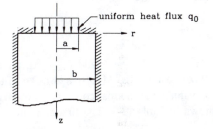

Figure 12.14 (*b*) Semi-infinite insulated cylinder with constant disk heat source.

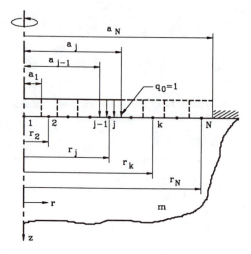

Figure 12.15 Geometry describing the influence functions for semi-infinite body.

resistances for bodies 1 and 2, respectively. The total thermal constriction resistance for the two semi-infinite bodies is then determined by

$$R_c(t) = R_{c1}(t) + R_{c2}(t) = \frac{T_{20} - T_{10}}{Q_c(t)} \qquad (12.112)$$

The average contact area temperature T_c is obtained by summing the products of the elemental temperature and the fraction of the total contact area occupied by the element.

$$T_c(t_M) = \sum_{j=1}^{N} T_{jM} \left(\frac{A_j}{A_c} \right) \qquad (12.113)$$

where T_{jM} is the temperature at the center of element j at time t_M, and

$$A_c = \pi a^2 \qquad A_j = \pi(a_j^2 - a_{j-1}^2) \qquad (12.114a, b)$$

The total heat flow through the contact area, Q_c, is determined by summing up the heat flows over all elements,

$$Q_c(t_M) = \sum_{j=1}^{N} q_{jM} A_j \qquad (12.115)$$

Substituting Eq. (12.115) into Eq. (12.112) yields

$$R_c(t_M) = \frac{T_{20} - T_{10}}{\sum_{j=1}^{N} q_{jM} A_j} \qquad (12.116)$$

With the values of k's = 1, α's = 1, a = 1, T_{20} = 2, and T_{10} = 0, the first problem was solved for elemental heat fluxes using Eqs. (12.109a, b). The fluxes were then introduced into Eq. (12.116) to evaluate the thermal constriction resistance across the contact area. The results are shown in Table 12.1. The first column in this table is the dimensionless time ($t^+ = \alpha t/a^2$) which extends over many decades. The results from the finite-difference solution of Schneider et al. (1977) are provided in the second column which are most accurate at the late times and least accurate at the early times. The third column comes from an exact solution given by Sadhel (1980) which is claimed to be valid for only large times. The next two columns are for the T-based and the q-based single-node USE solutions. The T-based solution is appropriate for late times and the q-based solution is accurate at early times. The results from the multinode USE solution are displayed in the sixth column. The last column is from a one-dimensional approximate solution given by Keltner (1973) which closely matches the multinode USE solution at short times and retains good accuracy for the mid to late times.

For the second problem, the intrinsic thermocouple problem, the USE solutions are compared with other existing solutions (Keltner, 1973; Henning and Parker, 1967; Shewen, 1976), based on the normalized area averaged interface temperature, defined as

$$T_c^+ = \frac{T_c - T_{10}}{T_{20} - T_{10}} \tag{12.117}$$

where T_c is the area averaged interface temperature given by Eq. (12.112), and T_{10} and T_{20} are the initial temperatures of the wire (semi-infinite cylinder) and the substrate (semi-infinite body), respectively. Note that at the initial moment when the substrate undergoes a step change of temperature, there is no spatial variation in the interface temperature. The normalized value of this instantaneous initial interface temperature is given by Keltner and Beck (1981) as

$$T_{c0}^+ = \frac{T_{c0} - T_{10}}{T_{20} - T_{10}} = (1 + \beta')^{-1} \tag{12.118}$$

Table 12.1 Results for dimensionless constriction resistance, for an isothermal disk region on the surface of a semi-infinite body. $R_c^+ = R_c$. a.k.

t^+	SSY (1977)	Sadhal (1980)	Beck and Keltner, 1982			Keltner (1973)
			Eq. (56)	Eq. (22)	MUSE	
0.001	0.0386		0.0202	0.0172	0.0166	0.0162
0.002	0.0409		0.0277	0.0240	0.0230	0.0223
0.005	0.0463		0.0411	0.0368	0.0349	0.0340
0.01	0.0532		0.0544	0.0503	0.0471	0.0473
0.02	0.0637		0.0706	0.0678	0.0625	0.0641
0.05	0.0851	4.8988	0.0959	0.0972	0.0879	0.0914
0.1	0.1074	0.2029	0.1171	0.1226	0.1102	0.1142
0.2	0.1336	0.1685	0.1386	0.1468	0.1336	0.1382
0.5	0.1695	0.1752	0.1658	0.1634	0.1631	0.1685
1	0.1933	0.1879	0.1839	0.1500	0.1824	0.1895
2	0.2120	0.2010	0.1994	0.1071	0.1984	0.2074
10	0.2368	0.2247	0.2245		0.2242	0.2347
100	0.2475	0.2413	0.2413		0.2414	0.2477
1000	0.2495	0.2472	0.2472		0.2472	
10000	0.2499	0.2491	0.2491		0.2491	
∞	0.2500	0.2500	0.2500		0.2500	

where T_{c0} is the instantaneous initial interface temperature and β' is the reciprocal of β defined by Eq. (12.93b); that is,

$$\beta' = \left(\frac{k_1 \rho_1 c_1}{k_2 \rho_2 c_2}\right)^{1/2} \tag{12.119}$$

Using the thermal properties of a chromel substrate ($k = 19.21$ W/m k, $\alpha = 0.492 \times 10^{-5}$ m^2/s) and an alumel wire ($k = 29.76$ W/m k, $\alpha = 0.663 \times 10^{-5}$ m^2/s), the USE results for normalized area averaged interface temperatures compared with other existing values are as shown in Table 12.2. For a chromel and an alumel combination, the normalized instantaneous initial interface temperature, given by Eq. (12.118), is equal to 0.4285. The results presented are actually valid for any combination of the materials with the ratios of $k_2/k_1 = 0.645$ and $\alpha_2/\alpha_1 = 0.742$. The first column in this table is the dimensionless time which extends from $t^+ = 0.001$ to $t^+ = 500$. The dimensionless time is based on the thermal diffusivity of the substrate (body 2),

$$t^+ = \frac{\alpha_2 t}{a^2} \tag{12.120}$$

The second and the third columns give the results of the finite-difference solutions given by Keltner (1973) and Shewen (1967), respectively. The fourth column is evaluated from the analytical solution given by Henning and Parker (1967) which is only good for late times ($t^+ > 20$). The early and late times results of the T-based and the q-based single-node USE solutions are displayed in the next four columns. The last column represents the multinode USE solution.

As can be seen from Table 12.2, there is a very good agreement between the finite-difference solutions and the USE solutions for the time range covered. However, note that both finite-difference solutions have difficulty regarding the computational effort and cost, particularly for the early times, $t^+ < 0.01$, and the late times, $t^+ > 10$. The T-based and the q-based single-node USE solutions are convenient in that the mathematics is not difficult and the expressions are simple to evaluate. Each solution provides two expressions; one for early times and the other for late times. The q-based solution is more appropriate for the early times. It approaches the exact solution (0.4285) as t^+ goes to zero, and closely matches the multinode USE solution up to dimensionless time $t^+ = 0.1$. It also provides relatively good results for the late times, $t^+ > 10$. The T-based solution does not approach the exact solution as t^+ goes to zero, and consequently, is less accurate than the q-based solution for early times. Because of the uniform interface temperature assumption, however, it yields very good results for the times of $t^+ > 0.1$. Even though neither the T-based solution nor the q-based solution is solely suitable for the complete time domain, a combination of the early-time q-based solution and the late-time T-based solution provides very good results over the entire time domain. These two solutions match very closely at the dimensionless time $t^+ = 0.1$.

A study of the Tables 12.1 and 12.2 shows that for each of the above problems, the USE method performed very well. The single-node solutions represent relatively accurate results for certain ranges of time. The advantage of this approach is in its

Table 12.2 Normalized area averaged interface temperature histories for chromel semi-infinite body and alumel semi-infinite cylinder

	FD Solutions		Henning and	T-based	(1981)	q-based		
				early	late	early	late	
t^+	1973	1976	Parker, 1967	time	time	time	time	MUSE
.001			.6084	.4489		.4342		.4335
.002		.4421	.6118	.4500		.4366		.4364
.005		.4480	.6185	.4521		.4413		.4422
.01	.4402	.4510	.6257	.4545		.4467		.4488
.02	.4505	.4599	.6356			.4546		.4581
.05	.4700	.4782	.6540			.4709		.4765
.1	.4916	.4991	.6731		.4907	.4904		.4973
.2	.5215	.5283	.6972		.5280			.5263
.5	.5770	.5826	.7373		.5910			.5805
1	.6302	.6338	.7729		.6452			.6328
2	.6896	.6921	.8109		.7042			.6915
5	.7688	.7714	.8602		.7810			.7700
10	.8202	.8246	.8933		.8327		.8091	.8236
20		.8694	.9207		.8757		.8614	.8687
50		.9139	.9482		.9186		.9108	.9137
100		.9382	.9689		.9417		.9365	.9381
200			.9786		.9585		.9550	.9559
500			.9832		.9737		.9715	.9719

Note: $k_{ch} = 19.21$ and $k_{al} = 29.76$ w/m-K, $\alpha_{ch} = .492 \times 10^{-5}$ and $\alpha_{al} = .663 \times 10^{-5}$ m^2/s.

simplicity. The multinode solution is superior to other analytical and numerical solutions in terms of accuracy and ability to treat the complete time range. Also, there is no restriction regarding the choice of the time step in the multinode approach. For instance, in the above problems, the elemental surface heat fluxes are determined for various values of t^+, in 20 time steps. This means that for larger times, larger time steps are considered. [To evaluate, $q_j(t^+)$'s, at times of $t^+ = 0.01$, 1, and 1000, the time steps of $t^+ = 0.0005$, 0.05, and 50, were used, respectively.] This substantially reduces the computational work compared with the case where a small constant time step is used for the entire time range.

REFERENCES

Bannerjee, P. K., and Butterfield, R. (eds.), 1979, *Developments in Boundary Element Methods-1*, Applied Science Publishers, London.

Bannerjee, P. K., and Mukherjee, S. (eds.), 1984, *Developments in Boundary Element Methods-3*, Applied Science Publishers, London.

Bannerjee, P. K., and Shaw, R. T. (eds.), 1982, *Developments in Boundary Element Methods-2*, Applied Science Publishers, London.

Bannerjee, P. K., and Watson, J. O. (eds.), 1985, *Developments in Boundary Element Methods-4*, Applied Science Publishers, London.

Beck, J. V., 1968, Surface Heat Flux Determination Using an Integral Method, *Nuclear Eng. Design*, vol. 7, pp. 170–178.

Beck, J. V., 1980, Average Transient Temperature Within a Body Heated by a Disk Heat Source, *Heat Transfer Thermal Control and Heat Pipes*, ed. W. B. Olstad, vol. 70, Progress in Astronautics and Aeronautics, pp. 3–24.

Beck, J. V., 1981a, Large Time Solutions for Temperatures in a Semi-infinite Body with a Disk Heat Source, *Int. J. Heat Mass Transfer*, vol. 24, pp. 155–164.

Beck, J. V., 1981b, Transient Temperatures in a Semi-Infinite Cylinder Heated by a Disk Heat Source, *Int. J. Heat Mass Transfer*, vol. 24, pp. 1631–1640.

Beck, J. V., and Keltner, N. R., 1987, Green's Function Partitioning Method Applied to Foil Heat Flux Gage, *J. Heat Transfer*, vol. 109, pp. 274–280.

Beck, J. V., Blackwell, B., and St. Clair, Jr., C. R., 1985a, *Inverse Heat Conduction*, Wiley, New York.

Beck, J. V., Keltner, N. R., and Schisler, I. P., 1985b, Influence Functions for the Unsteady Surface Element Method, *AIAA J.*, vol. 23, pp. 1978–1982.

Carslaw, H. S., and Jaeger, J. C., 1959, *Conduction of Heat in Solids*, 2d ed., Oxford University Press, London.

Chang, Y. P., Kang, C. S., and Chen, D. J., 1973, The Use of Fundamental Green's Functions for Solution of Problems of Heat Conduction in Anisotropic Media, *J. Heat Mass Transfer*, vol. 16, pp. 1905–1918.

Cole, K. D., and Beck, J. V., 1987, Conjugated Heat Transfer From a Strip Heater with the Unsteady Surface Element Method, *J. Thermophys. Heat Transfer*, vol. 1, pp. 348–354.

Cole, K. D., and Beck, J. V., 1988, Conjugated Heat Transfer from a Hot-film Probe for Transient Air Flow, *J. Heat Transfer*, vol. 110, pp. 290–296.

Eckert, E. R., and Drake, Jr., R. M., 1972, *Analysis of Heat and Mass Transfer*, McGraw-Hill, New York.

Henning, C. D., and Parker, R., 1967, Transient Response of an Intrinsic Thermocouple, *J. Heat Transfer, Trans. ASME, Ser. C*, vol. 39, p. 146.

Kays, W. M., and Crawford, M. E., 1980, *Convection Heat and Mass Transfer*, 2d ed., McGraw-Hill, New York.

Keltner, N. R., 1973, Heat Transfer in Intrinsic Thermocouples Applications to Transient Measurement Errors, Report SC-RR-72-0719, Sandia National Laboratory, Albuquerque, N. Mex.

Keltner, N. R., and Beck, J. V., 1981, Unsteady Surface Element Method, *J. Heat Transfer*, vol. 103, pp. 759–764.

Keltner, N. R., Bainbridge, B. L., and Beck, J. V., 1988, Rectangular Heat Source on a Semi-infinite Solid-an Analysis for Thin-film Heat Flux Gage Calibration, *J. Heat Transfer*, vol. 110, pp. 42–48.

Khader, M. S., 1980, Heat Conduction with Temperature Dependent Thermal Conductivity, presented in the 19th National Heat Transfer Conference, ASME paper No. 80-Ht-4.

Khader, M. S., and Hanna, M. C., 1981, An Iterative Boundary Integral Numerical Solution for General Steady State Heat Conduction Problems, *J. Heat Transfer*, vol. 103, pp. 26–31.

Litkouhi, B. L., and Beck, J. V., 1982, Temperatures in Semi-Infinite Body Heat by Constant Heat Flux Over Half Space, *Heat Transfer 1982*, ed. U. Grigull et al., pp. 21–27, Hemisphere, Washington, D.C.

Litkouhi, B. L., and Beck, J. V., 1985, Intrinsic Thermocouple Analysis Using Multinode Unsteady Surface Element Method, 1985, *AIAA J.*, vol. 23, no. 10, pp. 1609–1614.

Litkouhi, B. L., and Beck, J. V., 1986, Multinode Unsteady Surface Element Method With Application Contact Conductance Problem, *J. Heat Transfer*, vol. 108, pp. 257–263.

Luikov, A. V., 1968, *Analytical Heat Diffusion Theory*, Academic Press, New York.

Myers, G. E., 1987, *Analytical Methods in Conduction Heat Transfer*, Genium Publishing.

Ozisik, M. N., 1980, *Heat Conduction*, Wiley, New York.

Rizzo, F. J., and Shippy, D. J., 1970, A Method of Solution for Certain Problems of Transient Heat Conduction, *AIAA J.*, vol. 8, no. 11, pp. 2004–2009.

Sadhal, S. S., 1980, Transient Thermal Response of Two Solids in Contact Over a Circular Disk, *Int. J. Heat Mass Transfer*, vol. 23, pp. 731–733.

Schneider, G. E., 1979, Thermal Constriction Resistance Due to Arbitrary Contacts on a Half-Space-Numerical Solution of the Dirichlet Problem, in *Progress in Astronautics and Aeronautics: Thermophysics and Thermal Control*, vol. 65, ed. R. Viskanta, New York, pp. 103–119.

Schneider, G. E., and LeDain, B. L., 1979, The Boundary Integral Equation Method Applied to Steady

Heat Conduction with Special Attention Given to the Corner Problem, AIAA paper no. 79-0176, presented at the 17th Aerospace Sciences Meeting, New Orleans, La.

Schneider, G. E., Strong, A. B., and Yovanivich, M. M., 1977, Transient Thermal Response of Two Bodies Communicating Through a Small Circular Contact Area, *Int. J. Heat Mass Transfer*, vol. 23, pp. 731–733.

Shaw, R. P., 1974, An Integral Equation Approach to Diffusion, *Int. J. Heat Mass Transfer*, vol. 17, pp. 693–699.

Shewen, E. C., 1967, A Transient Numerical Analysis of Conduction between Contacting Circular Cylinders and Halfspaces applied to a Biosensor, M.S. thesis, University of Waterloo, Canada.

Wrobel, L. C., and Brebbia, C. A., 1981, A Formulation of the Boundary Element Method for Axisymmetric Transient Heat Conduction, *Int. J. Heat Mass Transfer*, vol. 24, no. 5, pp. 843–850.

Yovanovich, M. M., and Martin, K. A., 1980, Some Basic Three-Dimensional Influence Coefficients for the Surface Element Method, AIAA paper no. 80-1491, AIAA 15th Thermophysics Conference, Snowmass, Colo.

PROBLEMS

12.1 A semi-infinite body is initially at zero temperature. For times $t > 0$, the surface at $x = 0$ is subjected to a temperature which varies linearly with time as $T(0, x) = at$. Using the Duhamel's theorem, find the transient temperature distribution in the body.

12.2 Consider a semi-infinite body initially at zero temperature subjected to a uniform surface temperature over an infinite strip of width $2a$ with the rest of the surface being insulated. See Fig. 12.10a.

(a) By considering only two elements along the active part of the surface, give the appropriate USE equations for three time steps ($M = 3$).

(b) What is the required influence function for this problem?

12.3 Show that for Problem 12.2 the integration of the corresponding GF over the surface element and the time step results in the corresponding flux-based influence function.

12.4 Solve Problem 12.2 by utilizing the GF USE formulation.

12.5 Starting with Eq. (12.46a), derive the appropriate GF form of the flux-based USE Eq. (12.64) for two bodies in perfect contact over a portion of their boundaries with the rest of the boundaries being insulated.

12.6 Consider two semi-infinite bodies initially at different temperatures, T_{i1} and T_{i2}, brought together in perfect contact over an infinite strip of width $2a$ with the rest of the boundaries being insulated. Using the single-node USE method, obtain an approximate analytical solution for the interface heat flux.

12.7 A thermal property probe can be made by placing a small, thin rectangular electrical resistance heater on a large flat body, named body 1, for which k_1 and α_1 are known. If body 1 and the heater are put into good thermal contact with some body 2 with unknown thermal properties, a transient experiment may be carried out to find k_2 and α_2. The matching conditions between body 1 and body 2 are $\overline{T}_1(t) = \overline{T}_2(t)$ and

$$\overline{q}_1(t) + \overline{q}_2(t) = \begin{cases} q_0 & \text{for } t > 0 \text{ on heater} \\ 0 & \text{otherwise} \end{cases}$$

where the overbar ($^-$) denotes spatial average over the rectangular heater and where q_1 and q_2 are the surface heat flux into bodies 1 and 2, respectively. Initially body 1 and body 2 are at zero temperature.

(a) Formulate the flux-based one-node USE method for two bodies with a heater between them. The heater has negligible mass and negligible temperature gradients perpendicular to the interface.

(b) Using names $\overline{\phi}_1(t)$ and $\overline{\phi}_2(t)$ for the dimensionless spatial average influence functions on the heater, solve the one-node USE equation in the Laplace transform domain.

(c) If $\overline{\phi}_1 = 2(t^+/\pi)^{1/2} - (t^+/\pi)[1 + 1/(b/a) - 2/3(t^+/\pi)^{1/2}/b]$ where $t^+ = \alpha_1 t/a^2$ and b/a is the length/width of the rectangular heater, find $\overline{T}(t)$, the spatial average temperature on the heater at early time (an approximate inverse transform is required). If $\overline{T}(t)$ and q_0 are measured, is it possible to deduce k_2 and α_2?

NOTE 1: DERIVATION OF EQUATIONS (12.65a) AND (12.65b)

Equation (12.65a) can readily be obtained by considering that for the first time step ($M = 1$), the vectors $\bar{E}_M$ and $\bar{F}_M$ [given by Eqs. (12.62a and b), respectively] are zero.

$$\bar{E}_M = \bar{F}_M = 0 \tag{1}$$

Substituting Eq. (1) into Eq. (12.61b) and then its results into Eq. (12.64) yields

$$\bar{q}_1 = \bar{\bar{C}}^{-1}\bar{T}_0 \tag{2}$$

which is the same as Eq. (12.65a).

To show how Eq. (12.65b) is derived, Eq. (12.63) is expanded for different values of M. By introducing Eqs. (12.61b) and (12.62a, b) into Eq. (12.63), for $M = 1, 2, 3, \ldots$, one can write

for $M = 1$, $\qquad\qquad\qquad\qquad \bar{\bar{C}}\bar{q}_1 = \bar{T}_0 \tag{3.1}$

for $M = 2$, $\qquad\qquad\qquad\qquad \bar{\bar{C}}\bar{q}_2 = \bar{T}_0 + \bar{E}_2 - \bar{F}_2 \tag{3.2}$

for $M = 3$, $\qquad\qquad\qquad\qquad \bar{\bar{C}}\bar{q}_3 = \bar{T}_0 + \bar{E}_3 - \bar{F}_3 \tag{3.3}$

for $M - 1$, $\qquad\qquad\qquad \bar{\bar{C}}\bar{q}_{M-1} = \bar{T}_0 + \bar{E}_{M-1} - \bar{F}_{M-1} \tag{3.M-1}$

for M, $\qquad\qquad\qquad\qquad \bar{\bar{C}}\bar{q}_M = \bar{T}_0 + \bar{E}_M - \bar{F}_M \tag{3.M}$

By adding all M equations together, (3.1) through (3.M), and noticing that

$$\bar{E}_M = \bar{F}_{M-1} + \bar{\bar{\Phi}}_1 \bar{q}_{M-1} \tag{4}$$

it can be shown that

$$\bar{\bar{C}}\left\{ \sum_{i=1}^{M} \bar{q}_i \right\} = M\bar{T}_0 + \bar{\bar{\Phi}}_1 \sum_{i=1}^{M-1} \bar{q}_i - \bar{F}_M \tag{5}$$

or

$$\bar{q}_M + \sum_{i=1}^{M-1} \bar{q}_i = M\bar{\bar{C}}^{-1}\bar{T}_0 + \bar{\bar{C}}^{-1}\bar{\bar{\Phi}}_1 \sum_{i=1}^{M-1} \bar{q}_i - \bar{\bar{C}}^{-1}\bar{F}_M \tag{6}$$

Substituting for $\bar{\bar{C}}$, $\bar{F}_M$ and $\bar{\bar{C}}^{-1}\bar{T}_0$ from Eqs. (12.61a), (12.62b), and (2), respectively, yields

$$\bar{q}_M = M\bar{q}_1 + \bar{\bar{B}}\left\{ \sum_{i=1}^{M-1} \bar{q}_i \right\} - \bar{\bar{C}}^{-1}\bar{F}_M \tag{7}$$

where the matrix $\bar{\bar{B}}$ is defined as

$$\bar{\bar{B}} = \bar{\bar{H}}^{-1}\bar{\bar{\Phi}}_1 \tag{8}$$

Equation (7) is the same as Eq. (12.3.32b) and is valid for $M \geq 2$. Notice that the vector $\bar{F}_M$ is a function of time and should be calculated at each time step.

BESSEL FUNCTIONS

The differential equation

$$\frac{d^2R}{dz^2} + \frac{1}{z}\frac{dR}{dz} + \left(1 - \frac{\nu^2}{z^2}\right)R = 0 \tag{B.1}$$

is called the Bessel equation of order ν. Two linearly independent solutions of this equation for all values of ν are $J_\nu(z)$, the Bessel function of the first kind of order ν and $Y_\nu(z)$, the Bessel function of the second kind of order ν. Thus, the general solution of Eq. (B.1) is written as (Watson, 1966; McLachlan, 1961; Hildebrand, 1949)

$$R(z) = c_1 J_\nu(z) + c_2 Y_\nu(z) \tag{B.2}$$

The Bessel function $J_\nu(z)$ in series form is defined as

$$J_\nu(z) = \left(\frac{1}{2}z\right)^\nu \sum_{k=0}^{\infty} (-1)^k \frac{[(1/2)z]^{2k}}{k!\,\Gamma(\nu + k + 1)} \tag{B.3}$$

and

$$Y_\nu(z) = \frac{J_\nu(z)\cos(\nu\pi) - J_{-\nu}(z)}{\sin(\nu\pi)}$$

where $\Gamma(x)$ is the gamma function. The differential equation

$$\frac{d^2R}{dz^2} + \frac{1}{z}\frac{dR}{dz} - \left(1 + \frac{\nu^2}{z^2}\right)R = 0 \tag{B.4}$$

is called the modified Bessel equation of order ν. Two linearly independent solutions of this equation for all values of ν are $I_\nu(z)$ (the modified Bessel function of the first kind of order ν) and $K_\nu(z)$ (the modified Bessel function of the second kind of order ν). Thus, the general solution of Eq. (B.4) is written as

$$R(z) = c_1 I_\nu(z) + c_2 K_\nu(z) \tag{B.5}$$

$I_\nu(z)$ and $K_\nu(z)$ are real and positive when $\nu > -1$ and $z > 0$. The Bessel function

$I_\nu(z)$ in series form is given by

$$I_\nu(z) = \left(\frac{1}{2} z\right)^\nu \sum_{k=0}^{\infty} \frac{[(1/2)z]^{2k}}{k!\,\Gamma(\nu + k + 1)} \tag{B.6}$$

When ν is neither zero nor a positive integer, the general solutions (B.2) and (B.5) can be taken, respectively, in the form

$$R(z) = c_1 J_\nu(z) + c_2 J_{-\nu}(z) \tag{B.7a}$$

$$R(z) = c_1 I_\nu(z) + c_2 I_{-\nu}(z) \tag{B.7b}$$

When $\nu = n$ is a positive integer, the solutions $J_n(z)$ and $J_{-n}(z)$ are not independent (see Tables B.1–B.5); they are related by

$$J_n(z) = (-1)^n J_{-n}(z) \quad \text{and} \quad J_{-n}(z) = J_n(-z) \tag{B.8}$$

(n = integer). Similarly, when $\nu = n$ is a positive integer, the solutions $I_n(z)$ and $I_{-n}(z)$ are not independent.

We summarize various forms of solutions of Eq. (B.1) as

$$R(z) = c_1 J_\nu(z) + c_2 Y_\nu(z) \quad \text{always} \tag{B.9a}$$

$$R(z) = c_1 J_\nu(z) + c_2 J_{-\nu}(z) \quad \nu \text{ is not zero or a positive integer} \tag{B.9b}$$

and the solutions of Eq. (B.4) as

$$R(z) = c_1 I_\nu(z) + c_2 K_\nu(z) \quad \text{always} \tag{B.10a}$$

$$R(z) = c_1 I_\nu(z) + c_2 I_{-\nu}(z) \quad \nu \text{ is not zero or a positive integer} \tag{B.10b}$$

B.1 GENERALIZED BESSEL EQUATION

Sometimes a given differential equation, after suitable transformation of the independent variable, yields a solution that is a linear combination of Bessel functions. A convenient way of finding out whether a given differential equation possesses a solution

Table B.1 First ten roots of $J_n(z) = 0$; $n = 0, 1, 2, 3, 4, 5$

n	J_0 (R01 case)	J_1 (R02 case)	J_2	J_3	J_4	J_5
1	2.4048	3.8317	5.1356	6.3802	7.5883	8.7715
2	5.5201	7.0156	8.4172	9.7610	11.0647	12.3386
3	8.6537	10.1735	11.6198	13.0152	14.3725	15.7002
4	11.7915	13.3237	14.7960	16.2235	17.6160	18.9801
5	14.9309	16.4706	17.9598	19.4094	20.8269	22.2178
6	18.0711	19.6159	21.1170	22.5827	24.0190	25.4303
7	21.2116	22.7601	24.2701	25.7482	27.1991	28.6266
8	24.3525	25.9037	27.4206	28.9084	30.3710	31.8117
9	27.4935	29.0468	30.5692	32.0649	33.5371	34.9888
10	30.6346	32.1897	33.7165	35.2187	36.6990	38.1599

Table B.2 **First six roots of** $\beta J_1(\beta) - cJ_0(\beta) = 0$
(case $R03$**, where** $c = h\,b/k)^\dagger$

c	β_1	β_2	β_3	β_4	β_5	β_6
0.00	0.0000	3.8317	7.0156	10.1735	13.3237	16.4706
0.01	0.1412	3.8343	7.0170	10.1745	13.3244	16.4712
0.02	0.1995	3.8369	7.0184	10.1754	13.3252	16,4718
0.04	0.2814	3.8421	7.0213	10.1774	13.3267	16.4731
0.06	0.3438	3.8473	7.0241	10.1794	13.3282	16.4743
0.08	0.3960	3.8525	7.0270	10.1813	13.3297	16.4755
0.10	0.4417	3.8577	7.0298	10.1833	13.3312	16.4767
0.15	0.5376	3.8706	7.0369	10.1882	13.3349	16.4797
0.20	0.6170	3.8835	7.0440	10.1931	13.3387	16.4828
0.30	0.7465	3.9091	7.0582	10.2029	13.3462	16.4888
0.40	0.8516	3.9344	7.0723	10.2127	13.3537	16.4949
0.50	0.9408	3.9594	7.0864	10.2225	13.3611	16.5010
0.60	1.0184	3.9841	7.1004	10.2322	13.3686	16.5070
0.70	1.0873	4.0085	7.1143	10.2419	13.3761	16.5131
0.80	1.1490	4.0325	7.1282	10.2516	13.3835	16.5191
0.90	1.2048	4.0562	7.1421	10.2613	13.3910	16.5251
1.00	1.2558	4.0795	7.1558	10.2710	13.3984	16.5312
1.50	1.4569	4.1902	7.2233	10.3188	13.4353	16.5612
2.00	1.5994	4.2910	7.2884	10.3658	13.4719	16.5910
3.00	1.7887	4.4634	7.4103	10.4566	13.5434	16.6499
4.00	1.9081	4.6018	7.5201	10.5423	13.6125	16.7073
5.00	1.9898	4.7131	7.6177	10.6223	13.6786	16.7630
6.00	2.0490	4.8033	7.7039	10.6964	13.7414	16.8168
7.00	2.0937	4.8772	7.7797	10.7646	13.8008	16.8684
8.00	2.1286	4.9384	7.8464	10.8271	13.8566	16.9179
9.00	2.1566	4.9897	7.9051	10.8842	13.9090	16.9650
10.00	2.1795	5.0332	7.9569	10.9363	13.9580	17.0099
15.00	2.2509	5.1773	8.1422	11.1367	14.1576	17.2008
20.00	2.2880	5.2568	8.2534	11.2677	14.2983	17.3442
30.00	2.3261	5.3410	8.3771	11.4221	14.4748	17.5348
40.00	2.3455	5.3846	8.4432	11.5081	14.5774	17.6508
50.00	2.3572	5.4112	8.4840	11.5621	14.6433	17.7272
60.00	2.3651	5.4291	8.5116	11.5990	14.6889	17.7807
80.00	2.3750	5.4516	8.5466	11.6461	14.7475	17.8502
100.00	2.3809	5.4652	8.5678	11.6747	14.7834	17.8931
∞	2.4048	5.5201	8.6537	11.7915	14.9309	18.0711

†From Carslaw and Jaeger, 1959.

in terms of Bessel functions is to compare it with the generalized Bessel equation (Sherwood and Reed, 1939, p. 65)

$$\frac{d^2R}{dx^2} + \left(\frac{1 - 2m}{x} - 2\alpha\right)\frac{dR}{dx}$$

$$+ \left[p^2 a^2 x^{2p-2} + \alpha^2 + \frac{\alpha(2m - 1)}{x} + \frac{m^2 - p^2 v^2}{x^2}\right]R = 0 \quad \text{(B.11a)}$$

Table B.3 First five roots of $J_0(\beta)Y_0(\lambda\beta) - Y_0(\beta)J_0(\lambda\beta)$
(case $R11$, where $\lambda = b/a$, $\lambda > 1$)

λ^{-1}	1	2	3	4	5
0.80	12.55847 031	25.12877	37.69646	50.26349	62.83026
0.60	4.69706 410	9.41690	14.13189	18.84558	23.55876
0.40	2.07322 886	4.17730	6.27537	8.37167	10.46723
0.20	0.76319 127	1.55710	2.34641	3.13403	3.92084
0.10	0.33139 387	0.68576	1.03774	1.38864	1.73896
0.08	0.25732 649	0.53485	0.81055	1.08536	1.35969
0.06	0.18699 458	0.39079	0.59334	0.79522	0.99673
0.04	0.12038 637	0.25340	0.38570	0.51759	0.64923
0.02	0.05768 450	0.12272	0.18751	0.25214	0.31666
0.00	0.00000 000	0.00000	0.00000	0.00000	0.00000

Table B.4 First five roots of $J_1(\beta)Y_0(\lambda\beta) - Y_1(\beta)J_0(\lambda\beta)$
(cases $R12$ or $R21$, where $\lambda = b/a$, $\lambda > 1$)

λ^{-1}	1	2	3	4	5
0.80	6.56973 310	18.94971	31.47626	44.02544	56.58224
0.60	2.60328 138	7.16213	11.83783	16.53413	21.23751
0.40	1.24266 626	3.22655	5.28885	7.36856	9.45462
0.20	0.51472 663	1.24657	2.00959	2.78326	3.56157
0.10	0.24481 004	0.57258	0.90956	1.25099	1.59489
0.08	0.19461 772	0.45251	0.71635	0.98327	1.25203
0.06	0.14523 798	0.33597	0.53005	0.72594	0.92301
0.04	0.09647 602	0.22226	0.34957	0.47768	0.60634
0.02	0.04813 209	0.11059	0.17353	0.23666	0.29991
0.00	0.00000 000	0.00000	0.00000	0.00000	0.00000

Table B.5 First five roots of $J_1(\beta)Y_1(\lambda\beta) - Y_1(\beta)J_1(\lambda\beta)$
(case $R22$, where $\lambda = b/a$, $\lambda > 1$)

λ^{-1}	1	2	3	4	5
0.80	12.59004 151	25.14465	37.70706	50.27145	62.83662
0.60	4.75805 426	9.44837	14.15300	18.86146	23.57148
0.40	2.15647 249	4.22309	6.30658	8.39528	10.48619
0.20	0.84714 961	1.61108	2.38532	3.16421	3.94541
0.10	0.39409 416	0.73306	1.07483	1.41886	1.76433
0.08	0.31223 576	0.57816	0.84552	1.11441	1.38440
0.06	0.23235 256	0.42843	0.62483	0.82207	1.02001
0.04	0.15400 729	0.28296	0.41157	0.54044	0.66961
0.02	0.07672 788	0.14062	0.20409	0.26752	0.33097
0.00	0.00000 000	0.00000	0.00000	0.00000	0.00000

and the corresponding solution of which is

$$R = x^m e^{\alpha x}[c_1 J_\nu(ax^p) + c_2 Y_\nu(ax^p)] \tag{B.11b}$$

where c_1 and c_2 are arbitrary constants. For example, by comparing the differential equation

$$\frac{d^2R}{dx^2} + \frac{1}{x}\frac{dR}{dx} - \frac{\beta}{x}R = 0 \tag{B.12}$$

with the above generalized Bessel equation, we find

$$\alpha = 0 \qquad m = 0 \qquad p = \frac{1}{2} \qquad a = 2i\sqrt{\beta} \qquad \nu = 0$$

Hence, the solution of differential equation (B.12) is in the form

$$R = c_1 J_0(2i\sqrt{\beta x}) + c_2 Y_0(2i\sqrt{\beta x}) \tag{B.13a}$$

or

$$R = c_1 I_0(2\sqrt{\beta x}) + c_2 K_0(2\sqrt{\beta x}) \tag{B.13b}$$

which involves Bessel functions.

B.2 LIMITING FORM FOR SMALL z

For small values of $z(z \to 0)$, the retention of the leading terms in the series results in the following approximations for the values of Bessel functions (Abramowitz and Stegun, 1964, p. 360)

$$J_\nu(z) \approx \left(\frac{1}{2}z\right)^\nu \frac{1}{\Gamma(\nu + 1)} \qquad \nu \neq -1, -2, -3, \dots \tag{B.14a}$$

$$Y_\nu(z) \approx -\frac{1}{\pi}\left(\frac{2}{z}\right)^\nu \Gamma(\nu) \qquad \nu \neq 0 \qquad \text{and} \qquad Y_0(z) = \frac{2}{\pi}\ln z \tag{B.14b}$$

$$I_\nu(z) \approx \left(\frac{1}{2}z\right)^\nu \frac{1}{\Gamma(\nu + 1)} \qquad \nu \neq -1, -2, -3, \dots \tag{B.15a}$$

$$K_\nu(z) \approx \frac{1}{2}\left(\frac{2}{z}\right)^\nu \Gamma(\nu) \qquad \nu \neq 0 \qquad K_0(z) \approx -\ln z \tag{B.15b}$$

B.3 LIMITING FORM FOR LARGE z

For large values of $z(z \to \infty)$, the values of Bessel functions can be approximated as (Abramowitz and Stegun, 1964, pp. 364 and 377)

$$J_\nu(z) \approx \left(\frac{2}{\pi z}\right)^{1/2}\cos\left(z - \frac{\pi}{4} - \frac{\nu\pi}{2}\right) \tag{B.16a}$$

$$Y_\nu(z) \approx \left(\frac{2}{\pi z}\right)^{1/2} \sin\left(z - \frac{\pi}{4} - \frac{\nu\pi}{4}\right) \tag{B.16b}$$

$$I_\nu(z) \approx \frac{e^z}{\sqrt{2\pi z}} \quad \text{and} \quad K_\nu(z) \approx \left(\frac{2}{\pi z}\right)^{1/2} e^{-z} \tag{B.16c}$$

B.4 DERIVATIVES OF BESSEL FUNCTIONS [Hildebrand, 1949, pp. 161–163]

$$\frac{d}{dz}[z^\nu W_\nu(\beta z)] = \begin{cases} \beta z^\nu W_{\nu-1}(\beta z) & \text{for } W = J, Y, I \quad \text{(B.17a)} \\ -\beta z^\nu W_{\nu-1}(\beta z) & \text{for } W = K \quad \text{(B.17b)} \end{cases}$$

$$\frac{d}{dz}[z^{-\nu} W_\nu(\beta z)] = \begin{cases} -\beta z^{-\nu} W_{\nu+1}(\beta z) & \text{for } W = J, Y, K \quad \text{(B.18a)} \\ \beta z^{-\nu} W_{\nu+1}(\beta z) & \text{for } W = I \quad \text{(B.18b)} \end{cases}$$

For example, by setting $\nu = 0$, we obtain

$$\frac{d}{dz}[W_0(\beta z)] = \begin{cases} -\beta W_1(\beta z) & \text{for } W = J, Y, K \quad \text{(B.19a)} \\ \beta W_1(\beta z) & \text{for } W = I \quad \text{(B.19b)} \end{cases}$$

B.5 RECURRENCE RELATIONS

The recurrence formulas for the Bessel functions are given as [Watson, 1966, pp. 45 and 66; Abramowitz and Stegun, 1964, p. 361]

$$W_{\nu-1}(z) + W_{\nu+1}(z) = \frac{2\nu}{z} W_\nu(z) \tag{B.20a}$$

$$W_{\nu-1}(z) - W_{\nu+1}(z) = 2W_\nu'(z) \tag{B.20b}$$

$$W_{\nu-1}(z) - \frac{\nu}{z} W_\nu(z) = W_\nu(z) \tag{B.20c}$$

$$-W_{\nu+1}(z) + \frac{\nu}{z} W_\nu(z) = W_\nu'(z) \tag{B.20d}$$

where $W = J$ or Y or any linear combination of these functions the coefficients in which are independent of z and ν.

B.6 INTEGRALS OF BESSEL FUNCTIONS

$$\int xJ_0(x)\, dx = xJ_1(x) \tag{B.21}$$

$$\int J_1(x)\, dx = -J_0(x) \tag{B.22}$$

$$\int_0^z J_0(x)\, dx = z \sum_{k=0}^{\infty} J_{2k+1}(z) \tag{B.23}$$

Note: Numerical values for $\int J_0(x)$ and $\int Y_0(x)$ are tabulated in Abramowitz and Stegun (1964, pp. 491–493).

$$\int xY_0(x)\, dx = xY_1(x) \tag{B.24}$$

$$\int Y_1(x)\, dx = -Y_0(x) \tag{B.25}$$

$$(\beta^2 - \alpha^2) \int_0^1 xJ_k(\alpha x)J_k(\beta x)\, dx = \frac{\alpha\beta}{2k} [J_{k-1}(\alpha)J_{k+1}(\beta) - J_{k+1}(\alpha)J_{k-1}(\beta)] \tag{B.26}$$

$$\int xJ_k^2(\alpha x)\, dx = \frac{x^2}{2} [J_k^2(\alpha x) - J_{k-1}(\alpha x)J_{k+1}(\alpha x)] \tag{B.27}$$

In the following formulas, $C_k(x)$ and $\overline{C}_k(x)$ denote two general Bessel functions, (i.e., linear combinations):

$$C_k(x) = aJ_k(x) + bY_k(x) \qquad \overline{C}_k(x) = \overline{a}J_k(x) + \overline{b}Y_k(x)$$

with arbitrary constants, $a, b, \overline{a}, \overline{b}$.

$$\int x^{k+1}C_k(x)\, dx = x^{k+1}C_{k+1}(x) \tag{B.28}$$

$$\int x^{1-k}C_k(x)\, dx = -x^{1-k}C_{k-1}(x) \tag{B.29}$$

$$\int xC_k(hx)\overline{C}_k(gx)\, dx = (h^2 - g^2)^{-1}x[hC_{k+1}(hx)\overline{C}_k(gx) - gC_k(hx)\overline{C}_{k+1}(gx)] \tag{B.30}$$

$$\int xC_k(hx)\overline{C}_k(hx)\, dx = -\frac{1}{4}x^2[C_{k-1}(hx)\overline{C}_{k+1}(hx) - 2C_k(hx)\overline{C}_k(hx) + C_{k+1}(hx)\overline{C}_{k-1}(hx)] \tag{B.31}$$

$$\int x^{-1}C_m(hx)\overline{C}_k(hx)\, dx = (m^2 - k^2)^{-1}[(m - k)C_m(hx)\overline{C}_{k+1}(hx) - hxC_{m+1}(hx)\overline{C}_k(hx) + hxC_m(hx)\overline{C}_{k+1}(hx)] \tag{B.32}$$

REFERENCES

Abramowitz, M., and Stegun, I. A., 1964, *Handbook of Mathematical Functions*, National Bureau of Standards, Applied Mathematic Series 55, U.S. Government Printing Office, Washington, D.C.

Carslaw, H. S., and Jaeger, J. C., 1959, *Conduction of Heat in Solids*, 2d ed., Oxford University Press, New York.

Hildebrand, F. B., 1949, *Advanced Calculus for Engineers*, Prentice-Hall, Englewood Cliffs, N.J.

McLachlan, N. W., 1961, *Bessel Functions for Engineers*, 2d ed., Oxford-Clarendon Press, London.

Sherwood, T. K., and Reed, C. E., 1939, *Applied Mathematics in Chemical Engineering*, McGraw-Hill, New York.

Watson, G. N., 1966, *A Treatise on the Theory of Bessel Functions*, 2d ed., Cambridge University Press, London.

ERROR FUNCTION AND RELATED FUNCTIONS

E.1 DEFINITION

The error function is denoted erf (x) and is defined by

$$\text{erf } (x) = \frac{2}{\pi^{1/2}} \int_0^x e^{-t^2} \, dt \qquad (E.1)$$

and the complementary error function, erfc (x), is defined by

$$\text{erfc } (x) = \frac{2}{\pi^{1/2}} \int_x^\infty e^{-t^2} \, dt \qquad (E.2)$$

It can be shown that

$$\text{erf } (\infty) = 1 \qquad (E.3a)$$

$$\text{erf } (-x) = - \text{ erf } (x) \qquad (E.3b)$$

$$\text{erf } (x) + \text{erfc } (x) = 1 \qquad (E.3c)$$

$$\text{erfc } (-x) = 1 + \text{erf } (x) = 2 - \text{erfc } (x) \qquad (E.3d)$$

Alternative definitions of erf (x) are

$$\text{erf } (x) = \frac{\text{sgn } (x)}{\pi^{1/2}} \int_0^{x^2} t^{-1/2} e^{-t} \, dt \qquad (E.4a)$$

$$\text{erf } (x) = \frac{2}{\pi} \int_0^\infty t^{-1} e^{-t^2} \sin (2xt) \, dt \qquad (E.4b)$$

$$\text{erf } (x) = \frac{2x}{\pi^{1/2}} \int_0^1 e^{-x^2 t^2} \, dt \qquad (E.4c)$$

E.2 SERIES EXPRESSIONS

One way to expand erf (x) is

$$\text{erf}\,(x) = \frac{2}{\pi^{1/2}}\left[x - \frac{x^3}{3} + \frac{x^5}{10} - \cdots\right] = \frac{2x}{\pi^{1/2}}\sum_{j=0}^{\infty}\frac{(-x^2)^j}{j!(2j+1)} \tag{E.5a}$$

$$\text{erf}\,(x) = \frac{2}{\pi^{1/2}}e^{-x^2}\left(x + \frac{2x^3}{3} + \frac{4x^5}{15} + \cdots\right) = e^{-x^2}\sum_{j=0}^{\infty}\frac{x^{2j+1}}{\Gamma[(2j+3)/2]} \tag{E.5b}$$

These expressions are most useful for small values of x. For large values of x, erfc (x) is expansible asymptotically as

$$\text{erfc}\,(x) \sim \frac{\exp(-x^2)}{x\pi^{1/2}}\left(1 - \frac{1}{2x^2} + \frac{1\cdot 3}{2^2 x^4} - \frac{1\cdot 3\cdot 5}{2^3 x^6} + \cdots\right) \tag{E.6}$$

Care must be exercised in using Eq. (E.6) numerically because the error is only less than the absolute value of the last term retained. Also for large x, the continued fraction expression given below may be used:

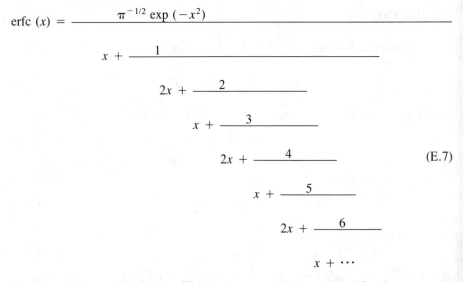

erfc $(x) = \cfrac{\pi^{-1/2}\exp(-x^2)}{x + \cfrac{1}{2x + \cfrac{2}{x + \cfrac{3}{2x + \cfrac{4}{x + \cfrac{5}{2x + \cfrac{6}{x + \cdots}}}}}}}$ (E.7)

See Press et al., p. 135 for an efficient way to evaluate continued fractions.

E.3 NUMERICAL EVALUATION SUBROUTINES

Table E.1 contains a main program and two subroutines, *ERFCG* and *ERFCP*, with the former being more accurate.

Table E.1 Program ERRORF which calls two different subroutines for evaluating erf (x), erfc (x), and ierfc (x).[†]

```
            PROGRAM ERRORF
            DOUBLE PRECISION X,ERF,ERFC,FIERFC
            X = 1.5D + 0
            CALL ERFCG(X,ERF,ERFC,FIERFC)
            WRITE(*,*)'X,ERF,ERFC,FIERFC'
            WRITE(*,5)X,ERF,ERFC,FIERFC
     5      FORMAT(5F15.10)
     6      FORMAT(5F15.8)
            XX = X
            CALL ERFCP(XX,PERF,PERFC,PIERFC)
            WRITE(*,5)XX,PERF,PERFC,PIERFC
            END
            SUBROUTINE ERFCG(X,ERF,ERFC,FIERFC)
     C      COEFFICIENTS TAKEN FROM R. GHEZ, A PRIMER OF DIFFUSION PROBLEMS
     C      WILEY-INTERSCIENCE, 1988 P. 108
            DOUBLE PRECISION P,Q,ERFC,S1,S2,X,ERF,XS,PI,FIERFC
            DIMENSION P(7),Q(8)
            DATA P/440.4137358247522D + 0,625.686535769683D + 0,
     I      448.3171659914834D + 0,191.8401405879669D + 0,50.9916976282532D + 0,
     I      7.923929828741545D + 0,0.5641999455982575D + 0/
            DATA Q/440.4137358247522D + 0,1122.640220177047D + 0,
     I      1274.667266757662D + 0,838.8103654784064D + 0,347.1229288117843D + 0,
     I      90.87271120353703D + 0,14.04533730546114D + 0,1.0D + 0/
            PI = 4.0D + 0*DATAN(1.0D + 0)
            Z = ABS(X)
            ERFC = 0.0D + 0
            IF(Z .GT. 20.0D + 0)GOTO 1000
            IF(Z .NE. 0.0D + 0)GOTO 50
            ERFC = 1.0D + 0
            GOTO 1000
     50     CONTINUE
            S1 = 0.0D + 0
            S2 = 0.0D + 0
            XS = Z
            DO 100 I = 1,7
            S1 = S1 + P(I)*XS
     100    XS = XS*Z
            XS = Z
            DO 200 J = 1,8
            S2 = S2 + Q(J)*XS
     200    XS = XS*Z
            ERFC = DEXP(-X*X)*S1/S2
     1000   IF(X .LT. 0.0D + 0)ERFC = 2.0D + 0-ERFC
            ERF  = 1.0D + 0-ERFC
            FIERFC = 0.0D + 0
            IF(Z .GT. 20.0D + 0)GOTO 2000
            FIERFC = ((1.0D + 0/DSQRT(PI))/DEXP(X*X))-X*ERFC
     2000   CONTINUE
            RETURN
            END
```

Table E.1 Program ERRORF which calls two different subroutines for evaluating erf (x), erfc (x), and ierfc (x).[†] (Continued)

```
          SUBROUTINE ERFCP(XX,PERF,PERFC,PIERFC)
C         FUNCTION TAKEN FROM PRESS, FLANNERY, TEUKOLSKY, AND
C         VETTERLING, "NUMERICAL RECIPES", CAMBRIDGE, 1986
C         SOME MODIFICATIONS
C         ERF = PERF, PERFC = ERFC, IERFC = PIERFC
C         ERFC SAID TO BE ACCURATE WITH A FRACTIONAL ERROR EVERYWHERE
C         LESS THAN 1.2E-7
          PI = 4.0*ATAN(1.0)
          Z = ABS(XX)
          PERFC = 0.0
          IF(Z .GT. 10.0)GOTO 200
          T = 1.0/(1.0 + 0.5*Z)
          PERFC = T*EXP(-Z*Z-1.26551223 + T*(1.000023368 + T*(0.37409196 +
     1    T*(0.09678418 + T*(-0.18628806 + T*(0.27886807 + T*(-1.13520398 +
     2    T*(1.48851587 + T*(-0.82215223 + T*0.17087277)))))))))
200       CONTINUE
          IF(XX .LT. 0.0) PERFC = 2.0-PERFC
1000      PERF = 1.0-PERFC
          PIERFC = 0.0
          IF(XX .GT. 10.0)GOTO 2000
          PIERFC = ((1.0/SQRT(PI))/EXP(XX*XX))-XX*PERFC
2000      CONTINUE
          RETURN
          END
```

[†] Subroutine ERFCG is more accurate and uses double precision.

E.4 RELATED FUNCTIONS

A set of functions is defined by the integral

$$i^n \text{ erfc } (x) = \frac{2}{\pi^{1/2}} \int_x^\infty \frac{(t - x)^n}{n!} e^{-t^2} dt \qquad n = 2, 3, 4, \ldots \qquad (E.8)$$

The notation is usually extended to embrace

$$i^1 \text{ erfc } (x) = \text{ierfc } (x) = \int_x^\infty \text{erfc } (t) \, dt \qquad (E.9a)$$

$$i^0 \text{ erfc } (x) = \text{erfc } (x) \qquad (E.9b)$$

$$i^{-1} \text{ erfc } (x) = \frac{2}{\pi^{1/2}} e^{-x^2} \qquad (E.9c)$$

The ierfc and i^n erfc functions are known as the complementary error function integral, and the repeated integrals of the error function complement, respectively. Plots of some of these functions are given in Fig. E.1.

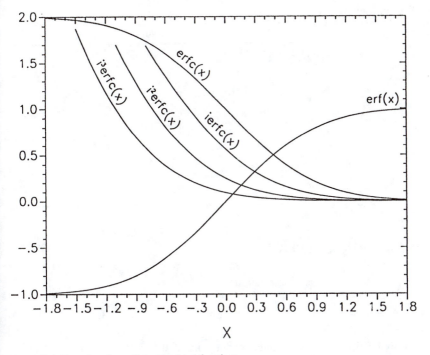

Figure E.1 Error function erf(x) and related functions.

The power series expansion is

$$i^n \text{ erfc } (x) = \frac{1}{2^n} \sum_{j=0}^{\infty} \frac{(-2x)^j}{j!\,\Gamma[(2+n-j)/2]} \tag{E.10}$$

which shows that

$$i^n \text{ erfc } (0) = \left[2^n \Gamma \left(1 + \frac{n}{2} \right) \right]^{-1} \tag{E.11}$$

of which the first few values are shown in Table E.2. The behavior of i^n erfc (x) for large values of x is described by

$$i^n \text{ erfc } (x) \sim \frac{2 \exp (-x^2)}{\pi^{1/2}(2x)^{n+1}} \left[1 - \frac{(n+1)(n+2)}{4x^2} \right.$$

$$+ \frac{(n+1)(n+2)(n+3)(n+4)}{32x^4}$$

$$\left. - \cdots + \frac{(n+2j)!}{n!j!(-4x^2)^j} + \cdots \right] \tag{E.12}$$

See also the tables of numerical values of erfc and related functions in Table 6.4.

Table E.2 Values of i^n erfc (0)

n	i^n erfc (0)
-1	$2\pi^{-1/2}$
0	1
1	$\pi^{-1/2}$
2	$\dfrac{1}{4}$
3	$\dfrac{\pi^{-1/2}}{6}$
4	$\dfrac{1}{32}$

E.5 RECURSION RELATION

The i^n erfc (x) functions obey the relation

$$i^n \text{ erfc } (x) = -\frac{x}{n} i^{n-1} \text{ erfc } (x) + \frac{1}{2n} i^{n-2} \text{ erfc } (x) \qquad n = 1, 2, 3, \ldots, \quad \text{(E.13)}$$

Sufficient applications of this formula permits any of the i^n erfc (x) functions to be expressed in terms of erfc (x) and exp $(-x^2)$ and hence evaluated. Some examples are

$$\text{ierfc } (x) = \pi^{-1/2} e^{-x^2} - x \text{ erfc } (x) \tag{E.14a}$$

$$i^2 \text{ erfc } (x) = \frac{1 + 2x^2}{4} \text{ erfc } (x) - \frac{x}{2\pi^{1/2}} e^{-x^2} \tag{E.14b}$$

$$i^3 \text{ erfc } (x) = \frac{1 + x^2}{6\pi^{1/2}} e^{-x^2} - \frac{3x + 2x^3}{12} \text{ erfc } (x) \tag{E.14c}$$

E.6 INTEGRALS AND DERIVATIVES

Differentiation gives

$$\frac{d}{dx} \text{ erf } (bx + c) = -\frac{d}{dx} \text{ erfc } (bx + c) = \frac{2b}{\pi^{1/2}} e^{-(bx+c)^2} \tag{E.15a}$$

Integration gives

$$\int_0^x \text{ erf } (bt) \, dt = x \text{ erf } (bx) - \frac{1 - \exp (-b^2 x^2)}{b\pi^{1/2}} \tag{E.15b}$$

$$\int_x^\infty \text{ erfc } (bt) \, dt = \frac{1}{b} \text{ ierfc } (bx) \tag{E.15c}$$

Also for the i^n erfc (x) function, the relations

$$\frac{d}{dx} i^n \text{ erfc } (x) = - i^{n-1} \text{ erfc } (x) \tag{E.16a}$$

$$\int_x^\infty i^n \text{ erfc } (t) \, dt = i^{n+1} \text{ erfc } (x) \tag{E.16b}$$

are valid. Notice that

$$\frac{d}{dx} \text{ erfc } (x) = - \frac{2}{\pi^{1/2}} e^{-x^2} \tag{E.16c}$$

E.7 COMPLEX ARGUMENT

For x replaced by $x + iy$, the complex function is given by

$$\text{erf } (x + iy) = \frac{2}{\pi^{1/2}} \left[e^{y^2} \int_0^x e^{-t^2} \cos (2yt) \, dt \right.$$

$$+ e^{-x^2} \int_0^y e^{t^2} \sin (2xt) \, dt \bigg]$$

$$- \frac{2i}{\pi^{1/2}} \left[e^{y^2} \int_0^x e^{-t^2} \sin (2yt) \, dt \right.$$

$$- e^{-x^2} \int_0^y e^{t^2} \cos (2xt) \, dt \bigg] \tag{E.17}$$

These four integrals cannot be evaluated in simpler terms. Confusingly, the function generally known as the "error function of the complex argument" is denoted

$$W(x + iy) = W(z) = e^{-z^2} \left[1 + \frac{2i}{\pi^{1/2}} \int_0^z e^{t^2} \, dt \right] = e^{-z^2} \text{ erfc } (-iz) \quad \text{(E.18)}$$

FUNCTIONS AND INTEGRALS

Table F.1 Expressions involving exponential functions and integrals

No.	Equation	Reference
0	$E_0(z) = z^{-1}e^{-z}$	(A + S)
1	$E_1(z) = \displaystyle\int_z^\infty \frac{e^{-t}}{t}\,dt = \int_{z/c^2}^\infty \frac{e^{-c^2 t}}{t}\,dt \ (\lvert \arg z \rvert < \pi)$	
2	$E_1(x) = -Ei(-x)$	(A + S)
3	$E_n(z) = \displaystyle\int_1^\infty \frac{e^{-zt}}{t^n}\,dt \ (Rz > 0)$	
3.1	$E_n(x) = x^{n-1} \displaystyle\int_x^\infty \frac{e^{-w}}{w^n}\,dw$	
3.2	$\displaystyle\int_0^t u^j e^{-1/u}\,du = t^{j+1} E_{j+2}(t^{-1})$	
3.3	$\displaystyle\int_x^\infty \frac{1}{t^{i+1}}\,e^{-c^2 t}\,dt = x^{-i}\,E_{i+1}(c^2 x)$	
4	$E_{n+1}(z) = \dfrac{1}{n}\left[e^{-z} - zE_n(z)\right] \quad n = 1, 2, 3, \ldots$	(A + S)
4.1	$E_2(z) = e^{-z} - zE_1(z)$	
4.2	$E_3(z) = \tfrac{1}{2}\left[z^2 E_1(z) + e^{-z}(1 - z)\right]$	
5	$\displaystyle\int \frac{1}{r^{2m-1}}\,e^{-r^2/(4t)}\,dr = -\frac{1}{2}\,r^{2(1-m)} E_m\left(\frac{r^2}{4t}\right) \quad \text{for } m = 1, 2, 3, \ldots$	
5.1	$\displaystyle\int \frac{1}{r}\,e^{-r^2/(4t)}\,dr = -\frac{1}{2}\,E_1\left(\frac{r^2}{4t}\right)$	
5.2	$\displaystyle\int \frac{1}{r^3}\,e^{-r^2/(4t)}\,dr = -\frac{1}{2}\,r^{-2} E_2\left(\frac{r^2}{4t}\right)$	
6	$\displaystyle\int r^{2n+1}\,e^{-r^2/(4t)}\,dr = \frac{1}{2}\,(4t)^{n+1} \int^{r^2/4t} v^n e^{-v}\,dv$	
	$\quad = -\dfrac{1}{2}\,(4t)^{n+1}\,e^{-r^2/(4t)} \displaystyle\sum_{i=0}^{n} \frac{n!}{(n-i)!}\left(\frac{r^2}{4t}\right)^{n-i}$	
	$\qquad \text{for } n = 0, 1, 2, \ldots$	

6.1 $\displaystyle\int re^{-r^2/(4t)}\, dr = -2te^{-r^2/(4t)}$

6.2 $\displaystyle\int r^3 e^{-r^2/(4t)}\, dr = -8t^2\left(1 + \frac{r^2}{4t}\right)e^{-r^2/(4t)}$

6.3 $\displaystyle\int r^5 e^{-r^2/(4t)}\, dr = -32t^3 e^{-r^2/(4t)}\left[2 + \frac{2r^2}{4t} + \left(\frac{r^2}{4t}\right)^2\right]$

7 $\displaystyle\int E_n(u)\, du = -E_{n+1}(u)$

8 $\displaystyle\int_0^b x E_1(x^2)\, dx = \frac{1}{2}\int_0^{b^2} E_1(u)\, du = \frac{1}{2}[1 - E_2(b^2)]$

9 $\displaystyle\int_a^\infty x^2 E_1(x^2)\, dx = \frac{1}{6}[\sqrt{\pi}\ \mathrm{erfc}\,(a) + 2a E_2(a^2)]$

10a $\displaystyle\int_z^\infty E_1(u^2)\, du = \sqrt{\pi}\ \mathrm{erfc}\,(z) - z E_1(z^2)$

10b $\displaystyle\int_0^\infty E_1(u^2)\, du = \sqrt{\pi}$

11 $E_{3/2}(x^2) = 2\sqrt{\pi}\ \mathrm{ierfc}\,(x)$

12 $\displaystyle\int_a^\infty \frac{1}{w(1 + bw)}\, e^{-w}\, dw = E_1(a) - e^{b^{-1}}E_1(a + b^{-1})$

Table F.2 Integrals involving erf (x) and erfc (x)

No.	Equation	Reference
1	$\displaystyle\int \mathrm{erf}\,(ax)\, dx = x\,\mathrm{erf}\,(ax) + \frac{1}{a\sqrt{\pi}}e^{-a^2x^2} = x + \frac{1}{a}\mathrm{ierfc}\,(ax)$	(N + G 4.1.1)
2	$\displaystyle\int \mathrm{erfc}\,(ax)\, dx = x\,\mathrm{erfc}\,(ax) - \frac{1}{a\sqrt{\pi}}e^{-a^2x^2} = -\frac{1}{a}\mathrm{ierfc}\,(ax)$	(N + G 4.1.2)
3	$\displaystyle\int x\,\mathrm{erf}\,(ax)\, dx = \left(\frac{x^2}{2} - \frac{1}{4a^2}\right)\mathrm{erf}\,(ax) + \frac{x}{2a\sqrt{\pi}}e^{-a^2x^2}$	
	$\displaystyle\qquad = \frac{x^2}{2} - \frac{1}{4a^2}\,\mathrm{erf}\,(ax) + \frac{x}{2a}\mathrm{ierfc}\,(ax)$	(N + G 4.1.4)
4	$\displaystyle\int x\,\mathrm{erfc}\,(ax)\, dx = \frac{1}{4a^2}\,\mathrm{erf}\,(ax) + \frac{x^2}{2}\,\mathrm{erfc}\,(ax) - \frac{x}{2a\sqrt{\pi}}e^{-x^2a^2}$	
	$\displaystyle\qquad = \frac{1}{4a^2}\,\mathrm{erf}\,(ax) - \frac{x}{2a}\mathrm{ierfc}\,(ax)$	(N + G 4.1.5)
5	$\displaystyle\int x^{-1}\,\mathrm{erf}\,(ax)\, dx = ln\,(x)\,\mathrm{erf}\,(ax) - \frac{2a}{\sqrt{\pi}}\int ln\,(x)e^{-a^2x^2}\, dx$	(N + G 4.1.12)
6	$\displaystyle\int x^{-1}\,\mathrm{erfc}\,(ax)\, dx = ln\,(x)\,\mathrm{erfc}\,(ax) + \frac{2a}{\sqrt{\pi}}\int ln\,(x)e^{-a^2x^2}\, dx$	(N + G 4.1.13)
7	$\displaystyle\int x^{-n}\,\mathrm{erf}\,(ax)\, dx = \frac{-\mathrm{erf}\,(ax)}{(n-1)x^{n-1}} + \frac{2a}{(n-1)\sqrt{\pi}}$	
	$\displaystyle\qquad \times \int \frac{e^{-a^2x^2}}{x^{n-1}}\, dx \qquad n \geq 2$	(N + G 4.1.14)

8 $$\int x^{-n} \operatorname{erfc}(ax)\,dx = \frac{-\operatorname{erfc}(ax)}{(n-1)\,x^{n-1}} - \frac{2a}{(n-1)\,\sqrt{\pi}}$$

$$\times \int \frac{e^{-a^2x^2}}{x^{n-1}}\,dx \qquad n \ge 2 \qquad\qquad \text{(N + G 4.1.15)}$$

9 $$\int e^{bx} \operatorname{erf}(ax)\,dx = \frac{e^{bx}}{b} \operatorname{erf}(ax) - \frac{e^{b^2/4a^2}}{b} \operatorname{erf}\left(ax - \frac{b}{2a}\right) \qquad \text{(N + G 4.2.1)}$$

10 $$\int e^{bx} \operatorname{erfc}(ax)\,dx = \frac{e^{bx}}{b} \operatorname{erfc}(ax) + \frac{e^{b^2/4a^2}}{b} \operatorname{erf}\left(ax - \frac{b}{2a}\right) \qquad \text{(N + G 4.2.2)}$$

11 $$\int x \operatorname{erf}^3(x)\,dx = \left(\frac{x^2}{2} - \frac{1}{4}\right) \operatorname{erf}^3(x) + \frac{3x}{2\sqrt{\pi}}\, e^{-x^2} \operatorname{erf}^2(x)$$

$$+ \frac{3}{2\pi}\, e^{-2x^2} \operatorname{erf}(x) - \frac{\sqrt{3}}{2\pi} \operatorname{erf}(x\sqrt{3}) \qquad\qquad \text{(Cho 2.3.6)}$$

12 $$\int_z^\infty \frac{1}{x^4} \operatorname{ierfc}(ax)\,dx = \frac{2}{z^3}\, i^3\, \operatorname{erfc}(az)$$

13 $$\int_z^\infty e^{-a^2x^2} \operatorname{erf}(x)\,dx = \frac{\pi^{1/2}}{2a}\, H(z, a) \qquad \text{(See Section 6.8 and Eq. [6.151])}$$

Table F.3 Integrals involving erf $(a\sqrt{x})$ or erfc $(a\sqrt{x})$

No.	Equation	Reference
1	$$\int \operatorname{erf}(a\sqrt{x})\,dx = \left(x - \frac{1}{2a^2}\right) \operatorname{erf}(a\sqrt{x}) + \frac{1}{a}\sqrt{\frac{x}{\pi}}\, e^{-a^2x} \qquad a \ne 0$$	(Cho 2.4.1)
	$$= x - \frac{1}{2a^2}(\operatorname{erf}(a\sqrt{x}) - 1) + \sqrt{\frac{x}{a^2}}\, \operatorname{ierfc}(a\sqrt{x})$$	
2	$$\int x \operatorname{erf}(a\sqrt{x})\,dx = \frac{x^2}{2} \operatorname{erf}(a\sqrt{x}) - \frac{3}{8a^4} \operatorname{erf}(a\sqrt{x})$$	
	$$+ \left(\frac{x^{3/2}}{2a\sqrt{\pi}} + \frac{3x^{1/2}}{4a^3\sqrt{\pi}}\right) e^{-a^2x} \qquad a \ne 0$$	(Cho 2.4.2)
3	$$\int \frac{\operatorname{erf}(a\sqrt{x})}{\sqrt{x}}\,dx = 2\sqrt{x} \operatorname{erf}(a\sqrt{x}) + \frac{2}{a\sqrt{\pi}}\, e^{-a^2x}$$	
	$$= 2\sqrt{x} + \frac{2}{a} \operatorname{ierfc}(a\sqrt{x}) \qquad a \ne 0$$	(Cho 2.4.3)
4	$$\int \frac{x-b}{\sqrt{x}\,(x+b)^2} \operatorname{erf}(a\sqrt{x})\,dx = \frac{2\sqrt{x}}{x+b} \operatorname{erf}(a\sqrt{x})$$	
	$$- 2e^{a^2b}\left(\frac{a}{\sqrt{\pi}}\right) E_1(a^2b + a^2x)$$	
5	$$\int e^x \operatorname{erf}(\sqrt{x})\,dx = e^x \operatorname{erf}\sqrt{x} - 2(x/\pi)^{1/2}$$	(Cho 2.4.4)
5.1	$$\int e^x \operatorname{erfc}(x^{1/2})\,dx = e^x \operatorname{erfc}(x^{1/2}) + 2(x/\pi)^{1/2}$$	
6	$$\int e^{ax} \operatorname{erf}(\sqrt{bx}) = \frac{1}{a}\, e^{ax} \operatorname{erf}(\sqrt{bx}) - \frac{1}{a}\sqrt{\frac{b}{b-a}}$$	

$$\times \text{ erf } \{[(b - a)x]^{1/2}\} \qquad a \neq 0 \qquad b \neq 0 \qquad a \neq b \qquad \text{(Cho 2.4.5)}$$

7 $\displaystyle\int_0^t u^{a/2} e^u \text{ erf } (u^{1/2}) \, du$

$$= \frac{2}{\pi^{1/2}} \sum_{i=0}^{\infty} 2^i \frac{t^{i+(a+3)/2}}{1 \cdot 3 \cdot 5 \cdots (2i+1) \, [i + (a+3)/2]} \qquad a > -3$$

$$\approx \frac{2}{\pi^{1/2}} \left(\frac{t^2}{2} + \frac{2t^3}{9} + \frac{1}{15} t^4 \right) \qquad \text{for } a = 1 \text{ and small } t \text{ values}$$

$$\approx \frac{2}{\pi^{1/2}} \left(\frac{t^3}{3} + \frac{t^4}{6} + \frac{4t^5}{75} \right) \qquad \text{for } a = 3 \text{ and small } t \text{ values}$$

8 $\displaystyle\int_0^t u^{a/2} e^u \text{ erfc } (u^{1/2}) \, du \approx t^{(a+2)/2} \left[\frac{2}{a+2} - \frac{4}{a+3} \left(\frac{t}{\pi} \right)^{1/2} + \cdots \right]$

for small t values and $a > -2$

9 $\displaystyle\int_0^t u^{-1/2} e^u \text{ erfc } (u^{1/2}) \, du = 2 \int_0^{t^{1/2}} e^{v^2} \text{ erfc } (v) \, dv \approx 2t^{1/2} \left[1 - \left(\frac{t}{\pi} \right)^{1/2} + \frac{1}{3} t \right]$

for small t values

10 $\displaystyle\int u^{1/2} e^u \text{ erfc } (u^{1/2}) \, du = 2 \int v^2 e^{v^2} \text{ erfc } (v) \, dv$

$$= \frac{u}{\pi^{1/2}} + u^{1/2} e^u \text{ erfc } (u^{1/2}) \qquad \text{(Barber)}$$

$$- \frac{1}{2} \int u^{-1/2} e^u \text{ erfc}(u^{1/2}) \, du$$

11 $\displaystyle\int_0^t (1 - e^{-1/4u}) e^{B^2 u} \text{ erfc } (B u^{1/2}) \, du$

$$\approx t \left[1 - E_2 \left(\frac{1}{4t} \right) \right]$$

$$- \frac{B}{\pi^{1/2}} \left[\frac{4}{3} t^{3/2} - \frac{1}{3} \pi^{1/2} \text{ erfc } \left(\frac{1}{2t^{1/2}} \right) + \frac{2}{3} t^{1/2} e^{-1/4t} (1 - 2t) \right]$$

$$+ \frac{1}{2} B^2 t^2 \left[1 - 2E_3 \left(\frac{1}{4t} \right) \right] \qquad \text{for } Bt^{1/2} \ll 1$$

12 $\displaystyle\int_0^t \frac{1}{(t-u)^{1/2}} \left\{ \frac{1}{u^{1/2}} - \frac{(\pi \alpha)^{1/2}}{a} e^{\alpha u/a^2} \text{ erfc } \left[\frac{(\alpha u)^{1/2}}{a} \right] \right\} \, du = \pi e^{\alpha t/a^2} \text{ erfc } \left[\frac{(\alpha t)^{1/2}}{a} \right] \qquad \text{(Levine)}$

Table F.4 Integrals involving erf $(a/\sqrt{x})$ or erfc $(a/\sqrt{x})$

No.	Equation	Reference
1	$\displaystyle\int \text{erfc} \left(\frac{a}{\sqrt{x}} \right) dx = (x + 2a^2) \text{ erfc } \left(\frac{a}{\sqrt{x}} \right) - 2a \sqrt{\frac{x}{\pi}} e^{-a^2/x}$	
	$\qquad\qquad = x \text{ erfc } \left(\frac{a}{\sqrt{x}} \right) - 2a\sqrt{x} \text{ ierfc } \left(\frac{a}{\sqrt{x}} \right)$	(Cho 2.5.1)
2	$\displaystyle\int x \text{ erfc } (a/\sqrt{x}) \, dx = \left(\frac{x^2}{2} - \frac{2a^4}{3} \right) \text{ erfc } (a\sqrt{x})$	

$$- (x^{3/2} - 2a^2 x^{1/2}) \frac{a}{3\sqrt{\pi}} e^{-a^2/x}$$

$$= \frac{x^2}{2} \operatorname{erfc}\left(\frac{a}{\sqrt{x}}\right) - \frac{a}{3} \sqrt{\frac{x^3}{\pi}} e^{-a^2/x}$$

$$+ \frac{2a^3 \sqrt{x}}{3} \operatorname{ierfc}\left(\frac{a}{\sqrt{x}}\right) \qquad \text{(Cho 2.5.2)}$$

3a
$$\int_{u=0}^{t} \frac{1}{u^2} \operatorname{erf}\left(\frac{x}{(4\alpha u)^{1/2}}\right) e^{-y^2/4\alpha u} \, du = \frac{4\alpha}{y^2} \left\{ e^{-y^2/4\alpha t} \operatorname{erf}\left[\frac{x}{(4\alpha t)^{1/2}}\right]\right.$$

$$+ \frac{x}{(x^2 + y^2)^{1/2}}$$

$$\times \operatorname{erfc}\left[\frac{(x^2 + y^2)^{1/2}}{(4\alpha t)^{1/2}}\right]\right\} \qquad y \neq 0$$

3b
$$\int_{u=0}^{t} \frac{1}{u^2} \operatorname{erfc}\left(\frac{x}{(4\alpha u)^{1/2}}\right) e^{-y^2/4\alpha u} \, du = \frac{4\alpha}{y^2} \left\{ e^{-y^2/4\alpha t} \operatorname{erfc}\left[\frac{x}{(4\alpha t)^{1/2}}\right]\right.$$

$$- \frac{x}{(x^2 + y^2)^{1/2}}$$

$$\times \operatorname{erfc}\left[\frac{(x^2 + y^2)^{1/2}}{(4\alpha t)^{1/2}}\right]\right\} \qquad y \neq 0$$

4
$$\int_{u=0}^{t} \frac{1}{u^{3/2}} e^{-y^2/4\alpha u} \operatorname{erf}\left[\frac{x}{(4\alpha u)^{1/2}}\right] du = \frac{2\sqrt{\alpha\pi}}{y} H\left(\frac{x}{\sqrt{4\alpha t}}, \frac{y}{x}\right)$$

$$\text{(see Section 6.8)}$$

5
$$\int_{u=0}^{t} \frac{1}{(\pi\alpha u)^{1/2}} \operatorname{erfc}\left[\frac{C_1}{(4\alpha u)^{1/2}}\right] \operatorname{erfc}\left[\frac{C_2}{(4\alpha u)^{1/2}}\right] du$$

$$\approx \frac{1}{\pi^{3/2}} \frac{(C_1^2 + C_2^2)^{3/2}}{2\alpha C_1 C_2} \left\{ \Gamma\left(-\frac{3}{2}, \frac{C_1^2 + C_2^2}{4\alpha t}\right)\right.$$

$$- \frac{1}{2} \left[\frac{C_1^2 + C_2^2}{C_1 C_2}\right]^2 \Gamma\left(-\frac{5}{2}, \frac{C_1^2 + C_2^2}{4\alpha t}\right)\right\}$$

$$\approx \frac{1}{\pi^{3/2}} \frac{(C_1^2 + C_2^2)^{1/2}}{C_1 C_2} \exp\left[-\frac{C_1^2 + C_2^2}{4\alpha t}\right]$$

$$\times \left(\frac{4\alpha t}{C_1^2 + C_2^2}\right)^{5/2} \left\{ 1 - \frac{5}{2} \frac{4\alpha t}{C_1^2 + C_2^2}\right.$$

$$- \frac{1}{2} \left[\frac{C_1^2 + C_2^2}{C_1 C_2}\right]^2 \left[\frac{4\alpha t}{C_1^2 + C_2^2}\right]\right\}$$

$$\text{for small values of } \frac{4\alpha t}{C_1^2} \text{ and } \frac{4\alpha t}{C_2^2}$$

6
$$\int_{u=0}^{t} [4\pi\alpha u]^{-1/2} \exp\left(-\frac{z^2}{4\alpha u}\right) \operatorname{erf}\left[\frac{a}{(4\alpha u)^{1/2}}\right] \operatorname{erf}\left[\frac{b}{(4\alpha u)^{1/2}}\right] du$$

$$\approx \left(\frac{t}{\alpha}\right)^{1/2} \operatorname{ierfc}\left[\frac{|z|}{(4\alpha t)^{1/2}}\right] - \frac{t}{\pi}\left[\frac{1}{a} E_2\left(\frac{z^2 + a^2}{4\alpha t}\right)\right.$$

$$+ \frac{1}{b} E_2\left(\frac{z^2 + b^2}{4\alpha t}\right)\right] \qquad \text{for small } \frac{\alpha t}{a^2} \text{ and } \frac{\alpha t}{b^2} \text{ values}$$

7a
$$\int_0^t \frac{1}{(\pi\alpha u)^{1/2}} \frac{(4\alpha u)^{(m+n+2)/2}}{C_1^{m+1} C_2^{n+1}} e^{-(C_1^2+C_2^2)/(4\alpha u)} \, du$$

$$= \frac{1}{2\alpha\pi^{3/2}} \frac{(C_1^2 + C_2^2)^{(m+n+3)/2}}{C_1^{m+1} C_2^{n+1}} \Gamma\left(-\frac{m+n+3}{2}, \frac{C_1^2 + C_2^2}{4\alpha t}\right)$$

7b
$$\int_0^t \frac{1}{(\pi\alpha u)^{1/2}} (4\alpha u)^{(n+2)/2} e^{-C^2/(4\alpha u)} \, du = \frac{1}{2\alpha\pi^{3/2}} C^{n+3} \Gamma\left(-\frac{n+3}{2}, \frac{C^2}{4\alpha t}\right)$$

8
$$\frac{\alpha}{a^2} \int_{u=0}^t \left(\frac{\alpha u}{a^2}\right)^{n/2} \text{ierfc}\left[\frac{a}{(\alpha u)^{1/2}}\right] du$$

$$= 2 \int_{a/(\alpha t)^{1/2}}^\infty \frac{1}{w^{n+3}} \text{ierfc}(w) \, dw$$

$$= \frac{2}{n+2} \left\{ \left(\frac{\alpha t}{a^2}\right)^{n/2+1} \text{ierfc}\left[\frac{a}{(\alpha t)^{1/2}}\right] - \frac{1}{n+1} \left(\frac{\alpha t}{a^2}\right)^{n/2+1/2} \right.$$

$$\left. \times \text{erfc}\left[\frac{a}{(\alpha t)^{1/2}}\right] + \frac{1}{n+1} \frac{1}{\pi^{1/2}} \Gamma\left(-\frac{n}{2}, \frac{a^2}{\alpha t}\right) \right\} \qquad n = 0, 1, 2, \ldots$$

Below are some $\Gamma(-m, x^2)$ relations:

$$\Gamma(-m, x^2) = x^{-2m} E_{m+1}(x^2) \qquad m = 0, 1, 2, 3, \ldots$$

$$\Gamma\left(-\frac{1}{2}, x^2\right) = \frac{2\pi^{1/2}}{x} \text{ierfc}(x)$$

$$\Gamma\left(-\frac{3}{2}, x^2\right) = \frac{4}{3}\left[\pi^{1/2} \text{erfc}(x) - \frac{1}{x} e^{-x^2}\left(1 - \frac{1}{2x^2}\right)\right]$$

(See also Appendix E)

Table F.5 Integrals involving erfc $(a\sqrt{x} + b/\sqrt{x})$

No.	Equation	Reference
1	$\int \text{erfc}\left(a\sqrt{x} + \frac{b}{\sqrt{x}}\right) dx = -\left[\frac{1}{4a^2}\left(\text{erfc}(a\sqrt{x} + b/\sqrt{x})\right.\right.$ $+ e^{-4ab} \text{erfc}(a\sqrt{x} - b/\sqrt{x})$ $\left.\left. + \frac{\sqrt{x}}{a} \text{ierfc}\left(a\sqrt{x} + \frac{b}{\sqrt{x}}\right)\right]\right. \qquad a \neq 0$	(Cho 2.6.1)
2	$\int e^x \text{erfc}\left(a\sqrt{x} + \frac{b}{\sqrt{x}}\right) dx = e^x \text{erfc}\left(a\sqrt{x} + \frac{b}{\sqrt{x}}\right)$ $- \frac{1}{2}\left(1 + \frac{a}{(a^2-1)^{1/2}}\right) e^{-2b|a-(a^2-1)^{1/2}|}$ $\times \text{erfc}\left[(a^2-1)^{1/2} x^{1/2} + \frac{b}{\sqrt{x}}\right]$ $+ \frac{1}{2}\left(1 - \frac{a}{(a^2-1)^{1/2}}\right) e^{-2b|a+(a^2-1)^{1/2}|}$ $\times \text{erfc}\left[(a^2-1)^{1/2} x^{1/2} - \frac{b}{\sqrt{x}}\right] \qquad a > 1$	(Cho 2.6.3)
3	$\int e^{(a^2-b^2)x} \text{erfc}\left(a\sqrt{x} + \frac{c}{\sqrt{x}}\right) dx = \frac{1}{a^2-b^2} e^{(a^2-b^2)x} \text{erfc}\left(a\sqrt{x} + \frac{c}{\sqrt{x}}\right)$	

$$- \frac{1}{2(a^2 - b^2)} \left(1 + \frac{a}{b}\right) e^{-2(a-b)c}$$

$$\times \operatorname{erfc}\left(b\sqrt{x} + \frac{c}{\sqrt{x}}\right) + \frac{1}{2(a^2 - b^2)}$$

$$\times \left(1 - \frac{a}{b}\right) e^{-2(a+b)c}$$

$$\times \operatorname{erfc}\left(b\sqrt{x} - \frac{c}{\sqrt{x}}\right) \qquad a^2 \neq b^2 \qquad \text{(Cho 2.6.4)}$$

Table F.6 Integrals involving the exponential function

No.	Equation	Reference
1	$\int e^{-(ax^2 + 2bx + c)}\, dx = \frac{1}{2}\sqrt{\frac{\pi}{a}}\, e^{(b^2 - ac)/a} \operatorname{erfc}\left(\sqrt{a}\, x + \frac{b}{\sqrt{a}}\right) \qquad a > 0$	(A + S 7.4.32)
2	$\int e^{-a^2 x^2 + bx}\, dx = \frac{\sqrt{\pi}}{2a}\, e^{b^2/4a^2} \operatorname{erf}\left(ax - \frac{b}{2a}\right)$	
3	$\int x^{3/2} e^{-a^2 x}\, dx = \frac{3\sqrt{\pi}}{4a^5} \operatorname{erf}(a\sqrt{x}) - \frac{3\sqrt{x}}{2a^4} e^{-a^2 x} - \frac{x^{3/2}}{a^2} e^{-a^2 x}$	
	$\qquad = \frac{3\sqrt{\pi}}{4a^5} \operatorname{erf}(a\sqrt{x}) - \left(\frac{3\sqrt{x}}{2a^4} + \frac{x^{3/2}}{a^2}\right) e^{-a^2 x} \qquad a \neq 0$	(Cho 2.7.2)
4	$\int \sqrt{x}\, e^{-a^2 x}\, dx = \frac{\sqrt{\pi}}{2a^3} \operatorname{erf}(a\sqrt{x}) - \frac{\sqrt{x}}{a^2} e^{-a^2 x} \qquad a \neq 0$	(Cho 2.7.3)
5	$\int x^{-1/2} e^{-a^2 x}\, dx = \frac{\sqrt{\pi}}{a} \operatorname{erf}(a\sqrt{x}) \qquad a \neq 0$	(Cho 2.7.4)
6	$\int x^{-3/2} e^{-a^2 x}\, dx = -2a\sqrt{\pi} \operatorname{erf}(a\sqrt{x}) - \frac{2}{\sqrt{x}} e^{-a^2 x}$	
	$\qquad - 2a\sqrt{\pi}\left(1 + \frac{1}{a\sqrt{x}} \operatorname{ierfc}(a\sqrt{x})\right) \qquad a \neq 0$	(Cho 2.7.5)
7	$\int \sqrt{x}\, e^{-a^2/x}\, dx = \left(\frac{2x^{3/2}}{3} - \frac{4a^2\sqrt{x}}{3}\right) e^{-a^2/x} - \frac{4a^3\sqrt{\pi}}{3} \operatorname{erf}\left(\frac{a}{\sqrt{x}}\right)$	
	$\qquad = \frac{2}{3} x^{3/2} e^{-a^2/x} - \frac{4}{3} a^2 \sqrt{\pi}\left[a + \sqrt{x} \operatorname{ierfc}\left(\frac{a}{\sqrt{x}}\right)\right]$	(Cho 2.8.2)
8	$\int x^{3/2} e^{-a^2/x}\, dx = \left(\frac{2x^{5/2}}{5} - \frac{4a^2 x^{3/2}}{15} + \frac{8a^4 x^{1/2}}{15}\right) e^{-a^2/x}$	
	$\qquad + \frac{8a^5\sqrt{\pi}}{15} \operatorname{erf}\left(\frac{a}{\sqrt{x}}\right)$	
	$\qquad = \frac{x^{3/2}}{15} (6x - 4a^2) e^{-a^2/x} + \frac{8a^4\sqrt{\pi}}{15}$	
	$\qquad \times \left[a + \sqrt{x} \operatorname{ierfc}\left(\frac{a}{\sqrt{x}}\right)\right]$	(Cho 2.8.1)

9 $\displaystyle\int x^{-1/2}\, e^{-a^2/x}\, dx = 2a\,\sqrt{\pi}\ \text{erf}\left(\frac{a}{\sqrt{x}}\right) + 2\sqrt{x}\, e^{-a^2/x}$

$\displaystyle\qquad\qquad = 2\pi^{1/2}\left[a + \sqrt{x}\ \text{ierfc}\left(\frac{a}{\sqrt{x}}\right)\right]$ (Cho 2.8.3)

10 $\displaystyle\int x^{-3/2}\, e^{-a^2/x}\, dx = -\frac{\sqrt{\pi}}{a}\ \text{erf}\left(\frac{a}{\sqrt{x}}\right)\qquad a \neq 0$ (Cho 2.8.4)

11 $\displaystyle\int e^{-a^2 x^2 - b^2/x^2}\, dx = \frac{\sqrt{\pi}}{4a}\left[e^{2ab}\,\text{erf}\left(ax + \frac{b}{x}\right)\right.$

$\displaystyle\qquad\qquad \left. + e^{-2ab}\,\text{erf}\left(ax - \frac{b}{x}\right)\right]\qquad a \neq 0$ (Cho 2.9.1)

12 $\displaystyle\int x^{-1/2}\, e^{-a^2 x - b^2/x}\, dx = \frac{\sqrt{\pi}}{2a}\left[e^{2ab}\,\text{erf}\left(a\sqrt{x} + \frac{b}{\sqrt{x}}\right)\right.$

$\displaystyle\qquad\qquad \left. + e^{-2ab}\,\text{erf}\left(a\sqrt{x} - \frac{b}{\sqrt{x}}\right)\right]\qquad a \neq 0$ (Cho 2.9.4)

13 $\displaystyle\int x^{-3/2}\, e^{-a^2 x - b^2/x}\, dx = \frac{\sqrt{\pi}}{2b}\left[e^{-2ab}\,\text{erf}\left(a\sqrt{x} - \frac{b}{\sqrt{x}}\right)\right.$

$\displaystyle\qquad\qquad \left. - e^{2ab}\,\text{erf}\left(a\sqrt{x} + \frac{b}{\sqrt{x}}\right)\right]\qquad b \neq 0$ (Cho 2.9.5)

14 $\displaystyle\int x^{1/2}\, e^{-a^2 x - b^2/x}\, dx = \frac{\sqrt{\pi}}{2a^2}\left(\frac{1}{2a} - b\right) e^{2ab}\,\text{erf}\left(a\sqrt{x} + \frac{b}{\sqrt{x}}\right)$

$\displaystyle\qquad\qquad - \frac{\sqrt{x}}{a^2}\, e^{-a^2 x - b^2/x} + \frac{\sqrt{\pi}}{2a^2}\left(\frac{1}{2a} + b\right) e^{-2ab}$

$\displaystyle\qquad\qquad \times\,\text{erf}\left(a\sqrt{x} - \frac{b}{\sqrt{x}}\right)\qquad a \neq 0$ (Cho 2.9.3)

15 $\displaystyle\int x^{3/2}\, e^{-a^2 x - b^2/x}\, dx = \left(b^2 + \frac{3}{4a^2} - \frac{3b}{2a}\right)\frac{\sqrt{\pi}}{2a^3}\, e^{2ab}\,\text{erf}\left(a\sqrt{x} + \frac{b}{\sqrt{x}}\right)$

$\displaystyle\qquad\qquad - \frac{3\sqrt{x}}{2a^4}\, e^{-a^2 x - b^2/x} + \left(b^2 + \frac{3}{4a^2} + \frac{3b}{2a}\right)\frac{\sqrt{\pi}}{2a^3}\, e^{-2ab}$

$\displaystyle\qquad\qquad \times\,\text{erf}\left(a\sqrt{x} - \frac{b}{\sqrt{x}}\right) - \frac{x^{3/2}}{a^2}\, e^{-a^2 x - b^2/x}\qquad a \neq 0$ (Cho 2.9.2)

REFERENCES

Abramowitz, M., and Stegun, I. A., (Eds.), 1964, *Handbook of Mathematical Functions with Formulas, Graphs, and Mathematical Tables*, Applied Mathematics Series 55, National Bureau of Standards, Wash., D.C.

Cho, S. H., 1971, *A Short Table of Integrals Involving the Error Function*, Dept. of Mechanics, Korean Military Academy, Seoul, Korea.

Levine, H., 1987, On the Solution of an Unsteady Diffusion Problem and its Ultimate Behavior, *Q.J. Mech. Applied Math.*, vol. 40 (Pt. 2), pp. 213–223.

Ng, E. W., and Geller, M., 1969, A Table of Integrals of the Error Functions, *Journal of Research of the National Bureau of Standards—B. Mathematical Sciences*, vol. 73B, pp. 1–20.

Ozisik, M. N., 1980, *Heat Conduction*, Wiley-Interscience, New York.

SERIES EXPRESSIONS

No.	Equation	Reference

1 $\displaystyle\sum_{n=1}^{\infty} \frac{\cos(n\pi x)}{n} = -\ln\left(2\sin\frac{\pi x}{2}\right)$ $(0 < x < 1)$ (A + S, p. 1005)

$\displaystyle = \frac{1}{2}\ln\frac{1}{2[1-\cos(\pi x)]}$ $(0 < x < 1)$ (G + R, p. 38)

2 $\displaystyle\sum_{n=1}^{\infty}\frac{\cos(n\pi x)}{n^2} = \frac{\pi^2}{2}\left[\frac{1}{2}x^2 - x + \frac{1}{3}\right]$ $(0 \le x \le 1)$ (A + S, p. 1005)

3 $\displaystyle\sum_{n=1}^{\infty}(-1)^{n-1}\frac{\cos(n\pi x)}{n^2} = \frac{\pi^2}{4}\left(\frac{1}{3}-x^2\right)$ $-1 \le x \le 1$ (G + R, p. 38)

4 $\displaystyle\frac{2}{\pi^3}\sum_{n=1}^{\infty}\frac{1}{n^3}\cos(n\pi x)\sin(n\pi\delta)$

$$= \begin{cases} \left[-\dfrac{1}{2}x^2 + \dfrac{\delta}{3} - \dfrac{\delta^2}{6}\right](1-\delta) & 0 \le x \le \delta \\[2ex] \dfrac{1}{2}\delta x(x-2) + \dfrac{\delta^3}{6} + \dfrac{1}{3}\delta & \delta \le x \le 1 \end{cases}$$

5.1 $\displaystyle\frac{2}{\pi}\sum_{n=1}^{\infty}\frac{\cos(n\pi x)\sin(n\pi\delta)}{n} = \begin{cases} 1-\delta & 0 \le x < \delta \\ -\delta & \delta < x \le 1 \end{cases}$

5.2 $\displaystyle\frac{2}{\pi}\sum_{n=1}^{\infty}\frac{\sin(n\pi x)\cos(n\pi\delta)}{n} = \begin{cases} -x & 0 \le x < \delta \\ 1-x & \delta < x \le 1 \end{cases}$

6 $\displaystyle\frac{4}{\pi^3}\sum_{m=1}^{\infty}\sum_{n=1}^{\infty}\frac{1}{m^3}\left[1+\left(\frac{n}{m}\frac{L_x}{L_y}\right)^2\right]^{-1}\cos\left(m\pi\frac{x}{L_x}\right)\cos\left(n\pi\frac{y}{L_y}\right)\sin\left(m\pi\frac{\delta}{L_x}\right)(-1)^n$

$$= -\frac{2}{\pi^3}\sum_{m=1}^{\infty}\frac{1}{m^3}\cos\left(m\pi\frac{x}{L_x}\right)\sin\left(m\pi\frac{\delta}{L_x}\right)$$

$$+ \frac{L_y}{L_x}\frac{2}{\pi^3}\sum_{n=1}^{\infty}\frac{\sin(n\pi\delta/L_x)\cos(n\pi x/L_x)\cosh(n\pi y/L_x)}{n^2\sinh(n\pi L_y/L_x)}$$

$$0 \le x \le L_x$$

$$0 \le y \le L_y$$

$$0 \le \delta \le L_x$$

7 $$\frac{2}{\pi} \sum_{m=1}^{\infty} \frac{\sin (m\pi y/L_y)}{m\{1 + [(n/m)(L_y/L_x)]^2\}} = \frac{\sinh [n\pi(L_y - y)/L_x]}{\sinh (n\pi L_y/L_x)}$$

8 $$\sum_{n=-\infty}^{\infty} e^{-(x-x'+2nL)^2/4\alpha t} = \frac{(\pi\alpha t)^{1/2}}{L} \left[1 + 2 \sum_{n=1}^{\infty} \cos \frac{n\pi(x - x')}{L} e^{-n^2\pi^2\alpha t/L^2} \right]$$

(Poisson's summation formula, Carslaw and Jaeger, 1959, p. 275)
x' can be positive or negative

9 $$\sum_{n=1}^{\infty} (-1)^n n \sin \left(n\pi \frac{x}{L} \right) e^{-n^2\pi^2\alpha t/L^2}$$

$$= -\frac{L}{2\pi^{3/2}(\alpha t/L^2)^{1/2}} \frac{d}{dx} \sum_{n=-\infty}^{\infty} \exp \left\{ -\frac{[x - (2n + 1)L]^2}{4\alpha t} \right\}$$ (M + F, p. 1587)

10 $$\sum_{n=1}^{\infty} \frac{\sin (n\pi x)}{n} = \frac{\pi}{2} (1 - x) \qquad 0 < x < 2$$ (A + S, p. 1005)

11 $$\sum_{n=1}^{\infty} (-1)^n \frac{\sin (n\pi x)}{n} = -\frac{\pi}{2} x$$ (Ozisik, p. 203)

12 $$\sum_{n=1}^{\infty} \frac{(-1)^n \cos (n\pi y/L_y)}{[(m/L_x)^2 + (n/L_y)^2]} = -\frac{1}{2} (L_x/m)^2 + L_y \frac{\pi}{2} (L_x/m) \frac{\cosh (m\pi y/L_x)}{\sinh (m\pi L_y/L_x)}$$

13 $$\sum_{n=1}^{\infty} (-1)^n \{1 + [(n/m)(L_x/L_y)]^2\}^{-1} \cos \left(n\pi \frac{y}{L_y} \right) = \frac{\pi}{2} \frac{mL_y}{L_x} \frac{\cosh (m\pi y/L_x)}{\sinh (m\pi L_y/L_x)}$$

REFERENCES

Abramowitz, M., and Stegun, I. A. (Eds), 1964, *Handbook of Mathematical Functions with Formulas, Graphs, and Mathematical Tables*, Applied Mathematics Series 55, National Bureau of Standards, Wash., D.C.

Gradshteyn, I. S., and Ryzhik, I. M., 1965, *Tables of Integrals, Series, and Products*, 4th ed., Academic Press, New York.

Morse, P. M., and Feshbach, H., 1953, *Methods of Theoretical Physics*, McGraw-Hill, New York.

Ozisik, M. N., 1980, *Heat Conduction*, Wiley-Interscience, New York.

RADIAL DIRECTION GREEN'S FUNCTIONS
CYLINDRICAL COORDINATES:
$$dv' = 2\pi r' \, dr'$$
$$ds' = 2\pi a \text{ or } 2\pi b$$

The partial differential equation for transient, cylindrical radial heat conduction is

$$\frac{1}{r} \frac{\partial}{\partial r} \left(r \frac{\partial T}{\partial r} \right) = \frac{1}{\alpha} \frac{\partial T}{\partial t}$$

For steady GFs in cylindrical coordinates see Table R.1. The Green's function (GF) is

$$G_{R00}(r, t|r', \tau) = \frac{1}{4\pi\alpha(t - \tau)} \exp\left[-\frac{r^2 + r'^2}{4\alpha(t - \tau)} \right] I_0\left[\frac{rr'}{2\alpha(t - \tau)} \right] \quad \text{(R00.1a)}$$

$$G_{R00}(r, t|r', \tau) = \frac{1}{4\pi\alpha(t - \tau)} \exp\left[-\frac{(r - r')^2}{4\alpha(t - \tau)} \right]$$

$$\times \exp\left[-\frac{rr'}{2\alpha(t - \tau)} \right] I_0\left[\frac{rr'}{2\alpha(t - \tau)} \right] \quad \text{(R00.1b)}$$

(units of $1/m^2$) (Carslaw and Jaeger, 1959, pp. 259 and 368). See Figure R00.1 which shows $r'^2 G_{R00}(\cdot)$ versus r/r' for fixed values of $\alpha(t - \tau)/r'^2$. A similar plot is given by Fig. R00.2 for $r'^2 G_{R00}(\cdot)$ versus r/r' for fixed values of $\alpha(t - \tau)/rr'$. The integral of $G_{R00}2\pi r \, dr'$ for $r' = 0$ to ∞ is unity,

$$\int_0^\infty G_{R00}(r, t|r', \tau)2\pi r' \, dr' = 1 \quad \text{(R00.2)}$$

A special case is for the source at $r' = 0$ (or equivalently for a source at $r' = r$ and the observation at $r = 0$):

$$G_{R00}(r, t|0, \tau) = \frac{1}{4\pi\alpha(t - \tau)} e^{-r^2/[4\alpha(t - \tau)]} \quad \text{(R00.3a)}$$

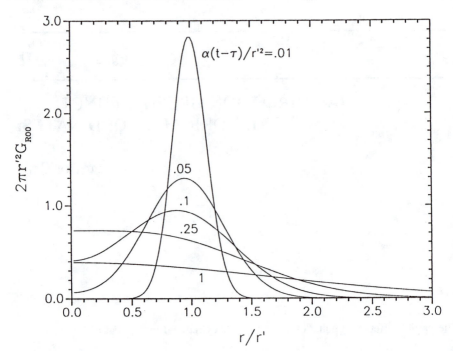

Figure R00.1 Green's function G_{R00} (multiplied by $2\pi r'^2$) versus r/r' for several values of $\alpha(t - \tau)/r'^2$.

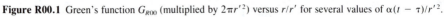

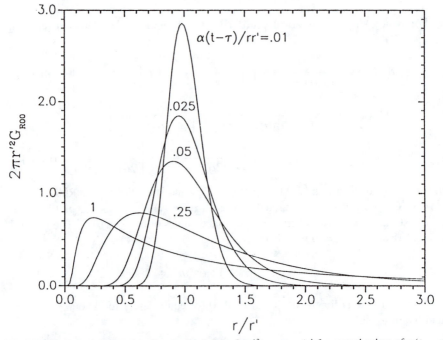

Figure R00.2 Green's function G_{R00} (multiplied by $2\pi r'^2$) versus r/r' for several values of $\alpha(t - \tau)/(rr')$.

$$G_{R00}(r, t|0, \tau) = G_{X00}(x, t|0, \tau)G_{Y00}(y, t|0, t) \qquad \text{(R00.3b)}$$

since $r^2 = x^2 + y^2$.

R.1 SMALL- AND LARGE-TIME APPROXIMATIONS FOR $G_{R00}(\cdot)$

For small values of $\alpha(t - \tau)/(rr')$, $G_{R00}(r, t|r', t)$ can be approximated by

$$G_{R00}(r, t|r', \tau) \approx \frac{1}{2\pi[4\pi rr'\alpha(t - \tau)]^{1/2}} \exp\left[-\frac{(r - r')^2}{4\alpha(t - \tau)}\right]$$

$$\times \left\{1 + \frac{\alpha(t - \tau)}{4rr'} + \frac{9[\alpha(t - \tau)]^2}{32(rr')^2}\right.$$

$$\left. + 1.5\frac{75[\alpha(t - \tau)]^3}{128(rr')^3}\right\} \qquad \text{(R00.4)}$$

The coefficient 1.5 in the last term is used to improve the accuracy. (The series expansion has a unity coefficient instead of 1.5.) The percent errors for Eq. (R00.4) are given by:

$\frac{\alpha(t - \tau)}{rr'}$	0	0.05	0.1	0.125	0.167	0.25
% error	0	0.0022	−0.005	−0.038	−0.14	0.016

For large values of $\alpha(t - \tau)/rr'$, $G_{R00}(\cdot)$ can be approximated by:

$$G_{R00}(r, t|r', \tau) \approx \frac{1}{4\alpha(t - \tau)} e^{-r'^2/[4\alpha(t-\tau)]}\left\{1 + \frac{(rr')^2}{[4\alpha(t - \tau)]^2}\right.$$

$$\left. + \frac{1}{4}\frac{(rr')^4}{[4\alpha(t - \tau)]^4} + \frac{1}{36}\frac{(rr')^6}{[4\alpha(t - \tau)]^6}\right\} \qquad \text{(R00.5)}$$

where the errors are given by

$\frac{\alpha(t - \tau)}{rr'}$	0.25	0.33	0.5	0.625	∞
% error	−0.08	−0.012	−0.0005	−0.00009	0

For large values of both $\alpha(t - \tau)/r^2$ and $\alpha(t - \tau)/r'^2$, $G(r, t|r', \tau)$ can be approximated by

$$G_{R00}(r, t|r', \tau) \approx \frac{1}{4\pi\alpha(t - \tau)}\left[1 - \frac{r^2 + r'^2}{4\alpha(t - \tau)} + \frac{r^4 + 3r^2r'^2 + r'^4}{[4\alpha(t - \tau)]^2}\right] \qquad \text{(R00.6)}$$

R.2 INTEGRAL FROM $r' = 0$ TO $r' = a$

For small times, an approximate result for the integral over r' from $r' = 0$ to a for $r^+ \geq 1$ is

$$\int_0^a G_{R00}(r, t|r', \tau)2\pi r'\, dr'$$

$$\approx \frac{1}{2}(r^+)^{-1/2}\left\{ \text{erfc}\left[\frac{r^+ - 1}{(4u)^{1/2}}\right] - \frac{1}{4}\left(\frac{1}{r^+} + 3\right)u^{1/2}\,\text{ierfc}\left[\frac{r^+ - 1}{(4u)^{1/2}}\right]\right.$$

$$\left. + \frac{3}{32}\left[\frac{3}{(r^+)^2} + \frac{2}{r^+} - 5\right]u\,\text{i}^2\,\text{erfc}\,\frac{r^+ - 1}{(4u)^{1/2}}\right\} \tag{R00.7}$$

where

$$u \equiv \frac{\alpha(t - \tau)}{a^2} \qquad r^+ \equiv \frac{r}{a} \tag{R00.8a, b}$$

This expression is accurate for $u < 0.1$. For $u < (r^+ - 1)^2/36$, the integral is nearly zero.

For $r^+ = 1$, Eq. (R00.7) gives

$$\int_0^a G_{R00}(a, t|r', \tau)\, 2\pi r'\, dr \approx \frac{1}{2}\left[1 - \left(\frac{u}{\pi}\right)^{1/2} - \frac{1}{4\pi^{1/2}}u^{3/2}\right] \tag{R00.9}$$

At $u = 0.25$, this gives the value of 0.35014 which is 1.3% high and is more accurate for smaller u values.

For $a/2 < r < a$ and for small times u, the integral over r' is

$$\int_0^a G_{R00}(r, t|r', \tau)2\pi r'\, dr' \approx 1 - \frac{1}{2}(r^+)^{-1/2}\left\{\text{erfc}\left[\frac{1 - r^+}{(4u)^{1/2}}\right]\right.$$

$$+ \frac{1}{4}\left(\frac{1}{r^+} + 3\right)u^{1/2}\,\text{ierfc}\,\frac{1 - r^+}{(4u)^{1/2}}$$

$$+ \frac{3}{32}\left(\frac{3}{r^{+2}} + \frac{2}{r^+} - 5\right)$$

$$\left. \times u\,\text{i}^2\,\text{erfc}\,\frac{1 - r^+}{(4u)^{1/2}}\right\} \tag{R00.10}$$

Note the similarity with Eq. (R00.7). Equation (R00.10) is accurate for $u < 0.01$. For $0 < r < a/2$, the value of the integral is nearly unity. For $r^+ = 0.5$ and $u = 0.01$, Eq. (R00.10) gives 0.9997. For large times u, an approximate result for the integral over r' from $r' = 0$ to a is

$$\int_0^a G_{R00}(r, t|r', \tau) 2\pi r' \, dr'$$

$$\approx e^{-r^{+2}/4u} \left[P\left(1, \frac{1}{4u}\right) + \frac{r^{+2}}{4u} P\left(2, \frac{1}{4u}\right) \right.$$

$$\left. + \frac{1}{2!} \frac{r^{+4}}{(4u)^2} P\left(3, \frac{1}{4u}\right) + \frac{1}{3!} \frac{r^{+6}}{(4u)^3} P\left(4, \frac{1}{4u}\right) \right] \qquad \text{(R00.11)}$$

where $P(n, x)$ is given by

$$P(n, x) = 1 - e_{n-1}(x) e^{-x} \qquad n = 1, 2, \ldots \qquad \text{(R00.12a)}$$

$$e_{n-1}(x) = \sum_{j=0}^{n-1} \frac{x^j}{j!} \qquad \text{(R00.12b)}$$

Equations (R00.12a, b) were derived using the four-term expression for $I_0(x)$ for large x for $x = \alpha(t - \tau)/rr'$.

For the center point, $r^+ = 0$, the integral is

$$\int_0^a G_{R00}(0, t|r', \tau) 2\pi r' \, dr' = 1 - e^{-1/4u} = P\left(1, \frac{1}{4u}\right) \qquad \text{(R00.13)}$$

which is exact. For $r^+ = 1$ and $u = 0.25$, the value of given by Eq. (R00.11) is 0.34569 which is -0.016% in error. Some values of the integral versus u for $r^+ = 0$, $0.5^{1/2}$, 1, 2, and 4 are given in Table R00.1.

Another expression for large times is (Beck, 1981)

$$\int_0^a G_{R00}(r, t|r', \tau) 2\pi r' \, dr'$$

$$\approx \frac{1}{4u} \left\{ 1 - \frac{1 + 2r^{+2}}{2!} \frac{1}{4u} + \frac{1 + 6r^{+2} + 3r^{+4}}{3!} \frac{1}{(4u)^2} \right\} \qquad \text{(R00.14)}$$

which requires larger u values as r^+ increases.

Table R00.1 Integral of $G_{R00}(r, t|r', \tau)$ from $r' = 0$ to a for various radii

$u^\dagger$	Green's function values for radii				
	$r^+ = 0$	$r^+ = 2^{1/2}$	$r^+ = 1$	$r^+ = 2$	$r^+ = 4$
0.020	0.999996	0.910694	0.459902	0.000000	0.000000
0.100	0.917915	0.646447	0.408230	0.008333	0.000000
0.200	0.713495	0.515406	0.364977	0.035206	0.000000
0.500	0.393469	0.324351	0.267120	0.081892	0.000590
1.000	0.221199	0.198146	0.177482	0.091529	0.006337
10.000	0.024690	0.024387	0.024088	0.022368	0.016633
100.000	0.002497	0.002494	0.002491	0.002472	0.002399

$^\dagger u \equiv \alpha(t - \tau)/a^2$.

R.3 AVERAGE GREEN'S FUNCTION FOR CIRCULAR REGION

The exact expression for the average GF for a circular region is (Amos, 1979)

$$\overline{G}_{R00}(t, \tau) \equiv \frac{4}{a^4} \int_{r=0}^{a} \int_{r'=0}^{a} G_{R00}(r, t|r', \tau) r' r \, dr' \, dr$$

$$= \frac{1}{\pi a^2} \left\{ 1 - e^{-1/2u} \left[I_0\left(\frac{1}{2u}\right) + I_1\left(\frac{1}{2u}\right) \right] \right\} \quad \text{(R00.15)}$$

where $u \equiv \alpha(t - \tau)/a^2$. Because this expression is not an easy one for subsequent integration, some approximate relations are given. One expression is

$$\pi a^2 \overline{G}_{R00}(t, \tau) = 1 - \left(\frac{u}{\pi}\right)^{1/2} \left(2 - \frac{u}{2} - \frac{u^2}{4}\right) \quad \text{for } 0 < u < 0.5 \quad \text{(R00.16a)}$$

$$\pi a^2 \overline{G}_{R00}(t, \tau) = \frac{1}{4u} \left(1 - \frac{1}{4u} + \frac{5}{96u^2} - \frac{1}{128u^3}\right) \quad \text{for } u > 0.5 \quad \text{(R00.16b)}$$

This is about 0.5% in error for $u = 0.4$ and more accurate elsewhere. Much more accurate relations are, for $0 < u < 0.15$:

$$\overline{G}_{R00}(t, \tau) \approx \frac{1}{\pi a^2} \left[1 - \left(\frac{u}{\pi}\right)^{1/2} \left(2 - \frac{u}{2} - \frac{3u^2}{16}\right. \right.$$

$$\left. \left. - \frac{15u^3}{64} - \frac{525u^4}{1024} - \frac{6615u^5}{2048}\right) \right] \quad \text{(R00.17a)}$$

for $0.15 < u < 0.55$:

$$\overline{G}_{R00}(t, \tau) \approx \frac{1}{\pi a^2} \sum_{n=1}^{6} A_n e^{-\beta_n^2 u/9} \quad \text{(R00.17b)}$$

where the A_n and β_n values are:

n	A_n	β_n
1	0.350307417	2.4048256
2	0.383866200	5.5200781
3	0.10146524	8.6537279
4	0.0008193642	11.7915344
5	0.044399184	14.9309177
6	0.0257058527	18.0710640

or $u > 0.55$:

$$\overline{G}_{R00}(t, \tau) \approx \frac{v}{\pi a^2} \left(1 - v + \frac{5v^2}{6} - \frac{7v^3}{12}\right.$$

$$\left. + \frac{7v^4}{20} - \frac{11v^5}{60} + \frac{143v^6}{1680} - \frac{143v^7}{8064}\right) \quad \text{(R00.17c)}$$

where $v \equiv 1/4u = a^2/[4\alpha(t - \tau)]$.

The Eq. (R00.17a) expression is only -0.001% in error at $u = 0.15$ and is better for smaller u values. The Eq. (R00.17b) expression is accurate to six significant figures at $u = 0.15$ and is -0.009% in error at $u = 0.55$. At $u = 0.55$, Eq. (R00.17c) is 0.007% in error and the accuracy improves with increasing u.

Values of $\pi a^2 \overline{G}_{R00}$ are given in Table R00.2 along with values for $\pi a^2 \overline{G}_{R0I}$, $I = 1, 2, 3$ for $b^+ = b/a = 2$.

E.4 DERIVATIVE OF G_{R00} WITH RESPECT TO r

Using Eq. (R00.4), for small values of $\alpha(t - \tau)/rr'$, the derivative of G_{R00} with respect to r is

$$\frac{\partial G_{R00}}{\partial r} \approx -\frac{1}{2\pi[4\pi rr'\alpha(t - \tau)]^{1/2}} \frac{1}{2\alpha(t - \tau)}$$

$$\times \exp\left[-\frac{(r - r')^2}{4\alpha(t - \tau)}\right]\left[(r - r') + \frac{\alpha(t - \tau)}{4rr'}(r + 3r')\right.$$

$$\left. + \frac{[\alpha(t - \tau)]^2}{32(rr')^2}(9r + 63r')\right] \tag{R00.18}$$

If $r = r'$, the first term inside the braces disappears and makes no contribution for $t - \tau$ greater than zero. However, as $t - \tau$ goes to zero, there can be a Dirac delta function at $r = r'$. See the X00 case.

R01 Solid cylinder, $G = 0$ at $r = b$

$$G_{R01}(r, t|r', \tau) = \frac{1}{\pi b^2}\sum_{m=1}^{\infty} e^{-\beta_m^2\alpha(t-\tau)/b^2}\frac{J_0(\beta_m r/b)J_0(\beta_m r'/b)}{J_1^2(\beta_m)} \tag{R01.1}$$

Eigenvalues are found from

$$J_0(\beta_m) = 0 \tag{R01.2}$$

Table R00.2 Comparison of $\pi a^2 \overline{G}_{R0I}$ values for $b^+ = 2$ with values of $\pi a^2 \overline{G}_{R00}$

u	$\pi a^2 \overline{G}_{R00}$	$\pi a^2 \overline{G}_{R01}$	$\pi a^2 \overline{G}_{R02}$	$\pi a^2 \overline{G}_{R03}$ $B = 0.0002$	$B = 10^4$
0.001	0.964326				
0.01	0.887445				
0.1	0.652487	0.652480	0.652487	0.652488	0.652279
0.2	0.523369			0.523620	
1	0.198544	0.150571	0.264423	0.264415	0.150534
10	0.024388	0.00000034	0.25	0.249769	3.3E-7
100	0.002494		0.25	0.247531	
1000	0.000250		0.25	0.226227	

(Ozisik, 1980, p. 92). The derivative of G with respect to n' and evaluated at $r' = b$ is

$$-\left.\frac{\partial G_{R01}}{\partial n'}\right|_{r'=b} = \frac{1}{\pi b^3} \sum_{m=1}^{\infty} e^{-\beta_m^2 \alpha(t-\tau)/b^2} \frac{\beta_m J_0(\beta_m r/b)}{J_1(\beta_m)} \qquad (R01.3)$$

Also the cross derivative at $r = r' = b$ is

$$-\left.\frac{\partial^2 G_{R01}}{\partial r\,\partial r'}\right|_{r=r'=b} = \frac{1}{2\pi b}\frac{1}{\{4\pi[\alpha(t-\tau)]^3\}^{1/2}}\left[1 + \frac{\alpha(t-\tau)}{8b^2}\right] \qquad (R01.4)$$

For small values of $\alpha(t-\tau)/b^2$ and r and r' not near b,

$$G_{R01}(r, t|r', \tau) \approx G_{R00}(r, t|r', \tau)$$

For small $\alpha(t-\tau)/b^2$ values and r and r' not near zero, $G(\cdot)$ is

$$G_{R01}(r, t|r', \tau) \approx \frac{1}{4\pi[\pi\alpha rr'(t-\tau)]^{1/2}}\left\{\exp\left[-\frac{(r-r')^2}{4\alpha(t-\tau)}\right]\right.$$
$$\left. -\exp\left[-\frac{(2b-r-r')^2}{4\alpha(t-\tau)}\right]\right\}$$

$$+\frac{1}{32\pi(rr')^{1/2}}\left(\left(\frac{1}{r'} - \frac{1}{r}\right)\text{erfc}\left\{\frac{r-r'}{[4\alpha(t-\tau)]^{1/2}}\right\}\right.$$

$$\left. -\left(\frac{1}{r'} + \frac{1}{r} - \frac{2}{b}\right)\text{erfc}\left\{\frac{2b-r-r'}{[4\alpha(t-\tau)]^{1/2}}\right\}\right) \qquad (R01.5)$$

for $0 < r' < r$. For $0 < r < r'$, exchange r and r'. For small $\alpha(t-\tau)/b^2$ values and r not near zero, the derivative is

$$-\left.\frac{\partial G_{R01}}{\partial n'}\right|_{r'=b} \approx \frac{1}{4\pi b^2\sqrt{rb}}\frac{r-b}{[\pi\alpha(t-\tau)]^{1/2}}e^{-(b-r)^2/[4\alpha(t-\tau)]}\left[\frac{b^2}{\alpha(t-\tau)} + \frac{1}{8}\frac{b}{r}\right]$$

$$(R01.6)$$

The average GF for a circular region of radius a is given by Eq. (R00.19) for $u \equiv \alpha(t-\tau)/a^2$ less than $(b^+ - 1)^2/12$ (with $b^+ \equiv b/a$) and for larger u values by

$$\overline{G}_{R01}(t, \tau) \equiv \frac{4}{a^2}\int_{r=0}^{a}\int_{r'=0}^{a} G_{R01}(r, t|r', \tau)rr'\,dr\,dr'$$

$$= \frac{4}{\pi a^2}\sum_{m=1}^{\infty} e^{-\beta_m^2\alpha(t-\tau)/b^2}\left[\frac{J_1(\beta_m a/b)}{\beta_m J_1(\beta_m)}\right]^2 \qquad (R01.7)$$

R02 Solid cylinder, $\partial G/\partial r = 0$ at $r = b$

$$G_{R02}(r, t|r', \tau) = \frac{1}{\pi b^2}\left[1 + \sum_{m=1}^{\infty} e^{-\beta_m^2\alpha(t-\tau)/b^2}\frac{J_0(\beta_m r/b)J_0(\beta_m r'/b)}{J_0^2(\beta_m)}\right] \qquad (R02.1)$$

Eigenvalues are found from

$$J_1(\beta_m) = 0 \tag{R02.2}$$

Some values are given in Table R02.

For small times ($\alpha t/b^2 < .01$) and r/b and r'/b not near zero, $G_{R02}(\cdot)$ is

$$G_{R02}(r, t|r', \tau) \approx G_{R00}(r, t|r', \tau) + \frac{1}{4\pi b(rr')^{1/2}} \left(\frac{b}{[\pi\alpha(t - \tau)]^{1/2}} \right.$$

$$\times \exp\left[-\frac{(2b - r - r')^2}{4\alpha(t - \tau)} \right]$$

$$+ \frac{1}{8}\left(6 + \frac{b}{r} + \frac{b}{r'} \right) \operatorname{erfc}\left\{ \frac{2b - r - r'}{[4\alpha(t - \tau)]^{1/2}} \right\}$$

$$+ \frac{1}{64}\left[36 + 12\frac{b}{r} + 12\frac{b}{r'} + 9\left(\frac{b}{r}\right)^2 + 9\left(\frac{b}{r'}\right)^2 \right.$$

$$\left. + 2\frac{b}{r}\frac{b}{r'} \right] \left[\frac{\alpha(t - \tau)}{b^2} \right]^{1/2}$$

$$\left. \times \operatorname{ierfc}\left\{ \frac{2b - r - r'}{[4\alpha(t - \tau)]^{1/2}} \right\} \right) \tag{R02.3}$$

For $r = r' = b$, this expression with $\tau = 0$ reduces to

$$G_{R02}(b, t|b, 0) \approx \frac{1}{2\pi b^2}\left[\frac{b}{(\pi\alpha t)^{1/2}} + \frac{1}{2} + \frac{3}{4(\pi)^{1/2}}\left(\frac{\alpha t}{b^2}\right)^{1/2} \right] \tag{R02.4}$$

A more accurate expression is given by Eq. (R02.5). For $\alpha t/b^2 < 0.1$ and at $r = b' = b$ (-0.1% at $\alpha t/b^2 = 0.04$ and -0.8% at 0.1),

$$G(b, t|b, 0) \approx \frac{1}{2\pi b^2}\left[\frac{b}{(\pi\alpha t)^{1/2}} + \frac{1}{2} + \frac{3}{4\sqrt{\pi}}\left(\frac{\alpha t}{b^2}\right)^{1/2} + \frac{3}{8}\left(\frac{\alpha t}{b^2}\right) \right] \tag{R02.5}$$

For $\alpha t/b^2 > 0.04$ at $r = r' = b$ (0.002% error at $\alpha t/b^2 = 0.04$),

$$G_{R02}(b, t|b, 0) \approx \frac{1}{\pi b^2}\left(1 + \sum_{m=1}^{4} e^{-\beta_m^2 \alpha t/b^2} \right) \tag{R02.6}$$

The β_m values are given in Table R02.

For $\alpha t/b^2 < 0.02$ for $r = 0$ and $r' = b$ (about 0.0015% error at $\alpha t/b^2 = 0.02$),

$$b^2 G(0, t|b, 0) \approx b^2[4\pi\alpha(t - \tau)]^{-1} \exp\left[-\frac{b^2}{4\alpha(t - \tau)} \right] \tag{R02.7}$$

and for $\alpha t/b^2 > 0.02$ for $r = 0$ and $r' = b$ (about 0.005% error at $\alpha t/b^2 = 0.02$)

$$b^2 G(0, t|b, 0) = \frac{1}{\pi}\left[1 + \sum_{m=1}^{6}\frac{1}{J_0(\beta_m)}\exp\left(-\beta_m^2\frac{\alpha t}{b^2}\right)\right] \qquad (R02.8)$$

where the β_m and $J_0(\beta_m)$ values are given in Table R02.

Table R02

m	β_m	$J_0(\beta_m)$
1	3.83170597	−0.40275940
2	7.01558667	0.30011575
3	10.17346814	−0.24970488
4	13.32369194	0.21835941
5	16.47063005	−0.19646537
6	19.61585851	0.18006338

The integral of $G_{R02}(r, t|r', \tau)\, 2\pi r'\, dr'$ from $r' = 0$ to a is

$$\int_0^a G_{R02}(r, t|r', \tau)\, 2\pi r'\, dr'$$

$$= \left(\frac{a}{b}\right)^2 + \frac{a}{b}\sum_{m=1}^{\infty} e^{-\beta_m^2\alpha(t-\tau)/b^2}\frac{J_0(\beta_m r/b)J_1(\beta_m a/b)}{\beta_m J_0^2(\beta_m)} \qquad (R02.9)$$

Though this expression applies for all $\alpha(t - \tau)/b^2$ values equal to or greater than zero, for small dimensionless times the expression for the integral over G_{R00} can be used. The average GF for a circular region of radius a is given by Eq. (R00.15) for $u \equiv \alpha(t - \tau)/a^2$ less than $(b^+ - 1)^2/12$ (with $b^+ \equiv b/a$) and for larger values of u by

$$\overline{G}_{R02}(t, \tau) \equiv \frac{4}{a^4}\int_{r=0}^{a}\int_{r'=0}^{a} G_{R02}(r, t|r', \tau)rr'\, dr'\, dr$$

$$= \frac{1}{\pi b^2}\left\{1 + 4\left(\frac{b}{a}\right)^2\sum_{m=1}^{\infty} e^{-\beta_m^2\alpha(t-\tau)/b^2}\left[\frac{J_1(\beta_m a/b)}{\beta_m J_0(\beta_m)}\right]^2\right\} \qquad (R02.10)$$

For $a/b = 1$, $\overline{G}_{R02}(t, \tau)$ is given by

$$\overline{G}_{R02}(t, \tau) = \frac{1}{\pi b^2} \qquad (R02.11)$$

The integral of $G_{R02}\, 2\pi r'\, dr'$ for $r' = 0$ to b is unity:

$$\int_0^b G_{R02}(r, t|r', \tau)2\pi r'\, dr' = 1 \qquad (R02.12)$$

R03 Solid cylinder, $k\partial G/\partial r + hG = 0$ at $r = b$

$$G_{R03}(r, t|r', \tau) = \frac{1}{\pi b^2}\sum_{m=1}^{\infty} e^{-\beta_m^2\alpha(t-\tau)/b^2}\frac{\beta_m^2 J_0(\beta_m r/b)J_0(\beta_m r'/b)}{J_0^2(\beta_m)(B^2 + \beta_m^2)} \qquad (R03.1)$$

The eigencondition is

$$-\beta_m J_1(\beta_m) + B J_0(\beta_m) = 0 \tag{R03.2}$$

$$B = \frac{hb}{k} \tag{R03.3}$$

The average GF for a circular region of radius a is given by Eq. (R00.15) for $u \equiv \alpha(t - \tau)/a^2$ less than $(b^+ - 1)^2/12$ (with $b^+ \equiv b/a$) and for larger values of u by

$$\overline{G}_{R03}(t, \tau) = \frac{4}{\pi a^2} \sum_{m=1}^{\infty} e^{-\beta_m^2 \alpha(t-\tau)/b^2} \frac{J_1^2(\beta_m a/b)}{J_0^2(\beta_m)(B^2 + \beta_m^2)} \tag{R03.4}$$

R10 Outside the cylindrical region $r = a$, $G = 0$ at $r = a$

$$G_{R10}(r, t|r', \tau) = \frac{1}{2\pi a^2} \int_{\beta=0}^{\infty} e^{-\beta^2(t-\tau)/a^2}$$

$$\times \ \beta[J_0(\beta r/a)Y_0(\beta) - Y_0(\beta r/a)J_0(\beta)]$$

$$\times \ \frac{[J_0(\beta r'/a)Y_0(\beta) - Y_0(\beta r'/a)J_0(\beta)]}{J_0^2(\beta) + Y_0^2(\beta)} \, d\beta \tag{R10.1}$$

$$-\frac{\partial G_{R10}}{\partial n'}\bigg|_{r'=a} = -\frac{1}{\pi^2 a^3} \int_{\beta=0}^{\infty} e^{-\beta^2 \alpha(t-\tau)/a^2}$$

$$\times \ \frac{\beta[J_0(\beta r/a)Y_0(\beta) - Y_0(\beta r/a)J_0(\beta)]}{J_0^2(\beta) + Y_0^2(\beta)} \, d\beta \tag{R10.2}$$

$$-\frac{\partial^2 G_{R10}}{\partial r \, \partial n'}\bigg|_{r'=r=a} = \frac{2}{\pi^3 a^4} \int_{\beta=0}^{\infty} e^{-\beta^2 \alpha(t-\tau)/a^2} \frac{\beta}{J_0^2(\beta) + Y_0^2(\beta)} \, d\beta \tag{R10.3}$$

R11 Hollow cylinder, $G = 0$ at $r = a$ and b

$$G_{R11}(r, t|r', \tau) = \frac{\pi}{4a^2} \sum_{m=1}^{\infty} e^{-\beta_m^2 \alpha(t-\tau)/a^2}$$

$$\times \ \frac{\beta_m^2 J_0^2(\beta_m)[J_0(\beta_m r/a)Y_0(\beta_m b/a) - J_0(\beta_m b/a)Y_0(\beta_m r/a)]}{J_0^2(\beta_m) - J_0^2(\beta_m b/a)}$$

$$\times \ \left[J_0\left(\beta_m \frac{r'}{a}\right) Y_0\left(\beta_m \frac{b}{a}\right) \right.$$

$$\left. - J_0\left(\beta_m \frac{b}{a}\right) Y_0\left(\beta_m \frac{r'}{a}\right) \right] \tag{R11.1}$$

The eigenvalues are found from

$$J_0(\beta_m) Y_0\left(\beta_m \frac{b}{a}\right) - J_0\left(\beta_m \frac{b}{a}\right) Y_0(\beta_m) = 0 \tag{R11.2}$$

The normal derivatives at $r' = a$ and b are

$$-\left.\frac{\partial G_{R11}}{\partial n'}\right|_{r'=a} = \frac{\pi}{4a^3} \sum_{m=1}^{\infty} e^{-\beta_m^2 \alpha (t-\tau)/a^2}$$

$$\times \; \beta_m^3 J_0^2(\beta_m)\,[J_0(\beta_m r/a)Y_0(\beta_m b/a) - J_0(\beta_m b/a)Y_0(\beta_m r/a)] \tag{R11.3}$$

$$\times \; \frac{[J_1(\beta_m)Y_0(\beta_m b/a) - J_0(\beta_m b/a)Y_1(\beta_m)]}{J_0^2(\beta_m) - J_0^2(\beta_m b/a)}$$

$$-\left.\frac{\partial G_{R11}}{\partial n'}\right|_{r'=b} = \frac{1}{2a^3}\frac{a}{b} \sum_{m=1}^{\infty} e^{-\beta_m^2 \alpha (t-\tau)/a^2}$$

$$\times \; \frac{\beta_m^2 J_0^2(\beta_m)\,[J_0(\beta_m r/a)Y_0(\beta_m b/a) - J_0(\beta_m b/a)Y_0(\beta_m r/a)]}{J_0^2(\beta_m) - J_0^2(\beta_m b/a)} \tag{R11.4}$$

R12 Hollow cylinder, $G = 0$ at $r = a$, $\partial G/\partial r = 0$ at $r = b$

$$G_{R12}(r, t|r', \tau) = \frac{\pi}{4a^2} \sum_{m=1}^{\infty} e^{-\beta_m^2 \alpha (t-\tau)/a^2} \frac{\beta_m^2 J_0^2(\beta_m)}{J_0^2(\beta_m) - J_1^2(\beta_m b/a)}$$

$$\times \; [J_0(\beta_m r/a)Y_1(\beta_m b/a) - J_1(\beta_m b/a)Y_0(\beta_m r/a)]$$

$$\times \; [J_0(\beta_m r'/a)Y_1(\beta_m b/a) - J_1(\beta_m b/a)Y_0(\beta_m r'/a)] \tag{R12.1}$$

where the eigenvalues are found from

$$J_0(\beta_m)Y_1(\beta_m b/a) - J_1(\beta_m b/a)Y_0(\beta_m) = 0 \tag{R12.2}$$

The normal derivative at $r' = a$ is

$$-\left.\frac{\partial G_{R12}}{\partial n'}\right|_{r'=a} = \frac{\pi}{4a^3} \sum_{m=1}^{\infty} e^{-\beta_m^2 \alpha (t-\tau)/a^2} \frac{\beta_m^3 J_0^2(\beta_m)}{J_0^2(\beta_m) - J_1^2(\beta_m b/a)}$$

$$\times \left[J_0\left(\beta_m \frac{r}{a}\right)Y_1\left(\beta_m \frac{b}{a}\right) - J_1\left(\beta_m \frac{b}{a}\right)Y_0\left(\beta_m \frac{r}{a}\right)\right]$$

$$\times \left[J_1(\beta_m)Y_1\left(\beta_m \frac{b}{a}\right) - J_1\left(\beta_m \frac{b}{a}\right)Y_1(\beta_m)\right] \tag{R12.3}$$

*R*13 Hollow cylinder, $G = 0$ at $r = a$, $kGT/\partial r + hG = 0$ at $r = b$

$$G_{R13}(r, t|r', \tau) = \frac{\pi}{4a^2} \sum_{m=1}^{\infty} e^{-\beta_m^2 \alpha(t-\tau)/a^2}$$

$$\times \frac{\beta_m^2 J_0^2(\beta_m)}{(B^2 + \beta_m^2)J_0^2(\beta_m) - V_0^2}$$

$$\times \left[S_0 J_0\left(\beta_m \frac{r}{a}\right) - V_0 Y_0\left(\beta_m \frac{r}{a}\right) \right]$$

$$\times \left[S_0 J_0\left(\beta_m \frac{r'}{a}\right) - V_0 Y_0\left(\beta_m \frac{r'}{a}\right) \right] \qquad \text{(R13.1)}$$

where

$$V_0 = \beta_m J_1\left(\beta_m \frac{b}{a}\right) + B J_0\left(\beta_m \frac{b}{a}\right) \qquad B = \frac{ha}{k} \qquad \text{(R13.2a, b)}$$

$$S_0 = -\beta_m Y_1\left(\beta_m \frac{b}{a}\right) + B Y_0\left(\beta_m \frac{b}{a}\right) \qquad \text{(R13.3)}$$

and the eigencondition is

$$S_0 J_0(\beta_m) - V_0 Y_0(\beta_m) = 0 \qquad \text{(R13.4)}$$

The normal derivative at $r' = a$ is

$$-\left.\frac{\partial G_{R13}}{\partial n'}\right|_{r'=a} = \frac{\pi}{4a^3} \sum_{m=1}^{\infty} e^{-\beta_m^2 \alpha(t-\tau)/a^2} \frac{\beta_m^3 J_0^2(\beta_m)}{(B^2 + \beta_m^2)J_0^2(\beta_m) - V_0^2}$$

$$\times \left[S_0 J_0\left(\beta_m \frac{r}{a}\right) - V_0 Y_0\left(\beta_m \frac{r}{a}\right) \right] \qquad \text{(R13.5)}$$

$$\times [S_0 J_1(\beta_m) - V_0 Y_1(\beta_m)]$$

*R*20 Outside the cylindrical region $r = a$. $\partial G/\partial r = 0$ at $r = a$

$$G_{R20}(r, t|r', \tau) = \frac{1}{2\pi a^2} \int_{\beta=0}^{\infty} e^{-\beta^2 \alpha(t-\tau)/a^2} \, d\beta$$

$$\times \beta[J_0(\beta r/a)Y_1(\beta) - Y_0(\beta r/a)J_1(\beta)] \qquad \text{(R20.1)}$$

$$\times \frac{[J_0(\beta r'/a)Y_1(\beta) - Y_0(\beta r'/a)J_1(\beta)]}{J_1^2(\beta) + Y_1^2(\beta)}$$

Note that $J_0(z)Y_1(z) - Y_0(z)J_1(z) = -2/(\pi z)$ (see Carslaw and Jaeger, 1959, p. 489)

For $r = r' = a$

$$G_{R20}(a, t|a, \tau) = \frac{1}{2\pi a^2} \frac{4}{\pi^2} \int_0^\infty \frac{\exp[-\beta^2\alpha(t - \tau)/a^2] \, d\beta}{\beta[J_1^2(\beta) + Y_1^2(\beta)]} \quad (R20.2)$$

Approximate values for $G_{R20}(a, t|a, \tau)$:

For small $t^+ = \alpha(t - \tau)/a^2$ values,

$$G_{R20}(a, t|a, \tau) \approx \frac{1}{2\pi a^2} \left[(\pi t^+)^{-1/2} - \frac{1}{2} C_1 \right.$$

$$\left. + \frac{3}{4} \left(\frac{t^+}{\pi} \right)^{1/2} C_2 - \frac{3}{8} t^+ C_3 + \frac{21}{32} \frac{1}{\pi^{1/2}} (t^+)^{3/2} C_4 \right] \quad (R20.3)$$

The series expansion has $C_1 = C_2 = C_3 = C_4 = 1$ which is accurate only for $\alpha(t - \tau)/a^2 \ll 1.0$. If Euler's transformation is used (Abramowitz and Stegun, 1964, p. 16), $C_1 = 15/16$, $C_2 = 11/16$, $C_3 = 5/16$, and $C_4 = 1/16$ and the accuracy is much improved, better than 0.15% for $t^+ < 0.4$. See Table GR20 where (R20.3c) denotes Eq. (R20.3) for $C_1 = 15/16$ and so on. More accurate values are obtained using the polynomial fit of

$$G_{R20}(a, t|a, \tau) \approx \frac{1}{2\pi a^2} \left[(\pi t^+)^{-1/2} - 0.5 + 0.413434(t^+)^{1/2} \right.$$

$$- 0.299877 t^+ + 0.154483(t^+)^{3/2}$$

$$\left. - 0.045263(t^+)^2 + 0.005484(t^+)^{5/2} \right] \quad (R20.4)$$

For large t^+ values,

$$G_{R20}(a, t|a, \tau) \approx \frac{1}{4\pi a^2} \frac{1}{t^+} \left\{ 1 - \frac{1}{2t^+} L \left[1 + \frac{3}{4t^+} (1 - L) \right] \right.$$

$$\left. - (\pi^2 + 4) \frac{C}{16t^{+2}} \right\} \quad L = \ln 4t^+ - \gamma \quad (R20.5)$$

where $\gamma = $ Euler's constant $= 0.57722$. This equation with $C = 0.5$ is accurate to $+0.1\%$ for $t^+ > 10$, to -0.6% at $t^+ = 5$. A comparison of results is given in Table GR20. Equation (R20.4) is recommended for $t^+ < 6$ and Eq. (R20.5) for $t^+ > 6$.

The integral of $G_{R20}(r, t|r', \tau) \, 2\pi r' \, dr'$ from $r' = a$ to ∞ is unity,

$$\int_0^\infty G_{R20}(r, t|r', \tau) 2\pi r' \, dr' = 1 \quad (R20.6)$$

Table GR20 Exact $2\pi G_{R20}(a, t|a, \tau)$ and approximate Eqs. (R20.4), (R20.3), and (R20.5). $u \equiv \alpha(t - \tau)/a^2$

u	(R20.4)	(R20.3)	(R20.5)	Exact
0.5	0.4846	0.4844		0.484220
1.0	0.2924	0.2923		0.292633
1.5	0.2143	0.2149		0.214567
2.0	0.1708	0.1727		0.170938
2.5	0.1428	0.1465		0.142723
3.0	0.1230			0.122844
3.5	0.1082		0.1053	0.108019
4.0	0.0966		0.0950	0.096506
4.5	0.0872		0.0864	0.087288
5.0	0.0796		0.0792	0.079730
5.5	0.0733		0.0731	0.073414
6.0	0.0683		0.0679	0.068058
6.5	0.0646		0.0633	0.063440
7.0	0.0621		0.0594	0.059429
7.5	0.0609		0.0559	0.055907
		Percent Errors		
0.5	0.07	0.04		
1.0	-0.06	-0.11		
1.5	-0.13	0.16		
2.0	-0.08	1.01		
2.5	0.03	2.67		
3.0	0.15			
3.5	0.18		-2.51	
4.0	0.09		-1.56	
4.5	-0.07		-0.98	
5.0	-0.21		-0.63	
5.5	-0.15		-0.40	
6.0	0.38		-0.25	
6.5	1.77		-0.14	
7.0	4.46		-0.07	
7.5	9.00		-0.02	

R21 Hollow cylinder, $\partial G/\partial r = 0$ at $r = a$, $G = 0$ at $r = b$

$$G_{R21}(r, t|r', \tau) = \frac{\pi}{4a^2} \sum_{m=1}^{\infty} e^{-\beta_m^2 \alpha(t-\tau)/a^2} \frac{\beta_m^2 J_1^2(\beta_m)}{J_1^2(\beta_m) - J_0^2(\beta_m b/a)}$$

$$\times \left[J_0\left(\beta_m \frac{r}{a}\right) Y_0\left(\beta_m \frac{b}{a}\right) - J_0\left(\beta_m \frac{b}{a}\right) Y_0\left(\beta_m \frac{r}{a}\right) \right]$$

$$\times \left[J_0\left(\beta_m \frac{r'}{a}\right) Y_0\left(\beta_m \frac{b}{a}\right) - J_0\left(\beta_m \frac{b}{a}\right) Y_0\left(\beta_m \frac{r'}{a}\right) \right] \quad \text{(R21.1)}$$

Eigencondition

$$J_1(\beta_m)\, Y_0\!\left(\beta_m \frac{b}{a}\right) - J_0\!\left(\beta_m \frac{b}{a}\right) Y_1(\beta_m) = 0 \tag{R21.2}$$

$$-\left.\frac{\partial G_{R21}}{\partial n'}\right|_{r'=b} = -\frac{1}{2a^2}\frac{1}{b}\sum_{m=1}^{\infty} e^{-\beta_m^2 \alpha(t-\tau)/a^2} \frac{\beta_m^2 J_0^2(\beta_m)}{J_1^2(\beta_m) - J_0^2(\beta_m b/a)}$$

$$\times \left[J_0\!\left(\beta_m \frac{r}{a}\right) Y_0\!\left(\beta_m \frac{b}{a}\right) - J_0\!\left(\beta_m \frac{b}{a}\right) Y_0\!\left(\beta_m \frac{r}{a}\right) \right] \tag{R21.3}$$

R22 Hollow cylinder, $\partial G/\partial r = 0$ at $r = a$ and b

$$G(r, t|r', \tau) = \frac{1}{\pi(b^2 - a^2)} + \frac{\pi}{4a^2}\sum_{m=1}^{\infty} e^{-\beta_m^2 \alpha(t-\tau)/a^2}$$

$$\times \frac{\beta_m^2 J_1^2(\beta_m)}{J_1^2(\beta_m) - J_1^2(\beta_m b/a)}$$

$$\times \left[J_0\!\left(\beta_m \frac{r}{a}\right) Y_1\!\left(\beta_m \frac{b}{a}\right) - J_1\!\left(\beta_m \frac{b}{a}\right) Y_0\!\left(\beta_m \frac{r}{a}\right) \right]$$

$$\times \left[J_0\!\left(\beta_m \frac{r'}{a}\right) Y_1\!\left(\beta_m \frac{b}{a}\right) - J_1\!\left(\beta_m \frac{b}{a}\right) Y_0\!\left(\beta_m \frac{r'}{a}\right) \right] \tag{R22.1}$$

Eigencondition

$$J_1(\beta_m) Y_1\!\left(\beta_m \frac{b}{a}\right) - J_1\!\left(\beta_m \frac{b}{a}\right) Y_1(\beta_m) = 0 \tag{R22.2}$$

For $r = r' = b$ and $\alpha(t - \tau)/(b - a)^2 < 1/12$,

$$G(b, t|b, \tau) \approx \frac{1}{2\pi b^2}\left\{ \frac{b}{[\pi\alpha(t - \tau)]^{1/2}} + \frac{1}{2} \right.$$

$$\left. + \frac{3}{4\sqrt{\pi}}\left[\frac{\alpha(t - \tau)}{b^2}\right]^{1/2} + \frac{3}{8}\left[\frac{\alpha(t - \tau)}{b^2}\right] \right\} \tag{R22.3}$$

For $r = r' = a$ and $\alpha(t - \tau)/(b - a)^2 < 1/12$ and $\alpha(t - \tau)/a^2 < 0.4$ (or for $b/a > 3$, only $\alpha(t - \tau)/a^2 < 0.4$), use the small t^+ expression of the $R20$ case. The integral of $G_{R22}(r, t|r', \tau)2\pi r'\, dr'$ from $r' = a$ to b is unity,

$$\int_a^b G_{R22}(r, t|r', \tau)2\pi r'\, dr' = 1 \tag{R22.4}$$

R23 Hollow cylinder, $\partial G/\partial r = 0$ at $r = a$, $k\partial G/\partial r + hG = 0$ at $r = b$

$$G(r, t|r', \tau) = \frac{\pi}{4a^2} \sum_{m=1}^{\infty} e^{-\beta_m^2 \alpha(t-\tau)/a^2} \frac{\beta_m^2 J_1^2(\beta_m)}{(B^2 + \beta_m^2)J_1^2(\beta_m) - V_0^2}$$

$$\times \left[S_0 J_0\left(\beta_m \frac{r}{a}\right) - V_0 Y_0\left(\beta_m \frac{r}{a}\right) \right]$$

$$\times \left[S_0 J_0\left(\beta_m \frac{r'}{a}\right) - V_0 Y_0\left(\beta_m \frac{r'}{a}\right) \right] \tag{R23.1}$$

where S_0 and V_0 are defined for *R*13.

Eigencondition

$$S_0 J_1(\beta_m) - V_0 Y_1(\beta_m) = 0 \tag{R23.2}$$

For $r = r' = a$ and $\alpha(t - \tau)/(b - a)^2 < 1/12$ and $\alpha(t - \tau)/a^2 < 0.4$ [or for $b/a > 3$, only $\alpha(t - \tau)/a^2 < 0.4$], use the small t^+ expression of the *R*20 case.

R30 Region outside $r = a$, $-k\partial G/\partial r + hG = 0$ at $r = a$

$$G(r, t|r', \tau) = \frac{1}{2\pi a^2} \int_{\beta=0}^{\infty} e^{-\beta^2 \alpha(t-\tau)/a^2}$$

$$\times \beta[W_0 J_0(\beta r/a) - U_0 Y_0(\beta r/a)]$$

$$\times \frac{[W_0 J_0(\beta r'/a) - U_0 Y_0(\beta r'/a)]}{U_0^2 + W_0^2} d\beta \tag{R30.1}$$

where

$$W_0 = -\beta Y_1(\beta) - B Y_0(\beta) \tag{R30.2}$$

$$U_0 = -\beta J_1(\beta) - B J_0(\beta) \qquad B = \frac{ha}{k} \tag{R30.3a, b}$$

$$G(a, t|a, \tau) \approx \frac{kB}{\alpha} \left\{ \left[\frac{\alpha}{\pi a^2(t - \tau)} \right]^{1/2} + \left(B + \frac{1}{2} \right) \frac{\alpha}{a^2} \right.$$

$$\left. - 2\left(B^2 + B + \frac{3}{8} \right) \left[\frac{\alpha^3(t - \tau)}{a^6} \right]^{1/2} \right\} \tag{R30.4}$$

for small $\alpha(t - \tau)/a^2$ values and $B \le 1$.

Table R.1 Steady Green's functions, radial cylindrical coordinates

$$G \text{ satisfies: } \frac{1}{r}\frac{d}{dr}\left(r\frac{dG}{dr}\right) = -\delta(r - r')$$

Case	Homogeneous boundary conditions: $B_1 = h_1 a/k$, $B_2 = b_2 b/k$	$2\pi G(r\|r')$
R01	$\partial G(0\|r')/\partial r = 0$; $G(b\|r') = 0$	$\ln(b/r')$; $r < r'$ $\ln(b/r)$; $r > r'$
R03	$\partial G(0\|r')/\partial r = 0$ $k\partial G(b\|r')/\partial r + h_2 G(b\|r') = 0$	$\ln(b/r') + 1/B_2$; $r < r'$ $\ln(b/r) + 1/B_2$; $r > r'$
R11	$G(a\|r') = 0$; $G(b\|r') = 0$	$\ln(b/r')\ln(r/a)/\ln(b/a)$; $r < r'$ $\ln(b/r)\ln(r'/a)/\ln(b/a)$; $r > r'$
R12	$G(a\|r') = 0$; $\partial G(b\|r')/\partial r = 0$	$\ln(r/a)$; $r < r'$ $\ln(r'/a)$; $r > r'$
R13	$G(a\|r') = 0$ $k\partial G(b\|r')/\partial r + h_2 G(b\|r') = 0$	$(\ln r/a)(1 + B_2 \ln b/r')(1 + B_2 \ln b/a)$; $r < r'$ $(\ln r'/a)(1 + B_2 \ln b/r)(1 + B_2 \ln b/a)$; $r > r'$
R21	$\partial G(a\|r')/\partial r = 0$ $G(b\|r') = 0$	$\ln(b/r')$; $r < r'$ $\ln(b/r)$; $r > r'$
R23	$\partial G(a\|r')/\partial r = 0$ $k\partial G(b\|r')/\partial r + h_2 G(b\|r') = 0$	$(1/B_2 + \ln b/r')$; $r < r'$ $(1/B_2 + \ln b/r)$; $r > r'$
R31[†]	$k\partial G(a\|r')/\partial n + h_1 G(a\|r') = 0$ $G(b\|r') = 0$	$(\ln b/r')(1/B_1 + \ln r/a)/(1/B_1 + \ln b/a)$; $r < r'$ $(\ln b/r)(1/B_1 + \ln r'/a)/(1/B_1 + \ln b/a)$; $r > r'$
R32[†]	$k\partial G(a\|r')/\partial n + h_1 G(a\|r') = 0$ $\partial G(b\|r')/\partial r = 0$	$(1/B_1 + \ln r/a)$; $r < r'$ $(1/B_1 + \ln r'/a)$; $r > r'$
R33[†]	$k\partial G(a\|r')/\partial n + h_1 G(a\|r') = 0$ $k\partial G(b\|r')/\partial r + h_2 G(b\|r') = 0$	$[(B_2 B_2(\ln b/r')(\ln r/a) + B_1 \ln r/a + B_2 \ln b/r' + 1)]/$ $(B_1 + B_2 + B_1 B_2 \ln b/a)$, for $r < r'$ $[(B_2 B_2(\ln b/r)(\ln r'/a) + B_1 \ln r'/a + B_2 \ln b/r + 1)]/$ $(B_1 + B_2 + B_1 B_2 \ln b/a)$, for $r < r'$

[†] $\partial/\partial n|_{r=a} = -\partial/\partial r|_{r=a}$.

R31 Hollow cylinder, $-k\partial G/\partial r + hG = 0$ at $r = a$, $G = 0$ at $r = b$

$$G(r, t|r', \tau) = \frac{\pi}{4a^2} \sum_{m=1}^{\infty} e^{-\beta_m^2 \alpha(t-\tau)/a^2}$$

$$\times \frac{\beta_m^2 U_0^2}{U_0^2 - (B^2 + \beta_m^2)J_0^2(\beta_m b/a)}$$

$$\times \left[J_0\left(\beta_m \frac{r}{a}\right) Y_0\left(\beta_m \frac{b}{a}\right) - J_0\left(\beta_m \frac{b}{a}\right) Y_0\left(\beta_m \frac{r}{a}\right) \right]$$

$$\times \left[J_0\left(\beta_m \frac{r'}{a}\right) Y_0\left(\beta_m \frac{b}{a}\right) \right.$$

$$\left. - J_0\left(\beta_m \frac{b}{a}\right) Y_0\left(\beta_m \frac{r'}{a}\right) \right] \tag{R31.1}$$

where

$$U_0 = -\beta_m J_1(\beta_m) - BJ_0(\beta_m)$$

$$W_0 = -\beta_m Y_1(\beta_m) - BY_0(\beta_m) \qquad B = \frac{ha}{k} \tag{R31.2a, b, c}$$

Eigencondition (Ozisik, 1980, pp. 96, 97):

$$U_0 Y_0\left(\beta_m \frac{b}{a}\right) - W_0 J_0\left(\beta_m \frac{b}{a}\right) \tag{R31.3}$$

$$-\left. \frac{\partial G}{\partial n'} \right|_{r'=b} = \frac{1}{2a^2} \frac{1}{b} \sum_{m=1}^{\infty} e^{-\beta_m^2 \alpha(t-\tau)/a^2}$$

$$\times \frac{\beta_m^2 U_0^2 [J_0(\beta_m r/a) Y_0(\beta_m b/a) - J_0(\beta_m b/a) Y_0(\beta_m r/a)]}{U_0^2 - (B^2 + \beta_m^2)J_0^2(\beta_m b/a)} \tag{R31.4}$$

R32 Hollow cylinder, $-k\partial G/\partial r + hG = 0$ at $r = a$, $\partial G/\partial r = 0$ at $r = b$

$$G(r, t|r', \tau) = \frac{\pi}{4a^2} \sum_{m=1}^{\infty} e^{-\beta_m^2 \alpha(t-\tau)/a^2}$$

$$\times \frac{\beta_m^2 U_0^2}{U_0^2 - (B^2 + \beta_m^2)J_1^2(\beta_m b/a)}$$

$$\times \left[J_0\left(\beta_m \frac{r}{a}\right) Y_1\left(\beta_m \frac{b}{a}\right) - J_1\left(\beta_m \frac{b}{a}\right) Y_0\left(\beta_m \frac{r}{a}\right) \right]$$

$$\times \left[J_0\left(\beta_m \frac{r'}{a}\right) Y_1\left(\beta_m \frac{b}{a}\right) \right.$$

$$- J_1\left(\beta_m \frac{b}{a}\right) Y_0\left(\beta_m \frac{r'}{a}\right)\Bigg] \tag{R32.1}$$

For U_0 and W_0 see $R31$.

Eigencondition

$$U_0 Y_1\left(\beta_m \frac{b}{a}\right) - W_0 J_1\left(\beta_m \frac{b}{a}\right) = 0 \tag{R32.2}$$

R33 **Hollow cylinder,** $-k\partial G/\partial r + h_1 G = 0$ **at** $r = a$, $k\partial G/\partial r + h_2 G = 0$ **at** $r = b$

$$G(r, t|r', \tau) = \frac{\pi}{4a^2} \sum_{m=1}^{\infty} e^{-\beta_m^2 \alpha(t-\tau)/a^2}$$

$$\times \frac{\beta_m^2 U_0^2}{(B_2^2 + \beta_m^2)U_0^2 - (B_1^2 + \beta_m^2)V_0^2}$$

$$\times \left[S_0 J_0\left(\beta_m \frac{r}{a}\right) - V_0 Y_0\left(\beta_m \frac{r}{a}\right)\right]$$

$$\times \left[S_0 J_0\left(\beta_m \frac{r'}{a}\right) - V_0 Y_0\left(\beta_m \frac{r'}{a}\right)\right] \tag{R33.1}$$

where

$$S_0 \equiv -\beta_m Y_1\left(\beta_m \frac{b}{a}\right) + B_2 Y_0\left(\beta_m \frac{b}{a}\right) \tag{R33.2}$$

$$U_0 \equiv -\beta_m J_1(\beta_m) - B_1 J_0(\beta_m) \tag{R33.3}$$

$$V_0 \equiv -\beta_m J_1\left(\beta_m \frac{b}{a}\right) + B_2 J_0\left(\beta_m \frac{b}{a}\right) \tag{R33.4}$$

$$W_0 \equiv -\beta_m Y_1(\beta_m) - B_1 Y_0(\beta_m) \tag{R33.5}$$

$$B_1 \equiv \frac{h_1 a}{k} \qquad B_2 \equiv \frac{h_2 a}{k} \tag{R33.6a, b}$$

Eigencondition

$$S_0 U_0 - V_0 W_0 = 0 \tag{R33.7}$$

REFERENCES

Abramowitz, M., and Stegun, I. A. (Eds.), 1964, *Handbook of Mathematical Functions with Formulas, Graphs and Mathematical Tables*, Applied Mathematics Series 55, National Bureau of Standards, Wash., D.C.

Amos, D., 1979, Personal Communication, Sandia Laboratories, Albuquerque, NM.

Beck, J. V., 1981, Large Time Solutions for Temperature in a Semi-Infinite Body with a Disk Heat Source, *Int. J. Heat Mass Transfer*, vol. 24, pp. 155–164.

Carslaw, H. S., and Jaeger, J. C., 1959, *Conduction of Heat in Solids*, 2nd ed., Oxford University Press, New York.

GREEN'S FUNCTIONS FOR CYLINDRICAL COORDINATES (r, ϕ)

$$dv' = r' \, dr' \, d\phi'$$

$$ds' = a \, d\phi' \text{ at } r = a \text{ and } ds' = b \, d\phi' \text{ at } r = b$$

$$ds' = dr' \text{ at } \phi = 0 \text{ or } \phi_0$$

The partial differential equation for transient conduction with cylindrical coordinates (r, ϕ) is

$$\frac{1}{r} \frac{\partial}{\partial r} \left(r \frac{\partial T}{\partial r} \right) + \frac{1}{r^2} \frac{\partial^2 T}{\partial \phi^2} = \frac{1}{\alpha} \frac{\partial T}{\partial t}$$

R00Φ11 Wedge for $0 < \phi < \phi_0 < 2\pi$ and with $G = 0$ at $\phi = 0$ and $\phi = \phi_0$ $(dv' = r' \, dr' \, d\phi', \, ds' = dr')$

$$G(r, \phi, t | r', \phi', \tau) = \frac{1}{\phi_0 \, \alpha(t - \tau)} e^{-(r^2 + r'^2)/[4\alpha(t - \tau)]} \sum_{n=1}^{\infty}$$

$$\times \, n I_s \left(\frac{rr'}{2\alpha(t - \tau)} \right) \sin \left(n\pi \frac{\phi}{\phi_0} \right)$$

$$\times \sin \left(n\pi \frac{\phi'}{\phi_0} \right) \tag{R00Φ11.1}$$

where $s = n\pi/\phi_0$ (Carslaw and Jaeger, 1959, p. 379).

$$-\left. \frac{\partial G}{\partial n'} \right|_{\phi'=0} = \frac{\pi}{r' \phi_0^2 \, \alpha(t - \tau)} e^{-(r^2 + r'^2)/[4\alpha(t - \tau)]} \sum_{n=1}^{\infty}$$

$$\times \, n I_s \left[\frac{rr'}{2\alpha(t - \tau)} \right] \sin \left(n\pi \frac{\phi}{\phi_0} \right) \tag{R00Φ11.2}$$

451

$$-\left.\frac{\partial G}{\partial n'}\right|_{\phi'=\phi_0} = -\frac{\pi}{r'\phi_0^2\,\alpha(t-\tau)}\,e^{-(r^2+r'^2)/[4\alpha(t-\tau)]}$$

$$\times\sum_{n=1}^{\infty}nI_s\left[\frac{rr'}{2\alpha(t-\tau)}\right]\sin\left(n\pi\,\frac{\phi}{\phi_0}\right)(-1)^n \quad \text{(R00Φ11.3)}$$

R00Φ12 Wedge for $0 < \phi < \phi_0 < 2\pi$ and with $G = 0$ at $\phi = 0$ and $\partial G/\partial\phi = 0$ at $\phi = \phi_0$ ($dv' = r'\,dr'\,d\phi'$ and $ds' = dr'$)

$$G(r,\phi,t|r',\phi',\tau) = \frac{1}{\phi_0\,\alpha(t-\tau)}\,e^{-(r^2+r'^2)/[4\alpha(t-\tau)]}$$

$$\times\sum_{m=1}^{\infty}I_{\beta_m}\left(\frac{rr'}{2\alpha(t-\tau)}\right)\sin(\beta_m\phi)\sin(\beta_m\phi') \quad \text{(R00Φ12.1)}$$

where $\beta_m = (2m-1)\dfrac{\pi}{2\phi_0}$, $m = 1, 2, \ldots$.

R00Φ22 Wedge for $0 < \phi < \phi_0 < 2\pi$ and with $\partial G/\partial\phi$ at $\phi = 0$ and at $\phi = \phi_0$ ($dv' = r'\,dr'\,d\phi'$ and $ds' = dr'$)

$$G(r,\phi,t|r',\phi',\tau) = \frac{1}{2\phi_0\,\alpha(t-\tau)}\,e^{-(r^2+r'^2)/[4\alpha(t-\tau)]}\left\{I_0\left[\frac{rr'}{2\alpha(t-\tau)}\right]\right.$$

$$+2\sum_{n=1}^{\infty}\cos\left(n\pi\,\frac{\phi}{\phi_0}\right)\cos\left(n\pi\,\frac{\phi'}{\phi_0}\right)$$

$$\left.\times I_s\left[\frac{rr'}{2\alpha(t-\tau)}\right]\right\} \quad \text{(R00Φ22.1)}$$

where $s = n\pi/\phi_0$ (Carslaw and Jaeger, 1959, p. 379).

R01Φ00 Solid cylinder with radial and angular dependence; $G = 0$ at $r = a$

$$G(r,\phi,t|r',\phi',\tau) = \frac{2}{a^2}\sum_{n=0}^{\infty}\frac{1}{\pi}\cos[n(\phi-\phi')]\sum_{m=1}^{\infty}$$

$$\times e^{-\beta_{mn}^2\alpha(t-\tau)/a^2}\frac{J_n(\beta_{mn}r/a)\,J_n(\beta_{mn}r'/a)}{[J_n'(\beta_{mn})]^2} \quad \text{(R01Φ00.1)}$$

where β_{mn} for $m, n = 1, 2, \ldots$, are the positive roots of $J_n(\beta_{mn}) = 0$. Replace π for 2π for $n = 0$ [Carslaw and Jaeger, 1959, p. 377, Eq. (6); Ozisik, 1980, p. 119, Eq. (3-141)].

R01Φ11 Sector of radius b; $0 \leq \phi \leq \phi_0 < 2\pi$; $G = 0$ at $r = b$, $\phi = 0$ and $\phi = \phi_0$

$$G(r, \phi, t | r', \phi', \tau) = \frac{4}{b^2 \phi_0} \sum_{m=1}^{\infty} \sum_{\nu} e^{-\beta_{m\nu}^2 \alpha (t - \tau)/b^2}$$

$$\times \frac{J_\nu(\beta_{m\nu} r/b) \, J_\nu(\beta_{m\nu} r'/b) \, \sin{(\nu\phi)} \, \sin{(\nu\phi')}}{J_\nu'^2(\beta_{m\nu})} \quad \text{(R01Φ11.1)}$$

where $\qquad\qquad \nu = \dfrac{n\pi}{\phi_0} \qquad n = 1, 2, 3, \ldots \qquad$ (R01Φ11.2)

and the $\beta_{m\nu}$ eigenvalues are given by the positive roots of

$$J_\nu(\beta_{m\nu}) = 0 \qquad \text{for the above } \nu \text{ values} \qquad \text{(R01Φ11.3)}$$

R01Φ12 Sector of radius b; $0 \leq \phi \leq \phi_0 < 2\pi$; $G = 0$ at $r = b$ and $\phi = 0$, $\partial G/\partial \phi = 0$ at $\phi = \phi_0$.

$$G(r, \phi, t | r', \phi', \tau) = \frac{4}{b^2 \phi_0} \sum_{m=1}^{\infty} \sum_{\nu} e^{-\beta_{m\nu}^2 \alpha (t - \tau)/b^2}$$

$$\times \frac{J_\nu(\beta_{m\nu} r/b) \, J_\nu(\beta_{m\nu} r'/b) \, \sin{\nu\phi} \, \sin{\nu\phi'}}{J_\nu'^2(\beta_{m\nu})} \quad \text{(R01Φ12.1)}$$

where $\qquad\qquad \nu = (2n - 1)\dfrac{\pi}{2\phi_0} \qquad n = 1, 2, 3, \ldots \qquad$ (R01Φ12.2)

and the $\beta_{m\nu}$ eigenvalues are given by the positive roots of

$$J_\nu(\beta_{m\nu}) = 0 \qquad \text{for the above } \nu \text{ values} \qquad \text{(R01Φ12.3)}$$

R01Φ22 Sector of radius b; $0 \leq \phi \leq \phi_0 < 2\pi$; $G = 0$ at $r = b$, $\partial G/\partial \phi = 0$ at $\phi = 0$ and $\phi = \phi_0$.

$$G(r, \phi, t | r', \phi', \tau) = \frac{1}{b^2 \phi_0} \sum_{m=1}^{\infty} \sum_{\nu} 4 e^{-\beta_{m\nu}^2 \alpha (t - \tau)/b^2}$$

$$\times \frac{J_\nu(\beta_{m\nu} r/b) \, J_\nu(\beta_{m\nu} r'/b) \, \cos{(\nu\phi)} \, \cos{(\nu\phi')}}{J_\nu'^2(\beta_{m\nu})}$$

$$\text{(R01Φ22.1)}$$

where $\qquad\qquad \nu = \dfrac{n\pi}{\phi_0} \qquad n = 0, 1, 2, 3, \ldots \qquad$ (R01Φ22.2)

and the $\beta_{m\nu}$ eigenvalues are given by the positive roots of

$$J_\nu(\beta_{m\nu}) = 0 \qquad \text{for the above } \nu \text{ values.} \qquad \text{(R01Φ22.3)}$$

Also replace the 4 coefficient for $\nu = 0$ by the value of 2.

R02Φ00 Solid cylinder with radial and angular dependence; $\partial G/\partial r = 0$ at $r = a$

$$G(r, \phi, t|r', \phi', \tau) = \frac{2}{a^2} \left\{ \frac{1}{2\pi} + \sum_{n=0}^{\infty} \frac{1}{\pi} \cos [n(\phi - \phi')] \sum_{m=1}^{\infty} e^{-\beta_{mn}^2 \alpha(t-\tau)/a^2} \right.$$

$$\left. \times \frac{\beta_{mn}^2 J_n(\beta_{mn}r/a) J_n(\beta_{mn}r'/a)}{(\beta_{mn}^2 - n^2) J_n^2(\beta_{mn})} \right\}$$

where β_{mn} are the positive roots of $J_n'(\beta_{mn}) = 0$. (For $\beta = 0$, the r-direction equation from the separation of variables is $r^2 R'' + rR' - n^2 R = 0$ which has the solution $R = C_1 r^{-n} + C_2 r^n$, $n \neq 0$. The solution is $R = 0$. For $n = 0$, $R = C$.) Replace π inside the summation by 2π for $n = 0$ [Carslaw and Jaeger, 1959, p. 378, Eq. (7)].

R02Φ11 Sector of radius b; $\partial G/\partial r = 0$ at $r = b$, at $G = 0$ at $\phi = 0$ and ϕ_0

$$G(r, \phi, t|r', \phi', \tau) = \frac{4}{b^2 \phi_0} \left\{ \frac{1}{2} + \sum_{m=1}^{\infty} \sum_{\nu}^{\infty} e^{-\beta_{m\nu}^2 \alpha(t-\tau)/b^2} \right.$$

$$\left. \times \frac{\beta_{m\nu}^2 J_\nu(\beta_{m\nu}r/b) J_\nu(\beta_{m\nu}r'/b) \sin (\nu\phi) \sin (\nu\phi')}{(\beta_{m\nu}^2 - \nu^2) J_\nu^2(\beta_{m\nu})} \right\}$$

$$\text{(R02Φ11.1)}$$

where $\qquad \nu = \dfrac{n\pi}{\phi_0} \qquad n = 1, 2, 3, \ldots \qquad$ (R02Φ11.2)

and the $\beta_{m\nu}$ eigenvalues are given by the positive roots of

$$J_\nu'(\beta_{m\nu}) = 0 \qquad \text{(R02Φ11.3)}$$

R02Φ12 Sector of radius b; $\partial G/\partial r = 0$ at $r = b$, $G = 0$ at $\phi = 0$ and $\partial G/\partial r = 0$ at $\phi = \phi_0$

$$G(r, \phi, t|r', \phi', \tau) = \frac{4}{b^2 \phi_0} \left[\frac{1}{2} + \sum_{m=1}^{\infty} \sum_{\nu}^{\infty} e^{-\beta_{m\nu}^2 \alpha(t-\tau)/b^2} \right.$$

$$\left. \times \frac{\beta_{m\nu}^2 J_\nu(\beta_{m\nu}r/b) J_\nu(\beta_{m\nu}r'/b) \sin (\nu\phi) \sin (\nu\phi')}{(\beta_{m\nu}^2 - \nu^2) J_\nu^2(\beta_{m\nu})} \right]$$

$$\text{(R02Φ12.1)}$$

where $\qquad \nu = (2n - 1) \dfrac{\pi}{2\phi_0} \qquad n = 1, 2, 3, \ldots \qquad$ (R02Φ12.2)

and the $\beta_{m\nu}$ eigenvalues are given by the positive roots of

$$J_\nu'(\beta_{m\nu}) = 0 \qquad \text{(R02Φ12.3)}$$

R02Φ22 **Sector of radius b; $\partial G/\partial r = 0$ at $r = b$, $\partial G/\partial r = 0$ at $\phi = 0$ and ϕ_0**

$$G(r, \phi, t | r', \phi', \tau) = \frac{1}{b^2 \phi_0} \left[2 + \sum_{m=1}^{\infty} \sum_{\nu} 4 e^{-\beta_{m\nu}^2 \alpha(t-\tau)/b^2} \right.$$

$$\left. \times \frac{\beta_{m\nu}^2 J_\nu(\beta_{m\nu} r/b) J_\nu(\beta_{m\nu} r'/b) \cos(\nu\phi) \cos(\nu\phi')}{(\beta_{m\nu}^2 - \nu^2) J_\nu^2(\beta_{m\nu})} \right]$$

$$(R02\Phi22.1)$$

where $$\nu = \frac{n\pi}{\phi_0} \qquad n = 0, 1, 2, \ldots \qquad (R02\Phi22.2)$$

and the $\beta_{m\nu}$ eigenvalues are given by the positive roots of

$$J_\nu'(\beta_{m\nu}) = 0 \qquad \text{for the above } \nu \text{ values} \qquad (R02\Phi22.3)$$

Also replace the 4 coefficient for $\nu = 0$ by the value of 2.

R11Φ00 **Annulus with radial and angular dependence; $G = 0$ at $r = a$ and b**

$$G(r, \phi, t | r', \phi', \tau) = \frac{1}{b^2} \sum_{m=1}^{\infty} \sum_{n=0}^{\infty} \frac{e^{-\beta_{mn}^2 \alpha(t-\tau)/b^2}}{\pi N(\beta_{mn})} R_n(\beta_{mn}, r) R_n(\beta_{mn}, r')$$

$$\times \cos[n(\phi - \phi')]$$

Replace π by 2π for $n = 0$. Also the following relations are given:

$$R_n(\beta_{mn}, r) = J_n(\beta_{mn} r/b) Y_n(\beta_{mn}) - J_n(\beta_{mn}) Y_n(\beta_{mn} r/b)$$

$$\frac{1}{N(\beta_{mn})} = \frac{\pi^2}{2} \frac{\beta_{mn}^2 J_n^2(\beta_{mn} a/b)}{J_n^2(\beta_{mn} a/b) - J_n^2(\beta_{mn})}$$

and β_{mn}'s are the positive roots of

$$J_n(\beta_{mn} a/b) Y_n(\beta_{mn}) - J_n(\beta_{mn}) Y_n(\beta_{mn} a/b) = 0$$

R12Φ00 **Annulus with radial and angular dependence: $G = 0$ at $r = a$ and $\partial G/\partial x = 0$ at $r = b$**

$$G(r, \phi, t | r', \phi', \tau) = \frac{1}{b^2} \sum_{m=1}^{\infty} \sum_{n=0}^{\infty} \frac{e^{-\beta_{mn}^2 a(t-\tau)/b^2}}{\pi N(\beta_{mn})} R_n(\beta_{mn}, r) R_n(\beta_{mn}, r')$$

$$\times \cos[n(\phi - \phi')]$$

Replace π by 2π for $n = 0$. Also the following relations are given:

$$R_{mn}(\beta_{mn}, r) = J_n(\beta_{mn} r/b) Y_n'(\beta_{mn}) - J_n'(\beta_{mn}) Y_n(\beta_{mn} r/b)$$

$$\frac{1}{N(\beta_{mn})} = \frac{\pi^2}{2} \frac{\beta_{mn}^2 J_n^2(\beta_{mn} a/b)}{[1 - (n/\beta_{mn})^2] J_n^2(\beta_{mn} a/b) - J_n'^2(\beta_{mn})}$$

and the β_{mn}'s are the positive roots of

$$J_n(\beta_{mn}a/b)\, Y_n'(\beta_{mn}) - J_n'(\beta_{mn})\, Y_n(\beta_{mn}a/b) = 0$$

for $m = 1, 2, \ldots$, and $n = 0, 1, 2, \ldots$.

R13Φ00 Annulus with radial and angular dependence; $G = 0$ at $r = a$ and $k\partial G/\partial r + hG = 0$ at $r = b$

$$G(r, \phi, t|r', \phi', \tau) = \frac{1}{b^2} \sum_{m=1}^{\infty} \sum_{n=0}^{\infty} \frac{e^{-\beta_{mn}^2 \alpha(t-\tau)/b^2}}{\pi N(\beta_{mn})} R_n(\beta_{mn}, r)\, R_n(\beta_{mn}, r')$$

$$\times\, \cos[n(\phi - \phi')]$$

Replace π by 2π for $n = 0$. Also the following relations are given:

$$R_{mn}(\beta_{mn}, r) = S_{mn} J_n(\beta_{mn} r/b) - V_{mn} Y_n(\beta_{mn} r/b)$$

$$S_{mn} \equiv \beta_{mn} Y_n'(\beta_{mn}) + B Y_n(\beta_{mn}) \qquad B \equiv hb/k$$

$$V_{mn} \equiv \beta_{mn} J_n'(\beta_{mn}) + B J_n(\beta_{mn})$$

$$\frac{1}{N(\beta_{mn})} = \frac{\pi^2}{2} \frac{\beta_{mn}^2 J_n^2(\beta_{mn}a/b)}{C_{mn} J_n^2(\beta_{mn}a/b) - V_{mn}^2}$$

$$C_{mn} \equiv B^2 + \beta_{mn}^2 \left[1 - \left(\frac{n}{\beta_{mn}}\right)^2\right]$$

and the β_{mn}'s are the positive roots of

$$S_{mn} J_n(\beta_{mn}a/b) - V_{mn} Y_n(\beta_{mn}a/b) = 0$$

for $m = 1, 2, \ldots$, and $n = 0, 1, 2, \ldots$.

R21Φ00 Annulus with radial and angular dependence; $\partial G/\partial r = 0$ at $r = a$ and $G = 0$ at $r = b$ Same as R12Φ00 with $a \to b$ and $b \to a$.

R22Φ00 Annulus with radial and angular dependence; $\partial G/\partial r = 0$ at $r = a$ and $r = b$

$$G(r, \phi, t|r', \phi', \tau) = \frac{1}{b^2} \left\{ \frac{1}{\pi[1 - (a/b)^2]} + \sum_{m=1}^{\infty} \sum_{n=0}^{\infty} \frac{e^{-\beta_{mn}^2 \alpha(t-\tau)/b^2}}{\pi N(\beta_{mn})} \right.$$

$$\left. \times\, R_n(\beta_{mn}, r)\, R_n(\beta_{mn}, r')\, \cos\,[n(\phi - \phi')] \right\}$$

For $n = 0$, replace π inside the summation by 2π. Also the following relations are given:

$$R_n(\beta_{mn}, r) = J_n(\beta_{mn} r/b)\, Y_n'(\beta_{mn}) - J_n'(\beta_{mn})\, Y_n(\beta_{mn} r/b)$$

$$\frac{1}{N(\beta_{mn})} = \frac{\pi^2}{2} \frac{\beta_{mn}^2 J_n'^2(\beta_{mn}a/b)}{[1 - (n/\beta_{mn})^2] J_n'^2(\beta_{mn}a/b) - \{1 - [nb/(\beta_{mn}{}^a)]^2\} J_n'^2(\beta_{mn})}$$

and the β_{mn}'s are the positive roots of

$$J'_n(\beta_{mn}a/b) \, Y'_n(\beta_{mn}) - J'_n(\beta_{mn}) \, Y'_n(\beta_{mn}a/b) = 0$$

for $m = 1, 2, \ldots$, and $n = 0, 1, 2, \ldots$.

$R23\Phi00$ Annulus with radial and angular dependence; $\partial G/\partial r = 0$ at $r = a$ and $+ \, k\partial G/\partial r + hG = 0$ at $r = b$

$$G(r, \phi, t|r', \phi', \tau) = \frac{1}{b^2} \sum_{m=1}^{\infty} \sum_{n=0}^{\infty} \frac{e^{-\beta_{mn}^2 \alpha(t-\tau)/b^2}}{\pi N(\beta_{mn})} R_n(\beta_{mn}, r) R_n(\beta_{mn}, r')$$

$$\times \cos [n(\phi - \phi')]$$

Replace π by 2π for $n = 0$. Also the following relations are given:

$$R_{mn}(\beta_{mn}, r) = S_{mn} J_m(\beta_{mn} r/b) - V_{mn} Y(\beta_{mn} r/b)$$

See $R13\Phi00$ for S_{mn} and V_{mn}.

$$\frac{1}{N(\beta_{mn})} = \frac{\pi^2}{2} \frac{\beta_{mn}^2 \, J'^2_n(\beta_{mn}a/b)}{BJ'^2_n(\beta_{mn}a/b) - \{1 - [nb/(\beta_{mn}{}^g)]^2\} V_{mn}^2} \qquad B \equiv \frac{hb}{k}$$

and the β_{mn}'s are the positive roots of

$$S_{mn} J'_n(\beta_{mn}a/b) - V_{mn} Y'_n(\beta_{mn}a/b) = 0$$

for $m = 1, 2, \ldots$, and $n = 0, 1, 2, \ldots$.

$R31\Phi00$ Annulus with radial and angular dependence: $-k\partial G/\partial r + hG = 0$ at $r = a$ and $G = 0$ at $r = b$ Same as $R13\Phi00$ with $a \to b$, $b \to a$ and $h \to -h$.

$R32\Phi00$ Annulus with radial and angular dependence: $-k\partial G/\partial r + hG = 0$ at $r = a$ and $\partial G/\partial r = 0$ at $r = b$ Same as $R23\Phi00$ with $a \to b$, $b \to a$, and $h \to -h$.

$R33\Phi00$ Annulus with radial and angular dependence; $-k\partial G/\partial r + h_1 G = 0$ at $r = a$ and $k\partial G/\partial r + h_2 G = 0$ at $r = b$ (See Tables $R\Phi.1$–$R\Phi.4$ for a summary of the $R\Phi$ cases).

$$G(r, \phi, t|r', \phi', \tau) = \frac{1}{b^2} \sum_{m=1}^{\infty} \sum_{n=0}^{\infty} \frac{e^{-\beta_{mn}^2 \alpha(t-\tau)/b^2}}{\pi N(\beta_{mn})} R_n(\beta_{mn}, r) R_n(\beta_{mn}, r')$$

$$\times \cos [n(\phi - \phi')]$$

Replace π by 2π for $n = 0$. Also the following relations are given:

$$R_{mn}(\beta_{mn}, r) = S_{mn} J_n (\beta_{mn}{}^{r/b}) - V_{mn} Y_n (\beta_{mn}{}^{r/b})$$

$$S_{mn} = \beta_{mn} Y_n'(\beta_{mn}) + B_2 Y_n(\beta_{mn}) \qquad B_2 = \frac{h_2 b}{k}$$

$$V_{mn} = \beta_{mn} J_n'(\beta_{mn}) + B_2 J_n(\beta_{mn})$$

$$\frac{1}{N(\beta_{mn})} = \frac{\pi^2}{2} \frac{\beta_{mn}^2 U_{mn}^2}{C_2 U_{mn}^2 - C_1 V_{mn}^2}$$

$$C_1 = B_1^2 + \beta_{mn}^2 \left[1 - \left(\frac{n}{\beta_{mn} a/b} \right)^2 \right] \qquad B_1 = h_1 b/k$$

$$C_2 = B_2^2 + \beta_{mn}^2 \left[1 - \left(\frac{n}{\beta_{mn}} \right)^2 \right]$$

$$U_{mn} = \beta_{mn} J_n' (\beta_{mn}{}^{a/b}) - B_1 J_n (\beta_{mn}{}^{a/b})$$

and the β_{mn}'s are the positive roots of

$$S_{mn} U_{mn} - V_{mn} W_{mn} = 0$$

for $m = 1, 2, \ldots$, and $n = 0, 1, 2, \ldots$.

$$W_{mn} = \beta_{mn} Y_n' \left(\beta_{mn} \frac{a}{b} \right) - B_1 Y_n \left(\beta_{mn} \frac{a}{b} \right)$$

Summary for Cases $RIJ\Phi KL\ J \neq 0, K, L \neq 0$

$$G(r, \phi, t|r', \phi', \tau) = \sum_{m=0}^{\infty} \sum_{\nu}^{\infty} e^{-\beta_{m\nu}^2 \alpha(t-\tau)/b^2}$$

$$\times \frac{R_\nu(\beta_{m\nu}, r/b)\, R_\nu(\beta_{m\nu}, r'/b)}{N(\beta_{m\nu})} \frac{\Phi(\nu, \phi)\, \Phi(\nu, \phi')}{N(\nu)}$$

$$S_{m\nu} = \beta_{m\nu} Y_\nu'(\beta_{m\nu}) + B_2 Y_\nu(\beta_{m\nu})$$

$$V_{m\nu} = \beta_{m\nu} J_\nu'(\beta_{m\nu}) + B_2 J_\nu(\beta_{m\nu})$$

$$B_1 = h_1 b/k;\ B_2 = h_2 b/k;\ A_{1\nu} = 1 - \left(\frac{\nu}{\beta_{m\nu} a/b} \right)^2;\ A_{2\nu} = 1 - \left(\frac{\nu}{\beta_{m\nu}} \right)^2$$

$$V_{m\nu} = \beta_{m\nu} J_\nu'(\beta_{m\nu} a/b) - B_1 J_\nu(\beta_{m\nu} a/b);\ W_{m\nu} = \beta_{m\nu} Y_\nu'(\beta_{m\nu} a/b) - B_1 Y_\nu(\beta_m a/b)$$

Table Rϕ.1

IJ	$R_0(\beta_{00}, r/b)$	$R_\nu(\beta_{m\nu}, r/b)\ m \neq 0$
01	0	$J_\nu(\beta_{m\nu} r/b)$
02	1	$J_\nu(\beta_{m\nu} r/b)$
03	0	$J_\nu(\beta_{m\nu} r/b)$
11	0	$J_\nu\left(\beta_{m\nu}\dfrac{r}{b}\right) Y_\nu(\beta_{m\nu}) - J_\nu(\beta_{m\nu}) Y_\nu\left(\beta_{m\nu}\dfrac{r}{b}\right)$
12	0	$J_\nu\left(\beta_{m\nu}\dfrac{r}{b}\right) Y'_\nu(\beta_{m\nu}) - J'_\nu(\beta_{m\nu}) Y_\nu\left(\beta_{m\nu}\dfrac{r}{b}\right)$
13	0	$S_{m\nu} J_\nu\left(\beta_{m\nu}\dfrac{r}{b}\right) - V_{m\nu} Y_\nu\left(\beta_{m\nu}\dfrac{r}{b}\right)$
21	0	Same as $R11$
22	1	Same as $R12$
23	0	Same as $R13$
31	0	Same as $R11$
32	0	Same as $R12$
33	0	Same as $R13$

Table Rϕ.2

IJ	$1/N(\beta_{00})$	$1/N(\beta_{m\nu}),\ m \neq 0$
01	–	$2/[b^2 J'^2_\nu(\beta_{m\nu})]$
02	$2/b^2$	$2\beta^2_{m\nu}[b^2 J^2_\nu(\beta_{m\nu}) (\beta^2_{m\nu} - \nu^2)]^{-1}$
03	–	$2\beta^2_{m\nu}[b^2 J^2_\nu(\beta_{m\nu}) (B^2_2 + \beta^2_{m\nu} - \nu^2)]^{-1}$
11	–	$[\pi^2/(2b^2)]\,[\beta^2_{m\nu} J^2_\nu(\beta_{m\nu}a/b)]\,[J^2_\nu(\beta_{m\nu}ab) - J^2_\nu(\beta_{m\nu})]^{-1}$
12	–	$[\pi^2/(2b^2)]\,[\beta^2_{m\nu} J^2_\nu(\beta_{m\nu}a/b)]\,[A_{2\nu} J^2_\nu(\beta_{m\nu}a/b)] - J'^2_\nu(\beta_{m\nu})]^{-1}$
13	–	$[\pi^2/(2b^2)]\,[\beta^2_{m\nu} J^2_\nu(\beta_{m\nu}a/b)]\,[(B^2_2 + \beta^2_{m\nu}A_{2\nu}) J^2_\nu(\beta_{m\nu}a/b) - V^2_{m\nu}]^{-1}$
21	–	$[\pi^2/(2b^2)]\,[\beta^2_{m\nu} J'^2_m(\beta_{m\nu}a/b)]\,[J'^2_\nu(\beta_{m\nu}a/b) - A_{1\nu} J^2_\nu(\beta_{m\nu})]^{-1}$
22	$2/(b^2 - a^2)$	$[\pi^2/(2b^2)]\,[\beta^2_{m\nu} J'^2_\nu(\beta_{m\nu}a/b)]\,[A_{2\nu} J'^2_\nu(\beta_{m\nu}a/b) - A_{1\nu} J'^2_\nu(\beta_{m\nu})]^{-1}$
23	–	$[\pi^2/(2b^2)]\,[\beta^2_{m\nu} J'^2_\nu(\beta_{m\nu}a/b)]\,[(B^2_2 + A_{2\nu}\beta^2_{m\nu}) J'^2_\nu(\beta_{m\nu}a/b) - A_{1\nu} V^2_{m\nu}]^{-1}$
31	–	$[\pi^2/(2b^2)]\,[\beta^2_{m\nu} U^2_{m\nu}]\,[U^2_{m\nu} - (B^2_1 + A_{1\nu}\beta^2_{m\nu}) J^2_\nu(\beta_{m\nu})]^{-1}$
32	–	$[\pi^2/(2b^2)]\,[\beta^2_{m\nu} U^2_{m\nu}]\,[A_{2\nu} U^2_{m\nu} - (B^2_1 + A_{1\nu}\beta^2_{m\nu}) J'^2_\nu(\beta_{m\nu})]^{-1}$
33	–	$[\pi^2/(2b^2)]\,[\beta^2_{m\nu} U^2_{m\nu}]\,[(B^2_2 + A_{2\nu}\beta^2_{m\nu}) U^2_{m\nu} - (B^2_1 + A_{1\nu}\beta^2_{m\nu}) V^2_{m\nu}]^{-1}$

Table Rϕ.3

IJ	β_{00}	Eigencondition (positive roots of), $m \neq 0$
01	–	$J_\nu(\beta_{m\nu}) = 0$
02	0	$J'_\nu(\beta_{m\nu}) = 0$
03	–	$\beta_{m\nu} J'_\nu(\beta_{m\nu}) + H_2 J_\nu(\beta_{m\nu}) = 0$
11	–	$J_\nu(\beta_{m\nu}a/b) Y_\nu(\beta_{m\nu}) - J_\nu(\beta_{m\nu}) Y_\nu(\beta_{m\nu}a/b) = 0$
12	–	$J_\nu(\beta_{m\nu}a/b) Y'_\nu(\beta_{m\nu}) - J'_\nu(\beta_{m\nu}) Y_\nu(\beta_{m\nu}a/b) = 0$
13	–	$S_{m\nu} J_\nu(\beta_{m\nu}a/b) - V_{m\nu} Y_\nu(\beta_{m\nu}a/b) = 0$
21	–	$J'_\nu(\beta_{m\nu}a/b) Y_\nu(\beta_{m\nu}) - J_\nu(\beta_{m\nu}) Y'_\nu(\beta_{m\nu}a/b) = 0$
22	0	$J'_\nu(\beta_{m\nu}a/b) Y'_\nu(\beta_{m\nu}) - J'_\nu(\beta_{m\nu}) Y'_\nu(\beta_{m\nu}a/b) = 0$
23	–	$S_{m\nu} J'_\nu(\beta_{m\nu}a/b) - V_{m\nu} Y'_\nu(\beta_{m\nu}a/b) = 0$
31	–	$U_{m\nu} Y_\nu(\beta_{m\nu}) - W_{m\nu} J_\nu(\beta_{m\nu}) = 0$
32	–	$U_{m\nu} Y'_\nu(\beta_{m\nu}) - W_{m\nu} J'_\nu(\beta_{m\nu}) = 0$
33	–	$S_{m\nu} U_{m\nu} - V_{m\nu} W_{m\nu} = 0$

Table Rϕ.4

KL	$\Phi(\nu, \phi)$	$1/N(\nu)$	Eigencondition
11	$\sin \nu\phi$	$2/\phi_0$	$\sin \nu\phi_0 = 0$
12	$\sin \nu\phi$	$2/\phi_0$	$\cos \nu\phi_0 = 0$
13	$\cos \nu\phi$	$2/\phi_0$	$\cos \nu\phi_0 = 0$
22	$\nu = 0$: 1	$\nu = 0$: $1/\phi_0$	$\nu = 0$
	$\nu \neq 0$: $\cos(\nu\phi)$	$\nu \neq 0$: $2/\phi_0$	$\nu \neq 0$: $\sin(\nu\phi_0) = 0$

REFERENCES

Carslaw, H. S., and Jaeger, J. C., 1959, *Conduction of Heat in Solids*, 2d ed., Oxford University Press, New York.

Ozisik, M. N., 1980, *Heat Conduction*, Wiley, New York.

CYLINDRICAL POLAR COORDINATE, φ THIN SHELL CASE. $dv' = \delta a d\phi'.\ ds' = \delta.$

Partial differential equation:

$$\frac{1}{a^2}\frac{\partial^2 T}{\partial \phi^2} = \frac{1}{\alpha}\frac{\partial T}{\partial t}$$

Φ00 Complete cylindrical shell of radius a

$$G(\phi, t|\phi', \tau) = \frac{1}{\delta a}\sum_{m=0}^{\infty}\frac{1}{\pi}e^{-m^2\alpha(t-\tau)/a^2}\cos\left[m(\phi - \phi')\right]$$

where π is replaced by 2π for $m = 0$.

Φ11 Partial cylindrical shell of radius a

$$G(\phi, t|\phi', \tau) = \frac{2}{a\phi_0\delta}\sum_{m=1}^{\infty}e^{-\frac{m^2\pi^2\alpha(t-\tau)}{\phi_0^2 a^2}}\sin\left(m\pi\frac{\phi}{\phi_0}\right)\sin\left(m\pi\frac{\phi'}{\phi_0}\right)$$

The Φ12, Φ13, . . . , Φ33 cases are the same as the standard X12, X13, . . . , X33 cases with L replaced by $a\phi_0$, x by $a\phi$ and x' by $a\phi'$. Also the $G_{X\text{--}}$ expression is divided by δ.

GREEN'S FUNCTIONS FOR RADIAL SPHERICAL GEOMETRIES: $dv' = 4\pi r'^2\, dr',\ ds' = 4\pi a^2$ or $4\pi b^2$

The partial differential equation can be written as

$$\frac{1}{r}\frac{\partial^2(rT)}{\partial r^2} = \frac{1}{\alpha}\frac{\partial T}{\partial t} \text{ or } \frac{1}{r^2}\frac{\partial}{\partial r}\left(r^2\frac{\partial T}{\partial r}\right) = \frac{1}{\alpha}\frac{\partial T}{\partial t}$$

*RS*00 **Infinite region with radial spherical symmetry (see Table RS.1 for steady state radial spherical GFs.)**

$$G(r,t|r',t) = \frac{1}{8\pi rr'[\pi\alpha(t-\tau)]^{1/2}}\left\{\exp\left[-\frac{(r-r')^2}{4\alpha(t-\tau)}\right]\right.$$
$$\left. - \exp\left[-\frac{(r+r')^2}{4\alpha(t-\tau)}\right]\right\} \tag{RS00.1}$$

[Carslaw and Jaeger, 1959, p. 259, Eq. (6)].

*RS*01 **Solid sphere, $G = 0$ at $r = b$**

There are two expressions, one better for small times and one better for large times. For small times, the better expression to use is (see Carslaw and Jaeger, 1959, pp. 275 and 367)

$$G(r,t|r',\tau) = (4\pi rr')^{-1}[4\pi\alpha(t-\tau)]^{-1/2}\sum_{n=-\infty}^{\infty}$$
$$\times\left\{\exp\left[-\frac{(2nb+r-r')^2}{4\alpha(t-\tau)}\right]\right.$$
$$\left. - \exp\left[-\frac{(2nb+r+r')^2}{4\alpha(t-\tau)}\right]\right\} \tag{RS01.1}$$

Table RS.1 Steady Green's functions, radial spherical coordinates

G satisfies: $\dfrac{1}{r^2}\dfrac{d}{dr}\left(r^2\dfrac{dG}{dr}\right) = -\,\delta(r - r')$

Case[†]	Homogeneous Boundary Conditions	$4\pi G(r|r')\ (m^{-1})$
RS00	$\partial G(0|r')/\partial r = 0$ $G(r \to \infty, r') = 0$	$1/r';\ r < r'$ $1/r;\ r > r'$
RS01	$\partial G(0|r')/\partial r = 0$ $G(b, r') = 0$	$1/r' - 1/b;\ r < r'$ $1/r - 1/b;\ r > r'$
RS02	$\partial G(0|r')/\partial r = 0;\ \partial G(b|r')/\partial r = 0$	$-$[‡]
RS03	$\partial G(0|r')/\partial r = 0$ $k\partial G(b|r')/\partial r + h_2 G(b|r') = 0$	$1/r' + (1/B_2 - 1)/b;\ r < r'$ $1/r + (1/B_2 - 1)/b;\ r > r'$; where $B_2 = h_2 b/k$
RS10	$G(a|r') = 0$ $G(r \to \infty|r') = 0$	$(1 - a/r)/r';\ r < r'$ $(1 - a/r')/r;\ r > r'$
RS11	$G(a|r') = 0$ $G(b|r') = 0$	$(b - r')\,(1 - a/r)/[r'(b - a)];\ r < r'$ $(b - r)\,(1 - a/r')/[r(b - a)];\ r > r'$
RS12	$G(a|r') = 0$ $\partial G(b|r')/\partial r = 0$	$1/a - 1/r;\ r < r'$ $1/a - 1/r';\ r > r'$
RS20	$\partial G(a|r')/\partial r = 0$ $G(r \to \infty|r') = 0$	$1/r';\ r < r'$ $1/r;\ r > r'$
RS21	$\partial G(a|r')/\partial r = 0$ $G(b|r') = 0$	$1/r' - 1/b;\ r < r'$ $1/r - 1/b;\ r > r'$
RS22	$\partial G(a|r')/\partial r = 0$ $\partial G(b|r')/\partial r = 0$	$-$[‡]
RS30	$k\partial G(a|r')/\partial r - h_1 G(a|r') = 0$ $G(r \to \infty|r') = 0$	$1/r' + (1/B_1 - 1)/b;\ r < r'$ $1/r' + (1/B_1 - 1);\ r > r'$; where $B_1 = h_1 a/k$

[†]Cases RS13, RS31, RS23, RS32, RS33 exist but are not listed.

[‡]Homogeneous boundary conditions cannot be satisfied; no steady Green's function. (However, non-homogeneous steady solutions are possible.)

$$-\left.\frac{\partial G}{\partial n'}\right|_{r' = b} = \frac{1}{rb}\,[4\pi\alpha(t - \tau)]^{-3/2}\sum_{n=-\infty}^{\infty}|(2n - 1)\,b + r|$$

$$\times \exp\left[-\frac{(2nb + r - b)^2}{4\alpha(t - \tau)}\right] \tag{RS01.2}$$

For large times, the better expression is (Carslaw and Jaeger, 1959, pp. 233 and 366)

$$G(r, t|r', \tau) = \frac{1}{2\pi brr'}\sum_{m=1}^{\infty}e^{-m^2\pi^2\alpha(t-\tau)/b^2}$$

$$\times \sin\left(m\pi\frac{r}{b}\right)\sin\left(m\pi\frac{r'}{b}\right) \tag{RS01.3}$$

$$-\left.\frac{\partial G}{\partial n'}\right|_{r'=b} = -\frac{1}{2b^3 r}\sum_{m=1}^{\infty} e^{-m^2\pi^2\alpha(t-\tau)/b^2}$$

$$\times (-1)^m\, m\, \sin\left(m\pi\,\frac{r}{b}\right) \tag{RS01.4}$$

Note the similarity with the $X11$ case with $L_x \to b$, $x \to r$, $x' \to r'$ and the G_{X11} expression divided by $4\pi rr'$.

RS02 Solid sphere, $\partial G/\partial r = 0$ at $r = b$

$$G(r, t|r', \tau) = \frac{3}{4\pi b^3} + \frac{1}{2\pi brr'}$$

$$\times \sum_{m=1}^{\infty} e^{-\beta_m^2\alpha(t-\tau)/b^2}\,\frac{\beta_m^2 + 1}{\beta_m^2}$$

$$\times \sin\left(\beta_m\,\frac{r}{b}\right)\sin\left(\beta_m\,\frac{r'}{b}\right) \tag{RS02.1}$$

The eigenvalues are found from the positive roots of

$$\beta_m \cot \beta_m = 1 \tag{RS02.2}$$

See case $X13$ for approximate relations for the eigenvalues (Carslaw and Jaeger, 1959, p. 367). For the small times of $\alpha(t - \tau)/b^2 \le 0.022$

$$G(r, t|r', \tau) \approx \frac{1}{4\pi rr'}\frac{1}{[4\pi\alpha(t - \tau)]^{1/2}}\left\{\exp\left[-\frac{(r - r')^2}{4\alpha(t - \tau)}\right]\right.$$

$$-\exp\left[-\frac{(r + r')^2}{4\alpha(t - \tau)}\right] + \exp\left[-\frac{(2b - r - r')^2}{4\alpha(t - \tau)}\right]\right\}$$

$$-\frac{B_2}{4\pi brr'}\exp\left[B_2\frac{(2b - r - r')}{b} + B_2^2\frac{\alpha(t - \tau)}{b^2}\right]$$

$$\times \operatorname{erfc}\left\{\frac{2b - r - r}{[4\alpha(t - \tau)]^{1/2}} + \frac{B_2}{b}[\alpha(t - \tau)]^{1/2}\right\} \tag{RS02.3}$$

$$B_2 = -1 \tag{RS02.4}$$

RS03 Solid sphere, $k\partial G/\partial r + h_2 G = 0\ r = b$

$$G(r, t|r', \tau) = \frac{1}{2\pi brr'}\sum_{m=1}^{\infty} e^{-\beta_m^2\alpha(t-\tau)/b^2}\,\frac{\beta_m^2 + B_2^2}{\beta_m^2 + B_2^2 + B_2}$$

$$\times \sin\left(\beta_m\,\frac{r}{b}\right)\sin\left(\beta_m\,\frac{r'}{b}\right) \tag{RS03.1}$$

where
$$B_2 = \frac{h_2 b}{k} - 1 \tag{RS03.2}$$

and β_m, $m = 1, 2, \ldots$, are the positive roots of

$$\beta_m \cot \beta_m = - B_2 \tag{RS03.3}$$

(Ozisik, 1980, p. 164; Carslaw and Jaeger, 1959, p. 367). For small times $\alpha(t - \tau)/b^2 \leq 0.022$, $G(r, t|r', \tau)$ is approximated by Eq. (RS02.3) with B_2 defined by Eq. (RS03.2).

*RS*10 Infinite region outside the spherical cavity, $r = a$; $G = 0$ at $r = a$

$$G(r, t|r', \tau) = \frac{1}{4\pi rr'\,[4\pi\alpha(t - \tau)]^{1/2}}$$
$$\times \left(e^{-(r-r')^2/[4\alpha(t-\tau)]} - e^{-(r+r'-2a)^2/[4\alpha(t-\tau)]}\right) \tag{RS10.1}$$

(Carslaw and Jaeger, 1959, p. 247)

$$-\frac{\partial G}{\partial n'}\bigg|_{r'=a} = \frac{r - a}{ra[4\pi\alpha(t-\tau)]^{3/2}}\, e^{-(r-a)^2/[4\alpha(t-\tau)]} \tag{RS10.2}$$

*RS*11 Hollow sphere with $T = 0$ at $r = a$ and b

The better expression for small times is:

$$G(r, t|r', \tau) = \frac{1}{4\pi rr'[4\pi\alpha(t - \tau)]^{1/2}}$$
$$\times \sum_{n=-\infty}^{\infty} \left\{ \exp\left[-\frac{2n(b - a) + r - r')^2}{4\alpha(t - \tau)} \right]\right.$$
$$\left. - \exp\left[-\frac{(2n(b - a) + r + r' - 2a)^2}{4\alpha(t - \tau)} \right]\right\} \tag{RS11.1}$$

$$-\frac{\partial G}{\partial n'}\bigg|_{r'=a} = \frac{1}{ra[4\pi\alpha(t - \tau)]^{3/2}} \sum_{n=-\infty}^{\infty} |2n(b - a) + r - a|$$
$$\times \exp\left[-\frac{(2n(b - a) + r - a)^2}{4\alpha(t - \tau)} \right] \tag{RS11.2}$$

$$-\frac{\partial G}{\partial n'}\bigg|_{r'=b} = \frac{1}{rb[4\pi\alpha(t - \tau)]^{3/2}} \sum_{n=-\infty}^{\infty} |2n(b - a) + r - b|$$
$$\times \exp\left[-\frac{(2n(b - a) + r - b)^2}{4\alpha(t - \tau)} \right] \tag{RS11.3}$$

The better expression for large times is

$$G(r, t|r', \tau) = \frac{1}{2\pi(b - a)rr'} \sum_{m=1}^{\infty} e^{-m^2\pi^2\alpha(t-\tau)/(b-a)^2}$$

$$\times \sin\left(m\pi\frac{r - a}{b - a}\right) \sin\left(m\pi\frac{r' - a}{b - a}\right) \qquad \text{(RS11.4)}$$

$$-\left.\frac{\partial G}{\partial n'}\right|_{r'=a} = \frac{1}{2(a - b)^2 ra} \sum_{m=1}^{\infty} e^{-m^2\pi^2\alpha(t-\tau)/(b-a)^2}$$

$$\times m \sin\left(m\pi\frac{r - a}{b - a}\right) \qquad \text{(RS11.5)}$$

$$-\left.\frac{\partial G}{\partial n'}\right|_{r'=b} = \frac{1}{2(b - a)^2 rb} \sum_{m=1}^{\infty} e^{-m^2\pi^2\alpha(t-\tau)/(b-a)^2}$$

$$\times m \sin\left(m\pi\frac{r - a}{b - a}\right)(-1)^m \qquad \text{(RS11.6)}$$

*RS*12 Hollow sphere with $G = 0$ at $r = a$ and $\partial G/\partial r = 0$ at $r = b$

For large times, a convenient expression is (Shakir, 1982)

$$G(r, t|r', \tau) = \frac{1}{2\pi(b - a)rr'} \sum_{m=1}^{\infty} e^{-\beta_m^2\alpha(t-\tau)/(b-a)^2}$$

$$\times \frac{\beta_m^2 + H_2^2}{\beta_m^2 + H_2^2 + H_2} \sin\left(\beta_m\frac{r - a}{b - a}\right)$$

$$\times \sin\left(\beta_m\frac{r' - a}{b - a}\right) \qquad \text{(RS12.1)}$$

where β_m are the positive roots of

$$\beta_m \cot \beta_m = - H_2 \qquad H_2 = B_2 R_2 \qquad \text{(RS12.2a, b)}$$

$$B_2 = - 1 \qquad R_2 = 1 - \frac{a}{b} \qquad \text{(RS12.3a, b)}$$

See the *X*13 case for approximate eigenvalues.

$$-\left.\frac{\partial G}{\partial n'}\right|_{r'=a} = \frac{1}{2\pi(b - a)^2 ra} \sum_{m=1}^{\infty} e^{-\beta_m^2\alpha(t-\tau)/(b-a)^2}$$

$$\times \frac{\beta_m(\beta_m^2 + H_2^2)}{\beta_m^2 + H_2^2 + H_2} \sin\left(\beta_m\frac{r - a}{b - a}\right) \qquad \text{(RS12.4)}$$

For small times, $\alpha(t - \tau)/(b - a)^2 \leq 0.022$, $G(r, t|r', \tau)$ is efficiently given by

$$G(r, t|r', \tau) \approx \frac{1}{4\pi rr'} \frac{1}{[4\pi\alpha(t - \tau)]^{1/2}} \left\{ \exp\left[-\frac{(r - r')^2}{4\alpha(t - \tau)} \right] \right.$$
$$\left. - \exp\left[-\frac{(r + r' - 2a)^2}{4\alpha(t - \tau)} \right] + \exp\left[-\frac{(2b - r - r')^2}{4\alpha(t - \tau)} \right] \right\}$$
$$- \frac{B_2}{4\pi rr'b} \exp\left[B_2 \frac{(2b - r - r)}{b} + B_2^2 \frac{\alpha(t - \tau)}{b^2} \right]$$
$$\times \operatorname{erfc} \left\{ \frac{2b - r - r'}{[4\alpha(t - \tau)]^{1/2}} + \frac{B_2}{b} [\alpha(t - \tau)]^{1/2} \right\} \qquad \text{(RS12.5)}$$

RS13 Hollow sphere with $G = 0$ at $r = a$ and $k\,\partial G/\partial r + h_2 G = 0$ at $r = b$

For large times, the $G(r, t|r', \tau)$ relations are found using Eq. (RS12.1)–(RS12.3) with

$$B_2 = \frac{h_2 b}{k} - 1 \qquad \text{(RS13.1)}$$

For small times, $G(r, t|r', \tau)$ is approximated by Eq. (RS12.5) with B_2 given by Eq. (RS13.1).

RS20 Infinite region outside a spherical cavity at $r = a$ with $\partial G/\partial r = 0$ at $r = a$

$$G(r, t|r', \tau) = \frac{1}{4\pi rr'[4\pi\alpha(t - \tau)]^{1/2}} \left\{ \exp\left[-\frac{(r - r')^2}{4\alpha(t - \tau)} \right] \right.$$
$$\left. + \exp\left[-\frac{(r + r' - 2a)^2}{4\alpha(t - \tau)} \right] \right\} - \frac{B_1}{4\pi rr'a}$$
$$\times \exp\left[B_1 \frac{r + r' - 2a}{a} + B_1^2 \frac{\alpha(t - \tau)}{a^2} \right]$$
$$\times \operatorname{erfc} \left\{ \frac{r + r' - 2a}{[4\alpha(t - \tau)]^{1/2}} + \frac{B_1}{a} [\alpha(t - \tau)]^{1/2} \right\} \qquad \text{(RS20.1)}$$

where B_1 is equal to 1. See $X30$ case for approximate values.

RS21 Hollow sphere with $\partial G/\partial r = 0$ at $r = a$ and $G = 0$ at $r = b$

$$G(r, t|r', \tau) = \frac{1}{2\pi(b - a)rr'} \sum_{m=1}^{\infty} e^{-\beta_m^2 \alpha(t - \tau)/(b - a)^2}$$
$$\times \frac{(\beta_m^2 + H_1^2) \sin\left[\beta_m(b - r)/(b - a)\right] \sin\left[\beta_m(b - r')/(b - a)\right]}{\beta_m^2 + H_1^2 + H_1}$$
$$\text{(RS12.1)}$$

where β_m are the positive roots of

$$\beta_m \cot \beta_m = -H_1 \qquad H_1 = B_1 R_1 \qquad \text{(RS21.2a, b)}$$

$$B_1 = 1 \qquad R_1 = \frac{b}{a} - 1 \qquad \text{(RS21.3a, b)}$$

See case $X13$ for approximations for eigenvalues.

$$-\left.\frac{\partial G}{\partial n'}\right|_{r'=b} = \frac{1}{2\pi(b-a)^2 br} \sum_{m=1}^{\infty} e^{-\beta_m^2 \alpha(t-\tau)/(b-a)^2}$$

$$\times \frac{\beta_m(\beta_m^2 + H_1^2) \sin [\beta_m(b-r)/(b-a)]}{B_m^2 + H_1^2 + H_1} \qquad \text{(RS21.4)}$$

For small times, $\alpha(t - \tau)/(b - a)^2 \leq 0.022$, $G(\cdot)$ is approximated by

$$G(r, t | r', \tau) \approx \frac{1}{4\pi rr'[4\pi\alpha(t-\tau)]^{1/2}} \left\{ \exp\left[-\frac{(r-r')^2}{4\alpha(t-\tau)} \right] \right.$$

$$+ \exp\left[-\frac{(r+r'-2a)^2}{4\alpha(t-\tau)} \right] - \left.\exp\left[-\frac{(2b-r-r')^2}{4\alpha(t-\tau)} \right] \right\}$$

$$- \frac{B_1}{4\pi rr'a} \exp\left[B_1 \frac{r+r'-2a}{a} + B_1^2 \frac{\alpha(t-\tau)}{a^2} \right]$$

$$\times \text{erfc} \left\{ \frac{r+r'-2a}{[4\alpha(t-\tau)]^{1/2}} + \frac{B_1}{a} [\alpha(t-\tau)]^{1/2} \right\} \qquad \text{(RS21.5)}$$

RS22 Hollow sphere with $\partial G/\partial r = 0$ at $r = a$ and b

For large times (Shakir, 1982)

$$G(r, t | r', \tau) = \frac{3B_0}{4\pi(b^3 - a^3)} + \frac{1}{2\pi(b-a)rr'} \sum_{m=1}^{\infty} e^{-\beta_m^2 \alpha(t-\tau)/(b-a)^2}$$

$$\times \frac{\{\beta_m \cos [\beta_m(r-a)/(b-a)] + H_1 \sin [\beta_m(r'-a)/(b-a)]\}}{(\beta_m^2 + H_1^2) [1 + H_2/(\beta_m^2 + H_2^2)] + H_1}$$

$$\qquad \text{(RS22.1)}$$

$$B_0 = 1 \qquad H_1 = B_1 R_1 \qquad H_2 = B_2 R_2 \qquad \text{(RS22.2a, b)}$$

$$B_1 = 1 \qquad B_2 = -1$$

$$R_1 = \frac{b}{a} - 1 \qquad R_2 = \frac{a}{b} - 1 \qquad \text{(RS22.3a, b, c, d)}$$

The eigenvalues β_m are the positive roots of

$$\tan \beta_m = \frac{\beta_m(H_1 + H_2)}{\beta_m^2 - H_1 H_2} \qquad \text{(RS22.4)}$$

For small times such that $\alpha(t - \tau)/(b - a)^2 \le 0.022$, $G(\cdot)$ is approximated by

$$
\begin{aligned}
G(r, t|r', \tau) \approx \frac{1}{4\pi r r'[4\pi\alpha(t - \tau)]^{1/2}} &\left\{ \exp\left[-\frac{(r - r')^2}{4\alpha(t - \tau)} \right] \right. \\
&\left. + \exp\left[-\frac{(r + r' - 2a)^2}{4\alpha(t - \tau)} \right] + \exp\left[-\frac{(2b - r - r')^2}{4\alpha(t - \tau)} \right] \right\} \\
&- \frac{B_1}{4\pi r r' a} \exp\left[B_1\frac{r + r' - 2a}{a} + B_1^2\frac{\alpha(t - \tau)}{a^2} \right] \\
&\times \operatorname{erfc}\left\{ \frac{r + r' - 2a}{[4\alpha(t - \tau)]^{1/2}} + \frac{B_1}{a}[\alpha(t - \tau)]^{1/2} \right\} \\
&- \frac{B_2}{4\pi r r' b} \exp\left[B_2\frac{2b - r - r'}{b} + B_2^2\frac{\alpha(t - \tau)}{b^2} \right] \\
&\times \operatorname{erfc}\left\{ \frac{(2b - r - r')}{[4\alpha(t - \tau)]^{1/2}} + \frac{B_2}{b}[\alpha(t - \tau)]^{1/2} \right\}
\end{aligned}
\tag{RS22.5}
$$

RS23 Hollow sphere with $\partial G/\partial r = 0$ at $r = a$ and $k\partial G/\partial r + h_2 G = 0$ at $r = b$ For large times, $G(r, t|\theta', \tau)$ is found using Eqs. (RS22.1)–(RS22.4) with

$$
B_0 = 0 \qquad B_1 = 1 \qquad B_2 = \frac{h_2 b}{k} - 1
\tag{RS23.1}
$$

For small times, $G(r, t|r', \tau)$ is given by Eq. (RS22.5).

RS30 Infinite region outside a spherical cavity ($r \ge a$) with $k\partial G/\partial r - h_1 G = 0$ at $r = a$ $G(r, t|r', \tau)$ is given by Eq. (RS20.1) with B_1 given by

$$
B_1 = \frac{h_1 a}{k} + 1
\tag{RS30.1}
$$

See $X30$ case for approximate values.

RS31 Spherical shell with $-k\partial G/\partial r + h_1 G = 0$ at $r = a$ and $G = 0$ at $r = b$ For large times, the $G(r, t|r', \tau)$ relations are found from using Eqs. (RS21.1)–(RS21.4) with

$$
B_1 = \frac{ha}{k} + 1
\tag{RS31.1}
$$

For small times, $G(r, t|r', \tau)$ is found from Eq. (RS21.5).

***RS32* Spherical shell with** $-k\partial G/\partial r + h_1 G = 0$ **at** $r = a$ **and** $\partial G/\partial r = 0$ **at** $r = b$ For large times, the $G(r, t|r', \tau)$ relations are given by Eqs. (RS22.1)–(RS22.4) with

$$B_0 = 0 \qquad B_1 = \frac{h_1 a}{k} + 1 \qquad B_2 = -1 \qquad \text{(RS32.1a, b, c)}$$

For small times, $\alpha(t - \tau)/(b - a)^2 \le 0.022$, $G(r, t|r', \tau)$ is approximated by Eq. (RS22.5).

***RS33* Spherical shell with** $-k\partial G/\partial r + h_1 G = 0$ **at** $r = a$ **and** $k\partial G/\partial r + h_2 G = 0$ **at** $r = b$ For large times, the $G(r, t|r', \tau)$ relations are given by Eqs. (RS22.1)–(RS22.4) with

$$B_0 = 1 \qquad B_1 = \frac{h_1 a}{k} + 1 \qquad B_2 = \frac{h_2 b}{k} - 1 \qquad \text{(RS33.1a, b, c)}$$

For small times, $\alpha(t - \tau)/(a - b)^2 \le 0.022$, $G(r, t|r', \tau)$ is approximated by Eq. (RS22.5).

***RS01θ00* Solid sphere with radial and azimuthal dependence;** $G = 0$ **at** $r = b$

$$dv' = 2\pi r^2\, dr'\, d\mu' \qquad \mu' = \cos\theta' \qquad -1 < \mu < 1$$
$$ds' = 2\pi b^2\, d\mu'$$

$$G_{RS01\theta00}(r, \theta, t|r', \theta', \tau) = \frac{1}{2\pi(rr')^{1/2}b^2} \sum_{n=0}^{\infty} \sum_{p=1}^{\infty} e^{-\beta_{np}^2 \alpha(t-\tau)/b^2}$$

$$\times \frac{(2n + 1) J_{n+1/2}(\beta_{np} r/b)\, J_{n+1/2}(\beta_{np} r'/b)\, P_n(\mu)\, P_n(\mu')}{[J'_{n+1/2}(\beta_{np})]^2} \qquad \text{(RS01θ00.1)}$$

where the β_{np}'s are the positive roots of

$$J_{n+1/2}(\beta_{np}) = 0 \qquad \text{(RS01θ00.2)}$$

and $P_n(\mu)$ is the nth Legendre polynomial.

Note that

$$J_{1/2}(x) = \left(\frac{2}{\pi x}\right)^{1/2} \sin x \qquad J_{-1/2}(x) = \left(\frac{2}{\pi x}\right)^{1/2} \cos x \qquad \text{(RS01θ00.3a, b)}$$

$$J_{n+1/2}(x) = \frac{2n - 1}{x} J_{n-1/2}(x) - J_{n-3/2}(x) \qquad \text{(RS01θ00.4)}$$

RS01θ01 Hemisphere with radial and azimuthal dependence; $G = 0$ at $r = b$ and $G = 0$ at $\mu = 0$ (or $\theta = \pi/2$) ($\mu = 0$ to 1)

$$G_{RS01\theta01}(r, \theta, t|r', \theta', \tau) = \frac{1}{\pi(rr')^{1/2}b^2} \sum_{n=1,3,\ldots}^{\infty} \sum_{p=1}^{\infty}$$

$$\times\ e^{-\beta_{np}^2 \alpha(t-\tau)/b^2} \frac{2n+1}{-J_{n-1/2}(\beta_{np})\,J_{n+3/2}(\beta_{np})}$$

$$\times\ J_{n+1/2}\left(\beta_{np}\frac{r}{b}\right) J_{n+1/2}\left(\beta_{np}\frac{r'}{b}\right)$$

$$\times\ P_n(\mu)\,P_n(\mu') \tag{RS01θ01.1}$$

where the eigenvalues β_{np} are the positive roots of

$$J_{n+1/2}(\beta_{np}) = 0 \tag{RS01θ01.2}$$

RS00Φ00θ00 (Butkovskiy, p. 171)

$$G(r, \phi, \theta, t|r', 0, 0, \tau) = \frac{1}{[4\pi\alpha(t-\tau)]^{3/2}} \exp\left[-\frac{r^2 + r'^2 - 2rr'\cos\theta}{4\alpha(t-\tau)}\right]$$

RS00Φ00 (Butkovskiy, p. 140)

$$G(r, \phi, t|r', \phi', \tau) = \frac{1}{4\pi\alpha(t-\tau)} \exp\left[-\frac{r^2 - r'^2 + 2rr'\cos(\phi-\phi')}{4\alpha(t-\tau)}\right]$$

RSIJΦ00, $J \neq 0$

$$G(r, \phi, t|r', \phi', \tau) = \sum_{m=0}^{\infty} \sum_{n=0}^{\infty} e^{-\beta_{mn}^2 \alpha(t-\tau)/b^2}$$

$$\times\ \frac{R_n(\beta_{mn}, r/b)\,R_n(\beta_{mn}, r'/b)}{\pi N(\beta_{mn})} \cos\left[n(\phi - \phi')\right]$$

Replace π by 2π for $n = 0$. See Table R0.1 for $R_n(\beta_{mn}, r/b)$, Table R0.2 for $N(\beta_{mn})$ and Table R0.3 for eigenconditions.

REFERENCES

Butkovskiy, A. G., 1982, *Green's Functions and Transfer Functions Handbook*, Ellis Horwood Limited, Halsted Press, division of John Wiley and Sons, New York.

Carslaw, H. S., and Jaeger, J. C., 1959, *Conduction of Heat in Solids*, 2nd ed., Oxford University Press, New York.

Shakir, S., 1982, Personal Communication, Dept. of Mech. Eng., Mich. State University, East Lansing, Michigan.

GREEN'S FUNCTIONS: RECTANGULAR COORDINATES

Rectangular coordinates, x, y, z

For a one-dimensional case, $dv' = dx'$.
For a two-dimensional case, $dv' = dx'\, dy'$.
For a three-dimensional case, $dv' = dx'\, dy'\, dz'$.

The transient Green's functions (GFs) are for the equation

$$\frac{\partial^2 T}{\partial x^2} = \frac{1}{\alpha} \frac{\partial T}{\partial t}$$

where α = constant. See Tables X.1 and X.2 for steady Green's functions.

The solutions are arranged using a numbering system for the x coordinate with an X being the first letter. The X is followed by two numbers, the first is for the $x = 0$ boundary and the second is for the $x = L$ boundary. If the boundary goes to infinity, the digit zero is used.

For a homogeneous temperature boundary condition, the "first kind," the digit 1 is used (Dirichlet condition).
For an insulation condition ($\partial T/\partial x = 0$), the "second kind," the digit 2 is used (Neumann condition).
For a convective boundary condition, the "third kind," the digit 3 is used (Robin condition).
For a thin film of finite heat capacity, the digit 4 is used (Carslaw condition).
For a combination of a convective and thin film boundary conditions, the digit 5 is used (Jaeger condition).

Products of two or three one-dimensional GFs for boundary conditions of the zeroth, first, second, and third kinds can be used to get two- and three-dimensional GFs in rectangular coordinates. Hence, it is necessary to give the GFs for only one rectangular coordinate for these cases.

In most *finite-body* cases (XIJ, I, and J not equal to zero) two forms of the GFs are given, one said to be best for small values of $\alpha(t - \tau)/L^2$, sometimes referred to simply as "small times," and one for large values of $\alpha(t - \tau)/L^2$. If an *infinite* summation is used in an expression, it is actually valid for all times, both small and large. However, the small time expression needs only a few terms for small values of $\alpha(t - \tau)/L^2$ (since it approaches a semi-infinite body, which does not require a semi-infinite number of terms). On the other hand, the large time expression requires only a few terms, i.e., eigenvalues, for large values of $\alpha(t - \tau)/L^2$. If the large time expression is used for all times, the required number of terms can be very large, even for large times, because the convolution integral over time is over all times, both small and large. The large time expressions are usually much easier to manipulate mathematically and thus are preferred if only one is used. If the required integrations are performed numerically rather than analytically, then the complexity is reduced and the convergence is speeded by using both types of expressions. This is discussed more fully in connection with time partitioning in Chapter 6.

X00 Infinite region

$$G_{X00}(x, t|x', \tau) = G_{X00}(x - x', t - \tau)$$

$$= [4\pi\alpha(t - \tau)]^{-1/2} \exp\left[-\frac{(x - x')^2}{4\alpha(t - \tau)} \right] \quad \text{(X00.1)}$$

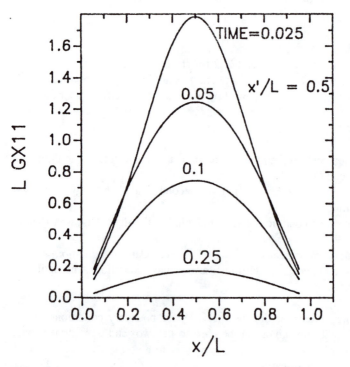

Figure X.1 $L\, G_{X11}(x, t|x'/L = 0.5, \tau)$ versus x/L for $\alpha(t - \tau)/L^2 = 0.025$, 0.05, 0.10, and 0.25.

Notice that

$$\int_{-\infty}^{\infty} G_{X00}(x, t|x', \tau) \, dx' = 1 \qquad \text{(X00.2a)}$$

$$\frac{\partial G_{X00}(x - x', t - \tau)}{\partial x} = - \frac{\partial G_{X00}(x - x', t - \tau)}{\partial x'} \qquad \text{(X00.2b)}$$

Notice that the integral over x' from a to b is

$$[4\pi\alpha(t - \tau)]^{-1/2} \int_a^b \exp\left[- \frac{(x - x')^2}{4\alpha(t - \tau)} \right] dx'$$

$$= \frac{1}{2} \left(\text{erfc}\left\{ \frac{x - b}{[4\alpha(t - \tau)]^{1/2}} \right\} - \text{erfc}\left\{ \frac{x - a}{[4\alpha(t - \tau)]^{1/2}} \right\} \right) \qquad \text{(X00.3)}$$

and thus

$$\int_0^{\infty} G_{X00}(x, t|x', \tau) \, dx' = 1 - \frac{1}{2} \, \text{erfc} \frac{x}{[4\alpha(t - \tau)]^{1/2}} \qquad \text{(X00.4)}$$

A relation involving differentiation and integration is

$$\frac{\partial}{\partial x} \int_a^b G_{X00}(x - x', t - \tau) \, dx'$$

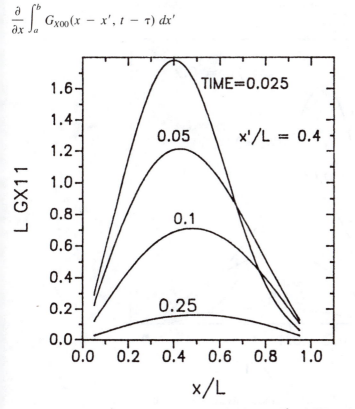

Figure X.2 $L \, G_{X11}(x, t|x'/L = 0.4, \tau)$ versus x/L for $\alpha(t - \tau)/L^2 = 0.025$, 0.05, 0.10, and 0.25.

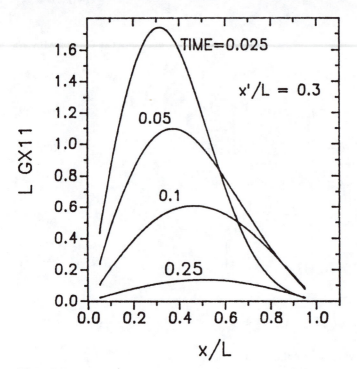

Figure X.3 $L\,G_{X11}(x,\,t\,|\,x'/L = 0.3,\,\tau)$ versus x/L for $\alpha(t - \tau)/L^2 = 0.025$, 0.05, 0.10, and 0.25.

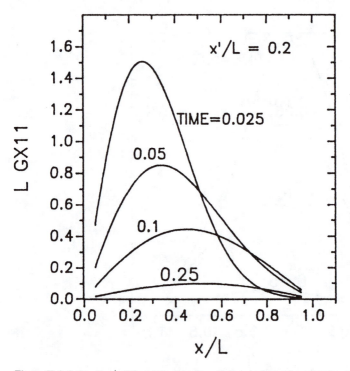

Figure X.4 $L\,G_{X11}(x,\,t\,|\,x'/L = 0.2,\,\tau)$ versus x/L for $\alpha(t - \tau)/L^2 = 0.025$, 0.05, 0.10, and 0.25.

$$= G_{X00}(x - a, t - \tau) - G_{X00}(x - b, t - \tau) \quad \text{(X00.5)}$$

The average from $x = c$ to d for integration over x' from a to b is

$$\frac{1}{d - c} \int_{x=c}^{d} \int_{x'=a}^{b} G_{X00}(x, t|x', \tau) \, dx' \, dx$$

$$= \frac{[\alpha(t - \tau)]^{1/2}}{d - c} \left\{ \text{ierfc} \frac{c - b}{[4\alpha(t - \tau)]^{1/2}} - \text{ierfc} \frac{d - b}{[4\alpha(t - \tau)]^{1/2}} \right.$$

$$\left. - \text{ierfc} \frac{c - a}{[4\alpha(t - \tau)]^{1/2}} + \text{ierfc} \frac{d - a}{[4\alpha(t - \tau)]^{1/2}} \right\} \quad \text{(X00.6)}$$

The average over $a < x < b$ is

$$\overline{G}_{X00} = \frac{1}{b - a} \int_{x=a}^{b} \int_{x'=a}^{b} G_{X00}(x - x', t - \tau) \, dx' \, dx$$

$$= 1 - \frac{[4\alpha(t - \tau)]^{1/2}}{b - a} \left(\pi^{-1/2} - \text{ierfc} \left\{ \frac{b - a}{[4\alpha(t - \tau)]^{1/2}} \right\} \right) \quad \text{(X00.7)}$$

For accurate, approximate expressions, see the $X20$ case.
Let $4\alpha(t - \tau)/(b - a)^2$ for the $X00$ case be equal to u in the $\overline{G}_{X20}$ approximations.

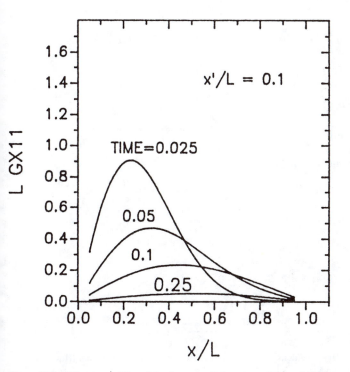

Figure X.5 $L*G_{X11}(x, t|x'/L = 0.1, \tau)$ versus x/L for $\alpha(t - \tau)/L^2 = 0.025$, 0.05, 0.10, and 0.25.

The integral of G_{X00} over τ is

$$\int_0^t G_{X00}(x - x', t - \tau)\, d\tau = \frac{(\alpha t)^{1/2}}{\alpha} \text{ ierfc} \left[\frac{|x - x'|}{(4\alpha t)^{1/2}} \right] \tag{X00.8a}$$

Table X.1 Steady Green's functions, one-dimensional rectangular coordinates

$$G \text{ satisfies: } \frac{d^2G}{dx^2} = -\delta(x - x'),\ 0 \le x \le L$$

Case	Homogeneous boundary conditions	$G(x\|x')$ (in meters)
$X11$	$G(0\|x') = 0$ $G(L\|x') = 0$	$x(1 - x'/L);\ x < x'$ $x'(1 - x/L);\ x > x'$
$X12$	$G(0\|x') = 0$ $\dfrac{\partial G}{\partial x}(L\|x') = 0$	$x;\ x < x'$ $x';\ x > x'$
$X13^†$	$G(0\|x') = 0$ $k\dfrac{\partial G}{\partial n}(L\|x') + hG(L\|x') = 0$	$x[1 - B(x'/L)/(1 + B)];\ x < x'$ $x'[1 - B(x/L)/(1 + B)];\ x > x'$ where $B = hL/k$
$X21$	$\dfrac{\partial G(0\|x')}{\partial x} = 0$ $G(L\|x') = 0$	$L - x';\ x < x'$ $L - x;\ x > x'$
$X23^†$	$\dfrac{\partial G(0\|x')}{\partial x} = 0$ $k\dfrac{\partial G(L\|x')}{\partial n} + hG(L\|x') = 0$	$L(1 + 1/B - x'/L);\ x < x'$ $L(1 + 1/B - x/L);\ x > x'$ where $B = hL/k$
$X32^†$	$k\dfrac{\partial G(0\|x')}{\partial n} + hG(0\|x') = 0$ $\dfrac{\partial G(L\|x')}{\partial x} = 0$	$L(1/B + x/L);\ x < x'$ $L(1/B + x'/L);\ x > x'$
$X33^†$	$k\dfrac{\partial G(0\|x')}{\partial n} + h_1G(0\|x') = 0$ $k\dfrac{\partial G(L\|x')}{\partial n} + h_2G(L\|x') = 0$	$(B_1B_2x + B_1x - B_1B_2xx'/L - B_2x'$ $+ B_2L + L)/(B_1B_2 + B_1 + B_2);\ x < x'$ $(B_1B_2x' + B_1x' - B_1B_2xx'/L - B_2x$ $+ B_2L + L)/(B_1B_2 + B_1 + B_2);\ x > x',$ where $B_1 = h_1L/k;\ B_2 = h_2L/k$
$X00$	$G(-\infty\|x') = 0;\ G(+\infty\|x') = 0$	$-^‡$
$X10$	$G(0\|x') = 0;\ G(\infty\|x') = 0$	$-^‡$
$X20$	$\dfrac{\partial G(0\|x')}{\partial x} = 0;\ G(\infty\|x') = 0$	$-^‡$
$X30^†$	$k\dfrac{\partial G(0\|x')}{\partial n} + hG(0\|x') = 0;\ G(\infty\|x') = 0$	$-^‡$
$X22$	$\dfrac{\partial G(0\|x')}{\partial x} = 0;\ \dfrac{\partial G(L\|x')}{\partial x} = 0$	$-^‡$

† Note $\partial/\partial n|_{x=0} = -\partial/\partial x|_{x=0};\ \partial/\partial n|_{x=L} = +\partial/\partial x|_{x=L}$.

‡ Homogeneous boundary conditions cannot be satisfied; no steady Green's function. (However, non-homogeneous steady solutions are possible.)

Table X.2 Steady Green's function with fin term (m^2 = constant), one-dimensional rectangular coordinates

$$G \text{ satisfies: } \frac{d^2G}{dx^2} - m^2G = -\delta(x - x')$$

Case[†]	Homogeneous boundary conditions	$G(x|x')$ (in meters)
X00	$G(-\infty|x') = 0$ $G(+\infty|x') = 0$	$e^{-mx'}e^{mx}/(2m); \; x < x'$ $e^{mx'}e^{-mx}/(2m); \; x > x'$
X10	$G(0|x') = 0$ $G(+\infty|x') = 0$	$e^{-mx'} \sinh mx/m; \; x < x'$ $e^{-mx} \sinh mx'/m; \; x > x'$
X11	$G(0|x') = 0$ $G(L|x') = 0$	$\sinh [m(L - x')] \sinh mx/(m \sinh mL); \; x < x'$ $\sinh [m(L - x)] \sinh mx'/(m \sinh mL); \; x > x'$
X12	$G(0|x') = 0$ $\partial G(L, x')/\partial x = 0$	$\cosh [m(L - x')] \sinh mx/(m \cosh mL) \text{ for } x < x'$ $\cosh [m(L - x)] \sinh mx'/(m \cosh mL) \text{ for } x > x'$
X13	$G(0|x') = 0$ $k\partial G(L|x')/\partial x + hG(L|x') = 0$	$([\cosh m(L - x') + (h/km) \sinh m(L - x')]/$ $\quad [m(\cosh mL + (h/km) \sinh mL)]) \sinh mx \text{ for } x<x'$ $([\cosh m(L - x) + (h/km) \sinh m(L - x)]/$ $\quad [m(\cosh mL + (h/km) \sinh mL)]) \sinh mx' \text{ for } x>x'$
X20	$\partial G(0|x')/\partial x = 0$ $G(+\infty|x') = 0$	$e^{-mx'} \cosh mx/m; \; x < x'$ $e^{-mx} \cosh mx'/m; \; x > x'$
X21	$\partial G(0|x')/\partial x = 0$ $G(L|x') = 0$	$\cosh (mx) \sinh [m(L - x')]/(m \cosh mL); \; x < x'$ $\cosh (mx') \sinh [m(L - x)]/(m \cosh mL); \; x > x'$
X22	$\partial G(0|x')/\partial x = 0$ $\partial G(L|x')/\partial x = 0$	$\cosh (mx) \cosh [m(L - x')]/(m \sinh mL); \; x < x'$ $\cosh (mx') \cosh [m(L - x)]/(m \sinh mL); \; x > x'$

[†]Cases $X30$, $X23$, $X31$, $X32$, $X33$, exist but are not listed.

An integral from t_1 to t_2 over τ and at $x = x'$ is

$$\int_{t_1}^{t_2} G_{X00}(x, t|x, \tau) \, d\tau$$

$$= (\pi\alpha)^{-1/2} \left[(t - t_1)^{1/2} - (t - t_2)^{1/2} \right] \qquad t_1 < t_2 \leq t \quad \text{(X00.8b)}$$

A general integral is

$$\int_0^t \tau^{n/2} G_{X00}(x - x', t - \tau) \, d\tau$$

$$= \Gamma\left(\frac{n}{2} + 1\right) \frac{1}{2\alpha^{1/2}} (4t)^{(n+1)/2} \, i^{n+1} \, \text{erfc} \left[\frac{|x - x'|}{(4\alpha t)^{1/2}} \right] \quad \text{(X00.9)}$$

The integral over τ from 0 to t of $\partial G/\partial x$ is

$$\int_0^t \frac{\partial G_{X00}}{\partial x} (x, t|x', \tau) \, d\tau = -\int_0^t \frac{2(x - x')}{\pi^{1/2}[4\alpha(t - \tau)]^{3/2}} e^{-(x-x')^2/[4\alpha(t-\tau)]} \, d\tau$$

$$= -\text{sgn} \, (x - x') \frac{1}{2\alpha} \, \text{erfc} \left[\frac{|x - x'|}{(4\alpha t)^{1/2}} \right] \quad \text{(X00.10)}$$

where sgn $(x - x')$ means the sign of $(x - x')$. Note that

$$-k \int_0^t \frac{\partial G_{X00}(x, t|x', \tau)}{\partial x} d\tau \bigg|_{x \to x'^-}^{x \to x'^+} = -k\left(-\frac{1}{2\alpha}\right)[1 - (-1)] = \rho c \quad \text{(X00.11)}$$

$X10$ Semi-infinite region with $G = 0$ at $x = 0$

$$G_{X10}(x, t|x', \tau) = \frac{1}{[4\pi\alpha(t - \tau)]^{1/2}}$$

$$\times \left\{ \exp\left[-\frac{(x - x')^2}{4\alpha(t - \tau)}\right]\right.$$

$$\left. - \exp\left[-\frac{(x + x')^2}{4\alpha(t - \tau)}\right]\right\} \quad \text{(X10.1)}$$

$$-\frac{\partial G_{X10}}{\partial n'}\bigg|_{x'=0} = \frac{x}{\{4\pi[\alpha(t - \tau)]^3\}^{1/2}} \exp\left[-\frac{x^2}{4\alpha(t - \tau)}\right] \quad \text{(X10.2)}$$

$$\alpha \int_0^t \left(-\frac{\partial^2 G_{X10}}{\partial x \, \partial n'}\bigg|_{x'=0}\right) d\tau \bigg|_{x=0} = -(\pi\alpha t)^{-1/2} \quad \text{(X10.3)}$$

A relation between the $X00$ and $X10$ GFs is

$$G_{X10}(x, t|x', \tau) = G_{X10}(x, x', t - \tau)$$

$$= G_{X00}(x - x', t - \tau) - G_{X00}(x + x', t - \tau) \quad \text{(X10.4)}$$

A relation between the $X00$, $X10$, and $X20$ GFs is

$$2G_{X00}(x - x', t - \tau) = G_{X10}(x, x', t - \tau) + G_{X20}(x, x', t - \tau) \quad \text{(X10.5)}$$

An integral from $x' = 0$ to b gives

$$\int_0^b G_{X10}(x, t|x', \tau) \, dx' = \frac{1}{2}\left(\text{erfc}\left\{\frac{x - b}{[4\alpha(t - \tau)]^{1/2}}\right\}\right.$$

$$- 2 \, \text{erfc}\left\{\frac{x}{[4\alpha(t - \tau)]^{1/2}}\right\}$$

$$\left. + \text{erfc}\left\{\frac{x + b}{[4\alpha(t - \tau)]^{1/2}}\right\}\right) \quad \text{(X10.6)}$$

and for $b \to \infty$,

$$\int_0^\infty G_{X10}(x, t|x', \tau) \, dx' = 1 - \text{erfc}\left\{\frac{x}{[4\alpha(t - \tau)]^{1/2}}\right\}$$

$$= \text{erf}\left\{\frac{x}{[4\alpha(t - \tau)]^{1/2}}\right\} \quad \text{(X10.7)}$$

The average of the integral over x of the integral over x' is

$$\bar{G}_{X10}(t|\tau) = \frac{1}{b} \int_0^b \int_0^b G_{X10}(x, t|x', \tau)\, dx'\, dx$$

$$= 1 - \left[\frac{\alpha(t - \tau)}{b^2} \right]^{1/2} \left(\frac{3}{\sqrt{\pi}} + \mathrm{ierfc}\left\{ \frac{b}{[\alpha(t - \tau)]^{1/2}} \right\} \right.$$

$$\left. - 4\, \mathrm{ierfc}\left\{ \frac{b}{[4\alpha(t - \tau)]^{1/2}} \right\} \right) \tag{X10.8}$$

For $\alpha(t - \tau)/b^2$ less than 0.0625, the error in $\bar{G}(t|\tau)$ is less than 0.05% using

$$\bar{G}_{X10}(t|\tau) \approx 1 - 3 \left[\frac{\alpha(t - \tau)}{\pi b^2} \right]^{1/2} + \frac{8}{\pi^{1/2}} \left[\frac{\alpha(t - \tau)}{b^2} \right]^{3/2} e^{-b^2/[4\alpha(t-\tau)]} \tag{X10.9}$$

and for $\alpha(t - \tau)/b^2 > 1$, $\bar{G}(t|\tau)$ is within 0.1% using

$$\bar{G}_{X10}(t|\tau) \approx \frac{1}{\pi^{1/2}} \frac{1}{8} \left[\frac{b^2}{\alpha(t - \tau)} \right]^{3/2} \left\{ 1 - \frac{1}{4} \frac{b^2}{\alpha(t - \tau)} \right.$$

$$\left. + \frac{3}{64} \left[\frac{b^2}{\alpha(t - \tau)} \right]^2 - \frac{17}{2304} \left[\frac{b^2}{\alpha(t - \tau)} \right]^3 \right\} \tag{X10.10}$$

X11 Plate with $G = 0$ at $x = 0$ and L

Two expressions are available: one is more computationally efficient for "small" $\alpha(t - \tau)/L^2$ values and the other for "large" values. See Figs. X.1–X.5 for plots of G_{X11}. The expression best for small times [see Eq. (X10.10) for long-time expression] is

$$G_{X11}(x, t|x', \tau) = [4\pi\alpha(t - \tau)]^{-1/2} \sum_{n = -\infty}^{\infty}$$

$$\times \left\{ \exp\left[-\frac{(2nL + x - x')^2}{4\alpha(t - \tau)} \right] \right.$$

$$\left. - \exp\left[-\frac{(2nL + x + x')^2}{4\alpha(t - \tau)} \right] \right\} \tag{X11.1}$$

(see Carslaw and Jaeger, 1959, p. 274). For $\alpha(t - \tau)/L^2 < 0.022$ use

$$G_{X11}(x, t|x', \tau) \approx [4\pi\alpha(t - \tau)]^{-1/2} \left\{ \exp\left[-\frac{(x - x')^2}{4\alpha(t - \tau)} \right] \right.$$

$$\left. - \exp\left[-\frac{(x + x')^2}{4\alpha(t - \tau)} \right] - \exp\left[-\frac{(2L - x - x')^2}{4\alpha(t - \tau)} \right] \right\} \tag{X11.2}$$

Some important derivatives are

$$-\frac{\partial G_{X11}}{\partial n'}\bigg|_{x' = 0} = \{4\pi[\alpha(t - \tau)]^3\}^{-1/2} \sum_{n = -\infty}^{\infty} (2nL + x)$$

$$\times \exp\left[-\frac{(2nL + x)^2}{4\alpha(t - \tau)}\right] \tag{X11.3}$$

$$-\frac{\partial G_{X11}}{\partial n'}\bigg|_{x'=L} = \{4\pi[\alpha(t - \tau)]^3\}^{-1/2} \sum_{n=-\infty}^{\infty} [(2n - 1)L + x]$$

$$\times \exp\left[-\frac{[(2n - 1)L + x]^2}{4\alpha(t - \tau)}\right] \tag{X11.4}$$

$$-\frac{\partial^2 G_{X11}}{\partial x \, \partial n'}\bigg|_{x'=x=0} = \{4\pi[\alpha(t - \tau)]^3\}^{-1/2} \sum_{n=-\infty}^{\infty} \left(1 - \frac{2n^2L^2}{\alpha(t - \tau)}\right)$$

$$\times e^{-n^2L^2/|\alpha(t-\tau)|} \qquad t - \tau > 0 \tag{X11.5}$$

For $\alpha(t - \tau)/L^2 < 0.05$, the approximations

$$-\frac{\partial G_{X11}}{\partial n'}\bigg|_{x'=0} \approx x\{4\pi[\alpha(t - \tau)]^3\}^{-1/2} e^{-x^2/|4\alpha(t-\tau)|} \tag{X11.6}$$

$$-\frac{\partial G_{X11}}{\partial n'}\bigg|_{x'=L} \approx (L - x)\{4\pi[\alpha(t - \tau)]^3\}^{-1/2} \exp\left[-\frac{(L - x)^2}{4\alpha(t - \tau)}\right] \tag{X11.7}$$

are quite accurate. For $\alpha(t - \tau)/L^2 < 0.2$, with errors less than $(1E-6)/L^3$, the cross-derivative at $x = x' = 0$ is

$$-\frac{\partial^2 G_{X11}}{\partial x \, \partial n'}\bigg|_{x'=x=0} \approx \{4\pi[\alpha(t - \tau)]^3\}^{-1/2}$$

$$\times \left\{1 + \left[2 - \frac{4L^2}{\alpha(t-\tau)}\right] e^{-L^2/|\alpha(t-\tau)|}\right\}, \, t - \tau > 0 \tag{X11.8}$$

and for $\alpha(t - \tau)/L^2 < 0.067$, the error is less than 0.002% using

$$-\frac{\partial^2 G_{X11}}{\partial x \, \partial n'}\bigg|_{x'=x=0} \approx \{4\pi[\alpha(t - \tau)]^3\}^{-1/2}, \qquad t - \tau > 0 \tag{X11.9}$$

The expression best for large times is

$$G_{X11}(x, t|x', \tau) = \frac{2}{L} \sum_{m=1}^{\infty} e^{-m^2\pi^2\alpha(t-\tau)/L^2} \sin\left(m\pi\frac{x}{L}\right) \sin\left(m\pi\frac{x'}{L}\right) \tag{X11.10}$$

For $\alpha(t - \tau)/L^2 > 0.1$, the errors are less than about $0.0003/L$ for the maximum m value of 2; for the maximum $m = 3$, the error is less than $(3E-7)/L$. For $\alpha(t - \tau)/L^2 > 0.05$ and the maximum $m = 5$, the error is less than $(4E-8)/L$. Some important derivatives are

$$-\frac{\partial G_{X11}}{\partial n'}\bigg|_{x'=0} = \frac{2\pi}{L^2} \sum_{m=1}^{\infty} e^{-m^2\pi^2\alpha(t-\tau)/L^2} m \sin\left(m\pi\frac{x}{L}\right) \tag{X11.11}$$

$$-\left.\frac{\partial G_{X11}}{\partial n'}\right|_{x'=L} = -\frac{2\pi}{L^2} \sum_{m=1}^{\infty} e^{-m^2\pi^2\alpha(t-\tau)/L^2} m(-1)^m \sin\left(m\pi\frac{x}{L}\right) \qquad (X11.12)$$

$$-\left.\frac{\partial^2 G_{X11}}{\partial x\,\partial n'}\right|_{x'=x=0} = \frac{2\pi^2}{L^3} \sum_{m=1}^{\infty} m^2 e^{-m^2\pi^2\alpha(t-\tau)/L^2} \qquad t-\tau>0 \qquad (X11.13)$$

For $\alpha(t-\tau)/L^2 > 0.2$, with errors less than 0.0002%, the cross-derivative is

$$-\left.\frac{\partial^2 G_{X11}}{\partial x\,\partial n'}\right|_{x'=x=0} \approx \frac{2\pi^2}{L^3}\left[e^{-\pi^2\alpha(t-\tau)/L^2} + 4\,e^{-4\pi^2\alpha(t-\tau)/L^2}\right], \qquad t-\tau>0 \quad (X11.14)$$

and for $\alpha(t-\tau)/L^2 > 0.067$, with errors less than 0.0004%, the maximum m needed is 4. An integral of $G_{X11}(\cdot)$ from $x' = 0$ to b for small times is best given by

$$\int_0^b G_{X11}(x,t|x',\tau)\,dx' = \frac{1}{2}\sum_{n=-\infty}^{\infty}\left(\text{erfc}\left\{\frac{2nL+x-b}{[4\alpha(t-\tau)]^{1/2}}\right\}\right.$$

$$- 2\,\text{erfc}\left\{\frac{2nL+x}{[4\alpha(t-\tau)]^{1/2}}\right\}$$

$$\left.+ \text{erfc}\left\{\frac{2nL+x+b}{[4\alpha(t-\tau)]^{1/2}}\right\}\right) \qquad (X11.15)$$

and for large times by

$$\int_0^b G_{X11}(\cdot)\,dx' = \frac{2}{\pi}\sum_{m=1}^{\infty} e^{-m^2\pi^2\alpha(t-\tau)/L^2}\frac{1}{m}\sin\left(m\pi\frac{x}{L}\right)\left[1-\cos\left(m\pi\frac{b}{L}\right)\right]$$

$$(X11.16)$$

For $b = L$, the integral for small times is

$$\int_0^L G_{X11}(x,t|x',\tau)\,dx'$$

$$= \frac{1}{2}\sum_{n=-\infty}^{\infty}\left(\text{erfc}\left\{\frac{(2n-1)L+x}{[4\alpha(t-\tau)]^{1/2}}\right\}\right.$$

$$- 2\,\text{erfc}\left\{\frac{2nL+x}{[4\alpha(t-\tau)]^{1/2}}\right\}$$

$$\left.+ \text{erfc}\left\{\frac{(2n+1)L+x}{[4\alpha(t-\tau)]^{1/2}}\right\}\right) \qquad (X11.17a)$$

$$= \text{erf}\left\{\frac{x}{[4\alpha(t-\tau)]^{1/2}}\right\} - \sum_{n=1}^{\infty}\left(\text{erfc}\left\{\frac{2nL+x}{[4\alpha(t-\tau)]^{1/2}}\right\}\right.$$

$$+ \text{erfc}\left\{\frac{(2n-1)L-x}{[4\alpha(t-\tau)]^{1/2}}\right\}$$

$$\left.- \text{erfc}\left\{\frac{2nL-x}{[4\alpha(t-\tau)]^{1/2}}\right\} - \text{erfc}\left\{\frac{(2n-1)L+x}{[4\alpha(t-\tau)]^{1/2}}\right\}\right) \qquad (X11.17b)$$

For large times, use

$$\int_0^L G_{X11}(\cdot)\, dx' = \frac{4}{\pi} \sum_{m=1}^{\infty} \frac{\sin[(2m-1)\pi x/L]}{2m-1}\, e^{-(2m-1)^2 \pi^2 \alpha (t-\tau)/L^2} \quad \text{(X11.18)}$$

The average of the integral from $x' = 0$ to L is

$$\overline{G}_{X11}(t|\tau) \equiv \frac{1}{L} \int_{x=0}^{L} \int_{x'=0}^{L} G_{X11}(x, t|x', \tau)\, dx'\, dx$$

$$= 1 - 4 \left[\frac{\alpha(t-\tau)}{L^2} \right]^{1/2} \left[\frac{1}{\pi^{1/2}} \right.$$

$$- 2 \sum_{n=1}^{\infty} \left(\text{ierfc} \left\{ \frac{(2n-1)L}{[4\alpha(t-\tau)]^{1/2}} \right\} \right. \quad \text{(X11.19a)}$$

$$\left. - \text{ierfc} \left\{ \frac{2nL}{[4\alpha(t-\tau)]^{1/2}} \right\} \right) \right]$$

$$= \frac{8}{\pi^2} \sum_{m=1}^{\infty} \frac{1}{(2m-1)^2}\, e^{-(2m-1)^2 \pi^2 \alpha (t-\tau)/L^2} \quad \text{(X11.19b)}$$

where Eq. (X11.19a) is best for small times and Eq. (X11.19b) is best for large times. For $\alpha(t-\tau)/L^2 < 0.03$, an accurate expression is simply

$$\overline{G}_{X11}(t|\tau) \approx 1 - 4 \left[\frac{\alpha(t-\tau)}{\pi L^2} \right]^{1/2} \quad \text{(X11.19c)}$$

The error is less than 0.0016%. For $\alpha(t-\tau)/L^2 > 0.03$, only two terms in the large-time expression, Eq. (X11.19b), are needed for an error of less than 0.003%.

X12 Plate with $G = 0$ at $x = 0$ and $\partial G / \partial x = 0$ at L

Two expressions are available: one is more computationally efficient for small $\alpha(t-\tau)/L^2$ values and the other for large values.

Expression best for small times

$$G_{X12}(x, t|x', \tau) = [4\pi\alpha(t-\tau)]^{-1/2} \sum_{n=-\infty}^{\infty} (-1)^n$$

$$\times \left\{ \exp\left[-\frac{(2nL + x - x')^2}{4\alpha(t-\tau)} \right] \right.$$

$$\left. - \exp\left[-\frac{(2nL + x + x')^2}{4\alpha(t-\tau)} \right] \right\} \quad \text{(X12.1)}$$

For $\alpha(t-\tau)/L^2 < 0.2$ and for a maximum $n = 2$, the errors are less than $(2E-14)/L$. For $\alpha(t-\tau)/L^2 < 0.022$, use

$$G_{X12}(x, t|x', \tau) \approx [4\pi\alpha(t-\tau)]^{-1/2} \left\{ \exp\left[-\frac{(x-x')^2}{4\alpha(t-\tau)} \right] \right.$$

$$- \exp\left[-\frac{(x + x')^2}{4\alpha(t - \tau)}\right]$$

$$+ \exp\left[-\frac{(2L - x - x')^2}{4\alpha(t - \tau)}\right]\Big\}\Big] \quad \text{(X12.2)}$$

For $\alpha(t - \tau)/L^2 < 0.2$, with errors less than $(3E-9)/L$:

$$G_{X12}(L, t|L, \tau) = [\pi\alpha(t - \tau)]^{-1/2} (1 - 2e^{-L^2/[\alpha(t-\tau)]}) \quad \text{(X12.3)}$$

An expression for $-\partial G/\partial n'|_{x'=0}$ is

$$-\frac{\partial G_{X12}}{\partial n'}\bigg|_{x'=0} = \{4\pi[\alpha(t - \tau)]^3\}^{-1/2} \sum_{n=-\infty}^{\infty} (-1)^n (2nL + x)$$

$$\times \exp\left[-\frac{(2nL + x)^2}{4\alpha(t - \tau)}\right] \quad \text{(X12.4)}$$

For $\alpha(t - \tau)/L^2 < 0.022$, use

$$-\frac{\partial G_{X12}}{\partial n'}\bigg|_{x'=0} \approx \frac{1}{\{4\pi[\alpha(t - \tau)]^3\}^{1/2}} e^{-x^2/[4\alpha(t-\tau)]} \quad \text{(X12.5)}$$

The cross-derivative evaluated at $x = x' = 0$ is

$$-\frac{\partial^2 G_{X12}}{\partial x \, \partial n'}\bigg|_{x=x'=0} = \{4\pi[\alpha(t - \tau)]^3\}^{-1/2} \sum_{n=-\infty}^{\infty} (-1)^n$$

$$\times \left[1 - \frac{2n^2L^2}{\alpha(t - \tau)}\right] e^{-n^2L^2/[\alpha(t-\tau)]} \quad \text{(X12.6)}$$

For $\alpha(t - \tau)/L^2 < 0.2$, with errors less than $(5E-7)/L^3$,

$$-\frac{\partial^2 G_{X12}}{\partial x \, \partial n'}\bigg|_{x=x'=0} \approx \{4\pi[\alpha(t - \tau)]^3\}^{-1/2}$$

$$\times \left\{1 - 2\left[1 - \frac{2L^2}{\alpha(t - \tau)}\right] e^{-L^2/[\alpha(t-\tau)]}\right\} \quad \text{(X12.7)}$$

Expression best for large times:

$$G_{X12}(x, t|x', \tau) = \frac{2}{L} \sum_{m=1}^{\infty} e^{-\beta_m^2\alpha(t-\tau)/L^2} \sin\left(\beta_m \frac{x}{L}\right) \sin\left(\beta_m \frac{x'}{L}\right) \quad \text{(X12.8)}$$

where $\beta_m = (2m - 1)(\pi/2)$, $m = 1, 2, \ldots$.

For $\alpha(t - \tau)/L^2 > 0.2$ and for a maximum $m = 2$, the errors are less than $(9E-6)/L$. For $\alpha(t - \tau)/L^2 > 0.2$, with errors less than $(5E-6)/L$, at $x = x' = L$, G is

$$G_{X12}(L, t|L, \tau) \approx \frac{2}{L} (e^{-\pi^2\alpha(t-\tau)/4L^2} + e^{-9\pi^2\alpha(t-\tau)/4L^2}) \quad \text{(X12.9)}$$

An expression for $-\partial G/\partial n'\big|_{x'=0}$ is

$$-\frac{\partial G_{X12}}{\partial n'}\bigg|_{x'=0} = \frac{2}{L^2} \sum_{m=1}^{\infty} e^{-\beta_m^2 \alpha(t-\tau)/L^2} \beta_m \sin\left(\beta_m \frac{x}{L}\right) \qquad \text{(X12.10)}$$

$$-\frac{\partial^2 G_{X12}}{\partial x \partial n'}\bigg|_{x'=x=0} = \frac{2}{L^3} \sum_{m=1}^{\infty} e^{-\beta_m^2 \alpha(t-\tau)/L^2} \beta_m^2 \qquad \text{(X12.11)}$$

For $\alpha(t - \tau)/L^2 > 0.2$, with errors less than $0.0006/L^3$ (i.e., 0.02%) a cross-derivative is

$$-\frac{\partial^2 G_{X12}}{\partial x \partial n'}\bigg|_{x'=x} \approx \frac{\pi^2}{2L^3}\left(e^{-\pi^2\alpha(t-\tau)/4L^2} + 9e^{-9\pi^2\alpha(t-\tau)/4L^2}\right) \qquad \text{(X12.12)}$$

X13 Plate with $G = 0$ at $x = 0$ and $k\partial G/\partial x + hT = 0$ at $x = L$

For small values of $\alpha(t - \tau)/L^2$ (≤ 0.022) use

$$G_{X13}(x, t|x', \tau) \approx [4\pi\alpha(t - \tau)]^{-1/2} \left\{ \exp\left[-\frac{(x - x')^2}{4\alpha(t - \tau)}\right]\right.$$

$$- \exp\left[-\frac{(x + x')^2}{4\alpha(t - \tau)}\right]$$

$$\left. + \exp\left[-\frac{(2L - x - ')^2}{4\alpha(t - \tau)}\right]\right\}$$

$$- \frac{h}{k}\exp\left[\frac{h(2L - x - x')}{k} + \frac{h^2\alpha(t - \tau)}{k^2}\right]$$

$$\times \text{erfc}\left\{\frac{2L - x - x'}{[4\alpha(t - \tau)]^{1/2}} + \frac{h}{k}[\alpha(t - \tau)]^{1/2}\right\} \qquad \text{(X13.1)}$$

Also, for small $\alpha(t - \tau)/L^2$ values, use

$$-\frac{\partial G_{X13}}{\partial n'}\bigg|_{x'=0} \approx \frac{x}{\{4\pi[\alpha(t - \tau)]^3\}^{1/2}} \exp\left[-\frac{x^2}{4\alpha(t - \tau)}\right] \qquad \text{(X13.2)}$$

$$-\frac{\partial^2 G_{X13}}{\partial x \partial n'}\bigg|_{x=x'=0} \approx \{4\pi[\alpha(t - \tau)]^3\}^{-1/2} \qquad \text{(X13.3)}$$

For any time, but best for large $\alpha(t - \tau)/L^2$ values, use

$$G_{X13}(x, t|x', \tau) = \frac{2}{L} \sum_{m=1}^{\infty} e^{-\beta_m^2 \alpha(t-\tau)/L^2}$$

$$\times \frac{(\beta_m^2 + B^2) \sin(\beta_m x/L) \sin(\beta_m x'/L)}{\beta_m^2 + B^2 + B} \qquad \text{(X13.4)}$$

Eigencondition: $\qquad \beta_m \cot \beta_m = -B \qquad B \equiv \dfrac{hL}{k} \qquad \text{(X13.5a, b)}$

$$G_{X13}(L, t|L, \tau) = \frac{2}{L} \sum_{m=1}^{\infty} e^{-\beta_m^2 \alpha(t-\tau)/L^2} \frac{\beta_m^2}{\beta_m^2 + B^2 + B} \qquad \text{(X13.6)}$$

$$-\frac{\partial G_{X13}}{\partial n'}\bigg|_{x'=0} = \frac{2}{L^2} \sum_{m=1}^{\infty} e^{-\beta_m^2 \alpha(t-\tau)/L^2} \frac{\beta_m(\beta_m^2 + B^2) \sin(\beta_m x/L)}{\beta_m^2 + B^2 + B} \qquad \text{(X13.7)}$$

The eigenvalues of $\beta_m \cot \beta_m = -B$ can be found for the range of $-1 \leq B < \infty$. For $m = 1$ and $-1 \leq B < -0.6$,

$$\beta_1 \approx \{3 [1 + B - \tfrac{1}{5}(1 + B)^2]\}^{1/2} \qquad \text{(X13.8)}$$

For $m = 1, 2, \ldots$ and $-1 \leq B < 5$ (except for $m = 1$ and $-1 \leq B < -0.6$), use

$$\beta_m \approx \frac{\pi}{2}(2m - 1)\left(1 + \frac{3}{2(B + 3)}\left\{\left[1 + \frac{16B(B + 3)}{3(2m - 1)^2\pi^2}\right]^{1/2} - 1\right\}\right) \qquad \text{(X13.9)}$$

For $B > 5$, use, for $m = 1, 2, \ldots$,

$$\beta_m \approx m\pi - \left(A + \frac{3m\pi}{2B}\right)^{1/3} + \left(A - \frac{3m\pi}{2B}\right)^{1/3} \qquad \text{(X13.10)}$$

where

$$A \approx \left[\left(\frac{3m\pi}{2B}\right)^2 + \left(1 + \frac{1}{B}\right)^3\right]^{1/2} \qquad \text{(X13.11)}$$

To improve the accuracy of the values (except for $B = 0$), use

$$\beta_m \approx \beta'_m - \frac{B \tan \beta'_m + \beta'_m}{B/\cos^2 \beta'_m + 1} \qquad \text{(X13.12)}$$

where β'_m is found from Eqs. (X13.8), (X13.9), or (X13.10).

X14 Plate with $G = 0$ at $x = 0$ and $k\, \partial G/\partial x + (\rho cb)_2\, \partial G/\partial t = 0$ at $x = L$

For small values of $\alpha(t - \tau)/L^2 (\leq 0.022)$, use

$$G_{X14}(x, t|x', \tau) = [4\pi\alpha(t - \tau)]^{-1/2}\left\{\exp\left[-\frac{(x - x')^2}{4\alpha(t - \tau)}\right]\right.$$

$$- \exp\left[-\frac{(x + x')^2}{4\alpha(t - \tau)}\right] - \exp\left[-\frac{(2L - x - x')^2}{4\alpha(t - \tau)}\right]\right\}$$

$$+ \frac{1}{LC_2}\exp\left[\frac{1}{C_2}\frac{2L - x - x'}{L} + \frac{1}{C_2^2}\frac{\alpha(t - \tau)}{L^2}\right]$$

$$\times \operatorname{erfc}\left\{\frac{2L - x - x'}{[4\alpha(t - \tau)]^{1/2}} + \frac{1}{C_2}\frac{[\alpha(t - \tau)]^{1/2}}{L}\right\} \qquad \text{(X14.1)}$$

Also, for small $\alpha(t - \tau)/L^2$ values,

$$-\left.\frac{\partial G_{X14}}{\partial n'}\right|_{x'=0} \approx \frac{x}{\{4\pi[\alpha(t-\tau)]^3\}^{1/2}} \exp\left[-\frac{x^2}{4\alpha(t-\tau)}\right] \qquad (X14.2)$$

$$-\left.\frac{\partial^2 G_{X14}}{\partial x\,\partial n'}\right|_{x=x'=0} \approx \{4\pi[\alpha(t-\tau)]^3\}^{-1/2} \qquad (X14.3)$$

For any time, but best for large $\alpha(t-\tau)/L^2$ values, use

$$G_{X14}(x,\,t|x',\,\tau) = \frac{2}{L}\sum_{m=1}^{\infty} e^{-\beta_m^2\alpha(t-\tau)/L^2}$$

$$\times\ \frac{(C_2^2\,\beta_m^2 + 1)\sin(\beta_m x/L)\sin(\beta_m x'/L)}{C_2^2\beta_m^2 + C_2 + 1} \qquad (X14.4)$$

Eigencondition: $\quad \beta_m \tan \beta_m = \dfrac{1}{C_2} \qquad \beta_m > 0 \qquad m = 1, 2, \ldots \qquad (X14.5)$

$$C_2 \equiv \frac{(\rho cb)_2}{\rho cL} \qquad (X14.6)$$

For relations for the eigenvalues, see the $X23$ case. (Replace B_2 by $1/C_2$.)

X15 Plate with $G = 0$ at $x = 0$ and $k\,\partial G/\partial x + h_2 G + (\rho cb)_2\,\partial G/\partial t = 0$ at $x = L$ For small values of $\alpha(t-\tau)/L^2$ (≤ 0.022), use

$$G_{X15}(x,\,t|x',\,\tau) \approx [4\pi\alpha(t-\tau)]^{-1/2}\left\{\exp\left[-\frac{(x-x')^2}{4\alpha(t-\tau)}\right]\right.$$

$$-\exp\left[-\frac{(x+x')^2}{4\alpha(t-\tau)}\right] - \exp\left[-\frac{(2L-x-x')^2}{4\alpha(t-\tau)}\right]\right\}$$

$$+\frac{1}{L}\frac{1}{C_2(S_4-S_3)}\left\{\exp\left[\frac{1}{S_4}\frac{2L-x-x'}{L} + \frac{1}{S_4^2}\frac{\alpha(t-\tau)}{L^2}\right]\right.$$

$$\times\ \mathrm{erfc}\left[\frac{2L-x-x'}{[4\alpha(t-\tau)]^{1/2}} + \frac{1}{S_4}\frac{[\alpha(t-\tau)]^{1/2}}{L}\right]$$

$$-\exp\left[\frac{1}{S_3}\frac{2L-x-x'}{L} + \frac{1}{S_3^2}\frac{\alpha(t-\tau)}{L^2}\right]$$

$$\left.\times\ \mathrm{erfc}\left[\frac{2L-x-x'}{[4\alpha(t-\tau)]^{1/2}} + \frac{1}{S_3}\frac{[\alpha(t-\tau)]^{1/2}}{L}\right]\right\} \qquad (X15.1)$$

for $C_2 < \frac{1}{4}B_2$ and

$$S_3 = \frac{1}{2C_2}[1 - (1 - 4B_2C_2)^{1/2}] \qquad (X15.2a)$$

$$S_4 = \frac{1}{2C_2}[1 + (1 - 4B_2C_2)^{1/2}] \qquad (X15.2b)$$

Also, for small $\alpha(t - \tau)/L^2$ values,

$$-\left.\frac{\partial G_{X15}}{\partial n'}\right|_{x'=0} \approx \frac{x}{\{4\pi[\alpha(t - \tau)]^3\}^{1/2}} \exp\left[-\frac{x^2}{4\alpha(t - \tau)}\right] \qquad (X15.3)$$

$$-\left.\frac{\partial^2 G_{X15}}{\partial x\,\partial n'}\right|_{x=x'=0} \approx \{4\pi[\alpha(t - \tau)]^3\}^{-1/2} \qquad (X15.4)$$

For any time, but best for large $\alpha(t - \tau)/L^2$ values, use

$$G_{X15}(x, t|x', \tau) = \sum_{m=1}^{\infty} e^{-\beta_m^2\alpha(t-\tau)/L^2}\,\frac{\sin\,(\beta_m x/L)\,\sin\,(\beta_m x'/L)}{N_m}$$

where $$N_m = \frac{L}{2}\,\frac{(B_2 - C_2\beta_m^2)^2 + \beta_m^2 + B_2 + C_2\beta_m^2}{(B_2 - C_2\beta_m^2)^2 + \beta_m^2} \qquad (X15.5)$$

Eigencondition: $(B_2 - C_2\beta_m^2)\tan\beta_m = -\beta_m \qquad (X15.6)$
$$\beta_m > 0 \qquad m = 1, 2, \ldots$$

$$C_2 \equiv \frac{(\rho cb)_2}{\rho cL} \qquad B_2 = \frac{h_2 L}{k} \qquad (X15.7a, b)$$

X20 Semi-infinite body with $\partial G/\partial x = 0$ at $x = 0$

$$G_{X20}(x, t|x', \tau) = \frac{1}{[4\pi\alpha(t - \tau)]^{1/2}}\left\{\exp\left[-\frac{(x - x')^2}{4\alpha(t - \tau)}\right]\right.$$
$$\left. + \exp\left[-\frac{(x + x')^2}{4\alpha(t - \tau)}\right]\right\} \qquad (X20.1)$$

$$G_{X20}(0, t|0, \tau) = [\pi\alpha(t - \tau)]^{-1/2} \qquad (X20.2)$$

See Figs. $X20.1$ and $X20.2$. A relation between the $X00$ and $X20$ GFs is

$$G_{X20}(x, t|x', \tau) = G_{X20}(x, x', t - \tau)$$
$$= G_{X00}(x - x', t - \tau) + G_{X00}(x + x', t - \tau) \qquad (X20.3)$$

An integral from $x' = 0$ to ∞ is

$$\int_{x=0}^{\infty} G_{X20}(x, t|x', \tau)\,dx' = 1 \qquad (X20.4)$$

The integral of G_{X20} from $x' = 0$ to b is

$$\int_0^b G_{X20}(x, t|x', \tau)\,dx'$$

$$= \frac{1}{2}\left(\text{erfc}\left\{\frac{x - b}{[4\alpha(t - \tau)]^{1/2}}\right\} - \text{erfc}\left\{\frac{x + b}{[4\alpha(t - \tau)]^{1/2}}\right\}\right) \qquad (X20.5)$$

The average of this integral over $x = 0$ to b is

$$\overline{G}_{X20} \equiv \frac{1}{b} \int_{x=0}^{b} \int_{x'=0}^{b} G_{X20}(x, t|x', \tau) \, dx' \, dx$$

$$= 1 - \left[\frac{\alpha(t - \tau)}{\pi b^2} \right]^{1/2} \left(1 - \pi^{1/2} \, \text{ierfc} \left\{ \frac{b}{[\alpha(t - \tau)]^{1/2}} \right\} \right) \quad \text{(X20.6)}$$

Integrals of $\overline{G}_{X20}$ in this form can be difficult to evaluate analytically due to the ierfc $(\cdot)$ term. Expressions more amenable to analytical integrals are given next. For small values of $u \equiv \alpha(t - \tau)/b^2$, $\overline{G}_{X20}$ is approximated by

$$\overline{G}_{X20} \approx 1 - \left(\frac{u}{\pi} \right)^{1/2} + \frac{u^{3/2}}{2\pi^{1/2}} e^{-1/u} \left[1 - \frac{3}{2} u + \frac{15}{4} u^2 - \frac{105}{8} C_1 u^3 \right] \quad \text{(X20.7)}$$

where the greatest accuracy is found for C_1 near $1/3$. Hence, let $C_1 = 1/3$. For large values of u, $\overline{G}_{X20}$ is approximated by

$$\overline{G}_{X20} \approx \frac{1}{(\pi u)^{1/2}} \left(1 - \frac{1}{6u} + \frac{1}{30u^2} - \frac{1}{168u^3} + \frac{1}{1080u^4} - \frac{C_2}{7920u^5} \right) \quad \text{(X20.8)}$$

where $C_2 = 0.89$ improves accuracy over $C_2 = 1$ which comes from a series approximation. Table $X20$ provides a comparison of results. Eq. (X20.7) is an accurate approximation for $u \leq 0.5$ and Eq. (X20.8) for $u > 0.5$.

If desired, an even more accurate approximation in the intermediate range can be obtained from the $\overline{G}_{X22}$ equation for $b/L = 0.25$; the result is

$$\overline{G}_{X20} \approx \frac{1}{4} + \frac{8}{\pi^2} \sum_{m=1}^{7} \frac{A_m}{m^2} e^{-m^2 \pi^2 u/16} \quad \text{(X20.9)}$$

where $A_1 = A_3 = A_5 = A_7 = 0.5$, $A_2 = A_6 = 1$ and $A_4 = 0$. The answers are accurate to six significant figures for $u = 0.25$ to 0.75. An alternative set of equations can be obtained by restricting Eq. (X20.7) to the first two terms, namely

$$\overline{G}_{X20} \approx 1 - \left(\frac{u}{\pi} \right)^{1/2} \quad \text{(X20.10)}$$

for $u \leq 0.125$ and by using Eq. (X20.9) for an intermediate range but with the number of terms changed to 9 with $A_8 = 0$ and $A_9 = 0.5$. The errors would be less than 10^{-5}.

X21 Plate with $\partial G/\partial x = 0$ at $x = 0$ and $G = 0$ at $x = L$

General expressions best for small times:

$$G_{X21}(x, t|x', \tau) = [4\pi\alpha(t - \tau)]^{-1/2} \sum_{n=-\infty}^{\infty} (-1)^n$$

$$\times \left\{ \exp \left[-\frac{(2nL + x - x')^2}{4\alpha(t - \tau)} \right] \right.$$

$$+ \exp\left[-\frac{(2nL + x + x')^2}{4\alpha(t - \tau)}\right]\right\} \tag{X21.1}$$

$$-\left.\frac{\partial G_{X21}}{\partial n'}\right|_{x'=L} = [4\pi[\alpha(t - \tau)]^3]^{-1/2} \sum_{n=-\infty}^{\infty} (-1)^n[(2n + 1)L - x]$$

$$\times \exp\left[-\frac{[(2n + 1)L - x]^2}{4\alpha(t - \tau)}\right] \tag{X21.2}$$

For small values of $\alpha(t - \tau)/L^2$ (≤ 0.022), use

$$G_{X21}(x, t|x', \tau) \approx [4\pi\alpha(t - \tau)]^{-1/2}\left\{\exp\left[-\frac{(x - x')^2}{4\alpha(t - \tau)}\right]\right.$$

$$+ \exp\left[-\frac{(x + x')^2}{4\alpha(t - \tau)}\right]$$

$$\left.- \exp\left[-\frac{(2L - x - x')^2}{4\alpha(t - \tau)}\right]\right\} \tag{X21.3}$$

$$-\left.\frac{\partial G_{X21}}{\partial n'}\right|_{x'=L} \approx \frac{L - x}{[4\pi[\alpha(t - \tau)]^3]^{1/2}} \exp\left[-\frac{(L - x)^2}{4\alpha(t - \tau)}\right] \tag{X21.4}$$

$$-\left.\frac{\partial^2 G_{X21}}{\partial x\, \partial n'}\right|_{x=x'=L} \approx [4\pi[\alpha(t - \tau)]^3]^{-1/2} \tag{X21.5}$$

General expressions best for large times:

$$G_{X21}(x, t|x', \tau) = \frac{2}{L} \sum_{m=1}^{\infty} e^{-\beta_m^2\alpha(t-\tau)/L^2} \cos\left(\beta_m \frac{x}{L}\right) \cos\left(\beta_m \frac{x'}{L}\right) \tag{X21.6}$$

$$\beta_m = \pi\left(m - \frac{1}{2}\right)$$

Table X20 Comparison of results for $\overline{G}_{X20}$

u	Exact, Eq. (X20.6)	Approx., Eq. (X20.7)	% Error, Eq. (X20.7)	Approx., Eq. (X20.8)	% Error, Eq. (X20.8)
0	1	1			
0.25	0.718394	0.718416	+0.003		
0.4	0.647118	0.647393	+0.042	0.645724	
0.5	0.609548	0.609705	+0.026	0.609265	−0.046
0.6	0.577634	0.575486	−0.37	0.577562	−0.012
0.75	0.537721	0.518095	−3.6	0.537711	−0.002
1	0.486065			0.486065	
2	0.368746			0.368746	
4	0.270903				
10	0.1955				
100	0.0563				

$$
-\left.\frac{\partial G_{X21}}{\partial n'}\right|_{x'=L} = -\frac{2}{L^2} \sum_{m=1}^{\infty} e^{-\beta_m^2 \alpha(t-\tau)/L^2} \beta_m (-1)^m \cos\left(\beta_m \frac{x}{L}\right) \quad (X21.7)
$$

X22 Plate with $\partial G/\partial x = 0$ at $x = 0$ and L

Expression best for small times:

$$
G_{X22}(x, t|x', \tau) = [4\pi\alpha(t - \tau)]^{-1/2} \sum_{n=-\infty}^{\infty}
$$

$$
\times \left\{ \exp\left[-\frac{(2nL + x - x')^2}{4\alpha(t - \tau)} \right] \right.
$$

$$
\left. + \exp\left[-\frac{(2nL + x + x')^2}{4\alpha(t - \tau)} \right] \right\} \quad (X22.1)
$$

For $\alpha(t - \tau)/L^2 < 0.25$, the maximum n value needed for four significant figures is 1. For small values of $\alpha(t - \tau)/L^2$ (≤ 0.022), use

$$
G_{X22}(x, t|x', \tau) \approx [4\pi\alpha(t - \tau)]^{-1/2} \left\{ \exp\left[-\frac{(x - x')^2}{4\alpha(t - \tau)} \right] \right.
$$

$$
+ \exp\left[-\frac{(x + x')^2}{4\alpha(t - \tau)} \right]
$$

$$
\left. + \exp\left[-\frac{(2L - x - x')^2}{4\alpha(t - \tau)} \right] \right\} \quad (X22.2)
$$

Expression best for large times:

$$
G_{X22}(x, t|x', \tau) = \frac{1}{L}\left[1 + 2 \sum_{m=1}^{\infty} e^{-m^2\pi^2\alpha(t-\tau)/L^2} \right.
$$

$$
\left. \times \cos\left(\frac{m\pi x}{L}\right) \cos\left(\frac{m\pi x'}{L}\right) \right] \quad (X22.3)
$$

See Fig. X22.1 for $LG_{X22}(\cdot)$ for various values of x'/L and u ($\equiv \alpha(t - \tau)/L^2$) versus x'/L. Also see Fig. X22.2 for $LG_{X2I}(0, t/0, \tau)$ for $I = 0, 1, 2$, and 3.

For $\alpha(t - \tau)/L^2 > 0.25$, the maximum m value needed for four significant figures is 2. For the locations $x = x' = 0$ or $x = x' = L$,

$$
G_{X22}(0, t|0, \tau) = G(L, t|L, \tau)
$$

$$
\approx [\pi\alpha(t - \tau)]^{-1/2} (1 + 2e^{-L^2/[\alpha(t-\tau)]}) \quad (X22.4a)
$$

$$
\approx \frac{1}{L}(1 + 2e^{-\pi^2\alpha(t-\tau)/L^2}) \quad (X22.4b)
$$

where Eq. (X22.4a) is used for $\alpha(t - \tau)/L^2 < \pi^{-1}$ and Eq. (X22.4b) for larger values. The error is in the sixth significant figure or less. For example, at $\alpha(t - \tau)/L^2 = \pi^{-1}$, both give $LG_{X22}(0, t|0, \tau) = 1.086428$, while the exact value is 1.086435.

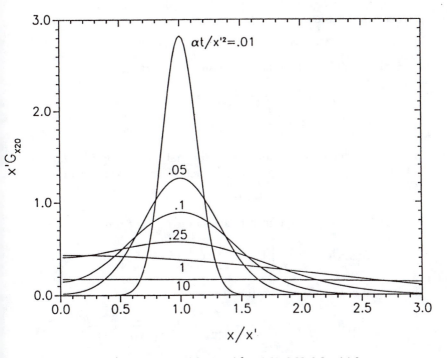

Figure X20.1 $x'G_{X20}(x, t|x', 0)$ versus x/x' for $\alpha t/(x')^2 = 0.01, 0.05, 0.5$ and 1.0.

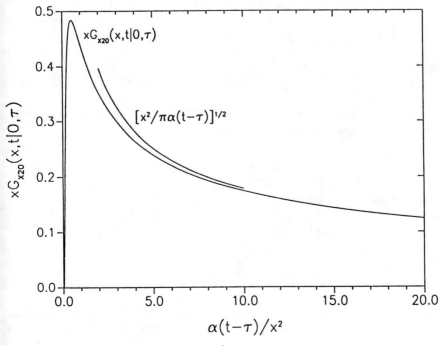

Figure X20.2 $x\, G_{X20}(x, t|0, \tau)$ versus $\alpha(t - \tau)/x^2$.

Alternative expressions are

$$G_{X22}(0, t|0, \tau) \approx [\pi\alpha(t - \tau)]^{-1/2} \qquad \text{(X22.5a)}$$

for $\alpha(t - \tau)/L^2 < 0.08$ and for larger values of $\alpha(t - \tau)/L^2$,

$$G_{X22}(0, t|0, \tau) \approx \frac{1}{L}[1 + 2(e^{-\pi^2\alpha(t-\tau)/L^2}$$
$$+ e^{-4\pi^2\alpha(t-\tau)/L^2} + e^{-9\pi^2\alpha(t-\tau)/L^2})] \qquad \text{(X22.5b)}$$

which are also accurate to about six decimal places. For $G(L, t|0, \tau)$ $[= G(0, t|L, \tau)]$ approximate expressions accurate to about six decimal places are

$$G(L, t|0, \tau) \approx 0 \qquad \text{(X22.6a)}$$

for $\alpha(t - \tau)/L^2 < 0.02$ and for greater values of $\alpha(t - \tau)/L^2$, use

$$G(L, t|0, \tau) \approx \frac{1}{L}\left[1 + 2 \sum_{m=1}^{7} (-1)^m e^{-m^2\pi^2\alpha(t-\tau)/L^2} \right] \qquad \text{(X22.6b)}$$

The integral of G_{X22} from $x' = 0$ to b is

$$\int_0^b G_{X22}(x, t|x', \tau)\, dx' = \frac{1}{2} \sum_{n=-\infty}^{\infty} \left(\text{erfc}\left\{ \frac{2nL + x - b}{[4\alpha(t - \tau)]^{1/2}} \right\} \right.$$
$$\left. - \text{erfc}\left\{ \frac{2nL + x + b}{[4\alpha(t - \tau)]^{1/2}} \right\} \right) \qquad \text{(X22.7a)}$$

$$\int_0^b G_{X22}(x, t|x', \tau)\, dx'$$
$$= \frac{b}{L} + \frac{2}{\pi} \sum_{m=1}^{\infty} \frac{1}{m} e^{-m^2\pi^2\alpha(t-\tau)/L^2} \cos\left(\frac{m\pi x}{L} \right) \sin\left(\frac{m\pi b}{L} \right) \qquad \text{(X22.7b)}$$

where Eq. (X22.7a) is better for small times, and Eq. (X22.7b) is better for large times. For $b = L$, the integral is

$$\int_0^L G_{X22}(x, t|x', \tau)\, dx' = 1 \qquad \text{(X22.8)}$$

At $x = 0$ for small times,

$$\int_0^b G_{X22}(0, t|x', \tau)\, dx'$$
$$\approx 1 - \text{erfc} \frac{b}{[4\alpha(t - \tau)]^{1/2}}$$
$$+ \text{erfc} \frac{2L - b}{[4\alpha(t - \tau)]^{1/2}} - \text{erfc} \frac{2L + b}{[4\alpha(t - \tau)]^{1/2}} \qquad \text{(X22.9)}$$

For $\alpha(t - \tau)/b^2 < 0.02$ and to six significant figures

$$\int_0^b G_{X22}(0, t|x', \tau) \, dx' \approx 1 \tag{X22.10}$$

For the average of the integral from $x' = 0$ to b, the result for small times is

$$\frac{1}{b} \int_{x=0}^b \int_{x'=0}^b G_{X22}(x, t|x', \tau) \, dx' \, dx$$

$$\approx 1 + \frac{[\alpha(t - \tau)]^{1/2}}{b} \left\{ -\frac{1}{\pi^{1/2}} + \text{ierfc} \frac{L - b}{[\alpha(t - \tau)]^{1/2}} \right.$$

$$+ \text{ierfc} \frac{b}{[\alpha(t - \tau)]^{1/2}} - 2 \, \text{ierfc} \frac{L}{[\alpha(t - \tau)]^{1/2}}$$

$$\left. + \text{ierfc} \frac{L + b}{[\alpha(t - \tau)]^{1/2}} \right\} \tag{X22.11a}$$

and for large times

$$= \frac{b}{L} + \frac{L}{b} \frac{2}{\pi^2} \sum_{m=1}^{\infty} \frac{1}{m^2} e^{-m^2 \pi^2 \alpha(t - \tau)/L^2} \left(\sin \frac{m\pi b}{L} \right)^2 \tag{X22.11b}$$

For $\alpha(t - \tau)/L^2 = 0.25$ and $b/L = 0.25$, the small time expression, Eq. (X22.11a), gives 0.2842 and the large time expression, Eq. (X22.11b), the value of 0.2844 which are in good agreement. For $0.1 < b/L < 0.9$ and $\alpha(t - \tau)/L^2 < 0.09$, only the first two ierfc $(\cdot)$ functions are needed in Eq. (X22.11a) with an error less than in the sixth significant digit.

X23 Plate with $\partial G/\partial x = 0$ at $x = 0$ and $k\partial G/\partial x + h_2 G = 0$ at $x = L$

For small values of $\alpha(t - \tau)/L^2 \, (\leq 0.022)$ use

$$G_{X23}(x, t|x', \tau) \approx [4\pi\alpha(t - \tau)]^{-1/2} \left\{ \exp \left[-\frac{(x - x')^2}{4\alpha(t - \tau)} \right] \right.$$

$$+ \exp \left[-\frac{(x + x')^2}{4\alpha(t - \tau)} \right] + \exp \left[-\frac{(2L - x - x')^2}{4\alpha(t - \tau)} \right] \right\}$$

$$- \frac{1}{L} B_2 \exp \left[B_2 \frac{2L - x - x'}{L} + B_2^2 \frac{\alpha(t - \tau)}{L^2} \right]$$

$$\times \text{erfc} \left\{ \frac{2L - x - x'}{[4\alpha(t - \tau)]^{1/2}} + B_2 \frac{[\alpha(t - \tau)]^{1/2}}{L} \right\} \tag{X23.1}$$

For any value of $\alpha(t - \tau)/L^2$ but best for $\alpha(t - \tau)/L^2 > 0.022$,

$$G_{X23}(x, t|x', \tau) = \frac{2}{L} \sum_{m=1}^{\infty} e^{-\beta_m^2 \alpha(t - \tau)/L^2} \frac{\beta_m^2 + B_2^2}{\beta_m^2 + B_2^2 + B_2}$$

$$\times \cos \left(\beta_m \frac{x}{L} \right) \cos \left(\beta_m \frac{x'}{L} \right) \tag{X23.2}$$

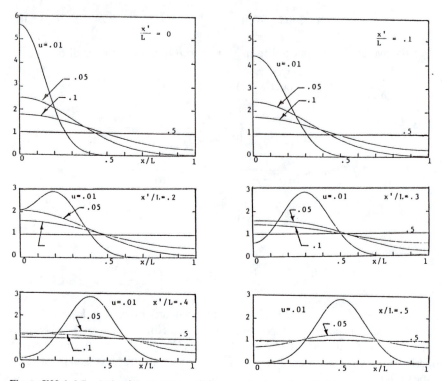

Figure X22.1 $LG_{X22}(x/L, x'/L, u)$ versus x/L for several values of u and x'/L.

Eigencondition: $\qquad\qquad \beta_m \tan \beta_m = B_2 \qquad B_2 \equiv \dfrac{h_2 L}{k}$ (X23.3)

The eigenvalues can be approximated as indicated below. For $0 \le B_2 \le 2$ to 0.08%:

$$\beta_1 \approx \left\{ \frac{3B_2}{3 + B_2} \left[1 - \frac{1}{45}\left(\frac{3B_2}{3 + B_2}\right)^2 \right] \right\}^{1/2}$$ (X23.4)

For $m \ge 2$ and small B_2:

$$\beta_m \approx \frac{(m-1)\pi}{2(B_2 + 3)} \left\{ 2B_2 + 3 + 3\left[1 + \frac{4B_2(B_2 + 3)}{3(m-1)^2\pi^2}\right]^{1/2} \right\}$$ (X23.5)

For $m = 2$, use for $B_2 \le 5$. 0.23% error at $B_2 = 5$.
For $m = 3$, use for $B_2 \le 8$. to 0.1%.
For $m = 4$, use for $B_2 \le 11$. to 0.07%.
For $m = 5$, use for $B_2 \le 13$. to 0.04%.

For $m \ge 1$ and large B_2:

$$\beta_m \approx \frac{(2m-1)\pi B_2}{2(B_2 + 1)} \left\{ 1 + \frac{[(2m-1)\pi]^2}{12(B_2 + 1)^3 + [(2m-1)\pi]^2(2B_2 - 1)} \right\}$$ (X23.6)

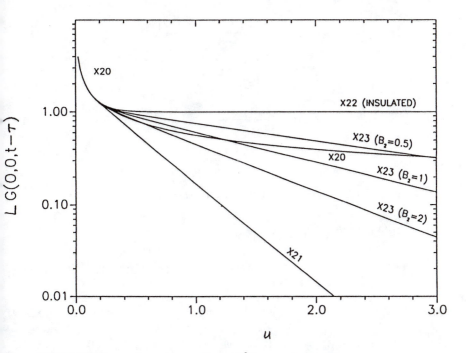

Figure X22.2 $L\,G(0, 0, t - \tau)$ versus $u = \alpha(t - \tau)/L^2$ for several geometries.

m	To within 0.1%	To within 0.01%
1	$B_2 \geq 2$	$B_2 \geq 3$
2	$B_2 \geq 5$	$B_2 \geq 9$
3	$B_2 \geq 8$	$B_2 \geq 14$
4	$B_2 \geq 11$	
5	$B_2 \geq 13$	

To obtain even more accurate values, use

$$\beta_m \approx \beta'_m - \frac{\beta'_m \tan (\beta'_m) - B_2}{\tan (\beta'_m) + \beta'_m \sec^2 \beta'_m}$$

where β'_m is a value from the above equations.

X24 Plate with $\partial G/\partial x = 0$ at $x = 0$ and film with finite heat capacity at $x = L$

The boundary condition at $x = L$ is $k\partial G/\partial x + (\rho cb)_2 \partial G/\partial t = 0$. For small values of $\alpha(t - \tau)/L^2$ (≤ 0.022) use

$$G_{X24}(x, t|x', \tau) \approx [4\pi\alpha(t - \tau)]^{-1/2} \left\{ \exp \left[-\frac{(x - x')^2}{4\alpha(t - \tau)} \right] \right.$$

$$\left. + \exp \left[-\frac{(x + x')^2}{4\alpha(t - \tau)} \right] \right.$$

$$- \exp\left[-\frac{(2L - x - x')^2}{4\alpha(t - \tau)}\right]\Bigg\}$$

$$+ \frac{1}{LC_2} \exp\left[\frac{1}{C_2}\frac{2L - x - x'}{L} + \frac{1}{C_2^2}\frac{\alpha(t - \tau)}{L^2}\right]$$

$$\times \operatorname{erfc}\left\{\frac{2L - x - x'}{[4\alpha(t - \tau)]^{1/2}} + \frac{1}{C_2}\frac{[\alpha(t - \tau)]^{1/2}}{L}\right\} \tag{X24.1}$$

where

$$C_2 \equiv \frac{(\rho c b)_2}{\rho c L} \tag{X24.2}$$

For any value of $\alpha(t - \tau)/L^2$ but best for $\alpha(t - \tau)/L^2 > 0.022$,

$$G_{X24}(x, t|x', \tau) = \frac{1}{1 + C_2}$$

$$+ \sum_{m=1}^{\infty} e^{-\beta_m^2 \alpha(t-\tau)/L^2} \frac{\cos(\beta_m x/L) \cos(\beta_m x'/L)}{N_m} \tag{X24.3}$$

where

$$N_m = \frac{L}{2}\frac{1 + C_2^2 \beta_m^2 + C_2}{1 + C_2^2 \beta_m^2} \tag{X24.4}$$

Eigencondition: $\tan \beta_m = -C_2 \beta_m \qquad m = 1, 2, \ldots \qquad \beta_m > 0$ \qquad (X24.5)

For $C_2 < 0.3$, the eigenvalues can be obtained using

$$\beta_m' \approx m\pi + \delta_m$$

$$\delta_m \approx \frac{3}{2m\pi C_2}\left\{1 - \left[1 + 4\frac{(m\pi C_2)^2}{3}\right]^{1/2}\right\}$$

and iterating,

$$\beta_m \approx \beta_m' - \frac{\tan(\beta_m') + C_2\beta_m'}{\sec^2(\beta_m') + C_2}$$

until convergence is obtained.

X25 Plate with $\partial G/\partial x = 0$ at $x = 0$ and film with finite heat capacity and convection coefficient at $x = L$

The boundary condition at $x = L$ is $k \, \partial G/\partial x + h_2 G + (\rho c b)_2 \partial G/\partial t = 0$. For small values of $\alpha(t - \tau)/L^2$ (≤ 0.022), use

$$G_{X25}(x, t|x', \tau) \approx [4\pi\alpha(t - \tau)]^{-1/2}\left\{\exp\left[-\frac{(x - x')^2}{4\alpha(t - \tau)}\right]\right.$$

$$+ \exp\left[-\frac{(x + x')^2}{4\alpha(t - \tau)}\right]$$

$$- \exp\left[-\frac{(2L - x - x')^2}{4\alpha(t - \tau)}\right]\Bigg\}$$

$$+ \frac{1}{L(1 - 4B_2C_2)^{1/2}}$$

$$\times \left(\exp \left[S_4 \frac{2L - x - x'}{L} + S_4^2 \frac{\alpha(t - \tau)}{L^2} \right] \right.$$

$$\times \operatorname{erfc} \left\{ \frac{2L - x - x'}{[4\alpha(t - \tau)]^{1/2}} + S_4 \frac{[\alpha(t - \tau)]^{1/2}}{L} \right\}$$

$$- \exp \left[S_3 \frac{2L - x - x'}{L} + S_3^2 \frac{\alpha(t - \tau)}{L^2} \right]$$

$$\left. \times \operatorname{erfc} \left\{ \frac{2L - x - x'}{[4\alpha(t - \tau)]^{1/2}} + S_3 \frac{[\alpha(t - \tau)]^{1/2}}{L} \right\} \right) \quad \text{(X25.1)}$$

where

$$C_2 \equiv \frac{(\rho c b)_2}{\rho c L} \qquad B_2 \equiv \frac{h_2 L}{k} \qquad \text{(X25.2a, b)}$$

For any value of $\alpha(t - \tau)/L^2$ but best for $\alpha(t - \tau)/L^2 > 0.022$,

$$G_{X25}(x, t|x', \tau) = \sum_{m=1}^{\infty} e^{-\beta_m^2 \alpha(t-\tau)/L^2} \frac{\cos (\beta_m x/L) \cos (\beta_m x'/L)}{N_m} \quad \text{(X25.3)}$$

where

$$N_m = \frac{L}{2} \frac{[\beta_m^2 + (B_2 - C_2\beta_m^2)^2](1 + 2C_2) + (B_2 - C_2\beta_m^2)[1 - 2C_2(B_2 - C_2\beta_m^2)]}{\beta_m^2 + (B_2 - C_2\beta_m^2)^2}$$

$$\text{(X25.4)}$$

Eigencondition: $\qquad \beta_m \tan \beta_m = B_2 - C_2\beta_m^2 \qquad \qquad \text{(X25.5)}$
$$m = 1, 2, \ldots, \qquad B_2 \neq 0$$

X30 Semi-infinite body with $-k\partial G/\partial x + hG = 0$ at $x = 0$

$$G_{X30}(x, t|x', \tau) = [4\pi\alpha(t - \tau)]^{-1/2} \left\{ \exp \left[-\frac{(x - x')^2}{4\alpha(t - \tau)} \right] \right.$$

$$+ \exp \left[-\frac{(x + x')^2}{4\alpha(t - \tau)} \right] \right\}$$

$$- \frac{h}{k} \exp \left[\alpha(t - \tau)h^2k^{-2} \right.$$

$$+ h(x + x')k^{-1}] \operatorname{erfc} \left\{ \frac{x + x'}{[4\alpha(t - \tau)]^{1/2}} \right.$$

$$\left. + \frac{h}{k} [\alpha(t - \tau)]^{1/2} \right\} \qquad \text{(X30.1)}$$

Then the GF can also be written as

$$G_{X30}(x, t|x', \tau) = G_{X20} - \frac{h}{k} \exp\left[-\frac{(x + x')^2}{4\alpha(t - \tau)}\right]$$

$$\times \operatorname{rerf}\left\{\frac{x + x'}{[4\alpha(t - \tau)]^{1/2}} + \frac{h}{k}[\alpha(t - \tau)]^{1/2}\right\} \quad (X30.2)$$

where

$$\operatorname{rerf}(z) \equiv e^{z^2} \operatorname{erfc}(z) \quad (X30.3)$$

Notice that the first two exp($\cdot$) terms in Eq. (X30.1) are equal to $G_{X20}(x, t|x', \tau)$.
For $x = x' = 0$,

$$G_{X30}(0, t|0, \tau) = [\pi\alpha(t - \tau)]^{-1/2}$$

$$- \frac{h}{k} \exp\left[\frac{\alpha(t - \tau)h^2}{k^2}\right] \operatorname{erfc}\left\{\frac{h}{k}[\alpha(t - \tau)]^{1/2}\right\} \quad (X30.4)$$

For small $(h/k)[\alpha(t - \tau)]^{1/2}$ values,

$$G_{X30}(0, t|0, \tau) \approx [\pi\alpha(t - \tau)]^{-1/2} - \frac{h}{k}\left\{1 - 2\frac{h}{k}\left[\frac{\alpha(t - \tau)}{\pi}\right]^{1/2}\right\} \quad (X30.5)$$

For large values of $(h/k)[\alpha(t - \tau)]^{1/2}$:

$$G_{X30}(0, t|0, \tau) \approx \frac{k^2}{2h^2\pi^{1/2}} \frac{1}{[\alpha(t - \tau)]^{3/2}}\left\{1 - \frac{3}{2}\frac{k^2}{h^2[\alpha(t - \tau)]}\right\} \quad (X30.6)$$

For small values of $(h/k)[\alpha(t - \tau)]^{1/2}$ and any x and x' values,

$$G_{X30}(x, t|x', \tau) \approx G_{X20}(x, t|x', \tau) - \frac{h}{k}\left(\operatorname{erfc}\left\{\frac{x + x'}{[4\alpha(t - \tau)]^{1/2}}\right\}\right.$$

$$\left. - \frac{h}{k}[4\alpha(t - \tau)]^{1/2} \operatorname{ierfc}\left\{\frac{x + x'}{[4\alpha(t - \tau)]^{1/2}}\right\}\right) \quad (X30.7)$$

For large values of $(h/k)[\alpha(t - \tau)]^{1/2}$ and any x and x' values,

$$G_{X30}(x, t|x', \tau) \approx G_{X10}(x, t|x', \tau)$$

$$+ \frac{x + x'}{2\pi^{1/2}[\alpha(t - \tau)]^{3/2}} \frac{k}{h} \exp\left[-\frac{(x + x')^2}{4\alpha(t - \tau)}\right] \quad (X30.8)$$

For any time with $h(x + x')/k$ large (about 10 or larger),

$$G_{X30}(x, t|x', \tau) \approx G_{X20}(x, t|x', \tau) - [\pi\alpha(t - \tau)]^{-1/2}$$

$$\times \left[1 + \frac{(x + x')^2}{2\alpha(t - \tau)} \frac{k}{h(x + x')}\right]^{-1} \exp\left[-\frac{(x + x')^2}{4\alpha(t - \tau)}\right] \quad (X30.9)$$

The integral of G_{X30} from $x' = 0$ to b is

$$\int_0^b G_{X30}(x, t|x', \tau) \, dx' = \frac{1}{2}\left(\text{erfc}\left[\frac{x - b}{4\alpha(t - \tau)^{1/2}}\right]\right.$$

$$+ \text{erfc}\left\{\frac{x + b}{[4\alpha(t - \tau)]^{1/2}}\right\}\right)$$

$$+ \exp\left[\frac{hx}{k} + \frac{h^2}{k^2}\alpha(t - \tau)\right]$$

$$\times \text{erfc}\left\{\frac{x}{[4\alpha(t - \tau)]^{1/2}}\right.$$

$$+ \frac{h}{k}[\alpha(t - \tau)]^{1/2}\Big\}$$

$$- \exp\left[\frac{h(x + b)}{k} + \frac{h^2}{k^2}\alpha(t - \tau)\right]$$

$$\times \text{erfc}\left\{\frac{x + b}{[4\alpha(t - \tau)]^{1/2}} + \frac{h}{k}[\alpha(t - \tau)]^{1/2}\right\}$$

$$(X30.10)$$

If $b \to \infty$, the integral becomes

$$\int_0^\infty G_{X30}(x, t|x', \tau) \, dx'$$

$$= \exp\left[\frac{hx}{k} + \frac{h^2}{k^2}\alpha(t - \tau)\right]$$

$$\times \text{erfc}\left[\frac{x}{[4\alpha(t - \tau)]^{1/2}} + \frac{h}{k}[\alpha(t - \tau)]^{1/2}\right]$$

$$(X30.11)$$

X31 Plate with $-k\partial G/\partial x + hG = 0$ at $x = 0$ and $G = 0$ at $x = L$

For small values of $\alpha(t - \tau)/L^2$ (≤ 0.022) use

$$G_{X31}(x, t|x', \tau) \approx [4\pi\alpha(t - \tau)]^{-1/2}\left\{\exp\left[-\frac{(x - x')^2}{4\alpha(t - \tau)}\right]\right.$$

$$+ \exp\left[-\frac{(x + x')^2}{4\alpha(t - \tau)}\right]$$

$$- \exp\left[-\frac{(2L - x - x')^2}{4\alpha(t - \tau)}\right]\Big\}$$

$$- \frac{h}{k}\exp\left[\frac{h(x + x')}{k} + \frac{h^2\alpha(t - \tau)}{k^2}\right]$$

$$\times \text{ erfc} \left\{ \frac{x + x'}{[4\alpha(t - \tau)]^{1/2}} + \frac{h}{k} [\alpha(t - \tau)]^{1/2} \right\} \qquad \text{(X31.1)}$$

$$-\left. \frac{\partial G_{X31}}{\partial n'} \right|_{x'=L} \approx \frac{L - x}{\{4\pi[\alpha(t - \tau)]^3\}^{1/2}} \exp\left[-\frac{(L - x)^2}{4\alpha(t - \tau)} \right] \qquad \text{(X31.2)}$$

$$-\left. \frac{\partial G_{X31}}{\partial n'} \right|_{x=x'=L} \approx \{4\pi[\alpha(t - \tau)]^3\}^{-1/2} \qquad \text{(X31.3)}$$

For larger values of $\alpha(t - \tau)/L^2$ (but valid for all times) use

$$G_{X31}(x, t|x', \tau) = \frac{2}{L} \sum_{m=1}^{\infty} e^{-\beta_m^2 \alpha(t-\tau)/L^2}$$

$$\times \frac{(\beta_m^2 + B^2) \sin[\beta_m(1 - x/L)] \sin[\beta_m(1 - x'/L)]}{\beta_m^2 + B^2 + B}$$

$$\text{(X31.4)}$$

Eigencondition: $\qquad \beta_m \cot \beta_m = -B \qquad B = \frac{hL}{k} \qquad \text{(X31.5a, b)}$

$$-\left. \frac{\partial G_{X31}}{\partial n'} \right|_{x'=L} = \frac{2}{L^2} \sum_{m=0}^{\infty} e^{-\beta_m^2 \alpha(t-\tau)/L^2} \frac{\beta_m(\beta_m^2 + B^2) \sin[\beta_m(1 - x/L)]}{\beta_m^2 + B^2 + B} \qquad \text{(X31.6)}$$

See $X13$ for equations for eigenvalues.

X32 Plate with $-k\partial G/\partial x + hG = 0$ at $x = 0$ and $\partial G/\partial x = 0$ at $x = L$

For small values of $\alpha(t - \tau)/L^2$ (≤ 0.022) use

$$G_{X32}(x, t|x', \tau) \approx [4\pi\alpha(t - \tau)]^{-1/2} \left\{ \exp\left[-\frac{(x - x')^2}{4\alpha(t - \tau)} \right] \right.$$

$$+ \exp\left[-\frac{(x + x')^2}{4\alpha(t - \tau)} \right]$$

$$\left. + \exp\left[-\frac{(2L - x - x')^2}{4\alpha(t - \tau)} \right] \right\}$$

$$- \frac{h}{k} \exp\left[\frac{h(x + x')}{k} + \frac{h^2\alpha(t - \tau)}{k^2} \right]$$

$$\times \text{ erfc} \left\{ \frac{x + x'}{[4\alpha(t - \tau)]^{1/2}} + \frac{h}{k} [\alpha(t - \tau)]^{1/2} \right\} \qquad \text{(X32.1)}$$

For larger values of $\alpha(t - \tau)/L^2$ (but valid for all times), use

$$G_{X32}(x, t|x', \tau) = \frac{2}{L} \sum_{m=1}^{\infty} e^{-\beta_m^2 \alpha(t-\tau)/L^2} \frac{\beta_m^2 + B^2}{\beta_m^2 + B^2 + B}$$

$$\times \cos\left[\beta_m\left(1 - \frac{x}{L}\right)\right] \cos\left[\beta_m\left(1 - \frac{x'}{L}\right)\right] \quad (X32.2)$$

Eigencondition: $\qquad \beta_m \tan \beta_m = B \qquad B = \dfrac{hL}{k} \qquad$ (X32.3a, b)

See $X23$ for equations for eigenvalues β_m.

X33 Plate with $-k\partial G/\partial x + h_1 G = 0$ at $x = 0$ and $k\partial G/\partial x + h_2 G = 0$ at $x = L$ For small values of $\alpha(t - \tau)/L^2$ (≤ 0.022) use

$$G_{X33}(x, t|x', \tau) \approx [4\pi\alpha(t - \tau)]^{-1/2}\left\{\exp\left[-\frac{(x - x')^2}{4\alpha(t - \tau)}\right]\right.$$

$$+ \exp\left[-\frac{(x + x')^2}{4\alpha(t - \tau)}\right]$$

$$\left. + \exp\left[-\frac{(2L - x - x')^2}{4\alpha(t - \tau)}\right]\right\}$$

$$- \frac{h_1}{k}\exp\left[\frac{h_1(x + x')}{k} + \frac{h_1^2\alpha(t - \tau)}{k^2}\right]$$

$$\times \operatorname{erfc}\left\{\frac{x + x'}{[4\alpha(t - \tau)]^{1/2}} + \frac{h_1}{k}[\alpha(t - \tau)]^{1/2}\right\}$$

$$- \frac{h_2}{k}\exp\left[\frac{h_2(2L - x - x')}{k} + \frac{h_2^2\alpha(t - \tau)}{k^2}\right]$$

$$\times \operatorname{erfc}\left\{\frac{2L - x - x'}{[4\alpha(t - \tau)]^{1/2}} + \frac{h_2}{k}[\alpha(t - \tau)]^{1/2}\right\} \quad (X33.1)$$

For larger values of $\alpha(t - \tau)/L^2$ (but valid for all times), use

$$G_{X33}(x, t|x', \tau) = \frac{2}{L}\sum_{m=1}^{\infty}e^{-\beta_m^2\alpha(t - \tau)/L^2}$$

$$\times [\beta_m \cos(\beta_m x/L) + B_1 \sin(\beta_m x/l)] \qquad (X33.2)$$

$$\times \frac{[\beta_m \cos(\beta_m x'/L) + B_1 \sin(\beta_m x'/L)]}{(\beta_m^2 + B_1^2)[1 + B_2/(\beta_m^2 + B_2^2)] + B_1}$$

where the β_m values are the positive eigenvalues (arranged in increasing order) of

$$\tan \beta_m = \frac{\beta_m(B_1 + B_2)}{\beta_m^2 - B_1 B_2} \qquad B_1 = \frac{h_1 L}{k} \qquad B_2 = \frac{h_2 L}{k} \quad (X33.3a, b, c)$$

An approximate relations for β_1 for $B_1 B_2 < 1$ is

$$\beta_1 \approx \left\{\frac{-45 - 15(B_1 + B_2) + [225(3 + B_1 + B_2)^2 + 180(B_1 + B_2)(B_1 B_2 + B_1 + B_2)]^{1/2}}{2(B_1 + B_2)}\right\}^{1/2} \quad (X33.4)$$

If $B_1B_2 < 1$ and $(B_1 + B_2)(B_1 + B_2 + B_1B_2)/(3 + B_1 + B_2)^2 < 0.04$, use

$$\beta_1 \approx \left[\frac{3(B_1 + B_2 + B_1 B_2)}{3 + B_1 + B_2} \right]^{1/2} \tag{X33.5}$$

For either (1) $1 < B_1B_2 < (\pi/2)^2$ or (2) $B_1B_2 > (\pi/2)^2$ if $B_1 + B_1 + 3 > 16 B_1B_2/3\pi^2$ use

$$\beta_1 \approx \frac{\pi/2 + [(\pi/2)^{1/2} + 4(B_1 + B_2 + 1)B_1B_2/(B_1 + B_2)^2]^{1/2}}{2(B_1 + B_2 + 1)/(B_1 + B_2)} \tag{X33.5}$$

For $B_1B_2 > (\pi/2)^2$ and $B_1 + B_2 + 3 < 16B_1B_2/3\pi^2$

$$\beta_1 \approx \frac{B_1B_2\pi}{B_1 + B_2 + B_1B_2} \tag{X33.6}$$

To improve the β_m values with the estimate of β_m', use

$$\beta_m \approx \beta_m' - \frac{(\beta_m'^2 - B_1B_2)\tan\beta_m' - \beta_m'(B_1 + B_2)}{2\beta_m'\tan\beta_m' + (\beta_m'^2 - B_1B_2)\sec^2\beta_m' - (B_1 + B_2)} \tag{X33.7}$$

provided that

$$\tan\beta_m' - \frac{\beta_m'(B_1 + B_2)}{\beta_m'^2 - B_1B_2} << 1 \tag{X33.8}$$

For very small values of B_1 and B_2 and $m \geq 2$,

$$\beta_m \approx (m - 1)\pi + \frac{B_1 + B_2}{(m - 1)\pi} \tag{X33.9}$$

For somewhat larger values of B_1 and B_2 and $m \geq 2$, use

$$\beta_m \approx (m - 1)\pi + \frac{3}{2(3 + B_1 + B_2)}\left[(m - 1)\pi - \frac{B_1B_2}{\beta_m'}\right]$$
$$\times \left(\left\{1 + \frac{4(B_1 + B_2)(3 + B_1 + B_2)}{3[(m - 1)\pi - B_1B_2/\beta_m']}\right\}^{1/2} - 1\right) \tag{X33.10}$$

where β_m' comes from Eq. (X33.9).

X34 Plate with $-k\partial G/\partial x + h_1G = 0$ at $x = 0$ and at $x = L$ the boundary condition is $k\partial G/\partial x + (\rho cb)_2 \, \partial G/\partial t = 0$

For small values of $\alpha(t - \tau)/L^2 (\leq 0.022)$ use

$$G_{X34}(x, t|x', \tau) \approx [4\pi\alpha(t - \tau)]^{-1/2}\left\{\exp\left[-\frac{(x - x')^2}{4\alpha(t - \tau)}\right]\right.$$
$$\left. + \exp\left[-\frac{(x + x')^2}{4\alpha(t - \tau)}\right] - \exp\left[-\frac{(2L - x - x')^2}{4\alpha(t - \tau)}\right]\right\}$$

$$-\frac{h_1}{k} \exp\left[\frac{h_1(x + x')}{k} + \frac{h_1^2\alpha(t - \tau)}{k^2}\right]$$

$$\times \operatorname{erfc}\left\{\frac{x + x'}{[4\alpha(t - \tau)]^{1/2}} + \frac{h_1}{k}[\alpha(t - \tau)]^{1/2}\right\}$$

$$+ \frac{\rho c}{(\rho cb)_2} \exp\left[\frac{\rho c(2L - x - x')}{(\rho cb)_2}\right.$$

$$+ \left.\frac{(\rho c)^2\alpha(t - \tau)}{(\rho cb)_2^2}\right] \operatorname{erfc}\left\{\frac{2L - x - x'}{[4\alpha(t - \tau)]^{1/2}}\right.$$

$$+ \left.\frac{\rho c}{(\rho cb)_2}[\alpha(t - \tau)]^{1/2}\right\} \tag{X34.1}$$

For larger values of $\alpha(t - \tau)/L^2$ (but valid for all times), use

$$G_{X34}(x, t|x', \tau) = \sum_{m=1}^{\infty} e^{-\beta_m^2\alpha(t-\tau)/L^2} \frac{X_m(x, \beta_m)X_m(x', \beta_m)}{N_m} \tag{X34.2}$$

where

$$X_m(x, \beta_m) = B_1 \sin\left(\beta_m \frac{x}{L}\right) + \beta_m \cos\left(\beta_m \frac{x}{L}\right) \tag{X34.3}$$

$$N_m = L\left(\frac{1}{2}(B_1^2 + \beta_m^2) + \beta_m^2 C_2 + \frac{\tan \beta_m}{1 + \tan^2\beta_m}\left\{\frac{1}{2\beta_m}(\beta_m^2 - B_1^2)\right.\right.$$

$$+ \left.\left. 2 C_2 B_1 \beta_m + \tan(\beta_m)[C_2(B_1^2 - \beta_m^2) + B_1]\right\}\right) \tag{X34.4}$$

The eigenvalues are the positive roots of

$$\tan \beta_m = \frac{B_1 - C_2\beta_m^2}{\beta_m(1 + B_1 C_2)} \tag{X34.5}$$

where

$$B_1 = \frac{h_1 L}{k} \qquad C_2 = \frac{(\rho cb)_2}{\rho cL} \tag{X34.6a, b}$$

X35 Plate with $-k\partial G/\partial x + h_1 G = 0$ **at** $x = 0$ **and at** $x = L$, **the boundary condition is** $k\partial G/\partial x + h_2 G + (\rho cb)_2\partial G/\partial t = 0$

For small values of $\alpha(t - \tau)/L^2$ (≤ 0.022) use

$$G_{X35}(x, t|x', \tau) \approx [4\pi\alpha(t - \tau)]^{-1/2}\left\{\exp\left[-\frac{(x - x')^2}{4\alpha(t - \tau)}\right]\right.$$

$$+ \exp\left[-\frac{(x + x')^2}{4\alpha(t - \tau)}\right]$$

$$- \exp\left[-\frac{(2L - x - x')^2}{4\alpha(t - \tau)}\right]\right\}$$

$$+ \frac{1}{L} \left\{ - B_1 \, ER(x + x', t - \tau, B_1) \right.$$

$$+ \frac{1}{(1 - 4B_2C_2)^{1/2}} \, [S_4 \, ER(2L - x - x', t - \tau, S_4)$$

$$\left. - S_3 \, ER(2L - x - x', t - \tau, S_3)] \right\} \tag{X35.1}$$

where for $C_2 < 1/4B_2$

$$S_3 = \frac{1}{2C_2} \, [1 - (1 - 4B_2C_2)^{1/2}] \tag{X35.2}$$

$$S_4 = \frac{1}{2C_2} \, [1 + (1 - 4B_2C_2)^{1/2}] \tag{X35.3}$$

$$ER(x, t - \tau, B) = \exp \left[\frac{Bx}{L} + \frac{B^2\alpha(t - \tau)}{L^2} \right]$$

$$\times \, \mathrm{erfc} \left\{ \frac{x}{[4\alpha(t - \tau)]^{1/2}} + B \frac{[\alpha(t - \tau)]^{1/2}}{L} \right\} \tag{X35.4}$$

For larger times of $\alpha(t - \tau)/L^2$ (but valid for all times), use

$$G_{X35}(x, t|x', \tau) = \sum_{m=1}^{\infty} e^{-\beta_m^2\alpha(t-\tau)/L^2} \frac{X_m(x, \beta_m)X_m(x', \beta_m)}{N_m} \tag{X35.5}$$

where

$$X_m(x, \beta_m) = B_1 \sin \left(\beta_m \frac{x}{L} \right) + \beta_m \cos \left(\beta_m \frac{x}{L} \right) \tag{X35.6}$$

$$N_m = L \left(\frac{1}{2} \, (B_1^2 + \beta_m^2) + \beta_m^2 C_2 + \frac{\tan \beta_m}{1 + \tan^2 \beta_m} \left\{ \frac{1}{2\beta_m} \, (\beta_m^2 - B_1^2) \right. \right.$$

$$\left. \left. + 2 \, C_2 B_1 \beta_m + \tan \beta_m \, [C_2(B_1^2 - \beta_m^2) + B_1] \right\} \right) \tag{X35.7}$$

The eigenvalues are the positive roots of

$$\tan \beta_m = \frac{\beta_m(B_1 + B_2 - C_2\beta_m^2)}{\beta_m^2 - B_1(B_2 - C_2\beta_m^2)} \tag{X35.8}$$

where

$$B_1 = \frac{h_1 L}{k} \qquad B_2 = \frac{h_2 L}{k} \qquad C_2 = \frac{(\rho c b)_2}{\rho c L} \tag{X35.9a, b, c}$$

$X40$ Semi-infinite body with $-k\partial G/\partial x + (\rho cb)_1\, \partial G/\partial t = 0$ at $x = 0$

$$G_{X40}(x, t|x', \tau) = [4\pi\alpha(t - \tau)]^{-1/2} \left\{ \exp\left[-\frac{(x - x')^2}{4\alpha(t - \tau)} \right] \right.$$

$$\left. - \exp\left[-\frac{(x + x')^2}{4\alpha(t - \tau)} \right] \right\}$$

$$+ \frac{1}{bP} \exp\left[-\frac{(x + x')^2}{4\alpha(t - \tau)} \right]$$

$$\times \text{rerf}\left\{ \frac{x + x'}{2[\alpha(t - \tau)]^{1/2}} + \frac{1}{P} \frac{[\alpha(t - \tau)]^{1/2}}{b} \right\} \quad \text{(X40.1)}$$

$$P = \frac{(\rho c)_1}{\rho c} \qquad \text{rerf}\,(z) = e^{z^2}\,\text{erfc}\,(z) \qquad \text{(X40.2a, b)}$$

$X41$ Plate with $-k\partial G/\partial x + (\rho cb)_1\partial G/\partial t = 0$ at $x = 0$ and $G = 0$ at $x = L$

For $\alpha(t - \tau)/L^2 < 0.1$, an accurate approximation is

$$G_{X41}(x, x'|t, \tau) \approx \frac{1}{L}\{EX(x - x', t - \tau) - EX(x + x', t - \tau)$$

$$- EX(2L - x - x', t - \tau)$$

$$+ EX(2L + x - x', t - \tau)$$

$$+ EX(2L - x + x', t - \tau)$$

$$+ C_1^{-1}[ER(x + x', t - \tau, C_1^{-1})$$

$$- ER(2L + x - x', t - \tau, C_1^{-1})$$

$$- ER(2L - x + x', t - \tau, C_1^{-1})]\} \quad \text{(X41.1)}$$

where $ER(\cdot)$ is defined by Eq. (X35.4), and

$$EX(z, t - \tau) = [4\pi\alpha(t - \tau)]^{-1/2} \exp\left[-\frac{z^2}{4\alpha(t - \tau)} \right] \quad \text{(X41.2)}$$

For all times but best for large times, $G_{X41}(\cdot)$ is

$$G_{X41}(x, t|x', \tau) = \frac{1}{L} \sum_{m=1}^{\infty} \frac{1}{N_m} e^{-\beta_m^2\alpha(t-\tau)/L^2} X_m(x)X_m(x') \quad \text{(X41.3)}$$

where

$$X_m(x) = \sin\frac{\beta_m(L - x)}{L} \quad \text{(X41.4)}$$

$$C_1 = \frac{(\rho cb)_1}{\rho cL} \quad \text{(X41.5)}$$

$$N_m = \frac{1}{2}[(C_1\beta_m)^2 + C_1 + 1] \tag{X41.6}$$

Eigencondition:
$$\beta_m \tan \beta_m = C_1^{-1} \tag{X41.7}$$

For eigenvalues, see case $X23$.

$X42$ Plate with $-k\partial G/\partial x + (\rho cb)_1 \, \partial G/\partial t = 0$ at $x = 0$ and $\partial G/\partial x = 0$ at $x = L$

$$G(x, t|x', \tau) = \frac{1}{L}\left[\frac{1}{N_0} + \sum_{m=1}^{\infty} \frac{1}{N_m} e^{-\beta_m^2 \alpha(t-\tau)/L^2} X_m(x)X_m(x')\right] \tag{X42.1}$$

where

$$X_m(x) = \cos\left(\beta_m \frac{x}{L}\right) - C_1\beta_m \sin\left(\beta_m \frac{x}{L}\right) \tag{X42.2}$$

$$C_1 = \frac{(\rho cb)_1}{\rho cL} \tag{X42.3}$$

$$N_0 = 1 + C_1 \tag{X42.4}$$

$$N_m = \frac{1}{2}[(C_1\beta_m)^2 + C_1 + 1] \qquad m = 1, 2, \ldots \tag{X42.5}$$

Eigencondition:
$$\beta_m \cot \beta_m = \frac{-1}{C_1} \tag{X42.6}$$

$X50$ Semi-infinite body with $-k\partial G/\partial x + hG + (\rho cb)_1 \partial G/\partial t = 0$ at $x = 0$

$$\begin{aligned}
G_{X50}(x, t|x', \tau) = [4\pi\alpha(t - \tau)]^{-1/2} &\left\{\exp\left[-\frac{(x - x')^2}{4\alpha(t - \tau)}\right]\right. \\
&\left. - \exp\left[-\frac{(x + x')^2}{4\alpha(t - \tau)}\right]\right\} \\
+ \frac{1}{2bAP} &\exp\left[-\frac{(x + x')^2}{4\alpha(t - \tau)}\right] \\
\times \left((1 + A) \, \text{rerf}\right. &\left\{\frac{x + x'}{2[\alpha(t - \tau)]^{1/2}}\right. \\
+ (1 + A) &\frac{[\alpha(t - \tau)]^{1/2}}{2bP}\right\} \\
- (1 - A) \, \text{rerf} &\left\{\frac{x + x'}{2[\alpha(t - \tau)]^{1/2}}\right.
\end{aligned}$$

$$+ (1 - A) \frac{[\alpha(t - \tau)]^{1/2}}{2bP} \Bigg\} \Bigg) \qquad \text{(X50.1)}$$

$$P = \frac{(\rho c)_1}{\rho c} \qquad B = \frac{hb}{k} \qquad A = (1 - 4BP)^{1/2} \qquad \text{for } 4BP < 1 \qquad \text{(X50.2a, b, c)}$$

X51 Plate with $-k\partial G/\partial x + hG + (\rho cb)_1 \, \partial G/\partial t = 0$ at $x = 0$ and $G = 0$ at $x = L$

For $\alpha(t - \tau)/L^2 < 0.1$, an approximate expression is

$$
\begin{aligned}
G_{X51}(x, x'|t, \tau) \approx \frac{1}{L} \{ & EX(x - x', t - \tau) - EX(x + x', t - \tau) \\
& - EX(2L - x - x', t - \tau) \\
& + EX(2L + x - x', t - \tau) \\
& + EX(2L - x + x', t - \tau) \\
& + \frac{1}{C_1(S_1 - S_2)} [ER(x + x', t - \tau, S_2) \\
& - ER(x + x', t - \tau, S_1) \\
& - ER(2L + x - x', t - \tau, S_2) \\
& + ER(2L + x - x', t - \tau, S_1) \\
& - ER(2L - x + x', t - \tau, S_2) \\
& + ER(2L - x + x', t - \tau, S_1)] \} \\
& \qquad \text{for } C_1 < (1/4B_1) \qquad \text{(X51.1)}
\end{aligned}
$$

where

$$S_1 = \frac{1}{2C_1} [-1 + (1 - 4B_1C_1)^{1/2}]$$

$$S_2 = \frac{1}{2C_1} [-1 - (1 - 4B_1C_1)^{1/2}] \qquad \text{(X51.2)}$$

For larger times but valid for any time, $G_{X51}(\cdot)$ is given by

$$G_{X51}(x, t|x', \tau) = \frac{1}{L} \sum_{m=1}^{\infty} \frac{1}{N_m} e^{-\beta_m^2 \alpha(t-\tau)/L^2} X_m(x) X_m(x') \qquad \text{(X51.3)}$$

where

$$X_m(x) = D_m \sin \left(\beta_m \frac{x}{L} \right) + \cos \left(\beta_m \frac{x}{L} \right) \qquad \text{(X51.4)}$$

$$D_m = \frac{B}{\beta_m} - C\beta_m \qquad B = \frac{hL}{k} \qquad C = \frac{(\rho c)_1 b}{\rho cL} \tag{X51.5}$$

$$N_m = \frac{1}{2}\left(D_m^2 + \frac{D_m}{\beta_m} + 2C + 1\right) \tag{X51.6}$$

Eigencondition: $\tan \beta_m = \dfrac{\beta_m}{C\beta_m^2 - B} \qquad m = 1, 2, \ldots \qquad \beta_m > 0$ $\qquad$ (X51.7)

For eigenvalues, see case $X33$.

$X52$ Plate with $-k\partial G/\partial x + hG + (\rho c b)_1 \partial G/\partial t = 0$ at $x = 0$ and $\partial G/\partial x = 0$ at $x = L$

$$G(x, t|x', \tau) = \frac{1}{L}\sum_{m=1}^{\infty}\frac{1}{N_m} e^{-\beta_m^2 \alpha(t-\tau)/L^2} X_m(x)X_m(x') \tag{X52.1}$$

where

$$X_m(x) = D_m \sin\left(\beta_m \frac{x}{L}\right) + \cos\left(\beta_m \frac{x}{L}\right) \tag{X52.2}$$

$$D_m = \frac{B}{\beta_m} - C\beta_m \qquad B = \frac{hL}{k} \qquad C = \frac{(\rho c)_1 b}{\rho cL} \tag{X52.3a, b, c}$$

$$N_m = \frac{1}{2}\left(D_m^2 + \frac{1}{\beta_m}D_m + 2C + 1\right) \tag{X52.4}$$

Eigencondition: $\qquad \tan \beta_m = D_m \qquad m = 1, 2, \ldots$ $\qquad$ (X52.5)

$$(\beta_m > 0 \text{ for } B > 0)$$

REFERENCES

Carslaw, H. S., and Jaeger, J. C., 1959, *Conduction of Heat in Solids*, 2d ed., Oxford University Press, New York.

INDEX OF SOLUTIONS BY NUMBERING SYSTEM

Number	Equation	Comments
$X00$	(1.4)	$X00$ GF
$X00T0Gx7t1$	Example 1.3	Plane source
$X00T0Gx7t7$	(1.25)	
$X00T5$	Example 1.1	
$X00T5$	Example 1.2	
$X00T7$	(1.21)	
$X00T-G-$	(1.6)	Integral form
$X00T0Y20Bx5$	(6.142)	Heated half-plane, integral expression
$X00T0Y20Bx5$	(6.143)	Heated half-plane, surface temperature
$X00T0Y20Bx5$	(6.152)	Heated half-plane, series expression
$X00T0Y20Bx5$	(6.158)	Heated strip, series expression
$X00T0Y20Bx5$	(6.160)	Heated strip, surface temperature
$X00Y21B(x5)0$	(6.172)	Heated strip, steady state
$X10B0T0Gt3$	(6.34a)	$g(t) = q_0(t/t_0)^{n/2}$ and $I = 1$
$X10B0T1$	Example 1.4	
$X10B0T1$	Table 6.2	
$X10B0T2$	(6.9)	
$X10B0T5$	(6.7)	$I = 1$
$X10B1T0$	(6.17)	
$X10B1T0$	Table 6.3	
$X10B1T0$	Example 1.5	
$X10B3T0$	(6.20)	Boundary temperature $T_0(t/t_0)^{n/2}$
$X11B00T-$	(4.75a)	Large-time form
$X11B00Gx7$	(6.66)	Steady state
$X11B00T(i + 1)$	(6.39)	$I = 1, J = 1, i = 0, 1,$ or 2
$X11B00T0Gx5t1$	(6.112)	$I = J = 1$, time partitioned
$X11B00T0Gx5t2$	(6.113)	$I = J = 1$, time partitioned
$X11B00T0Gx7t1$	(6.68)	Alternative solution
$X11B00T0Gx7t3$	(6.50)	$I = J = 1$, small-time form
$X11B00T1$	(6.43)	Small time
$X11B00T1$	(10.30)	Galerkin-based GF method
$X11B00T1$	(10.23)	Approximate solution
$X11B00T1$	(6.59)	Large-time form
$X11B00T5$	(6.56)	$J = 1$

$X11B06T0$	(10.83)	Galerkin-based GF method
$X11B06T0$	(10.84)	Exact, large-time form
$X11B00T0Y21B(x5)0$	(6.128)	Large time
$X11B00T0Y21B(x5)0$	(6.132)	Small time, approximate form
$X11Y11Z11$	(6.176)	Parallelpiped, one face at $T > 0$, steady
$X12B00T(i + 1)$	(6.39)	$I = 1, J = 2, i = 0, 1,$ or 2
$X12B00T0Gx5t1$	(6.112)	$I = 1, J = 2,$ time partitioned
$X12B00T0Gx5t2$	(6.113)	$I = 1, J = 2,$ time partitioned
$X12B00T0Gx7t1$	(6.69)	Alternative solution
$X12B00T0Gx7t3$	(6.50)	$I = 1, J = 2$
$X12B00T5$	(6.56)	$J = 2$
$X20B0T0Gt3$	(6.34)	$g = q_0(t/t_0)^{n/2}$ and $I = 2$
$X20B0T1$	Table 6.2	
$X20B0T2$	(6.9)	
$X20B0T5$	(6.7)	$I = 2$
$X20B1T0$	(6.24)	
$X20B1T0$	Table 6.3	
$X20B3T0$	(6.25)	Boundary heating $q_0(t/t_0)^{n/2}$
$X21B00G1$	(6.165)	Steady state
$X21B00G4$	(6.166)	Steady state
$X21B10$	(6.93)	Steady state
$X21B10T0$	(6.92)	Time partitioned
$X21B10T0$	(6.95)	Alternative solution
$X21B10Y21B01T0$	(6.121)	Large time
$X22B00T-$	(4.93)	Large time
$X22B00T(i + 1)$	(6.39)	$I = J = 2, i = 0, 1,$ or 2
$X22B00T0Gx5t1$	(6.112)	$I = J = 2,$ time partitioned
$X22B00T0Gx5t2$	(6.113)	$I = J = 2,$ time partitioned
$X22B00T0Gx7t3$	(6.50)	$I = J = 2$
$X22B10T0$	(6.71)	Large-time form
$X22B10T0$	(6.78)	Alternative solution
$X30B0T1$	Table 6.2	
$X30B1T0$	(6.29)	
$X30B1T0$	Table 6.3	
$X32B10T0$	(6.101)	Time partitioned
$X40B0T01$	Table 6.2	
$X40B0T10$	Table 6.2	
$X40B1T00$	Table 6.3	
$X50B0T01$	Table 6.2	
$X50B0T10$	Table 6.2	
$X50B1T00$	Table 6.3	
$R00T0Gx7t7$	(7.11)	$R00$ GF
$R00T5$	(7.14)	Integral forms
$R00T5$	(7.16)	Evaluated at $r = 0$
$R00T0Z20Br5$	(8.31)	Integral form
$R00Z20Br5$	(8.34)	Steady surface temperature
$R00T0Z20Br5$	(8.42)	Surface temperature, best for large time
$R00T0Z20Br5$	(8.48)	Average temperature on heated disk; large time
$R00T0Z20Br5$	(8.49)	Average temperature on heated disk; small time
$R01B0T-$	(7.37)	Integral form
$R01B0T1$	(7.43)	Large-time form
$R01B0T5$	(7.44)	Large-time form

SUBJECT INDEX

AUTHOR INDEX